Oligonucleotide-Based Drugs and Therapeutics

Oligonucleotide-Based Drugs and Therapeutics

Preclinical and Clinical Considerations for Development

Edited by

Nicolay Ferrari and Rosanne Seguin

WILEY

This edition first published 2018

Registered Office
John Wiley & Sons, Inc., 111 River Street, Hoboken, NJ 07030, USA

Editorial Office
111 River Street, Hoboken, NJ 07030, USA

For details of our global editorial offices, customer services, and more information about Wiley products visit us at www.wiley.com.

Wiley also publishes its books in a variety of electronic formats and by print-on-demand. Some content that appears in standard print versions of this book may not be available in other formats.

Library of Congress Cataloging-in-Publication Data

Names: Ferrari, Nicolay, 1969– editor. | Seguin, Rosanne, editor.
Title: Oligonucleotide-based drugs and therapeutics : preclinical and clinical considerations for development / edited by Nicolay Ferrari, Rosanne Seguin.
Description: Hoboken, NJ : John Wiley & Sons, 2018. | Includes bibliographical references and index. |
Identifiers: LCCN 2018006576 (print) | LCCN 2018009683 (ebook) | ISBN 9781119070290 (pdf) | ISBN 9781119070306 (epub) | ISBN 9781118537336 (cloth)
Subjects: LCSH: Oligonucleotides–Therapeutic use. | Antisense nucleic acids–Therapeutic use.
Classification: LCC RM666.N87 (ebook) | LCC RM666.N87 O445 2018 (print) | DDC 572.8/5–dc23
LC record available at https://lccn.loc.gov/2018006576

Cover Design: Wiley
Cover Images: © nechaev-kon/Getty Images; © Michal Sanca/Shutterstock

Set in 10/12pt Warnock by SPi Global, Pondicherry, India

10 9 8 7 6 5 4 3 2 1

MIX
Paper from responsible sources
FSC® C013604

Contents

List of Contributors

Annemieke Aartsma-Rus
Department of Human Genetics
Leiden University Medical Center
Leiden
The Netherlands

Scott A. Barros
Sage Therapeutics
Cambridge, MA
USA

Cindy L. Berman
Berman Consulting
Wayland, MA
USA

Kevin Brown
Fluidigm Corporation
South San Francisco, CA
USA

Joy Cavagnaro
Access BIO
Boyce, VA
USA

Scott Cormack
OncoGenex Pharmaceuticals
Bothell, WA
USA

Beth E. Davis
Department of Medicine
University of Saskatchewan
Saskatoon, Saskatchewan
Canada

Mehrdad Dirin
Department of Pharmaceutical
Chemistry
University of Vienna
Vienna, Austria

Sheila M. Galloway
MRL, Merck & Co., Inc.
West Point, PA
USA

Gail M. Gauvreau
Department of Medicine
McMaster University
Hamilton, Ontario
Canada

Isabella Gazzoli
Department of Human
Genetics
Leiden University Medical Center
Leiden
The Netherlands

Alain Guimond
InSymbiosis
Montreal
Quebec
Canada

Lydia Haile
Centre for Drug Evaluation and Research, Office of New Drugs
US FDA
Silver Spring, MD
USA

Jonathan Hall
ETH-Zurich
Zurich
Switzerland

Derek Ireland
Centre for Drug Evaluation and Research, Office of Biotechnology Products, US FDA
Silver Spring, MD
USA

Aimee L. Jackson
miRagen Therapeutics
Boulder, CO
USA

Cindy Jacobs
OncoGenex Pharmaceuticals
Bothell, WA
USA

Piotr J. Kamola
GlaxoSmithKline R&D
Ware
Hertfordshire
UK
Current address
RIKEN Center for Integrative Medical Sciences
Yokohama
Japan

Lauren Kane
GlaxoSmithKline R&D
Stevenage
Hertfordshire
UK
Current address
MRC Human Genetics Unit
Institute of Genetics and Molecular Medicine
University of Edinburgh
Scotland

Peter Kasper
Federal Institute for Drugs and Medical Devices (BfArM)
Bonn
Germany

Jeremy D. A. Kitson
GlaxoSmithKline R&D
Stevenage
Hertfordshire
UK

Doug Kornbrust
Preclinsight
Reno, NV
USA

Monica Krieger
OncoGenex Pharmaceuticals
Bothell, WA
USA

Arthur A. Levin
Avidity Biosciences
La Jolla, CA
USA

Helen Lightfoot
ETH-Zurich
Zurich
Switzerland

John Martucci
Centre for Drug Evaluation and Research, Office of Biotechnology Products, US FDA
Silver Spring, MD
USA

Xin Ming
Division of Molecular Pharmaceutics, UNC Eshelman School of Pharmacy
University of North Carolina
Chapel Hill, NC
USA

Nicolay Ferrari
Centre de recherche du CHUM – Tour Viger
Montreal, Quebec
Canada

Frederick B. Oleson
Independent Consultant
Concord, MA
USA

John Paul Oliveria
Department of Medicine
McMaster University
Hamilton, Ontario
Canada

Catherine C. Priestley
Innovative Medicines & Early Development
AstraZeneca
Cambridge
UK

Montserrat Puig
Centre for Drug Evaluation and Research Office of Biotechnology Products, US FDA
Silver Spring, MD
USA

Michael J. Schlosser
MSR Pharma Services, Inc.
Lincolnshire, IL
USA

Anneliese Schneider
Preclinical Services & Consulting
Munich
Germany

Rosanne Seguin
Montreal Neurological Institute
McGill University
Montreal
Quebec
Canada

Zhanna Sobol
Pfizer Inc.
Groton, CT
USA

Patricia S. Stewart
OncoGenex Pharmaceuticals
Bothell, MA
USA

Kevin S. Sweder
Forensic and National Security Sciences Institute
Syracuse University
Syracuse, NY
USA

Daniela Verthelyi
Centre for Drug Evaluation and Research, Office of Biotechnology Products, US FDA
Silver Spring, MD
USA

Stefan Vonhoff
NOXXON Pharma AG
Berlin
Germany

Jonathan K. Watts
RNA Therapeutics Institute and
Department of Biochemistry and
Molecular Pharmacology
University of Massachusetts
Medical School
Worcester, MA
USA

Tacey E.K. White
Aclairo Pharmaceutical Development
Group, Inc., Vienna, VA
USA

Johannes Winkler
Department of Cardiology
Medical University of Vienna
Vienna, Austria
Department of Pharmaceutical
Chemistry
University of Vienna
Vienna, Austria

Karen D. Wisont
OncoGenex Pharmaceuticals
Bothell, WA, USA

Preface

Development of oligonucleotide (ODN)-based therapeutics is being progressed for a wide range of indications and using various routes of administration. There is a diversity of structures, chemistries, and mechanisms of actions for ODN therapeutics, but most of the members of this class of drug candidates can be categorized on the basis of whether they target either mRNA or proteins. ODN-based therapy is distinct from gene therapy as it does not involve the modification of genes. Antisense ODN (ASO), short interfering RNA (siRNA), antagomirs, microRNA mimetics, and DNAzymes are part of the RNA-targeting group, while immunostimulatory sequences (ISS), aptamers, and decoys are members of the protein-targeting group.

Currently, six ODN-based pharmaceuticals, including four ASO, have achieved marketing authorization in Europe and/or United States, and many more are undergoing late-stage clinical testing. The first ASO drug, VITRAVENE (fomivirsen, Ionis Pharmaceuticals – formerly Isis), was approved in 1998 to treat CMV eye infections in HIV patients but within a few years was rendered obsolete by advances in antiretroviral cocktails for HIV therapy. The field waited 15 years for another approval. In 2013, the second ASO drug, KYNAMRO (mipomersen, Ionis Pharmaceuticals), was approved by the Food and Drug Administration (FDA) for the treatment of familial hypercholesterolemia. In 2016, out of 22 new drugs approved by FDA, 3 were for ODN therapeutics: DEFITELIO (defibrotide, Jazz Pharmaceuticals), a treatment for veno-occlusive disease of the liver in individuals who have undergone bone marrow transplants granted in March; EXONDYS 51 (eteplirsen, Sarepta Therapeutics), a treatment for Duchenne muscular dystrophy granted in August; and SPINRAZA (nusinersen, Biogen), a treatment for spinal muscular atrophy granted in December. In addition, Atlantic Pharmaceuticals is developing alicaforsen, an ASO targeting ICAM-1 for the treatment of pouchitis, and currently supplies alicaforsen in response to physicians' requests under international named patient supply regulations for patients with inflammatory bowel disease. In January 2017, Atlantic announced it received agreement from the FDA to initiate a rolling submission of its New Drug Application for alicaforsen to treat pouchitis ahead of data from an ongoing phase III study, which is expected at the end of 2018.

The recent ODN approvals are indicative of the enthusiasm, vigor, and vitality of the field observed in recent years. There are currently over 100 companies combining hundreds of ODN programs. In 2015 alone, there were more than 35 Investigational New Drug submissions for ODN candidates. More than 145 ODN clinical trials are listed on ClinicalTrials.gov, 31 of which are active/recruiting. The diverse types of indications for which ODN therapies have been approved and for those currently in clinical development demonstrate that these therapies are not a "one-off" development but rather are poised to claim their space in the apothecary of pharmaceuticals.

The advancement of a growing number of ODN programs, in particular ASO, in late stage of clinical development and the rapid pipeline expansion by various companies are testament of the progress, much of which was made in the 15 years between first and second drug approvals, in understanding the pharmacologic, pharmacokinetic, and toxicologic properties, as well as improving the delivery of ODN. There are now numerous examples of pharmacologic activity in animal models, and evidence of antisense activity in patients has been demonstrated in clinical trials.

The discovery of novel therapeutics is an inherently complex and interdisciplinary process, requiring close integration of scientists from several disciplines in an environment in which lessons are shared and taught across an organization.

The purpose of this book is to review the current state of knowledge of ODN and to examine the scientific principles and the tools utilized by scientists in preclinical and clinical settings as applied to ODN therapeutics.

Acknowledgments

We have embarked on this endeavor without anticipating the long twisting road that was ahead of us in putting this book together. We would like to give our heartfelt thanks to all authors. As editors, we were depending on their goodwill, commitment, and patience. We hope that their contribution will offer a useful review of the current understanding and recent advances in the field. In light of the challenges we are facing with this technology, we also hope the knowledge summarized in this book will provide guidance and will support those readers currently working in the field as well as the future developers that will further advance oligonucleotide therapeutics.

Nicolay Ferrari and Rosanne Seguin

1

Mechanisms of Oligonucleotide Actions

Annemieke Aartsma-Rus[1], Aimee L. Jackson[2], and Arthur A. Levin[3]

[1] *Department of Human Genetics, Leiden University Medical Center, Leiden, The Netherlands*
[2] *miRagen Therapeutics, Boulder, CO, USA*
[3] *Avidity Biosciences, La Jolla, CA, USA*

1.1 Introduction

The promise of antisense oligonucleotide (ASO) therapeutics is the ability to design drugs that are specific inhibitors of the expression solely on the basis of Watson and Crick base-pairing rules. The premise is that treatment of a patient with a DNA-like oligonucleotide complementary to a disease-related RNA (usually a messenger RNA) results in the formation of a heteroduplex that inhibits the function (generally translation) of that target RNA. Although antisense RNAs were first described in 1978 [1, 2], until recently the promise of selectivity and efficacy has always remained slightly out of reach for various reasons. Oligonucleotides are large molecules leading to synthesis and delivery issues. In addition, natural DNA and RNA oligonucleotides are rapidly degraded and cleared after systemic delivery. Over time many of the issues that have challenged developers of oligonucleotide-based therapeutics have been addressed: Synthesis costs have been reduced by orders of magnitude over the past two decades, allowing more investigators to use the technology. Stability issues were addressed partially with the introduction of phosphorothioate backbones (reviewed in Ref. [3]) and later sugar modifications (reviewed in Ref. [4]), and, as a result, oligonucleotides now used clinically and preclinically have more conventional drug-like properties [5]. In addition, fundamental discoveries have improved our understanding of the antisense mechanisms. We now know that target RNA structure and accessibility impacts activity of oligonucleotide therapeutics [6] and therefore pharmacologic activity. Apparently small changes in

Oligonucleotide-Based Drugs and Therapeutics: Preclinical and Clinical Considerations for Development, First Edition. Edited by Nicolay Ferrari and Rosanne Seguin.

oligonucleotide chemistry can also have large pharmacologic effects as analyzed at the phenomenological level [7] and at the quantum level [8].

Our ability to design effective oligonucleotide-based drugs has also been enhanced by studies of the molecular mechanisms of these agents. This chapter reviews the mechanisms of action, the chemistry, and the clinical applications of three broad categories of oligonucleotide therapeutics: antisense agents, splicing modifiers, and gene silencers that activate the RNA interference (RNAi) pathway.

Antisense technology has now produced dozens of clinical stage drugs and two approvals. That hybridization of an oligonucleotide to a pre-mRNA could modulate the splicing of that RNA was described in 1993 [9], and the therapeutic potential of that mechanism is being exploited to treat Duchenne muscular dystrophy and now other diseases (see below). Running at first behind but more recently in parallel with applications of single-stranded ASO agents is the use of double-stranded RNA-like molecules that activate the RNA-induced silencing complex (RISC) to cleave targeted mRNAs or interfere with their translation. Synthetic small interfering RNA (siRNA) therapeutics relies on the same mechanism that is used by eukaryotic cells to control mRNA translation by endogenous microRNAs (miRNAs).

1.2 Antisense Oligonucleotide Therapeutics

1.2.1 Antisense Activity Mediated by RNase H

Zamecnik and Stephenson [1, 2] were the first to describe that a DNA strand complementary to a sequence of an mRNA prevented translation. They observed that an ASO prevented the accumulation of Rous sarcoma virus by inhibiting the translation of proteins encoded by the viral mRNA. A whole new potential field of therapeutics was launched with a single (understated) sentence: "It might also be possible to inhibit the translation of a specific cell protein" [1]. That RNase H was responsible for the inhibitor effects on translation was a conclusion reached by multiple investigators over a period of time. An elegant proof of the role of this specific enzyme in antisense activity was provided by Wu et al. in 1999 [10]; these authors showed that modulation of RNase H levels in cells or animals produces a coordinate change in antisense activity.

1.2.2 The RNase H Mechanism

Members of the RNase H family are ubiquitously expressed. The endonuclease mechanism of action and the crystal structure have been reviewed [10–15]. RNase H is approximately 20 kDa, and the isoforms in mammalian cells are known to have distinct functions. RNase H1 is necessary for transcription, and

RNase H2 is thought to remove RNA primers in the replication of DNA [16]. The RNA binding domain of these enzymes is located at the N-terminus. The catalytic activity is located in a C-terminal domain and depends upon the presence of the 2′ hydroxyl on the ribose sugar for cleavage. The specificity of the enzyme is imparted by heteroduplex formation between a DNA and the targeted RNA. Thus, the enzyme does not cleave single-stranded RNA in the absence of a heteroduplex nor does it cleave DNA in a double strand because of the absence of the critical 2′ OH.

Binding of RNase H to the heteroduplex results in hydrolysis of the RNA at a site distal to the binding region. The enzyme has a DNA recognition site into which a phosphate fits. This phosphate on the DNA strand is two base pairs distal to the cleavage site on the RNA. This DNA binding and recognition is a factor in the recognition of the heteroduplex. The heteroduplex landing site must contain at least five 2′ OH groups, and the position of RNA cleavage is approximately one helical turn from the binding domain [15]. The distance of the cleavage site from the RNA binding site is determined by a spacer domain between the binding domain and the catalytic domain [11–13]. The enzyme extends across a groove in the helix formed by heteroduplex to cut the RNA. Catalysis requires the presence of two metal ions (Mg^{2+} or Mn^{2+}), which activate the nucleophile and stabilize the transition state during hydrolysis of the phosphodiester backbone of the RNA substrate. One metal ion serves to stabilize the transition state, and the other acts during strand transfer [15, 17].

Over a decade after RNase H antisense drugs had been in clinical trials, the identity of the specific RNase family member responsible for the mRNA cleavage remained unproven. By modulating the expression of human RNase H1 and RNase H2, Wu et al. [10] demonstrated that RNase H1 was associated with antisense activity *in vivo*. Antisense drug activity increased with RNase H1 overexpression and decreased with RNase H1 inhibition. The same was not true for RNase H2, demonstrating that the form of the enzyme associated with therapeutic activity is RNase H1.

1.2.3 Chemical Modifications to Enhance RNase H-mediated Antisense Activity

RNase H is rather intolerant to chemical modifications to the DNA strand, and, as a result, ASO drugs that work through the RNase H mechanism must have a DNA-like character in certain nucleotides. One modification tolerated by RNase H is the phosphorothioate linkage: a substitution of a nonbridging sulfur for the phosphodiester linkage between nucleotides. First described by Eckstein and due to the increased stability of the phosphorothioate linkage compared with the native phosphodiester linkage, the phosphorothioate is the most used chemical modification in ASO and siRNA agents. The substitution with sulfur increases the nuclease stability (reviewed in Ref. [3]) and has the added effect of increasing protein binding. This substitution also creates a chiral center at the phosphate. The increased nuclease resistance results from the fact that one of the two diastereomers is highly resistant to nuclease activity, probably as a result of the sulfur being in closer proximity to the metal ions of nucleases in the Sp configuration.

The phosphorothioate modification significantly alters the properties of an oligonucleotide compared with a native DNA oligonucleotide. Plasma half-lives are extended in the phosphorothioate-modified oligonucleotide due both to increased resistance to nucleases and to enhanced binding to plasma proteins. This later effect is both a blessing and a curse in that some of the acute toxicities of phosphorothioate oligonucleotides have been associated with binding to plasma proteins [18]. Ironically, whereas the phosphorothioate linkage is tolerated by RNase H, high concentrations of a phosphorothioate oligonucleotide are inhibitory to the enzyme's activity [19,20]. Thus phosphorothioate linkages must be used strategically to balance in delivery with toxicity to the organism and to the very enzyme that is responsible for the pharmacologic activity. A large number of chemical modifications to oligonucleotides have been tested with the goals of increasing potency to lower toxicity and reduce the potential for RNase H inhibition.

Because of the intolerance of the RNase mechanism for chemical modifications, a scheme was developed that ensured that the ASO retained a DNA-like character. In the so-called chimeric design [21], the central region has nucleotides with DNA-like character (usually natural bases and sugars and a phosphorothioate backbone), and the flanking regions are modified with the aim of increasing affinity to the mRNA target and enhancing nuclease resistance. This modification pattern is also dubbed the gapmer design for the deoxy characteristic of the region between the modified termini (the gap). The size of the region required for RNase recognition and binding must be at least five nucleotides [22]. The minimal binding site size may be larger depending on the nature of the modifications flanking these deoxynucleotides. Crooke and his group have demonstrated that the nature of the 2′ sugar modifications (e.g. 2′methoxy ethyl) influences RNase H activity by changing the conformation of the oligonucleotide–mRNA heteroduplex. The conformational change in a heteroduplex is transmitted for some distance from the 2′ modification. A typical gapmer design has approximately 8–12 central DNA-like residues. One of the factors that hamper the activity of phosphorothioate oligonucleotides that have been internalized by cells is protein binding and sequestration of the antisense molecule away from its target protein RNase H. Recent studies have begun the task of identifying these proteins, which is the first step to being able to exploit them for improving therapeutics [23].

Recognition and binding of the antisense drug to the RNA target are of course critical for the activity of antisense therapeutics. A host of different chemical modifications have been tested over the years with the goal of increasing binding affinity (reviewed in Ref. [24]). Addition of steric bulk at the 2′ position has the effect of producing a northern-type sugar conformation. This conformation is inhibitory to RNase H but may allow for better hydrogen bonding, thus resulting in increases in affinity for the target RNA. Conformationally restricted nucleic acids, such as LNA, or bicyclic nucleic acids (BNAs) are extreme examples of conformational restriction that result in high affinity for a complementary RNA strand.

Wengel et al. [25] described a modification that has the opposite effect in that the sugar no longer cyclizes but is acyclic (or unlocked), which promotes flexibility. These unlocked forms can be useful when it is in the drug designer's interest to reduce the potential for binding to an RNA target. These acyclic nucleotides support RNase H cleavage [26]. The 2′ arabino fluoro nucleotides also support RNase H binding and cleavage and are thus a potential modification that can be used anywhere in an ASO increase affinity to target mRNA [27].

1.3 Oligonucleotides that Sterically Block Translation

Single-stranded ASOs may also act independently of RNase H to block translation or processing of pre-mRNA. Subsequent sections of this chapter will discuss oligonucleotides designed to alter splicing. There are also reports of steric blockers that are inhibited in cell-free translation systems and in cells; ASO modified to inhibit RNase H activity that hybridizes with the region that includes the AUG start codon very effectively block protein synthesis. More recently an alternative strategy for blocking mRNA function through the inhibition of polyadenylation was proposed by Gunderson [28]. By selecting an oligonucleotide that has homology to the U1 adapter small nuclear RNA and homology to the sequence in the 3′ terminus of the target mRNA, it is possible to get a duplex formed where polyadenylation should occur and subsequently block the polyadenylation step that is critical for mRNA function. Without polyadenylation the nascent mRNA is degraded.

1.4 Oligonucleotides that Act Through the RNAi Pathway

1.4.1 The RISC Pathway

Small interfering RNAs (siRNAs) and miRNAs are duplexes of 20–30 base pairs that regulate gene expression and control a diverse array of biological processes. These small RNAs exert their function through the formation of ribonucleoprotein complexes called RISCs that are instrumental in target transcript regulation. Therapeutic modulation of target regulation by siRNAs and miRNAs has the potential to impact diverse disease indications including viral diseases, cardiovascular disease, fibrosis, and cancer. Understanding of the function and modulation of these small regulatory RNAs has progressed at a rapid pace, resulting in translation to therapeutic development in only 10 years since their initial characterization.

In 1993, the cloning of *lin-4* in *Caenorhabditis elegans* marked the discovery that a short (~22 nt) RNA could function as a regulatory molecule and regulate translation via an antisense RNA–RNA interaction [29]. Within a few years, it became clear that endogenously expressed miRNAs are abundant and evolutionarily conserved and play diverse roles in gene expression in species from worms to humans [29–32]. The discovery by Fire and Mello in 1998 that short double-stranded RNAs induce gene silencing in *C. elegans* [33] further revolutionized our understanding of gene regulation and the ability of RNAs to function as regulatory molecules. Shortly thereafter, short interfering RNAs (siRNAs) were shown to guide sequence-specific target silencing in plants [34], *Drosophila* [35, 36], and mammalian cells [37, 38] in a conserved process termed RNAi. The ability of miRNAs and siRNAs to trigger specific gene silencing generated significant excitement of these small RNAs as a therapeutic modality, particularly for targets that are considered to be "undruggable" with small molecules.

miRNAs bind target mRNAs with partial sequence complementarity in the 3′ UTR, mostly involving residues 2–8 (the seed sequence) at the 5′ end of the guide strand [39]. Seed pairing has been shown to be both necessary and sufficient for target regulation by miRNAs in some contexts [40–43], although sequences in addition to the seed can also be important [44–47]. Because miRNAs do not require perfect complementarity for target recognition, a single miRNA can regulate expression from numerous mRNAs [48–50]. It is estimated that miRNAs as a class regulate the expression of 60% of genes in the human genome [51] to control differentiation, development, and physiology. Altered expression or function of miRNAs is linked to human diseases, giving rise to the idea that selective therapeutic modulation of miRNAs could alter the course of disease. The therapeutic inhibition of a miRNA or addition of a miRNA mimetic might produce a phenotype that is derived from a complex set of gene expression changes. The regulation of coordinated gene networks distinguishes miRNAs and their modulation as a therapeutic modality and provides a therapeutic advantage suggestive of combination therapy. Therapeutically, miRNA mimetics can be utilized to restore activity of miRNAs whose loss of function is linked to disease, whereas miRNA inhibitors (called antimiRs or antagomirs) can be used to block activity of miRNAs whose gain of function is linked to disease. A miRNA mimetic is a duplex oligonucleotide analogous to the mature miRNA. An antimiR is a single-stranded oligonucleotide that is complementary to the miRNA and is designed to act as a steric block by binding with the miRNA to prevent it from interacting with target mRNA. Consequently, target transcripts are more highly expressed.

Both miRNAs and siRNAs are processed from double-stranded RNA precursors by the RNase III enzymes Drosha and Dicer to yield the mature, approximately 22-nt, double-stranded RNA [52–54]. Mature miRNAs and siRNAs catalyze gene regulation in complex with a ribonucleoprotein complex called the RISC. The catalytic component of RISC is a member of the Argonaute (Ago) family. Because small RNAs in RISC must anneal to their complementary target mRNAs, one strand, termed the guide strand, is retained in RISC and provides the sequence specificity to guide mRNA silencing. The other strand, termed the passenger strand, is cleaved. The process of strand selection is termed RISC loading. Strand selection is not random. Strand choice is partly encoded in the intrinsic structure of the small RNA duplex, with thermodynamic properties being a major determinant [55, 56]. Unwinding of the duplex, selection of the guide strand and cleavage of the passenger strand are facilitated by the Argonaute protein [57, 58] in an ATP-dependent process [59–63].

The most important domain of the guide strand is the seed sequence, which is the primary determinant of binding specificity for both siRNAs and miRNAs [39, 45, 49, 64–66]. Structural analysis of RNA associated with Argonaute provided insight into the role of the 5′ seed region of the guide strand in sequence-specific pairing with target mRNA [67–69]. The phosphorylated 5′ end of the guide RNA serves as the anchor and is buried within a highly conserved basic pocket in the Mid domain of Argonaute. In contrast, the seed region is exposed and displayed in a prehelical structure that favors the formation of a duplex with the target mRNA. Systematic mutation analysis of siRNA guide strands elucidated distinct siRNA guide domains within Argonaute [70]. Consistent with the structural analysis, mismatches between position 1 of the guide and the target RNA do not impair catalytic activity of Argonaute [66, 70], whereas mismatches within the seed regions reduce target binding and hinder target silencing [70, 71]. Mismatches at the center of the seed region (positions 4 and 5) are more detrimental than mismatches at the periphery (positions 2, 7, and 8), perhaps explaining how some small RNAs, including miRNAs, can regulate targets through imperfect seed matching [45, 72].

Of the four Argonaute proteins in mammals, only one, Ago2, has endonuclease activity [73]. Target cleavage occurs at the nucleotide opposite positions 10 and 11 of the siRNA guide strand, and mismatches or chemical modifications at these positions considerably decrease catalytic activity [37, 38, 74–76]. Pairing with the guide strand positions the scissile phosphate of the target near the catalytic residues in the PIWI domain of Ago2 [37, 66, 74, 77, 78]. siRNAs tend to be perfectly complementary to the target mRNA, and this pairing might enable Argonaute to achieve a catalytically competent conformation [66]. miRNAs typically lack significant pairing in the 3′ portion of the guide strand, although such supplemental base pairing can compensate for a weak seed region [79].

1.4.2 Mechanisms of RISC-mediated Gene Silencing

siRNAs and miRNAs guide RISC to target mRNAs in a sequence-dependent manner and subsequently affect one of three facets of mRNA metabolism: cleavage/destabilization, translation, or mRNA localization. In *Drosophila*, the ultimate fate of the target mRNA depends in part on the Argonaute protein and in part on the small RNA associated with RISC. There does not seem to be a strict small RNA sorting system in human RISC loading, perhaps because the four Ago proteins in humans have largely redundant functions.

siRNAs guide Ago2-containing RISC to complementary mRNA, whereupon the mRNA is degraded via endonucleolytic cleavage [80, 81]. The siRNA–RISC complex is subsequently released and able to bind and cleave another target mRNA molecule in a catalytic process. The power of RNAi arises from the discovery that the endogenous gene-silencing machinery can be conscripted by synthetic siRNAs for selective silencing of a gene of interest [38, 74]. In theory, siRNAs can be designed to silence any gene of interest based solely on the sequence of the target mRNA. Efficacy and potency of target silencing can be enhanced by leveraging thermodynamics, 5′ nucleotide identity, and structure to bias for guide strand selection [55, 82]. Well-designed siRNAs can achieve 95% silencing of the intended target.

Early reports suggested that siRNAs were absolutely specific for the target gene of interest. Target genes were silenced by complementary siRNAs but not unrelated siRNAs [83, 84], and silencing was abolished by single-nucleotide mismatches at the cleavage site of the siRNAs [74, 77, 78]. Subsequently, unbiased genome-scale expression profiling has revealed off-target activity of siRNAs [85]. Off-target silencing is mediated by limited target complementarity to the siRNA, primarily in the seed region [71], reminiscent of miRNA-based target repression. Sequence analysis of off-target transcripts revealed that the 3′ UTRs of these transcripts were complementary to the 5′ end of the siRNA guide strand containing the seed region [85]. Therefore, in addition to the intended, fully complementary target transcript, siRNAs can hybridize to and regulate the expression of transcripts with only partial sequence complementary to the siRNA. Interestingly, base mismatches in the 5′ end of a siRNA guide strand reduced silencing of the original set of off-target transcripts, but introduced a new set of off-target transcripts with complementarity to the mismatched guide strand [71]. This highlights the role of the seed sequence in nucleating RISC on complementary transcripts. As few as 10 nucleotides of sequence complementarity (including eight nucleotides in the seed region) are sufficient to trigger silencing of off-target transcripts [85].

Due to the limited sequence complementarity required for off-target silencing, off-target effects cannot be easily eliminated by siRNA sequence selection, but they can be mitigated by position-specific chemical modification [85]. A single 2′-*O*-methyl modification of position 2 of the seed region reduces the majority of off-target silencing while retaining silencing of the fully complementary target [71]. Modification of the siRNA seed with DNA at positions 1–8 reduces silencing of some off-target transcripts [86]. Modification of specific positions in the seed region with unlocked nucleobase analogs (UNAs), particularly at position 7, results in silencing of the intended target but not other tested mRNAs [87, 88]. Another approach to improving the specificity of target silencing is siRNA pooling. Combining multiple siRNAs to a single target mRNA can reduce the contribution of each individual siRNA to off-target regulation [89, 90]. This approach has demonstrated considerable improvement in RNAi specificity *in vitro*; however, the feasibility of this strategy for development of siRNA therapeutics is unclear.

miRNAs control posttranscriptional gene expression by inhibiting translation and/or initiating mRNA decay. miRNA-based target repression is distinct from siRNA-mediated target silencing in that miRNAs affect mRNA targets without the need for ribonuclease activity and miRNA-mediated repression is generally cleavage independent. miRNAs regulate gene expression by base pairing to partially complementary sequences in the 3′ UTRs of target mRNAs [91–93]. miRNAs interact with their targets through limited base-pairing interactions that mainly contain the seed but that are insufficient to place the target in the active site of Ago2 where cleavage can occur. miRNA-associated RISC, termed miRISC, contains one of the four Argonaute proteins and a glycine–tryptophan repeat-containing protein of 182 kDa (GW182). GW182 is essential for target silencing by miRNAs; it interacts directly with AGO proteins and serves as a molecular platform for binding of silencing effectors [94–99]. miRISC inhibits translation initiation by interfering with cap recognition or by interfering with ribosomal complex formation and might inhibit translation at post-initiation steps by inhibiting ribosome elongation. miRISC can promote mRNA decay by interacting with deadenylase complexes (CCR4-NOT and PAN2-PAN3) to facilitate deadenylation, which is followed by decapping and exonuclease decay of the mRNA ([100] and references therein).

The relative contributions of translational inhibition and mRNA decay to miRNA-mediated target repression remain unclear and might be influenced by the miRNA and the biological context. Some studies reported inhibition of translation in the absence of mRNA destabilization [29, 101], whereas others found significant correlations between mRNA and protein levels of miRNA targets in global analyses [48, 102, 103]. Data from several studies now demonstrate that miRNAs can function in a two-step mode of repression in which translation inhibition results in subsequent destabilization of the targeted mRNAs [104, 105]. However, it remains to be determined how these mechanisms contribute to target repression in different biological systems.

miRNAs impact a given phenotype through regulation of a single key target [106] or through coordinated regulation of a subset of targets [107–109]. miRNAs regulate each individual target mRNA only modestly (~30–50%), yet this degree of silencing is sufficient to induce phenotypic changes. Because a single miRNA can regulate hundreds of targets, it is not always clear which (or which combination) of the potential targets drive the biologic change of interest. Computer algorithms have been developed in an attempt to identify target mRNAs, but these algorithms predict only approximately 50% of regulated targets identified by global expression analysis [48]. Different prediction algorithms incorporate or emphasize different aspects of miRNA-target interactions (evolutionary conservation, target accessibility, sequence context), resulting in disparate sets of predicted targets. Further complicating prediction of miRNA targets and understanding of miRNA mechanism of action, a given miRNA might regulate different targets in different biological contexts [110]. For this reason, identification of target mRNAs and molecular mechanism of action is best measured using global mRNA expression methods in the biological setting of interest.

A unique feature of target regulation by miRNAs that is a consideration for therapeutic development of miRNA modulators is the potential for species-specific targeting. miRNAs are highly conserved across species, but the transcripts targeted by miRNAs are likely less conserved. The majority of functional binding sites for miRNAs reside in 3′ UTRs, which can be evolutionarily divergent [111]. As a result, the transcripts targeted by a miRNA and the resulting phenotypic consequences of miRNA modulation have the potential to differ across species. This has obvious consequences for the selection of appropriate preclinical models for drug development. However, in the best characterized example to date, inhibition of miR-122 has produced remarkably similar phenotypic changes in species from mouse to man [112–115]. As more miRNA-targeting drugs enter clinical trials, it will be instructive to compile cross-species comparisons and establish the factors that influence cross-species versus species-specific responses.

1.5 Chemical Modification of siRNAs and miRNAs

In order to realize the full potential of siRNA and miRNA therapeutics, strategies must be developed to overcome the challenges with RNA stability, specificity, immune modulation, and delivery. Chemical modifications to siRNAs, antimiRs, and miRNA mimetics can improve pharmacokinetic (PK) and pharmacodynamic (PD) properties and reduce immunogenicity. In general, the entire passenger strand as well as the 3′ proximal part of the guide strand is tolerant to chemical modification. The phosphorothioate modification provides resistance to nucleolytic degradation and increases affinity for plasma proteins [116–120]. Moderate modification of siRNAs with phosphorothioate linkages can support efficient RNAi, but tolerability is position dependent [77, 121–124]. For example, phosphorothioate linkages can reduce activity when located near the Ago2 cleavage site [121, 124].

Chemical modifications to the 2′ position of ribose are widely used to increase binding affinity, improve nuclease stability, and enhance specificity of siRNAs ([71, 124–127] and references therein) and have been incorporated to improve target affinity and activity of antimiRs ([128] and references therein). The ribose 2′-OH of siRNAs can be substituted with chemical groups, or the 2′ oxygen can be locked to the 4′-carbon in bridged nucleic acids such as LNAs. Electronegative modifications such as 2′-fluoro, 2′-*O*-methyl, and DNA (2′-H) enhance stability of the duplex between guide strand and target and enhance nuclease resistance. siRNAs containing alternating modifications of 2′-F and 2′-*O*-Me [129] or DNA [130] retain potency with nuclease resistance. Bulkier 2′-modifications, such as 2′-MOE and 2′-*O*-allyl, presumably distort the RNA helix structure necessary for Ago2 cleavage and therefore are only tolerated at certain positions in the siRNA [77, 88, 131]. The LNA modification provides enhanced thermostability, increases nuclease resistance *in vitro* [122] and *in vivo* [132, 133], and reduces immune modulation by siRNA duplexes [134].

Modifications based on sugar moieties other than ribose can also enhance hybridization affinity and/or specificity. Modifications including altritol nucleic acid (ANA), hexitol nucleic acid (HNA), 2′-deoxy-2′-fluoroarabinonucleic acid (2′-F-ANA), cyclohexenyl nucleic acid (CeNA), and unlocked nucleic acid (UNA) have been shown to support siRNA activity [87, 88, 124, 127, 135–137]. Ribose substitutions such as 2′-F-ANA can be combined with 2′-OH modifications such as 2′F or LNA to provide superior properties to siRNAs [138]. UNA, lacking the C2′—C3′ bond of the ribose ring, causes local destabilization of the siRNA duplex as well as interaction of the guide strand with the target mRNA. Therefore, modification with UNA must be limited to two to three nucleotides within the duplex. Modest UNA modification can enhance *in vivo* stability and function of siRNA when combined with other duplex stabilizing modifications such as LNA [139].

Duplex RNAs, including siRNAs and miRNA mimetics, modulate the immune response via pattern recognition receptors of the innate immune system, primarily the toll-like receptors (TLRs) 3, 7, and 8 [134, 140–142]. TLR3 is expressed on the cell surface and in endosomes of dendritic cells, epithelial cells, and endothelial linings and recognizes double-stranded RNA [141, 143–146]. TLR7 and TLR8 are found exclusively in endosomes of immune cells and recognize specific sequences in single-stranded RNA that can be exposed from RNA duplexes via random thermal fluctuations [147–149]. Activation of endosomal TLR7/8 is considered to be the major source of *in vivo* immunogenicity induced by siRNA [134, 142, 150–152].

Nucleobase and ribose modifications can reduce immunostimulation of siRNAs and miRNA mimetics [122, 153–157]. Nucleobase modification may reduce immunostimulation by siRNA and miRNA mimetics by preventing interaction with TLR and PKR receptors [156, 158]. Activation of PKR, a cytoplasmic sensor of double-stranded RNA, is reduced or abrogated by incorporation of purine N2-benzyl, 2′-deoxyuridine [159], 4-thiouridine, and 2-thiouridine. In contrast, the 2′F modification does not reduce PKR activation [160]. Modification of specific immunogenic sequences in siRNAs with small 2′-OH substitutions such as 2′-F, 2′-H, and 2′-O-Me abrogates interaction with TLR7/8 [161]. Uridine residues or U-rich regions are typically the focus of these 2′-ribose modifications, as uridine residues are critical for siRNA activation of TLR7/8 [134, 147, 157, 162]. Guanidine and adenosine modification have also been reported to reduce immunogenicity of siRNAs, whereas cytidine modifications have no effect [134, 153–155, 157]. Base modifications, including 5-methylcytidine (m5C), 5-methyluracil (m5U), N6-mehyladenosine (m6A), and pseudouridine (s2U), have been shown to reduce TLR7/8 activation [153] but are not commonly used due to the success of modifications such as 2′-O-Me that reduces immunogenicity and are compatible with siRNA activity [163, 164]. Immunogenicity has also been suggested to correlate negatively with the strength of hybridization between the siRNA strands. Therefore, modifications such as LNA can reduce exposure of immunostimulatory single-stranded RNA [134, 165]. Although nucleobase and sugar modifications can increase binding affinity, potency, and specificity if placed appropriately within the oligonucleotide, not all modifications are compatible with activity ([127, 161, 166] and references therein). Therefore, therapeutic siRNAs and miRNA modulators require optimization for binding affinity, nuclease stability, and avoidance of immune stimulation.

1.5.1 Delivery of Therapeutic siRNAs or miRNAs

Delivery of oligonucleotide-based therapeutics requires crossing multiple barriers, including serum instability; renal clearance; passage through the blood vessel wall, interstitium, and extracellular matrix; crossing the membrane of the target cell; and escape from the endosome. Systemic delivery is particularly challenging for duplex RNAs such as siRNA and miRNA mimetics because duplex RNA does not readily across the cell membrane and therefore relies heavily on delivery vehicles for cellular uptake. Liposomes containing cationic or neutral lipids are currently the dominant delivery technology. Lipid nanoparticles readily distribute to the liver, and other organs can be targeted by conjugating cell-specific ligands to the nanoparticle. Other delivery vehicles being developed for duplex RNAs include polymeric nanoparticles, metallic core nanoparticles, lipidoids, dendrimers, and polymeric micelles (reviewed in Refs. [167, 168]).

An additional delivery strategy employed for duplex RNAs is conjugation of cholesterol or a ligand (antibody, aptamer, small molecule, or peptide) directly to the oligonucleotide. Initial studies utilized cholesterol conjugated to the passenger strand of a siRNA and demonstrated knockdown of the endogenous target gene, *ApoB*, after systemic delivery *in vivo* [169, 170]. Conjugate chemistries such as cholesterol can improve cellular uptake of duplex RNAs, but the relatively high concentration required for efficacy has hindered their clinical development. Alnylam and Ionis are now using GalNAc conjugated oligonucleotides in multiple clinical trials in multiple indications. Alnylam has entered clinical trials for TTR-associated amyloidosis with a transthyretin-targeting siRNA that is conjugated with *N*-acetylgalactosamine (GalNac) for targeted delivery to hepatocytes after systemic subcutaneous administration (www.clinicaltrials.gov). Delivery vehicles and conjugates not only can improve cellular uptake of duplex RNAs but also have the potential to trigger immune modulation or non-specific effects [168, 171–173]. Therefore, delivery agents as well as therapeutic oligonucleotides must be selected and evaluated carefully for safety.

AntimiRs are typically delivered in saline and rely on chemical modifications including phosphorothioate backbones for enhanced uptake. Many peripheral tissues can be effectively targeted by systemically delivered chemically modified antimiRs. These single-stranded oligonucleotides show good PK properties along with serum and tissue stability *in vivo*. Systemic inhibition of miRNA function in mammals was first demonstrated with a cholesterol-conjugated, 2′-*O*-Me-modified oligonucleotide targeting miR-122 that produced derepression of miR-122 seed-matched transcripts in the liver [174]. Subsequently, several studies demonstrated efficient and long-lasting inhibition of miRNAs *in vivo* using unconjugated, phosphorothioated antimiRs with 2′ ribose modifications in species from mouse to human [112, 114, 115, 175–181]. Delivery strategies being developed for siRNAs can also be applied for targeted delivery of antimiRs. Results from the first phase II study of the effect of miRNA inhibition on HCV infection indicate that miRNA antagonists are well tolerated and provide long-lasting efficacy.

Local administration can avoid some of the challenges associated with systemic delivery by delivering high concentration of oligonucleotide in the direct vicinity of the target cells. Local administration reduces the overall dose of oligonucleotide needed for efficacy and limits toxicity that might accompany systemic exposure. Local delivery of siRNA and miRNA modulators in preclinical and clinical settings has been reported for the lung, vaginal epithelium, brain, eye, and skin ([167, 168, 171–173, 182]. Local delivery continues to be an area of intensive research for both formulated and unformulated oligonucleotides.

Exosome-mediated transfer of miRNAs has recently been identified as a novel mechanism of genetic exchange between cells [183]. Exosomes are small membrane vesicles of endocytic origin that are released into the extracellular environment when multivesicular bodies fuse with the plasma membrane [184]. miRNAs are found in multivesicular bodies, suggesting that these might be sites of miRISC accumulation and function. Furthermore, miRNAs have been found in secreted exosomes that derive from multivesicular bodies [185]. After fusion with the plasma membrane of the recipient cell, exosomes transfer their cargo to the recipient cell (for review see Ref. [186]). Exosomes may interact with recipient cells in a cell type-specific manner [183]. miRNA-loaded exosomes from T cells display antigen-driven, unidirectional transfer to antigen-presenting cells during immune synapse formation and modulate gene expression in recipient cells [187]. The miRNAs of the chromosome 19 miRNA cluster from placenta trophoblast-derived exosomes are transferred to recipient cells where they attenuate viral replication via autophagy [188]. In another example, exosomes from mesenchymal stem cells mediate the transfer of miR-133b to astrocytes and neurons, whereupon miR-133b regulates gene expression for neurite remodeling and functional recovery after stroke in rats [189]. Intercellular communication by exosome-derived miRNAs influences cancer progression via transfer of cancer-promoting contents within the tumor microenvironment or into the circulation to act at distant sites ([190] and references therein). Tumor cells of various cancer types secrete exosomes containing tumor-associated signaling molecules, including miRNAs, to modify angiogenesis, immune response, epigenetic reprogramming, migration, and invasion.

Exosomes consequently offer a novel strategy for delivery of cargo, including small RNAs, for targeted therapy. Alvarez-Erviti et al. were the first to utilize exosomes as a delivery vehicle for siRNA [191]. Targeted exosomes were produced by engineering dendritic cells to express a brain-targeting peptide fused to an exosomal membrane protein. Purified exosomes were loaded with siRNA via electroporation, and the loaded exosomes were delivered to mice via intravenous injection. siRNA was delivered specifically to neurons, microglia, and oligodendrocytes in the brain and produced silencing of the target mRNA *BACE1*. Subsequently, exosomal delivery and transfer of therapeutic miRNAs and siRNAs has been demonstrated in mouse hepatocytes [192], human monocytes and lymphocytes [193], and breast cancer cells [194]. Although much remains to be elucidated regarding purification, loading, cellular uptake, immune response, and toxicity of exosomes, these initial studies highlight the potential of these nanovesicles to deliver endogenous or exogenous small RNAs for therapeutic benefit.

1.6 Clinical Use of Oligonucleotides that Act through the RNAi Pathway

Small RNA-based therapeutics of each of the classes discussed here have entered clinical trials for a diverse array of indications and are demonstrating therapeutic benefit (see Table 1.1). These studies include siRNAs delivered via several different strategies as well as the first miRNA antagonist, antimiR-122 (miravirsen) to treat HCV infection, and the first miRNA mimetic, miR-34, being tested in hepatocellular carcinoma patients. The rapid translation of

Table 1.1 Therapeutics in clinical development that exploit the RISC mechanism.

Sponsor	Target	Mechanism	Indication	Phase
Alnylam	RSV nucleocapsid	siRNA	RSV infection	II
Alnylam	TTR	siRNA	TTR amyloidosis	II
Allergan/Sirna	VEGFR-1	siRNA	Age-related macular degeneration	II
Allergan/Sirna	VEGFR-1	siRNA	Choroidal neovascularization secondary to age-related macular degeneration	II
Opko Health	VEGF	siRNA	Diabetic macular edema	II
Sylentis	ADRB2	siRNA	Ocular hypertension	II
Sylentis	TRPV1R	siRNA	Ocular pain and dry eye syndrome	II
Quark	TP53	siRNA	Delayed graft function in kidney transplantation	II
Quark	RTP801	siRNA	Diabetic macular edema	II
Silence	PKN3	siRNA	Pancreatic ductal carcinoma	II
Santaris	miR-122	antimiR	HCV infection	II
Santaris	miR-122	antimiR	HCV infection (in combination with telaprevir and ribavirin)	II
Quark	TP53	siRNA	Acute renal failure	I
Transderm	Keratin 6A	siRNA	Pachyonichia congenita	I
Tekmira	ApoB	siRNA	Hypercholesterolemia	I
Tekmira	PLK1	siRNA	Cancer with liver involvement	I
Silence	PKN3	siRNA	Advanced solid tumors	I

(Continued)

Table 1.1 (Continued)

Sponsor	Target	Mechanism	Indication	Phase
Quark	Caspase 2	siRNA	Chronic optic nerve atrophy and acute nonarteritic anterior ischemic optic neuropathy	I
Calando	RNR1	siRNA	Solid tumors	I
Silenseed	K-RAS	siRNA-LODER	Pancreatic adenocarcinoma	I
Nitto Denko	Hsp47	siRNA	Liver fibrosis	I
Alnylam	TTR	siRNA	TTR amyloidosis	I
Alnylam	TTR	siRNA-GalNac	TTR amyloidosis (healthy volunteers)	I
Alnylam	PCSK9	siRNA	Hypercholesterolemia	I
Alnylam	KSP1/VEGFR	siRNA	Advanced solid tumors with liver involvement	I
MD Anderson Cancer Center	EphA2	siRNA	Advanced cancers	I
Mirna	miR-34	MicroRNA mimetic	Liver cancer	I
Miragen	miR-155	antimiR	Hematological malignancy	I
Miragen	miR-29	microRNA mimetic	Pathologic fibrosis	I
Miragen	miR-92a	antimiR	Revascularization	I

siRNA and miRNA modulators from discovery to the clinic highlights the excitement and enormous potential for this therapeutic modality. With more RNA therapeutics advancing through clinical development, data to address some of the remaining hurdles will be forthcoming and pave the way for exploiting these regulatory RNAs for therapeutic advantage.

1.7 Oligonucleotides that Modulate Splicing

1.7.1 Pre-mRNA Splicing and Disease

In most eukaryotic genes, genetic information that encodes proteins is dispersed over exons, which are interspersed by pieces of DNA that do not encode protein information (introns). After transcription, intronic sequences must be correctly removed from precursor RNA (pre-mRNA) to result in the mature mRNA copy that is translated into protein. This highly regulated processing of pre-mRNAs is known as splicing. Alternative splicing is a process where (part of an) exon (or intron) can be either included or excluded, allowing the production of different mRNAs (and thus proteins) from a single gene. This process is responsible for the fact that a mere 20 000 human genes encode more than 100 000 different proteins in humans.

As splicing and alternative splicing are such important processes, it is understandable that mutations interfering with splicing underlie many genetic diseases. Mutations that alter nucleotides in the splice sites, the consensus sites flanking each exon that are crucial for correct splicing, generally result in exon skipping. Mutations that introduce new splice signals may result in the aberrant inclusion of piece of an intron into the mRNA. Mutations have also been reported that affect the ratio of alternatively spliced transcripts and that interfere with general alternative splicing patterns due to sequestration of proteins normally involved in maintaining certain splicing patterns.

1.7.2 Mechanisms of Oligonucleotide-mediated Splicing Modulation

Over the past two decades, ASOs have been developed as tools to manipulate splicing. These agents also have promise in the clinic (Table 1.2). The approach can be exploited in several ways to address splicing mutations, as discussed in this section. For more detailed reviews see the following Refs. [195, 196].

Mutations that generate false splice sites may result in aberrant inclusion of intronic sequences in an mRNA. Generally this results into a disrupted reading frame and/or a premature stop codon and a nonfunctional protein. ASOs designed to hybridize to the cryptic splice site in the pre-mRNA compete for binding with the splicing machinery. Given a high-affinity oligonucleotide

Table 1.2 Oligonucleotides currently in clinical development that modulate splicing.

Sponsor	Drug name	Target	Mechanism	Indication	Phase
BioMarin	Drisapersen	*Dystrophin*	Induces exon skipping	DMD	Abolished
Isis Pharmaceuticals	ISIS-SMNRx	*SMN1*	Blocks intronic silencer motif	SMA	Approved
Sarepta Therapeutics	Eteplirsen	*SMN1*	Induces exon skipping	DMD	Approved

agent, the aberrant splice site is not used, and normal splicing is restored, allowing the production of a normal protein. This approach has therapeutic implications for many genetic diseases. Two examples are cystic fibrosis and beta-thalassemia; both commonly result from mutations that generate cryptic splice sites [9, 197, 198]. ASO-mediated restoration of splicing has been reported in patient-derived cell cultures, in mini-gene systems, and in some even in animal models [196, 199].

Alternative splicing allows the production of multiple proteins from one gene. Sometimes alternatively spliced mRNAs expressed from the same gene encode proteins that have opposite function such as apoptotic and anti-apoptotic or inflammatory and anti-inflammatory roles. Normally the ratios of alternatively spliced products are carefully regulated often in a tissue-specific manner, and, not surprisingly, defects in control of alternative splicing appear to be causative of diseases including certain cancers and chronic inflammatory syndromes. Antisense targeting of the splice sites of the alternatively spliced exon can prevent its inclusion into the mRNA. Antisense-mediated control of alternative splicing has been reported to induce spontaneous apoptosis or sensitivity to chemotherapy in cell lines and in a mouse model [200–202].

Mutations can also result in the exclusion of exons from the mRNA. When the splice sites are mutated, restoring correct splicing is challenging. However, when exon skipping is the result of mutations in exonic or intronic regions that inhibit exon recognition (so-called exonic or intronic splicing silencers), these can be targeted by ASOs to bring about exon inclusion. This has been demonstrated in models of spinal muscular atrophy. This disease is caused by knock-out mutations of the *SMN1* gene. The *SMN2* gene would produce a virtually identical protein except that in *SMN2* exon 7 is generally not included in the mRNA. Targeting splicing silencer motifs in intron 7 of the *SMN2* pre-mRNA induces exon 7 inclusion in cultured cells and animal models [203–205]. The ASO nusinersen targets the silencer motif and has received approval for treatment of spinal muscular atrophy patients by the Food and Drug Administration (FDA) and European Medicines Agency (EMA).

Rather than restoring splicing, ASOs can also be employed to intentionally induce exon skipping in order to remove an exon encoding a toxic domain or to inhibit production of a certain protein by blocking inclusion of an exon that encodes a functional domain or by causing production of an mRNA that is out of frame. Unlike the use of RNase H-mediated ASOs or siRNAs, which may almost complete inhibit production of protein from the targeted gene, the use of antisense to impact splicing may achieve controlled, partial inhibition. This may be desired when the disease causing gene is crucial for survival at certain doses. Furthermore, if a mutation that causes a dominant negative effect (e.g. a repeat expansion) is located in an in-frame exon, skipping that exon could produce a less toxic protein. This approach is being tested in models of spinocerebellar ataxia (SCA) type 3, a disease that is caused by an expanded CAG repeat in exon 10 of the *SCA3* gene that is translated into an ataxin protein that is toxic due to an expanded polyglutamine repeat. ASOs targeting exons 9 and 10 cause skipping of the repeat-encoding exon, while maintaining the reading frame to allow the production of a protein without the toxic expanded polyglutamine repeat that retains ubiquitin binding capacity [206]. Another advantage of exon skipping is that it allows isoform-specific inhibition of gene expression. Antisense-mediated gene inhibition has been demonstrated in cultured cells and an animal model with mutations in the gene that encodes ApoB100, the protein component of LDL-cholesterol that is involved in atherosclerosis; this oligonucleotide did not alter expression of another isoform, ApoB48, which is required for food absorption in the intestine [207, 208].

Finally, intentional exon skipping has been used to restore the reading frame of transcripts encoding modular proteins dystrophin and collagen [209, 210]. Modular proteins sometimes can function even when a part of the repeated domain is deleted. ASOs targeting in-frame exons containing nonsense mutations can induce the skipping of the targeted exon, thus bypassing the mutation without affecting the reading frame. Alternatively, ASOs have been used to bring about the skipping of an exon flanking an out-of-frame deletion to restore the open reading frame, allowing the production of partially functional proteins. This approach has been used to bypass mutations in the *COL7A1* gene in cultured cells from patients with dystrophic epidermolysis bullosa, a severe skin disease [210]. The ability of ASOs (drisapersen and eteplirsen) to restore the reading frame in the gene mutated in Duchenne muscular dystrophy patients, who suffer from a severe progressive neuromuscular disorder, has been tested by GlaxoSmithKline and BioMarin and Sarepta, respectively. [211]. Development of drisapersen has been abandoned, while eteplirsen received accellerated approval from FDA (see Chapter 17 for more details). Both splice sites and exonic sequences involved in exon recognition (exonic splicing enhancer sites) have successfully been targeted with ASOs to induce exon skipping. In Duchenne muscular dystrophy models, ASOs that target exonic enhancer sequences have outperformed those targeting splice sites. Exonic sequences may be thermodynamically more favorable targets due to the higher percentage of GC nucleotides in these sequences than in splice sites [212].

Finally, ASOs can be used to restore normal splicing in syndromes caused by disruption of alternative splicing patterns, such as is observed in myotonic dystrophy. This disease is caused by repeat expansions in the *DMPK* or the *CNBP* gene; these repeat regions form stable hairpin structures that sequester splicing factors to shift of splicing toward an embryonic pattern. Symptoms include myotonia as well as cardiac defects, insulin resistance, cataracts, and hypogonadism. ASOs targeting the repeats disrupt these secondary structures, freeing the sequestered splicing factors and normalizing normal splicing patterns [213–216].

1.7.3 Chemical Modifications that Enhance Activity of Oligonucleotide-based Splicing Modulators

Whereas natural DNA and RNA can bring about exon skipping *in vitro*, unmodified oligonucleotides have little activity in cells and in organisms due to degradation by DNases and RNases. Furthermore, for modulation of splicing, RNase-mediated cleavage of the RNA strand of the antisense RNA hybrids is not desired as this results in the inhibition of gene expression rather than splicing modulation. Thus, chemical modifications designed to improve stability, to render the hybrid RNase H resistant, and to improve PK properties have been incorporated into splice modulators (see Ref. [217] for an overview). The chemistries most often used are the 2′-*O*-methyl or 2′-*O*-methoxy RNAs with a phosphorothioate backbone (2OMePS and MOEPS, respectively), phosphorodiamidate morpholino oligomers (PMOs), peptide-conjugated PMOs, bridged nucleic acids (BNAs), and tricyclo RNAs. The 2OMePS and MOEPS chemistries are similar and impart RNase H resistance due to the ribose modification. The PS backbone improves uptake into cells and enables low-affinity serum protein binding, which prevents renal clearance. The drisapersen compound that was developed in Duchenne muscular dystrophy patients used the 2OMePS chemistry, while the compound approved for spinal muscular atrophy (nusinersen) is of the MOEPS chemistry.

PMOs are DNA homologues containing a morpholine ring; these oligonucleotide mimics are not recognized by any nuclease in the body. Since PMOs are neutral and do not bind serum proteins, these oligomers are quickly cleared by the kidney after systemic dosing. The ASO approved for Duchenne muscular dystrophy (eteplirsen from Sarapta Therapeutics) is of the PMO chemistry. To improve bioavailability, arginine-rich peptides have been linked to PMOs (PPMOs). Clinical development of PPMOs is currently hampered by toxicity [218, 219]. BNAs have a methylene, ethylene, or other type of bridge between the 2′ and 4′ positions of the ribose, forcing the nucleotide to adopt the endo-conformation and increasing affinity for the target. BNAs used *in vivo* usually have a PS backbone to prevent renal clearance. To prevent self-annealing of longer BNAs, generally BNA-2O-Me mixmers are used. BNAs show promise in cell and animal models for splicing modulation [220, 221]. In tricyclo-nucleosides, the 3′ C and the 5′ C of the deoxyribofuranose units are linked via an ethylene bridge that is rigid due to an annelated cyclopropane unit. Only limited cell studies have been reported with tricyclo ASOs [222].

Chemical compounds that can improve ASO-mediated exon skipping have been identified as well. The presence of 6-thioguanine (6TG) was reported to improve PMO-induced exon skipping *in vitro* and *in vivo*, although improvement was only observed *in vitro* in another study [204, 223]. Dantrolene has been shown to improve exon skipping levels for both PMOs and 2OMePS *in vitro* and *in vitro* [224]. Dantrolene is an FDA-approved drug.

1.7.4 Clinical Applications of Splicing Modulators

ASO-mediated splicing modulation holds promise for treatment of many genetic diseases. Two ASOs (nusinersen and eteplirsen) have now been approved for clinical use in spinal muscular atrophy and Duchenne muscular dystrophy, respectively. Oligonucleotide-based agents that induce splicing modulation often target single mutations or subsets of mutations and thus will find utility in only small groups of patients. This poses challenges on the clinical development as current regulatory models require large, placebo-controlled trials. A dialogue with regulatory agencies to discuss this issue has been initiated [225]. We anticipate that an antisense drug that modulates splicing will be approved for patient use in the near future, paving the way for the full exploitation of this approach.

1.8 Conclusions

There is an explosion of new knowledge on the mechanisms of action of oligonucleotides, and with the greater understanding of the regulatory roles of RNA, we may have many more processes and mechanisms to exploit for therapeutic purposes. Understanding mechanism is the best way to effectively utilize natural processes to achieve therapeutic ends.

References

1 Zamecnik, P.C. and Stephenson, M.L. (1978). Inhibition of Rous sarcoma virus replication and cell transformation by a specific oligodeoxynucleotide. *Proc. Natl. Acad. Sci. U. S. A.* 75 (1): 280–284.

2 Stephenson, M.L. and Zamecnik, P.C. (1978). Inhibition of Rous sarcoma viral RNA translation by a specific oligodeoxyribonucleotide. *Proc. Natl. Acad. Sci. U. S. A.* 75 (1): 285–288.

3 Eckstein, F. (2014). Phosphorothioates, Essential Components of Therapeutic Oligonucleotides. *Nucleic Acid Ther.* 141029131837003.

4 Swayze, E., Bhat, B., and Crooke, S.T. (2007). *The Medicinal Chemistry of Oligonucleotides* (ed. S.T. Crooke), 143–182. Boca Raton, FL: Taylor & Francis.

5 Altmann, K.-H., Dean, N.M., Fabbro, D. et al. (1996). Second generation of antisense oligonucleotides: from nuclease resistance to biological efficacy in animals. *Chimia* 50 (4): 168–176.

6 Vickers, T.A., Wyatt, J.R., and Freier, S.M. (2000). Effects of RNA secondary structure on cellular antisense activity. *Nucleic Acids Res.* 28 (6): 1340–1347.

7 Stanton, R., Sciabola, S., Salatto, C. et al. (2012). Chemical modification study of antisense gapmers. *Nucleic Acid Ther.* 22 (5): 344–359.

8 Koch, T., Shim, I., Lindow, M. et al. (2014). Quantum mechanical studies of DNA and LNA. *Nucleic Acid Ther.* 1–10.

9 Dominski, Z. and Kole, R. (1993). Restoration of correct splicing in thalassemic pre-mRNA by antisense oligonucleotides. *Proc. Natl. Acad. Sci. U. S. A.* 90 (18): 8673–8677.

10 Wu, H., Lima, W.F., Zhang, H. et al. (2004). Determination of the role of the human RNase H1 in the pharmacology of DNA-like antisense drugs. *Biochemistry* 279 (17): 17181–17189.

11 Lima, W.F., Rose, J.B., Nichols, J.G. et al. (2007). Human RNase H1 discriminates between subtle variations in the structure of the heteroduplex substrate. *Mol. Pharmacol.* 71 (1): 83–91.

12 Lima, W.F., Rose, J.B., Nichols, J.G. et al. (2007). The positional influence of the helical geometry of the heteroduplex substrate on human RNase H1 catalysis. *Mol. Pharmacol.* 71 (1): 73–82.

13 Lima, W.F., Nichols, J.G., Wu, H. et al. (2004). Structural requirements at the catalytic site of the heteroduplex substrate for human RNase H1 catalysis. *Biochemistry* 279 (35): 36317–36326.

14 Lima, W., Wu, H., and Crooke, S.T. (2007). The RNase H mechanism. In: *Antisense* (ed. S.T. Crooke), 47–74. Boca Raton, FL: CRC Press/Taylor & Francis Group.

15 Nowotny, M., Gaidamakov, S.A., Crouch, R.J., and Yang, W. (2005). Crystal structures of RNase H bound to an RNA/DNA hybrid: substrate specificity and metal-dependent catalysis. *Cell* 121 (7): 1005–1016.

16 Chapados, B.R., Chai, Q., Hosfield, D.J. et al. (2001). Structural biochemistry of a type 2 RNase H: RNA primer recognition and removal during DNA replication. *J. Mol. Biol.* 307 (2): 541–556.

17 Nowotny, M. and Yang, W. (2006). Stepwise analyses of metal ions in RNase H catalysis from substrate destabilization to product release. *EMBO J.* 25 (9): 1924–1933.

18 Levin, A.A. (1999). A review of the issues in the pharmacokinetics and toxicology of phosphorothioate antisense oligonucleotides. *Biochim. Biophys. Acta* 1489: 69–84.

19 Wu, H., Lima, W.F., and Crooke, S.T. (1999). Properties of cloned and expressed human RNase H1. *J. Biol. Chem.* 274 (40): 28270–28278.

20 Gao, W., Stein, C.A., Cohen, J.S. et al. (1989). Effect of phosphorothioate homo-oligodeoxynucleotides on herpes simplex virus type 2-induced DNA polymerase. *J. Biol. Chem.* 264 (19): 11521–11526.

21 Agrawal, S., Jiang, Z., Zhao, Q. et al. (1997). Mixed-backbone oligonucleotides as second generation antisense oligonucleotides: *in vitro* and *in vivo* studies. *Proc. Natl. Acad. Sci. U. S. A.* 94 (6): 2620–2625.

22 Crooke, S.T., Lemonidis, K.M., Neilson, L. et al. (1995). Kinetic characteristics of Escherichia coli RNase H1: cleavage of various antisense oligonucleotide-RNA duplexes. *Biochem. J.* 312 (Pt 2): 599–608.

23 Liang, X.-H., Sun, H., Shen, W., and Crooke, S.T. (2015). Identification and characterization of intracellular proteins that bind oligonucleotides with phosphorothioate linkages. *Nucleic Acids Res.* 43 (5): 2927–2945.

24 Bennett, C.F. and Swayze, E.E. (2010). RNA targeting therapeutics: molecular mechanisms of antisense oligonucleotides as a therapeutic platform. *Annu. Rev. Pharmacol. Toxicol.* 50: 259–293.

25 Pasternak, A. and Wengel, J. (2011). Unlocked nucleic acid – an RNA modification with broad potential. *Org. Biomol. Chem.* 9 (10): 3591–3597.

26 Fluiter, K., Mook, O.R.F., Vreijling, J. et al. (2009). Filling the gap in LNA antisense oligo gapmers: the effects of unlocked nucleic acid (UNA) and 4 0 - C - hydroxymethyl-DNA modifications on RNase H recruitment and efficacy of an LNA gapmer. *Mol. Biosyst.* 5: 838–843.

27 Kalota, A., Karabon, L., Swider, C.R. et al. (2006). 2′-Deoxy-2′-fluoro-beta-D-arabinonucleic acid (2′F-ANA) modified oligonucleotides (ON) effect highly efficient, and persistent, gene silencing. *Nucleic Acids Res.* 34 (2): 451–461.

28 Goraczniak, R., Behlke, M.A., and Gunderson, S.I. (2009). Gene silencing by synthetic U1 adaptors. *Nat. Biotechnol.* 27 (3): 257–263.

29 Lee, R.C., Feinbaum, R.L., and Ambros, V. (1993). The C. elegans heterochronic gene lin-4 encodes small RNAs with antisense complementarity to lin-14. *Cell* 75 (5): 843–854.

30 Reinhart, B.J., Slack, F.J., Basson, M. et al. (2000). The 21-nucleotide let-7 RNA regulates developmental timing in Caenorhabditis elegans. *Nature* 403 (6772): 901–906.

31 Lau, N.C., Lim, L.P., Weinstein, E.G., and Bartel, D.P. (2001). An abundant class of tiny RNAs with probable regulatory roles in Caenorhabditis elegans. *Science* 294 (5543): 858–862.

32 Lagos-Quintana, M., Rauhut, R., Lendeckel, W., and Tuschl, T. (2001). Identification of novel genes coding for small expressed RNAs. *Science* 294 (5543): 853–858.

33 Fire, A., Xu, S., Montgomery, M.K. et al. (1998). Potent and specific genetic interference by double-stranded RNA in Caenorhabditis elegans. *Nature* 391 (6669): 806–811.

34 Hamilton, A.J. and Baulcombe, D.C. (1999). A species of small antisense RNA in posttranscriptional gene silencing in plants. *Science* 286 (5441): 950–952.

35 Hammond, S.M., Bernstein, E., Beach, D., and Hannon, G.J. (2000). An RNA-directed nuclease mediates post-transcriptional gene silencing in Drosophila cells. *Nature* 404 (6775): 293–296.

36 Zamore, P.D., Tuschl, T., Sharp, P.A., and Bartel, D.P. (2000). RNAi: double-stranded RNA directs the ATP-dependent cleavage of mRNA at 21 to 23 nucleotide intervals. *Cell* 101 (1): 25–33.

37 Elbashir, S.M., Harborth, J., Lendeckel, W. et al. (2001). Duplexes of 21-nucleotide RNAs mediate RNA interference in cultured mammalian cells. *Nature* 411: 494–498.

38 Elbashir, S.M., Lendeckel, W., and Tuschl, T. (2001). RNA interference is mediated by 21- and 22-nucleotide RNAs. *Genes Dev.* 15 (2): 188–200.

39 Lai, E.C. (2002). Micro RNAs are complementary to 3′ UTR sequence motifs that mediate negative post-transcriptional regulation. *Nat. Genet.* 30 (4): 363–364.

40 Doench, J.G. and Sharp, P.A. (2004). Specificity of microRNA target selection in translational repression. *Genes Dev.* 18 (5): 504–511.

41 Krek, A., Grun, D., Poy, M.N. et al. (2005). Combinatorial microRNA target predictions. *Nat. Genet.* 37 (5): 495–500.

42 Lewis, B.P., Burge, C.B., and Bartel, D.P. (2005). Conserved seed pairing, often flanked by adenosines, indicates that thousands of human genes are microRNA targets. *Cell* 120 (1): 15–20.

43 Lewis, B.P., Shih, I.H., Jones-Rhoades, M.W. et al. (2003). Prediction of mammalian microRNA targets. *Cell* 115 (7): 787–798.

44 Didiano, D. and Hobert, O. (2006). Perfect seed pairing is not a generally reliable predictor for miRNA-target interactions. *Nat. Struct. Mol. Biol.* 13 (9): 849–851.

45 Ha, I., Wightman, B., and Ruvkun, G. (1996). A bulged lin-4/lin-14 RNA duplex is sufficient for Caenorhabditis elegans lin-14 temporal gradient formation. *Genes Dev.* 10 (23): 3041–3050.

46 Tay, Y., Zhang, J., Thomson, A.M. et al. (2008). MicroRNAs to Nanog, Oct4 and Sox2 coding regions modulate embryonic stem cell differentiation. *Nature* 455 (7216): 1124–1128.

47 Vella, M.C., Choi, E.Y., Lin, S.Y. et al. (2004). The C. elegans microRNA let-7 binds to imperfect let-7 complementary sites from the lin-41 3′ UTR. *Genes Dev.* 18 (2): 132–137.

48 Baek, D., Villen, J., Shin, C. et al. (2008). The impact of microRNAs on protein output. *Nature* 455 (7209): 64–71.

49 Lim, L.P., Lau, N.C., Garrett-Engele, P. et al. (2005). Microarray analysis shows that some microRNAs downregulate large numbers of target mRNAs. *Nature* 433: 769–773.

50 Selbach, M., Schwanhausser, B., Thierfelder, N. et al. (2008). Widespread changes in protein synthesis induced by microRNAs. *Nature* 455 (7209): 58–63.

51 Friedman, R.C., Farh, K.K., Burge, C.B., and Bartel, D.P. (2009). Most mammalian mRNAs are conserved targets of microRNAs. *Genome Res.* 19 (1): 92–105.

52 Carthew, R.W. and Sontheimer, E.J. (2009). Origins and mechanisms of miRNAs and siRNAs. *Cell* 136 (4): 642–655.

53 Ghildiyal, M. and Zamore, P.D. (2009). Small silencing RNAs: an expanding universe. *Nat. Rev. Genet.* 10 (2): 94–108.

54 Kim, D.H. and Rossi, J.J. (2007). Strategies for silencing human disease using RNA interference. *Nat. Rev. Genet.* 8 (3): 173–184.

55 Khvorova, A., Reynolds, A., and Jayasena, S.D. (2003). Functional siRNAs and miRNAs exhibit strand bias. *Cell* 115 (2): 209–216.
56 Schwarz, D.S., Hutvagner, G., Du, T. et al. (2003). Asymmetry in the assembly of the RNAi enzyme complex. *Cell* 115 (2): 199–208.
57 Rand, T.A., Petersen, S., Du, F., and Wang, X. (2005). Argonaute2 cleaves the anti-guide strand of siRNA during RISC activation. *Cell* 123 (4): 621–629.
58 Matranga, C., Tomari, Y., Shin, C. et al. (2005). Passenger-strand cleavage facilitates assembly of siRNA into Ago2-containing RNAi enzyme complexes. *Cell* 123 (4): 607–620.
59 Kawamata, T., Seitz, H., and Tomari, Y. (2009). Structural determinants of miRNAs for RISC loading and slicer-independent unwinding. *Nat. Struct. Mol. Biol.* 16 (9): 953–960.
60 Tomari, Y., Du, T., Haley, B. et al. (2004). RISC assembly defects in the Drosophila RNAi mutant armitage. *Cell* 116 (6): 831–841.
61 Yoda, M., Kawamata, T., Paroo, Z. et al. (2010). ATP-dependent human RISC assembly pathways. *Nat. Struct. Mol. Biol.* 17 (1): 17–23.
62 Liu, J., Carmell, M.A., Rivas, F.V. et al. (2004). Argonaute2 is the catalytic engine of mammalian RNAi. *Science* 305 (5689): 1437–1441.
63 Nykanen, A., Haley, B., and Zamore, P.D. (2001). ATP requirements and small interfering RNA structure in the RNA interference pathway. *Cell* 107: 309–321.
64 Brennecke, J., Stark, A., Russell, R.B., and Cohen, S.M. (2005). Principles of microRNA-target recognition. *PLoS Biol.* 3 (3): e85.
65 Wightman, B., Ha, I., and Ruvkun, G. (1993). Posttranscriptional regulation of the heterochronic gene lin-14 by lin-4 mediates temporal pattern formation in C. elegans. *Cell* 75: 855–862.
66 Haley, B. and Zamore, P.D. (2004). Kinetic analysis of the RNAi enzyme complex. *Nat. Struct. Mol. Biol.* 11 (7): 599–606.
67 Song, J.J., Liu, J., Tolia, N.H. et al. (2003). The crystal structure of the Argonaute2 PAZ domain reveals an RNA binding motif in RNAi effector complexes. *Nat. Struct. Biol.* 10 (12): 1026–1032.
68 Parker, J.S., Roe, M., and Barford, D. (2005). Structural insights into mRNA recognition from a PIWI domain-siRNA guide complex. *Nature* 434: 663–666.
69 Ma, J.-B., Yuan, Y.-R., Meister, G. et al. (2005). Structural basis for 5′-end-specific recognition of guide RNA by the A. fulgidus Piwi protein. *Nature* 434: 666–670.
70 Wee, L.M., Flores-Jasso, C.F., Salomon, W.E., and Zamore, P.D. (2012). Argonaute divides its RNA guide into domains with distinct functions and RNA-binding properties. *Cell* 151 (5): 1055–1067.
71 Jackson, A.L., Burchard, J., Schelter, J. et al. (2006). Widespread siRNA "off-target" transcript silencing mediated by seed region sequence complementarity. *RNA* 12: 1179–1187.
72 Yekta, S., Shih, I., and Bartel, D.P. (2004). MicroRNA-directed cleavage of HOXB8 mRNA. *Science* 304: 594–596.

73 Meister, G., Landthaler, M., Patkaniowska, A. et al. (2004). Human Argonaute2 mediates RNA cleavage targeted by miRNAs and siRNAs. *Mol. Cell* 15 (2): 185–197.
74 Elbashir, S.M., Martinez, J., Patkaniowska, A. et al. (2001). Functional anatomy of siRNAs for mediating efficient RNAi in Drosophila melanogaster embryo lysate. *EMBO J.* 20 (23): 6877–6888.
75 Martinez, J. and Tuschl, T. (2004). RISC is a 5′ phosphomonoester-producing RNA endonuclease. *Genes Dev.* 18 (9): 975–980.
76 Elbashir, S.M., Harborth, J., Weber, K., and Tuschl, T. (2002). Analysis of gene function in somatic mammalian cells using small interfering RNAs. *Methods* 26: 199–213.
77 Amarzguioui, M., Holen, T., Babaie, E., and Prydz, H. (2003). Tolerance for mutations and chemical modifications in a siRNA. *Nucleic Acids Res.* 31 (2): 589–595.
78 Holen, T., Amarzguioui, M., Wiiger, M.T. et al. (2002). Positional effects of short interfering RNAs targeting the human coagulation trigger tissue factor. *Nucleic Acids Res.* 30: 1757–1766.
79 Grimson, A., Farh, K.K., Johnston, W.K. et al. (2007). MicroRNA targeting specificity in mammals: determinants beyond seed pairing. *Mol. Cell* 27 (1): 91–105.
80 Hutvagner, G. and Simard, M.J. (2008). Argonaute proteins: key players in RNA silencing. *Nat. Rev. Mol. Cell Biol.* 9 (1): 22–32.
81 Martinez, L.A., Naguibneva, I., Lehrmann, H. et al. (2002). Synthetic small inhibiting RNAs: efficient tools to inactivate oncogenic mutations and restore p53 pathways. *Proc. Natl. Acad. Sci. U. S. A.* 99: 14849–14854.
82 Noland, C.L. and Doudna, J.A. (2013). Multiple sensors ensure guide strand selection in human RNAi pathways. *RNA* 19 (5): 639–648.
83 Caplen, N.J., Parrish, S., Imani, F. et al. (2001). Specific inhibition of gene expression by small double-stranded RNAs in invertebrate and vertebrate systems. *Proc. Natl. Acad. Sci. U. S. A.* 98: 9742–9747.
84 Tuschl, T., Zamore, P.D., Lehmann, R. et al. (1999). Targeted mRNA degradation by double-stranded RNA *in vitro*. *Genes Dev.* 13: 3191–3197.
85 Jackson, A.L., Bartz, S.R., Schelter, J. et al. (2003). Expression profiling reveals off-target gene regulation by RNAi. *Nat. Biotechnol.* 21: 635–637.
86 Ui-Tei, K., Naito, Y., Takahashi, F. et al. (2004). Guidelines for the selection of highly effective siRNA sequences for mammalian and chick RNA interference. *Nucleic Acids Res.* 32: 936–948.
87 Vaish, N., Chen, F., Seth, S. et al. (2011). Improved specificity of gene silencing by siRNAs containing unlocked nucleobase analogs. *Nucleic Acids Res.* 39 (5): 1823–1832.
88 Bramsen, J.B., Pakula, M.M., Hansen, T.B. et al. (2010). A screen of chemical modifications identifies position-specific modification by UNA to most potently reduce siRNA off-target effects. *Nucleic Acids Res.* 38 (17): 5761–5773.

89 Kittler, R., Surendranath, V., Heninger, A.K. et al. (2007). Genome-wide resources of endoribonuclease-prepared short interfering RNAs for specific loss-of-function studies. *Nat. Methods* 4 (4): 337–344.
90 Myers, J.W., Chi, J.-T., Gong, D. et al. (2006). Minimizing off-target effects by using diced siRNAs for RNA interference. *J. RNAi Gene Silencing* 2 (2): 181–194.
91 Eulalio, A., Huntzinger, E., and Izaurralde, E. (2008). Getting to the root of miRNA-mediated gene silencing. *Cell* 132 (1): 9–14.
92 Bartel, D.P. (2009). MicroRNAs: target recognition and regulatory functions. *Cell* 136 (2): 215–233.
93 Fabian, M.R., Sonenberg, N., and Filipowicz, W. (2010). Regulation of mRNA translation and stability by microRNAs. *Annu. Rev. Biochem.* 79: 351–379.
94 Chekulaeva, M., Mathys, H., Zipprich, J.T. et al. (2011). miRNA repression involves GW182-mediated recruitment of CCR4-NOT through conserved W-containing motifs. *Nat. Struct. Mol. Biol.* 18 (11): 1218–1226.
95 Jinek, M., Fabian, M.R., Coyle, S.M. et al. (2010). Structural insights into the human GW182-PABC interaction in microRNA-mediated deadenylation. *Nat. Struct. Mol. Biol.* 17 (2): 238–240.
96 Huntzinger, E., Braun, J.E., Heimstadt, S. et al. (2010). Two PABPC1-binding sites in GW182 proteins promote miRNA-mediated gene silencing. *EMBO J.* 29 (24): 4146–4160.
97 Braun, J.E., Huntzinger, E., Fauser, M., and Izaurralde, E. (2011). GW182 proteins directly recruit cytoplasmic deadenylase complexes to miRNA targets. *Mol. Cell* 44 (1): 120–133.
98 Fabian, M.R., Cieplak, M.K., Frank, F. et al. (2011). miRNA-mediated deadenylation is orchestrated by GW182 through two conserved motifs that interact with CCR4-NOT. *Nat. Struct. Mol. Biol.* 18 (11): 1211–1217.
99 Zekri, L., Huntzinger, E., Heimstadt, S., and Izaurralde, E. (2009). The silencing domain of GW182 interacts with PABPC1 to promote translational repression and degradation of microRNA targets and is required for target release. *Mol. Cell. Biol.* 29 (23): 6220–6231.
100 Fabian, M.R. and Sonenberg, N. (2012). The mechanics of miRNA-mediated gene silencing: a look under the hood of miRISC. *Nat. Struct. Mol. Biol.* 19 (6): 586–593.
101 Yang, Y., Chaerkady, R., Kandasamy, K. et al. (2010). Identifying targets of miR-143 using a SILAC-based proteomic approach. *Mol. Biosyst.* 6 (10): 1873–1882.
102 Hendrickson, D.G., Hogan, D.J., McCullough, H.L. et al. (2009). Concordant regulation of translation and mRNA abundance for hundreds of targets of a human microRNA. *PLoS Biol.* 7 (11): e1000238.
103 Guo, H., Ingolia, N.T., Weissman, J.S., and Bartel, D.P. (2010). Mammalian microRNAs predominantly act to decrease target mRNA levels. *Nature* 466 (7308): 835–840.

104 Djuranovic, S., Nahvi, A., and Green, R. (2012). miRNA-mediated gene silencing by translational repression followed by mRNA deadenylation and decay. *Science* 336 (6078): 237–240.

105 Bazzini, A.A., Lee, M.T., and Giraldez, A.J. (2012). Ribosome profiling shows that miR-430 reduces translation before causing mRNA decay in zebrafish. *Science* 336 (6078): 233–237.

106 Xiao, C., Calado, D.P., Galler, G. et al. (2007). MiR-150 controls B cell differentiation by targeting the transcription factor c-Myb. *Cell* 131 (1): 146–159.

107 Georges, S.A., Biery, M.C., Kim, S.Y. et al. (2008). Coordinated regulation of cell cycle transcripts by p53-Inducible microRNAs, miR-192 and miR-215. *Cancer Res.* 68 (24): 10105–10112.

108 Linsley, P.S., Schelter, J., Burchard, J. et al. (2007). Transcripts targeted by the microRNA-16 family cooperatively regulate cell cycle progression. *Mol. Cell. Biol.* 27 (6): 2240–2252.

109 Valastyan, S., Benaich, N., Chang, A. et al. (2009). Concomitant suppression of three target genes can explain the impact of a microRNA on metastasis. *Genes Dev.* 23 (22): 2592–2597.

110 Sood, P., Krek, A., Zavolan, M. et al. (2006). Cell-type-specific signatures of microRNAs on target mRNA expression. *Proc. Natl. Acad. Sci. U. S. A.* 103 (8): 2746–2751.

111 Chen, K. and Rajewsky, N. (2006). Natural selection on human microRNA binding sites inferred from SNP data. *Nat. Genet.* 38 (12): 1452–1456.

112 Esau, C., Davis, S., Murray, S.F. et al. (2006). miR-122 regulation of lipid metabolism revealed by *in vivo* antisense targeting. *Cell Metab.* 3: 87–98.

113 Elmen, J., Lindow, M., Silahtaroglu, A. et al. (2008). Antagonism of microRNA-122 in mice by systemically administered LNA-antimiR leads to up-regulation of a large set of predicted target mRNAs in the liver. *Nucleic Acids Res.* 36 (4): 1153–1162.

114 Elmen, J., Lindow, M., Schutz, S. et al. (2008). LNA-mediated microRNA silencing in non-human primates. *Nature* 452 (7189): 896–899.

115 Lanford, R.E., Hildebrandt-Eriksen, E.S., Petri, A. et al. (2010). Therapeutic silencing of microRNA-122 in primates with chronic hepatitis C virus infection. *Science* 327 (5962): 198–201.

116 Watanabe, T.A., Geary, R.S., and Levin, A.A. (2006). Plasma protein binding of an antisense oligonucleotide targeting human ICAM-1 (ISIS 2302). *Oligonucleotides* 16 (2): 169–180.

117 Eckstein, F. (1967). A dinucleoside phosphorothioate. *Tetrahedron Lett.* 8: 1157–1160.

118 Eckstein, F. (1966). Nucleoside phosphorothioates. *J. Am. Chem. Soc.* 88: 4292–4294.

119 Eckstein, F. (2000). Phosphorothioate oligodeoxynucleotides: what is their origin and what is unique about them. *Antisense Nucleic Acid Drug Dev.* 10: 117–121.

120 Geary, R.S., Yu, R.Z., and Levin, A.A. (2001). Pharmacokinetics of phosphorothioate antisense oligodeoxynucleotides. *Curr. Opin. Investig. Drugs* 2 (4): 562–573.
121 Leuschner, P.J., Ameres, S.L., Kueng, S., and Martinez, J. (2006). Cleavage of the siRNA passenger strand during RISC assembly in human cells. *EMBO Rep.* 7 (3): 314–320.
122 Braasch, D.A., Jensen, S., Liu, Y. et al. (2003). RNA interference in mammalian cells by chemically-modified RNA. *Biochemistry* 42 (26): 7967–7975.
123 Chiu, Y.L. and Rana, T.M. (2003). siRNA function in RNAi: a chemical modification analysis. *RNA* 9 (9): 1034–1048.
124 Deleavey, G.F., Watts, J.K., and Damha, M.J. (2009). Chemical modification of siRNA. *Curr. Protoc. Nucleic Acid Chem.* Chapter 16, Unit 16 3.
125 Corey, D.R. (2007). Chemical modification: the key to clinical application of RNA interference? *J. Clin. Invest.* 117 (12): 3615–3622.
126 Deleavey, G.F. and Damha, M.J. (2012). Designing chemically modified oligonucleotides for targeted gene silencing. *Chem. Biol.* 19 (8): 937–954.
127 Watts, J.K., Deleavey, G.F., and Damha, M.J. (2008). Chemically modified siRNA: tools and applications. *Drug Discov. Today* 13 (19–20): 842–855.
128 Stenvang, J., Petri, A., Lindow, M. et al. (2012). Inhibition of microRNA function by antimiR oligonucleotides. *Silence* 3 (1): 1.
129 Allerson, C.R., Sioufi, N., Jarres, R. et al. (2005). Fully 2′-modified oligonucleotide duplexes with improved *in vitro* potency and stability compared to unmodified small interfering RNA. *J. Med. Chem.* 48 (4): 901–904.
130 Blidner, R.A., Hammer, R.P., Lopez, M.J. et al. (2007). Fully 2′-deoxy-2′-fluoro substituted nucleic acids induce RNA interference in mammalian cell culture. *Chem. Biol. Drug Des.* 70 (2): 113–122.
131 Odadzic, D., Bramsen, J.B., Smicius, R. et al. (2008). Synthesis of 2′-O-modified adenosine building blocks and application for RNA interference. *Bioorg. Med. Chem.* 16 (1): 518–529.
132 Mook, O.R., Baas, F., de Wissel, M.B., and Fluiter, K. (2007). Evaluation of locked nucleic acid-modified small interfering RNA *in vitro* and *in vivo*. *Mol. Cancer Ther.* 6 (3): 833–843.
133 Glud, S.Z., Bramsen, J.B., Dagnaes-Hansen, F. et al. (2009). Naked siLNA-mediated gene silencing of lung bronchoepithelium EGFP expression after intravenous administration. *Oligonucleotides* 19 (2): 163–168.
134 Hornung, V., Guenthner-Biller, M., Bourquin, C. et al. (2005). Sequence-specific potent induction of IFN-alpha by short interfering RNA in plasmacytoid dendritic cells through TLR7. *Nat. Med.* 11 (3): 263–270.
135 Bramsen, J.B., Laursen, M.B., Nielsen, A.F. et al. (2009). A large-scale chemical modification screen identifies design rules to generate siRNAs with high activity, high stability and low toxicity. *Nucleic Acids Res.* 37 (9): 2867–2881.

136 Dowler, T., Bergeron, D., Tedeschi, A.-L. et al. (2006). Improvements in siRNA properties mediated by 2′-deoxy-2′-fluoro-beta-D-arabinonucleic acid (FANA). *Nucleic Acids Res.* 34 (6): 1669–1675.

137 Nauwelaerts, K., Fisher, M., Froeyen, M. et al. (2007). Structural characterization and biological evaluation of small interfering RNAs containing cyclohexenyl nucleosides. *J. Am. Chem. Soc.* 129 (30): 9340–9348.

138 Deleavey, G.F., Watts, J.K., Alain, T. et al. (2010). Synergistic effects between analogs of DNA and RNA improve the potency of siRNA-mediated gene silencing. *Nucleic Acids Res.* 38 (13): 4547–4557.

139 Laursen, M.B., Pakula, M.M., Gao, S. et al. (2010). Utilization of unlocked nucleic acid (UNA) to enhance siRNA performance *in vitro* and *in vivo*. *Mol. Biosyst.* 6 (5): 862–870.

140 Kariko, K., Bhuyan, P., Capodici, J., and Weissman, D. (2004). Small interfering RNAs mediate sequence-independent gene suppression and induce immune activation by signaling through toll-like receptor 3. *J. Immunol.* 172 (11): 6545–6549.

141 Kleinman, M.E., Yamada, K., Takeda, A. et al. (2008). Sequence- and target-independent angiogenesis suppression by siRNA via TLR3. *Nature* 452 (7187): 591–597.

142 Sioud, M. (2005). Induction of inflammatory cytokines and interferon responses by double-stranded and single-stranded siRNAs is sequence-dependent and requires endosomal localization. *J. Mol. Biol.* 348 (5): 1079–1090.

143 Cario, E. and Podolsky, D.K. (2000). Differential alteration in intestinal epithelial cell expression of toll-like receptor 3 (TLR3) and TLR4 in inflammatory bowel disease. *Infect. Immun.* 68 (12): 7010–7017.

144 Lim, D.M., Narasimhan, S., Michaylira, C.Z., and Wang, M.-L. (2009). TLR3-mediated NF-{kappa}B signaling in human esophageal epithelial cells. *Am. J. Physiol. Gastrointest. Liver Physiol.* 297 (6): G1172–G1180.

145 Taura, M., Eguma, A., Suico, M.A. et al. (2008). p53 Regulates Toll-like receptor 3 expression and function in human epithelial cell lines. *Mol. Cell. Biol.* 28 (21): 6557–6567.

146 Zimmer, S., Steinmetz, M., Asdonk, T. et al. (2011). Activation of endothelial toll-like receptor 3 impairs endothelial function. *Circ. Res.* 108 (11): 1358–1366.

147 Heil, F., Hemmi, H., Hochrein, H. et al. (2004). Species-specific recognition of single-stranded RNA via toll-like receptor 7 and 8. *Science* 303 (5663): 1526–1529.

148 Kokkinopoulos, I., Jordan, W.J., and Ritter, M.A. (2005). Toll-like receptor mRNA expression patterns in human dendritic cells and monocytes. *Mol. Immunol.* 42 (8): 957–968.

149 Hornung, V., Rothenfusser, S., Britsch, S. et al. (2002). Quantitative expression of toll-like receptor 1-10 mRNA in cellular subsets of human peripheral blood mononuclear cells and sensitivity to CpG oligodeoxynucleotides. *J. Immunol.* 168 (9): 4531–4537.

150 Judge, A.D., Sood, V., Shaw, J.R. et al. (2005). Sequence-dependent stimulation of the mammalian innate immune response by synthetic siRNA. *Nat. Biotechnol.* 23 (4): 457–462.

151 Robbins, M., Judge, A., Ambegia, E. et al. (2008). Misinterpreting the therapeutic effects of siRNA caused by immune stimulation. *Hum. Gene Ther.* 19 (10): 991–999.

152 Zamanian-Daryoush, M., Marques, J.T., Gantier, M.P. et al. (2008). Determinants of cytokine induction by small interfering RNA in human peripheral blood mononuclear cells. *J. Interferon Cytokine Res.* 28 (4): 221–233.

153 Kariko, K., Buckstein, M., Ni, H., and Weissman, D. (2005). Suppression of RNA recognition by Toll-like receptors: the impact of nucleoside modification and the evolutionary origin of RNA. *Immunity* 23 (2): 165–175.

154 Morrissey, D.V., Blanchard, K., Shaw, L. et al. (2005). Activity of stabilized short interfering RNA in a mouse model of hepatitis B virus replication. *Hepatology* 41 (6): 1349–1356.

155 Judge, A.D., Bola, G., Lee, A.C., and MacLachlan, I. (2006). Design of noninflammatory synthetic siRNA mediating potent gene silencing *in vivo*. *Mol. Ther.* 13 (3): 494–505.

156 Peacock, H., Kannan, A., Beal, P.A., and Burrows, C.J. (2011). Chemical modification of siRNA bases to probe and enhance RNA interference. *J. Organomet. Chem.* 76 (18): 7295–7300.

157 Eberle, F., Giessler, K., Deck, C. et al. (2008). Modifications in small interfering RNA that separate immunostimulation from RNA interference. *J. Immunol.* 180 (5): 3229–3237.

158 Peacock, H., Fucini, R.V., Jayalath, P. et al. (2011). Nucleobase and ribose modifications control immunostimulation by a microRNA-122-mimetic RNA. *J. Am. Chem. Soc.* 133 (24): 9200–9203.

159 Puthenveetil, S., Whitby, L., Ren, J. et al. (2006). Controlling activation of the RNA-dependent protein kinase by siRNAs using site-specific chemical modification. *Nucleic Acids Res.* 34 (17): 4900–4911.

160 Nallagatla, S.R. and Bevilacqua, P.C. (2008). Nucleoside modifications modulate activation of the protein kinase PKR in an RNA structure-specific manner. *RNA* 14 (6): 1201–1213.

161 Bramsen, J.B. and Kjems, J. (2012). Development of therapeutic-grade small interfering RNAs by chemical engineering. *Front. Genet.* 3: 154.

162 Diebold, S.S., Massacrier, C., Akira, S. et al. (2006). Nucleic acid agonists for Toll-like receptor 7 are defined by the presence of uridine ribonucleotides. *Eur. J. Immunol.* 36 (12): 3256–3267.

163 Fucini, R.V., Haringsma, H.J., Deng, P. et al. (2012). Adenosine modification may be preferred for reducing siRNA immune stimulation. *Nucleic Acid Ther.* 22 (3): 205–210.

164 Robbins, M., Judge, A., Liang, L. et al. (2007). 2′-O-methyl-modified RNAs act as TLR7 antagonists. *Mol. Ther.* 15 (9): 1663–1669.

165 Goodchild, A., Nopper, N., King, A. et al. (2009). Sequence determinants of innate immune activation by short interfering RNAs. *BMC Immunol.* 10: 40.

166 Jackson, A.L. and Linsley, P.S. (2010). Recognizing and avoiding siRNA off-target effects for target identification and therapeutic application. *Nat. Rev. Drug Discov.* 9 (1): 57–67.

167 Gavrilov, K. and Saltzman, W.M. (2012). Therapeutic siRNA: principles, challenges, and strategies. *Yale J. Biol. Med.* 85 (2): 187–200.

168 Whitehead, K.A., Langer, R., and Anderson, D.G. (2009). Knocking down barriers: advances in siRNA delivery. *Nat. Rev. Drug Discov.* 8 (2): 129–138.

169 Soutschek, J., Akinc, A., Bramlage, B. et al. (2004). Therapeutic silencing of an endogenous gene by systemic administration of modified siRNAs. *Nature* 432 (7014): 173–178.

170 Wolfrum, C., Shi, S., Jayaprakash, K.N. et al. (2007). Mechanisms and optimization of *in vivo* delivery of lipophilic siRNAs. *Nat. Biotechnol.* 25 (10): 1149–1157.

171 Hollins, A.J., Omidi, Y., Benter, I.F., and Akhtar, S. (2007). Toxicogenomics of drug delivery systems: exploiting delivery system-induced changes in target gene expression to enhance siRNA activity. *J. Drug Target.* 15 (1): 83–88.

172 Akhtar, S. (2010). Cationic nanosystems for the delivery of small interfering ribonucleic acid therapeutics: a focus on toxicogenomics. *Expert Opin. Drug Metab. Toxicol.* 6 (11): 1347–1362.

173 Omidi, Y., Hollins, A.J., Benboubetra, M. et al. (2003). Toxicogenomics of non-viral vectors for gene therapy: a microarray study of lipofectin- and oligofectamine-induced gene expression changes in human epithelial cells. *J. Drug Target.* 11 (6): 311–323.

174 Krutzfeldt, J., Rajewsky, N., Braich, R. et al. (2005). Silencing of microRNAs *in vivo* with "antagomirs". *Nature* 438 (7068): 685–689.

175 Montgomery, R.L., Hullinger, T.G., Semus, H.M. et al. (2011). Therapeutic inhibition of miR-208a improves cardiac function and survival during heart failure. *Circulation* 124 (14): 1537–1547.

176 Najafi-Shoushtari, S.H., Kristo, F., Li, Y. et al. (2010). MicroRNA-33 and the SREBP host genes cooperate to control cholesterol homeostasis. *Science* 328 (5985): 1566–1569.

177 Worm, J., Stenvang, J., Petri, A. et al. (2009). Silencing of microRNA-155 in mice during acute inflammatory response leads to derepression of c/ebp Beta and down-regulation of G-CSF. *Nucleic Acids Res.* 37 (17): 5784–5792.

178 Davis, S., Propp, S., Freier, S.M. et al. (2009). Potent inhibition of microRNA *in vivo* without degradation. *Nucleic Acids Res.* 37 (1): 70–77.

179 Hullinger, T.G., Montgomery, R.L., Seto, A.G. et al. (2012). Inhibition of miR-15 protects against cardiac ischemic injury. *Circ. Res.* 110 (1): 71–81.

180 Marquart, T.J., Allen, R.M., Ory, D.S., and Baldan, A. (2010). miR-33 links SREBP-2 induction to repression of sterol transporters. *Proc. Natl. Acad. Sci. U. S. A.* 107 (27): 12228–12232.

181 Janssen, H.L., Reesink, H.W., Lawitz, E.J. et al. (2013). Treatment of HCV infection by targeting MicroRNA. *N. Engl. J. Med.* 368: 1685–1694.

182 Vaishnaw, A.K., Gollob, J., Gamba-Vitalo, C. et al. (2010). A status report on RNAi therapeutics. *Silence* 1 (1): 14.

183 Valadi, H., Ekstrom, K., Bossios, A. et al. (2007). Exosome-mediated transfer of mRNAs and microRNAs is a novel mechanism of genetic exchange between cells. *Nat. Cell Biol.* 9 (6): 654–659.

184 Van Niel, G., Porto-Carreiro, I., Simoes, S., and Raposo, G. (2006). Exosomes: a common pathway for a specialized function. *J. Biochem.* 140 (1): 13–21.

185 Gibbings, D.J., Ciaudo, C., Erhardt, M., and Voinnet, O. (2009). Multivesicular bodies associate with components of miRNA effector complexes and modulate miRNA activity. *Nat. Cell Biol.* 11 (9): 1143–1149.

186 Pant, S., Hilton, H., and Burczynski, M.E. (2012). The multifaceted exosome: biogenesis, role in normal and aberrant cellular function, and frontiers for pharmacological and biomarker opportunities. *Biochem. Pharmacol.* 83 (11): 1484–1494.

187 Mittelbrunn, M., Gutierrez-Vazquez, C., Villarroya-Beltri, C. et al. (2011). Unidirectional transfer of microRNA-loaded exosomes from T cells to antigen-presenting cells. *Nat. Commun.* 2: 282.

188 Delorme-Axford, E., Donker, R.B., Mouillet, J.-F. et al. (2013). Human placental trophoblasts confer viral resistance to recipient cells. *Proc. Natl. Acad. Sci. U. S. A.* 110 (29): 12048–12053.

189 Xin, H., Li, Y., Liu, Z. et al. (2013). Mir-133b promotes neural plasticity and functional recovery after treatment of stroke with multipotent mesenchymal stromal cells in rats via transfer of exosome-enriched extracellular particles. *Stem Cells* 31: 2737–2746.

190 Hannafon, B.N. and Ding, W.Q. (2013). Intercellular communication by exosome-derived microRNAs in cancer. *Int. J. Mol. Sci.* 14 (7): 14240–14269.

191 Alvarez-Erviti, L., Seow, Y., Yin, H. et al. (2011). Delivery of siRNA to the mouse brain by systemic injection of targeted exosomes. *Nat. Biotechnol.* 29 (4): 341–345.

192 Pan, Q., Ramakrishnaiah, V., Henry, S. et al. (2012). Hepatic cell-to-cell transmission of small silencing RNA can extend the therapeutic reach of RNA interference (RNAi). *Gut* 61 (9): 1330–1339.

193 Wahlgren, J., De, L.K.T., Brisslert, M. et al. (2012). Plasma exosomes can deliver exogenous short interfering RNA to monocytes and lymphocytes. *Nucleic Acids Res.* 40 (17): e130.

194 Ohno, S., Takanashi, M., Sudo, K. et al. (2013). Systemically injected exosomes targeted to EGFR deliver antitumor microRNA to breast cancer cells. *Mol. Ther.* 21 (1): 185–191.

195 Hammond, S.M. and Wood, M.J.A. (2011). Genetic therapies for RNA mis-splicing diseases. *Trends Genet.* 27 (5): 196–205.

196 Van Roon-Mom, W.M.C. and Aartsma-Rus, A. (2012). Overview on applications of antisense-mediated exon skipping. *Methods Mol. Biol.* 867: 79–96.

197 Friedman, K.J., Kole, J., Cohn, J.A. et al. (1999). Correction of aberrant splicing of the cystic fibrosis transmembrane conductance regulator (CFTR) gene by antisense oligonucleotides. *J. Biol. Chem.* 274 (51): 36193–36199.

198 Svasti, S., Suwanmanee, T., Fucharoen, S. et al. (2009). RNA repair restores hemoglobin expression in IVS2-654 thalassemic mice. *Proc. Natl. Acad. Sci. U. S. A.* 106 (4): 1205–1210.

199 Spitali, P. and Aartsma-Rus, A. (2012). Splice modulating therapies for human disease. *Cell* 148 (6): 1085–1088.

200 Mercatante, D.R. and Kole, R. (2002). Control of alternative splicing by antisense oligonucleotides as a potential chemotherapy: effects on gene expression. *Biochim. Biophys. Acta* 1587 (2–3): 126–132.

201 Mercatante, D.R., Sazani, P., and Kole, R. (2001). Modification of alternative splicing by antisense oligonucleotides as a potential chemotherapy for cancer and other diseases. *Curr. Cancer Drug Targets* 1 (3): 211–230.

202 Zammarchi, F., de Stanchina, E., Bournazou, E. et al. (2011). Antitumorigenic potential of STAT3 alternative splicing modulation. *Proc. Natl. Acad. Sci.* 108 (43): 17779–17784.

203 Hua, Y., Sahashi, K., Rigo, F. et al. (2011). Peripheral SMN restoration is essential for long-term rescue of a severe spinal muscular atrophy mouse model. *Nature* 478 (7367): 123–126.

204 Hua, Y., Sahashi, K., Hung, G. et al. (2010). Antisense correction of SMN2 splicing in the CNS rescues necrosis in a type III SMA mouse model. *Genes Dev.* 24 (15): 1634–1644.

205 Passini, M.A., Bu, J., Richards, A.M. et al. (2011). Antisense oligonucleotides delivered to the mouse CNS ameliorate symptoms of severe spinal muscular atrophy. *Sci. Transl. Med.* 3 (72): 72ra18.

206 Evers, M.M., Tran, H.-D., Zalachoras, I. et al. (2013). Ataxin-3 protein modification as a treatment strategy for spinocerebellar ataxia type 3: removal of the CAG containing exon. *Neurobiol. Dis.* 58: 49–56.

207 Disterer, P., Al-Shawi, R., Ellmerich, S. et al. (2013). Exon skipping of hepatic APOB pre-mRNA with splice-switching oligonucleotides reduces LDL cholesterol *in vivo*. *Mol. Ther.* 21 (3): 602–609.

208 Khoo, B., Roca, X., Chew, S.L., and Krainer, A.R. (2007). Antisense oligonucleotide-induced alternative splicing of the APOB mRNA generates a novel isoform of APOB. *BMC Mol. Biol.* 8: 3.

209 Aartsma-Rus, A. (2014). Antisense-mediated modulation of splicing: therapeutic implications for Duchenne muscular dystrophy. *RNA Biol.* 7 (4): 453–461.

210 Goto, M., Sawamura, D., Nishie, W. et al. (2006). Targeted skipping of a single exon harboring a premature termination codon mutation: implications and potential for gene correction therapy for selective dystrophic epidermolysis bullosa patients. *J. Invest. Dermatol.* 126 (12): 2614–2620.

211 Verhaart, I.E.C. and Aartsma-Rus, A. (2012). Gene therapy for Duchenne muscular dystrophy. *Curr. Opin. Neurol.* 25 (5): 588–596.

212 Aartsma-Rus, A., Houlleberghs, H., van Deutekom, J.C.T. et al. (2010). Exonic sequences provide better targets for antisense oligonucleotides than splice site sequences in the modulation of Duchenne muscular dystrophy splicing. *Oligonucleotides* 20 (2): 69–77.

213 Wheeler, T.M., Sobczak, K., Lueck, J.D. et al. (2009). Reversal of RNA dominance by displacement of protein sequestered on triplet repeat RNA. *Science* 325 (5938): 336–339.

214 Wheeler, T.M., Lueck, J.D., Swanson, M.S. et al. (2007). Correction of ClC-1 splicing eliminates chloride channelopathy and myotonia in mouse models of myotonic dystrophy. *J. Clin. Invest.* 117 (12): 3952–3957.

215 Wheeler, T.M., Leger, A.J., Pandey, S.K. et al. (2012). Targeting nuclear RNA for *in vivo* correction of myotonic dystrophy. *Nature* 488 (7409): 111–115.

216 Mulders, S.A.M., Van Den Broek, W.J.A.A., Wheeler, T.M. et al. (2009). Triplet-repeat oligonucleotide-mediated reversal of RNA toxicity in myotonic dystrophy. *Proc. Natl. Acad. Sci. U. S. A.* 106 (33): 13915–13920.

217 Saleh, A.F., Arzumanov, A.A., and Gait, M.J. (2012). Overview of alternative oligonucleotide chemistries for exon skipping. *Methods Mol. Biol.* 867: 365–378.

218 El Andaloussi, S.A., Hammond, S.M., Mäger, I., and Wood, M.J.A. (2012). Use of cell-penetrating-peptides in oligonucleotide splice switching therapy. *Curr. Gene Ther.* 12 (3): 161–178.

219 Moulton, H.M. and Moulton, J.D. (2010). Morpholinos and their peptide conjugates: therapeutic promise and challenge for Duchenne muscular dystrophy. *Biochim. Biophys. Acta* 1798 (12): 2296–2303.

220 Graziewicz, M.A., Tarrant, T.K., Buckley, B. et al. (2008). An endogenous TNF-alpha antagonist induced by splice-switching oligonucleotides reduces inflammation in hepatitis and arthritis mouse models. *Mol. Ther.* 16 (7): 1316–1322.

221 Takagi, M., Yagi, M., Ishibashi, K. et al. (2004). Design of 2′-O-Me RNA/ENA chimera oligonucleotides to induce exon skipping in dystrophin pre-mRNA. *Nucleic Acids Symp. Ser. (Oxf).* (48): 297–298.

222 Ittig, D., Liu, S., Renneberg, D. et al. (2004). Nuclear antisense effects in cyclophilin A pre-mRNA splicing by oligonucleotides: a comparison of tricyclo-DNA with LNA. *Nucleic Acids Res.* 32 (1): 346–353.

223 Verhaart, I.E.C. and Aartsma-Rus, A. (2012). The effect of 6-thioguanine on alternative splicing and antisense-mediated exon skipping treatment for duchenne muscular dystrophy. *PLoS Curr.* 4. doi: 10.1371/currents.md.597d700f92eaa70de261ea0d91821377.

224 Kendall, G.C., Mokhonova, E.I., Moran, M. et al. (2012). Dantrolene enhances antisense-mediated exon skipping in human and mouse models of Duchenne muscular dystrophy. *Sci. Transl. Med.* 4 (164): 164ra160.
225 Muntoni, F. (2010). The development of antisense oligonucleotide therapies for Duchenne muscular dystrophy: report on a TREAT-NMD workshop hosted by the European Medicines Agency (EMA), on September 25th 2009. *Neuromuscul. Disord.* 20 (5): 355–362.

2

The Medicinal Chemistry of Antisense Oligonucleotides

Jonathan K. Watts

RNA Therapeutics Institute and Department of Biochemistry and Molecular Pharmacology, University of Massachusetts Medical School, Worcester, MA, USA

2.1 Introduction: The Antisense Approach and the Need for Chemical Modification

It has been 40 years since the seminal experiments with antisense DNA by Zamecnik and Stephenson in 1978 [1, 2]. The first antisense drug (fomivirsen) was approved in 1998 to treat CMV eye infections in HIV patients but within a few years was rendered obsolete by advances in antiretroviral cocktails for HIV therapy. At the time, it likely would have been unthinkable that the field would have to wait 15 years for another approval, but a second antisense drug (mipomersen) was not approved until 2013. A third approval (eteplirsen) was granted in August 2016, and a fourth (nusinersen) in December 2016.

What has taken the field so long to develop? The answer is that the enabling technologies have slowly but steadily evolved over all these years – namely, the chemistry of the oligonucleotide backbone and associated delivery strategies. As the chemical toolbox has grown in size and sophistication, improved drugs have been advanced into clinical trials, and the pace of approvals is likely to continue to accelerate.

Unmodified DNA makes a poor drug for a number of reasons. First, DNA is rapidly broken down by nucleases inside cells. Second, cellular mRNA targets are highly structured and often bound by protein chaperones. The binding affinity of DNA is generally insufficient to overcome this existing structure. Third, foreign nucleic acids can be recognized by the immune system, causing potentially serious side effects. Fourth, side effects can also arise if the antisense oligonucleotide (ASO) recognizes undesired RNA targets through

Oligonucleotide-Based Drugs and Therapeutics: Preclinical and Clinical Considerations for Development, First Edition. Edited by Nicolay Ferrari and Rosanne Seguin.

partial complementarity. And fifth, unmodified DNA and RNA are not taken up by cells efficiently enough to be used in the clinic.

2.1.1 How Does Medicinal Chemistry Apply to Oligonucleotides?

Medicinal chemists, no matter the project, are tasked with improving potency, specificity, and pharmacokinetics. These goals remain the same when medicinal chemistry is applied to oligonucleotides. However, there are also some differences.

While most traditional medicinal chemists work on a particular small molecule for a particular target, oligonucleotide medicinal chemists work at two levels. The *first level* could be called the platform level and involves developing chemistry to improve the properties of ASOs globally. The chemistry of the sugars and phosphates, the presence of a targeting ligand, single or double strandedness, and length all contribute to the tissue distribution, half-life, and potency. We have recently argued that the ensemble of molecular properties that determine these three properties should be termed the "dianophore," from the Greek word dianomi (διανομή), meaning distribution or delivery [3]. Development of an improved dianophore can lead to a step change in ASO technology. The biggest of these changes have been called "generations" – the development of the phosphorothioate (PS) modification led to the first generation of clinically practicable oligonucleotides and one approved drug (fomivirsen). In the late 1990s it became clear that a gapmer approach (see Section 2.2.3) offered better potency and pharmacokinetics and gapmers are often designated "second-generation" ASOs. These gapmers reduced the doses required by about an order of magnitude compared with first-generation ASOs, depending on the chemistry of the wings (details of sugar chemistry are described below).

A number of other changes can still be included in "platform" medicinal chemistry. Subtle changes in ASO chemistry can lead to improved uptake, RNase H activity, binding affinity, and potency. What is learned on one sequence will often be applicable to ASOs generally. Thus most of this review will focus on "platform" medicinal chemistry.

ASOs have been called "informational drugs," since the specificity of targeting is defined by the sequence – the information – in the drug [3, 4]. In no other class of drug can the pharmacophore be so easily represented by abstract information (a string of letters). This points to a tremendous advantage for oligonucleotide drugs: When a dianophore is established that allows clinical activity in a given tissue, the pharmacophore can be semirationally designed by screening a finite number of sequences complementary to the target gene sequence. This allows compounds to move rapidly from target validation into preclinical testing. Thus, the ability to separately optimize the dianophore from the pharmacophore is a unique advantage of oligonucleotide drugs.

However, biological information can never be fully divorced from molecular structure. Accordingly, small changes in chemistry or sequence can often lead to large changes in nonobvious parameters like off-target binding to proteins, subcellular localization, and gene silencing activity. Quantum mechanical calculations show dramatic changes in the global wavefunction of an oligonucleotide upon small changes in chemical modification [5, 6]. Some sequences, when chemically modified in certain ways, fold into inappropriate structures or cause high toxicity.

Because of this complex interplay between sequence and chemistry, the medicinal chemist must also take care to work at a *second level*: namely, the specific optimization of the modification pattern of each individual drug. Thus, even when a platform (dianophore) is highly effective and safe for some sequences, medicinal chemists need to make variants and screen a number of modified sequences before choosing a lead compound. Work on *predicting* the interplay of chemical modification and sequence (in terms of both toxicity and activity) is in early stages but has the potential to save a great deal of effort in the years ahead [7, 8].

2.1.2 Chemistry and Toxicity

One focus of this book is on the toxicology of oligonucleotides. The chemistry of an oligonucleotide can affect its toxicity in a number of distinct ways. In some cases, one issue is traded for another. For example, if a chemical modification lowers the required dose of oligonucleotide, it is likely to reduce adverse effects like immunostimulation and injection site reactions as well as class-dependent toxicities such as complement activation and nonspecific polyanion–protein toxic interactions. Modifications that increase the specificity of binding will also help reduce toxicity due to mismatched hybridization to unintended targets.

However, the introduction of chemical modifications can cause challenges as well. Chemical modifications may cause an ASO to fold into different structures or bind new proteins through sequence-independent effects [9]. Oligonucleotides are eventually metabolized into their component nucleotides inside cells, and these metabolites may come with their own toxicity issues. Finally, while increasing the binding affinity of an ASO for its target increases its potency, it is also likely to increase binding affinity and silencing potency toward mismatched targets. Over the past 2 years, three independent groups have shown that much of the hepatotoxicity of high-affinity gapmer ASOs results from RNase H-mediated cleavage of mismatched transcripts (often within introns) [10–13] (also see the Chapter 7 by Kamola, Kitson, and Kane in this volume).

Many of these issues are addressed in detail in other chapters of this book. As noted above, chemical modifications interact with each oligonucleotide sequence in a slightly different way [7, 8]. As discussed below, they also interact with other cellular factors in different ways. Thus the process of optimizing each ASO may benefit from having multiple options in terms of chemistry.

This chapter will briefly examine the rationale for giving careful consideration to medicinal chemistry in the context of ASOs. In this regard we examine the importance of nuclease stability, binding affinity, relationship with cellular proteins, immunostimulation, cleavage specificity, and trafficking to and within cells. In a second section, we will then examine the structures and properties of the most important chemical modifications, including analogues of the sugar, phosphate, and nucleobase.

In this chapter, we will focus mainly on the medicinal chemistry of single-stranded oligonucleotides but will make references to duplex RNAs when relevant. Approaches to chemical modification of duplex RNAs have been reviewed elsewhere [3, 14, 15].

2.2 Why Chemically Modify an Oligonucleotide?

2.2.1 Medicinal Chemistry Can Increase Nuclease Stability

One of the most important aspects of the pharmacokinetics of oligonucleotide drugs is their potential for digestion by nucleases. An unmodified single-stranded DNA or RNA oligonucleotide is rapidly digested by serum and cellular nucleases and thus has a half-life too short for clinical activity.

Almost all of the chemically modified nucleotides described in this chapter contribute to improved nuclease stability, though some modifications are more effective than others. Many of these examples will be cited when discussing the individual chemical modifications.

Another factor to consider is that duplex oligonucleotides tend to be more nuclease resistant than single-stranded oligomers. Thus analogues that increase self-structure or duplex stability will also tend to increase nuclease stability.

Understanding mechanisms of degradation can lead to targeted chemical modifications that improve clinical performance. For example, for many years it was assumed that 3′ exonucleases were predominant in the digestion of siRNAs, and so the 3′ ends of siRNAs were modified with phosphorothioate (PS) linkages [16]. But more recent studies showed that siRNA samples isolated from tissues after *in vivo* delivery were digested from their 5′ ends [17]. Accordingly, stabilization of the 5′ end of the siRNA guide strand by the inclusion of PS linkages led to an increase in activity and duration of effect [18]. This insight has influenced drugs that are now in advanced clinical trials, as exemplified by the PCSK9-targeted siRNA drug inclisiran, which can be administered as infrequently as every 6–9 months [19, 20].

2.2.2 Medicinal Chemistry Can Tune Binding Affinity and Specificity

Cellular RNAs are highly structured – folded onto themselves and bound to proteins and other nucleic acids. For an oligonucleotide to bind and exert a physiological effect in this context, it should show sufficient binding affinity to stably displace existing cellular binding partners.

Sugar modifications such as locked nucleic acid (LNA) (Section 2.3.1.4) can increase melting temperatures by ~5 °C per modified nucleotide. A single nucleobase modification like the G-clamp (Section 2.3.4.3) can increase melting temperatures by over 10 °C. These and other examples of chemical approaches are discussed below, but it is clear that even a small number of modifications can dramatically improve ASO target binding.

Higher affinity increases potency up to a point – but it is becoming increasingly clear that there is an optimal affinity above which potency begins to drop and off-target effects become more significant. In the context of LNA-modified, RNase H-dependent ASOs, for example, once an affinity "threshold" of about 55–60 °C was reached, further increases in length (and affinity) reduced potency [21]. An optimal T_m value of 60–65 °C was observed for a second bicyclic sugar modification (2′,4′-BNA^{NC}) for a single sequence [22]. It is not clear how widely applicable these values are to other sequences and other families of chemical modifications (such as 2′-modified unconstrained sugars), but the principle is likely to be widely applicable across RNase H-dependent ASOs. One factor in explaining these observations is the importance of multiple turnover: If ASO binding affinity is too high, then the release of cleaved RNA fragments is retarded, reducing the catalytic turnover of the ASO and thus reducing its potency [23, 24]. High on-target affinity also tends to correlate with high affinity toward mismatched transcripts: Thus higher-affinity ASOs tend to elicit more off-target cleavage events and thus higher toxicity [25].

The above observations concern RNase H-dependent ASOs. For splice-switching oligonucleotides (Section 2.2.3), the optimal T_m value may be higher – for example, in the context of LNA, one paper suggested an optimal T_m value of 80–90 °C [26], and in the context of 2′-*O*-methyl RNA, high affinity correlated with high activity with no upper ceiling specified [27].

The specificity of ASO binding (i.e. the ability of an ASO to clearly distinguish between fully matched and partially matched targets) is just as important as the affinity to minimize hybridization-mediated off-target effects. Fortunately, modifications of the sugar, phosphate, and nucleobase that increase binding affinity have generally been found to increase specificity of binding as well [28]. This may be related to the fact that analogues that bind tightly also tend to yield duplexes with increased rigidity. Conversely, destabilizing modifications tend to reduce specificity as well as affinity. Nevertheless, no matter the modifications used, if an oligonucleotide is designed with overly high affinity, it will be able to bind mismatched targets under cellular conditions. The related area of target *cleavage* specificity is discussed in Section 2.2.5.

2.2.3 Medicinal Chemistry Can Change Interactions with Cellular Factors

ASOs either serve to block or recruit cellular factors from/to their target RNA. The chemistry of the oligonucleotide plays a central role in determining these interactions.

The earliest antisense effects were presumed to occur by simply blocking translation of the target mRNA into protein. This mechanism is known as "steric blocking," and today's steric blocker ASOs also include many other applications besides inhibiting translation: They can be used to modulate splicing, to inhibit microRNAs (miRNAs), and to regulate the interactions of long noncoding RNAs. A steric blocker ASO needs to find its target and bind with high affinity and achieves its therapeutic effect by blocking the interaction of its target RNA with other cellular components, either proteins or nucleic acids – including notably the splicing machinery as well as regulatory RNAs. The chemical requirements for a steric blocker oligonucleotide are fairly simple: The ASO must distribute into cells and bind with high affinity and specificity to its RNA target.

The second class of ASOs is those that depend on a protein to achieve their activity. The classic example of such a protein is the enzyme RNase H. In the mid-1980s it became clear that DNA-based ASOs were not simply blocking translation. Rather, RNase H, which cleaves the RNA strand of RNA–DNA hybrid duplexes, can be recruited to a hybrid of an ASO with its target RNA [29–31]. RNase H can thus cleave the RNA strand, releasing the ASO to find and cleave another target in a catalytic fashion [29–31]. ASOs that function through an RNase H-dependent mechanism have an additional constraint on their chemistry besides those required of steric blockers: The oligonucleotide must be sufficiently DNA-like that RNase H will recognize the ASO/RNA hybrid duplex.

Chemically there are two ways to approach RNase H-dependent ASO design. The first option is to use a chemistry that mimics DNA and allows RNase H-mediated cleavage to occur [32, 33]. The second is to make a chimeric oligonucleotide such as a "gapmer," which contains a window of DNA surrounded on either side by a different chemistry to improve nuclease stability and binding affinity [34, 35]. Such gapmers dominate the current clinical pipeline for inhibitory ASOs.

Argonaute is another enzyme of key importance for oligonucleotide therapeutics. While Argonaute adopts a similar fold to RNase H [36, 37], its requirements in terms of trigger oligonucleotides are different. Argonaute naturally binds small RNA duplexes, from which it retains one strand to act as a guide [38].

However, recent work has identified a design for chemically modified single-stranded oligonucleotides that can be efficiently taken up by Argonaute without the need for a second strand. Key design elements of these oligomers include a nuclease-stable body and dipurine 3′-terminus and a metabolically stable 5′-phosphate [39–46]. Hybrid structures that are partly double stranded and partly single stranded, often in combination with a hydrophobic ligand, are promising designs that combine potent recruitment of Argonaute with effective delivery to a variety of tissues *in vivo* [47–51]. The most clinically advanced compound based on this asymmetric, self-delivering siRNA design is RXI-109, which targets connective tissue growth factor to reduce dermal scarring and is in phase 2 clinical trials. Finally, gapmer ASOs may also use Argonaute proteins to exert gene silencing through noncanonical mechanisms [52].

Two nuclease enzymes, RNase H and Argonaute, have been considered the most important factors for the protein-dependent class of gene silencing oligonucleotides. To recruit other proteins, researchers have sometimes made conjugates of ASOs with ligands or domains that recruit a protein of interest (for salient examples, see Refs. [53, 54]). Moreover, recent results point to the idea that the chemistry of the ASO backbone can *inherently* change recruitment of other important proteins: Rigo et al., while carrying out routine medicinal chemistry on a splice-switching ASO, observed that 2′-fluoro-RNA oligomers caused the splicing to be changed in the *opposite direction* compared with other chemistries [55]. They traced this phenomenon to the fact that interleukin enhancer-binding factors two and three (ILF2/3) were recruited to the ASO–target duplex when the ASO was made of 2′-fluoro-RNA (2′F-RNA) [55]. It is intriguing that the chemistry of the oligonucleotide could sufficiently change the proteins recruited to push splicing in opposite directions. This provocative finding suggests that we are only on the edge of understanding the mechanisms by which different oligonucleotide chemistries may contribute to differential interactions with other cellular factors and ultimately different biological activities. Protein binding can be a double-edged sword, and the same 2′F-RNA chemistry has been associated, in certain contexts, with toxic effects through inappropriate protein binding [9, 56].

More broadly, it is becoming increasingly clear that a large number of proteins not only bind oligonucleotides inside cells but also contribute to their activity [57]. The chemical nature of the oligonucleotide backbone affects these binding relationships – both for better and for worse.

2.2.4 Medicinal Chemistry Can Modulate Immunostimulation

Oligonucleotides are recognized by the innate immune system in a number of ways [58]. Toll-like receptors (TLRs) 3, 7, and 8 recognize RNA, while TLR9 recognizes single-stranded DNA with unmethylated CpG motifs (5′-CG-3′ dinucleotides) [59].

For clinical gene silencing applications, it is generally important to reduce or eliminate the immune response induced by oligonucleotides. This is particularly true for applications requiring repeat dosing, where an initial innate immune response to the oligonucleotide cargo can lead to such responses as antibody formation against the carrier. Immune responses are often observed as "flu-like symptoms" in ASO clinical trials.

Specific chemical modifications can reduce immune recognition of oligonucleotides. For example, 5-methylation of cytosine in CpG dinucleotides of DNA can reduce the TLR9 recognition of these sequences [60]. Similarly, 2′-*O*-methyl RNA (Section 2.3.1.1) inhibits recognition by TLR7 – even if only a subset of the sequence is chemically modified [61, 62].

On the other hand, for some applications it is desirable to induce an immune response. The use of immunostimulatory oligonucleotides based on CpG DNA in particular has become a major field of research and several candidates are in advanced clinical trials (e.g. as vaccine adjuvants) – see Refs. [63, 64] and Chapter 16 of this volume. Other innate immune receptors, including cytoplasmic receptors like MDA and RIG-I, show additional therapeutic promise by virtue of being expressed in all cells [58]. Chemical and structural modification of oligonucleotides can be used to increase immune responses [65–67].

2.2.5 Medicinal Chemistry Can Improve RNase H Cleavage Specificity

For diseases caused by a dominant negative mutation, it would often be desirable to cleave one allele while leaving another allele intact. Huntington's disease is one such example – the diseased allele produces a toxic protein, but the correct allele produces a functional protein that serves in multiple cellular processes. Silencing one while leaving the other untouched would be clinically attractive.

RNase H is inherently somewhat tolerant toward mismatches (see Chapter 7 by Kamola, Kitson, and Kane), which makes allele-selective inhibition challenging since an ASO that targets the mutant allele will also tend to lead to some cleavage of the wild-type allele. However, inclusion of particular modifications can improve the cleavage selectivity [68–71]. More details on these modifications are included later in this chapter. The common principle is not just found in increasing the difference in binding affinity, but rather in restricting the cleavage site, so that RNase H is limited to being able to cleave at a particular site, in combination with modification at or near the cleavage site that exaggerates the structural distortion produced by a mismatch. For example, in the case of 2-thiothymidine, the 2-thio group causes a tight contact with key amino acids near the active site of RNase H [68]. So far, these approaches focus on improving mismatch discrimination at a single position. It remains to be seen whether similar approaches can help reduce the more global problem of off-target cleavage of transcripts that are mismatched at any position.

2.2.6 Medicinal Chemistry Can Improve Cellular Uptake and Subcellular Trafficking

One of the main challenges for effective antisense therapeutics is the need to ensure effective cellular uptake of oligonucleotides. Once taken up, oligonucleotides can be degraded or sequestered without reaching their targets. However, chemical modification can play a major role in improving the bioavailability of oligonucleotides – both in terms of cell uptake and subcellular trafficking.

Single-stranded oligonucleotides can be taken up by cell surface receptors and are more readily taken up by cells than are (naked) double-stranded oligonucleotides. This is related to several factors: Single-stranded oligonucleotides are flexible, while duplexes are rigid. Single-stranded oligonucleotides are amphiphilic (with polar phosphates and hydrophobic nucleobases), while duplexes have a surface dominated by heavily hydrated phosphates. The flexible, amphiphilic nature of single-stranded oligonucleotides allows them to interact more readily with plasma proteins and cell surface proteins, including trafficking proteins, and thus be taken up more readily by cells. Thus most ASOs in clinical trials are delivered naked, while duplex RNAs typically need to be delivered as conjugates or formulated into liposomes.

Oligonucleotide conjugates can also dramatically improve cellular uptake, and conjugation is one of the best approaches to stimulate uptake by specific populations of cells. Recent work on such conjugates has been reviewed elsewhere [72, 73]. The most clinically advanced conjugates are based on multimers of the monosaccharide ligand *N*-acetylgalactosamine (GalNAc), which leads to selective uptake by hepatocytes [74–78]. Multiple companies are beginning to apply this to different oligonucleotide technologies with targets in the liver [18, 79, 80]; a combination of advanced siRNA chemistry with GalNAc conjugation can achieve potent clinical gene silencing for at least 6 months with a single 300 mg dose [19]. GalNAc is not the only promising conjugate, and many groups are investing energy to find ligands that will stimulate effective uptake in other tissues [73, 81].

Oligonucleotide chemistry can also have a large effect on subcellular trafficking. For example, a modified peptide nucleic acid with phosphonate ester side chains can be selectively localized in the mitochondria [82]. Sometimes an initially unexplained potency improvement can be traced to a more favorable subcellular distribution profile. For example, 11–15-mer tricyclo-DNA (tcDNA)-modified oligonucleotides with lower affinity than corresponding LNAs still showed superior potency as splice-switching oligonucleotides [83, 84]. Here the primary advantage of the tcDNA may be one of subcellular localization, as the tcDNA appears to localize more effectively to the nucleus [83]. As described in Section 2.2.3, this relationship is almost certainly mediated by changes in protein binding that are only beginning to be understood [57].

2.3 Chemical Modifications of Current Importance by Structural Class

2.3.1 Sugar Modifications

Almost all of the sugar-modified oligonucleotides in preclinical or clinical development fall into two categories – either 2′-modified sugars or constrained sugars. Another promising category of sugar modifications are the expanded ring sugars. These classes of modified oligonucleotides will be examined in turn.

2.3.1.1 2′-Modified Ribose Sugars

The most common sugar-modified nucleotides are 2′-*O*-substituted RNAs (Figure 2.1). For example, the naturally occurring 2′-*O*-methyl RNA improves binding affinity compared with normal DNA or RNA. It also improves stability toward endonuclease digestion for both single- and double-stranded oligomers [16, 85–87]. Its safety profile is considered excellent and indeed is one of the clinically approved chemical modifications, as a component of the approved aptamer pegaptanib [88].

Beginning from this point, medicinal chemists worked intensely since the mid-1990s to find the ideal 2′-*O*-substituent. Dozens of variants were tested, including alkyl, alkoxyalkyl, aminoalkyl, allyl, and others (reviewed in Refs. [28, 89–91]). Of this series, 2′-*O*-methoxyethyl (MOE) RNA [92] stands out from most others in terms of both nuclease stability and binding affinity ($\Delta T_m = 0.9–1.7\,°C$ per insert, Figure 2.1). MOE has been a main focus of development by Ionis Pharmaceuticals for the past 15 years and features in a number of clinical candidates including the approved drug mipomersen. Thus it is one of the best-studied modifications in terms of its influence on other drug properties such as pharmacokinetics. Two MOE drugs have been FDA approved: Mipomersen is an MOE gapmer that silences ApoB for treatment of homozygous familial hypercholesterolemia [93], while nusinersen is a uniformly modified MOE 18-mer used to modulate SMN2 splicing for treatment of spinal muscular atrophy [94].

An excellent 2005 study compared the crystal structures of MOE and 10 novel 2′-*O*-modified oligonucleotides, observing trends and correlations in terms of the effects on binding affinity and nuclease stability [95]. Interestingly, increased steric bulk of the 2′-substituent did not always correlate with improved nuclease stability. On the other hand, cationic substituents showed very high nuclease stability. Factors that contributed to high affinity included favorable electrostatic interactions with the phosphate, coordination of a water molecule in the vicinity of the 2′-substituent, and appropriate conformational preorganization of the substituent [95].

The 2′-oxygen can also be replaced altogether. The most advanced example of this is 2′F-RNA [35, 96], which is included in numerous clinically relevant candidates including the FDA-approved aptamer pegaptanib [88]. 2′F-RNA

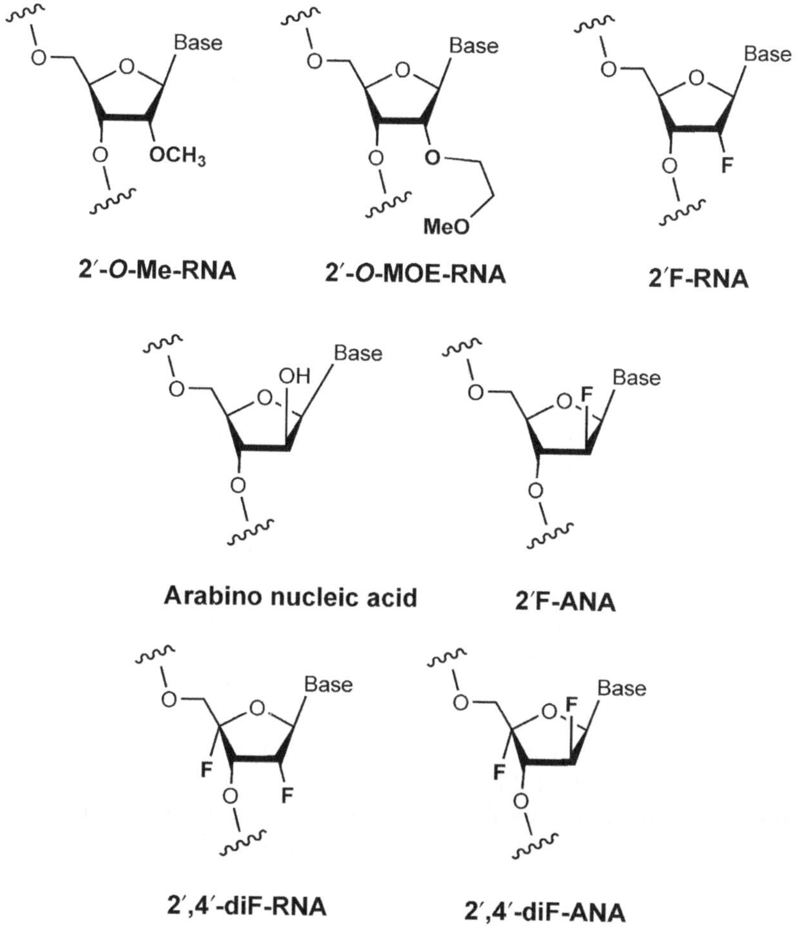

Figure 2.1 Some oligonucleotides based on 2′-modified sugars.

increases the binding affinity of ASOs more dramatically than the 2′-*O*-substituted derivatives ($\Delta T_m \sim 2.5\,°C$/modified nucleotide). The origin of this effect was long assumed to be an entropic advantage, through directing the conformation by stereoelectronic effects. However, recent work has shown that 2′F-RNA directly strengthens both base stacking and H-bonding [97, 98]. This may involve a direct effect of the fluorine on the electronics of the nucleobase through inductive effects [97, 98]. 2′F-RNA does not provide extensive protection from exonucleases [89], but does confer some endonuclease protection [16]. It can also reduce oligonucleotide immunogenicity [99, 100]. Furthermore, ASO modification with 2′F-RNA can change protein binding – for good or for ill – as discussed in Section 2.2.4.

2.3.1.2 2′-Modified Arabinose Sugars

Inverting the configuration of RNA at the 2′-position gives arabinonucleic acid – a nucleic acid based on an arabinose rather than ribose sugar (Figure 2.1). Because the top-face substituent points toward the base stack, only the smallest possible substituent (fluorine) is well tolerated here: 2′-Fluoroarabinonucleic acid (2′F-ANA) shows increased binding affinity to both DNA and RNA [101, 102], while the parent compound, arabinonucleic acid, shows reduced binding affinity to both DNA and RNA [103].

The inversion of configuration causes a change in ring conformation and associated enzyme recognition. 2′F-ANA·RNA duplexes adopt a helical structure similar to DNA·RNA hybrid duplexes. Accordingly, they are recognized by RNase H and the RNA strand is cleaved [32]. The increased rigidity of a 2′F-ANA duplex reduces the kinetics of RNase H cleavage compared with a DNA strand, but this can be mitigated by including one or more DNA residues or acyclic nucleotides in the oligomer [104, 105]. Thus unlike most RNA-like modifications (2′F-RNA, MOE, LNA), a very short DNA gap (as low as 1 nt) is compatible with potent activity in 2′F-ANA ASOs [104]. Gapmers and mixmers based on 2′F-ANA show improved potency [102, 104–108] and an acceptable toxicity profile [109]. 2′F-ANA modification also improves nuclease stability [107, 110], acid resistance [110], and cellular uptake [108] of oligonucleotides.

The increased binding affinity by 2′F-ANA is partly related to a pseudohydrogen bonding interaction between the fluorine and H8 of the neighboring purine base [111, 112]. This interaction occurs preferentially at pyrimidine–purine steps and can thus be used to influence sequence design [112]. Electronic effects like those described above for 2′F-RNA may also be in operation in 2′F-ANA. Moreover, 2′-fluorination activates hydrogens, allowing them to engage in FC—H···O hydrogen bonds that may contribute to high binding affinity [113].

2.3.1.3 2′,4′-Difluorinated Nucleosides

The difluorinated analogues 2′,4′-difluoro-RNA and 2′,4′-difluoro-ANA (Figure 2.1) have also recently been described [114–116]. Both stereoisomers show rigid north conformations. The difluoro-RNA monomer is neutral in its binding affinity in the context of an A-form RNA helix, in spite of its rigid north conformation [115]. The difluoro-ANA monomer is slightly stabilizing when in the DNA strand of a hybrid duplex and does not block but redirects the position of RNase H cleavage [116].

2.3.1.4 Constrained Nucleotides

Reducing the conformational flexibility of nucleotides can reduce the entropic penalty of binding and thus increase oligonucleotide binding affinity [117]. As such, locking the conformation of the ribose sugar has been a very successful approach in oligonucleotide medicinal chemistry.

LNA (Figure 2.2) [118–120] is the most advanced of the constrained nucleotides in terms of applications in antisense therapeutics [121, 122]. Adding LNA

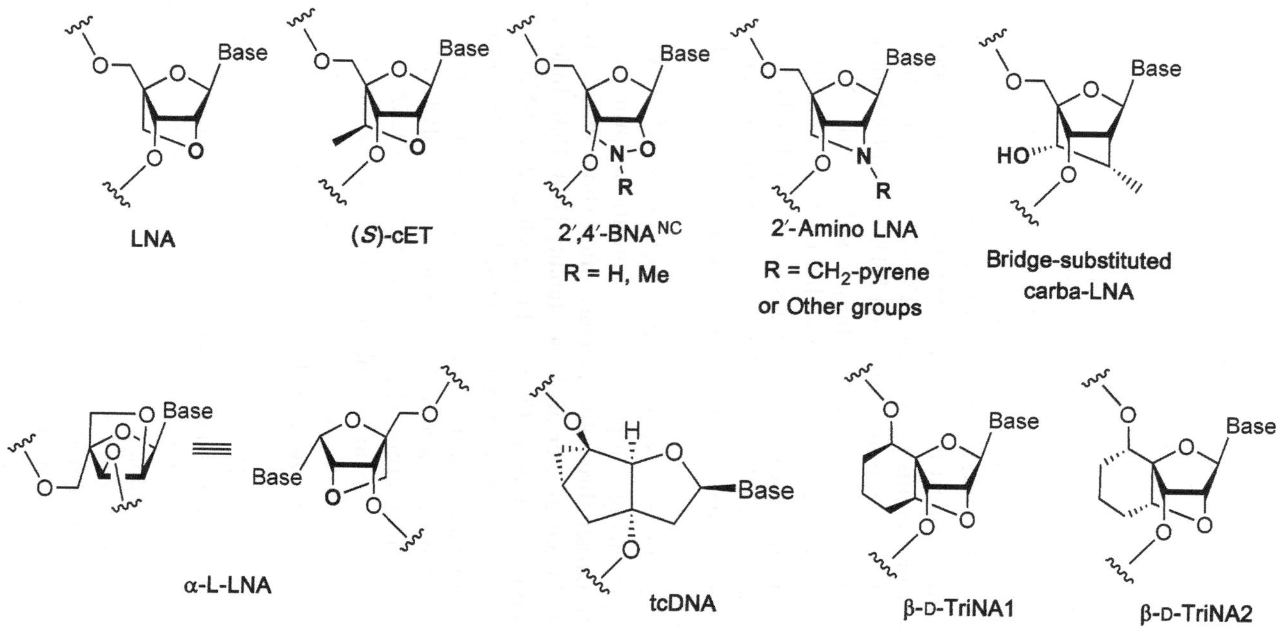

Figure 2.2 Oligonucleotides based on constrained sugars.

nucleotides to a DNA oligonucleotide produces a dramatic increase in binding affinity (3–8 °C per modification when binding RNA) [118, 119, 123]. This affinity advantage comes with a significant increase in binding specificity as well.

Adding an additional methyl group to the bridge in the S configuration produces "constrained ethyl" or cEt nucleotides (Figure 2.2). This modification provides a similar increase in binding affinity as LNA and may show reduced toxicity in some cases [124]. Both LNA- and cEt-modified ASOs have entered clinical trials.

Both LNA and cEt, by virtue of their very high binding affinity, can be used as the basis for shorter oligonucleotides than previously thought possible. For example, gapmers of 13–15 nt are now commonly used [21, 124, 125], whereas a decade ago, the field worked almost universally with ASOs of 18 nt or more. The shorter gapmers appear to have a potency advantage over longer oligonucleotides of the same chemistry [21, 124, 125], as long as their binding affinity does not drop below a minimum threshold [21]. For miRNA inhibition, even smaller oligonucleotides can be used if fully modified with LNA: Fully modified 8-mer "tiny LNAs" have been used to inhibit families of miRNAs sharing a common seed sequence (the 6–8 nucleotides at the 5′ end of the active miRNA) [126]. The seed sequence is key for miRNA recognition of target mRNAs [127], so a tiny LNA can prevent repression by a group of miRNAs with a common inhibitory profile.

The 2′-4′ bridge of LNA has been the object of intense effort by oligonucleotide chemists, producing a large number of additional promising analogues [124, 128–132]. Larger three-atom bridges often show similar hybridization properties as LNA but greater nuclease stability [128, 133]. Indeed, bridging moieties containing three-atom O–N–C bridges (e.g. 2′,4′-BNA^{NC}; Figure 2.2) show excellent properties including unsurpassed nuclease resistance, increased potency in the context of gapmers, increased selectivity for binding RNA over single-stranded DNA, and increased triplex-forming ability [22, 133, 134].

Analogues with a fully carbocyclic bridge, called carba-LNA (Figure 2.2), are also active [130] and can be further substituted. In one interesting example, the analogue shown in Figure 2.2 with the bridge OH group pointing back away from the phosphate improved nuclease stability, while analogues with the bridge OH pointing forward (i.e. toward the internucleotide phosphate) reduced nuclease stability [135]. Bridges containing nitrogen are also suitable for further functionalization. For example, the 2′-amino analogue of LNA has allowed functional pendant groups to be projected into the minor groove of the duplex (Figure 2.2) [136].

Of eight possible stereoisomers of LNA (i.e. taking all combinations of α/β, xylo/ribo, and D/L), it is remarkable that six of them bind RNA with high affinity [137]. One very interesting member of this family is α-L-LNA, which shows increased binding and preorganizes the ASO into a DNA-like conformation

[138, 139]. Given that most high-binding analogues are RNA mimics, this is a noteworthy exception and may be useful for both basic research and antisense therapeutics.

tcDNA is another promising nucleotide constrained by a two-ring bridge between the 3′ and 5′ positions (Figure 2.2). Its binding affinity ($\Delta T_m \sim +2\,°C$ compared with RNA) is smaller than that of LNA; however, it lends itself well to fully modified oligomers, while fully modified LNAs longer than about 8 nt tend to aggregate.

In experiments using PS tcDNA-based gapmer ASOs in mice, the tcDNA showed similar activity to an isosequential MOE gapmer for gene silencing in liver but outperformed the MOE gapmer in a number of extrahepatic tissues, including the heart, lung, fat, and muscle [125]. However, some of the most exciting applications of tcDNA have come in the area of splice-switching oligomers. For example, fully modified tcDNA oligomers showed four fold higher potency than LNA oligomers for redirecting splicing in a cyclophilin A model [83] and outperformed other modifications for redirecting splicing of dystrophin [84]. tcDNA ASOs may to be ideal for splice-switching applications in part because they preferentially localize to the nucleus [83].

Finally, another family of tricyclic modifications is based on conceptually joining the C5′ position to the bridge of LNA (TriNA1 and TriNA2; Figure 2.2) [140]. Of the two stereoisomers, TriNA2 shows slightly higher binding affinity ($\Delta T_m \sim +6\,°C$). Similar tricyclic modifications have also been developed based on α-L-LNA and are equally promising in terms of affinity ($\Delta T_m \sim +4–8\,°C$) [141, 142].

2.3.1.5 Sugars with Expanded Ring Size

Six-membered ring sugars are more rigid than the five-membered ring sugars found in natural nucleic acids. Therefore, if analogues can be found to adopt an *appropriate* rigid conformation, they too can stabilize oligonucleotide binding. Outstanding members of this group are hexitol nucleic acid (HNA), its fluorinated analogue (FHNA), and altritol nucleic acid (Figure 2.3).

The Herdewijn laboratory dedicated significant effort over two decades to studying the six-membered sugar-based nucleosides and nucleotides [143]. The initial targets were the pyranose nucleic acids (conceptually adding a methylene group between C1′ and C2′ of DNA; Figure 2.3), but these did not bind well to complementary nucleic acids. In contrast, the (anhydro)-HNA series is derived from conceptually adding a methylene group between the ring heteroatom and C1′. HNA shows a moderately increased binding affinity toward RNA ($\Delta T_m = 0.5–1.4\,°C$ per insert). The fluorinated analogue, FHNA, shows a further increase in binding affinity ($\Delta T_m = 0.6–2.6\,°C$ per insert) and outstanding nuclease stability (over 50-fold increased half-life compared with MOE or LNA) [144]. While HNA and FHNA showed similar potency for gene silencing in cultured cells, FHNA showed superior potency in animals,

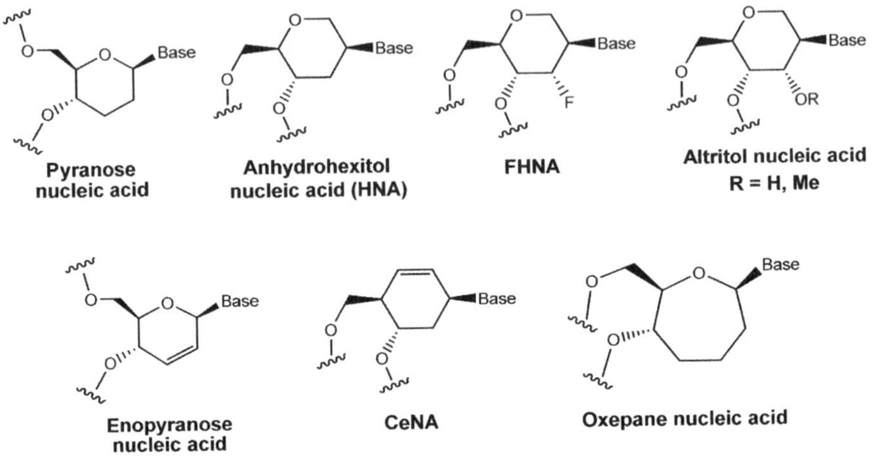

Figure 2.3 Oligonucleotides based on sugars with expanded ring size.

implying that it may also confer an advantage in terms of biodistribution [144]. The hydroxy analogue, altritol nucleic acid, and its *O*-methylated congener also show improved binding affinity relative to HNA [145, 146].

Including a double bond in the six-membered ring gives cyclohexene nucleic acid (CeNA) [33, 147] or 2′-enopyranose nucleic acid [148] (Figure 2.3). These unsaturated rings are more flexible than their saturated counterparts, and this has an influence on their enzymatic recognition. Both can be used to generate fully modified oligonucleotides that can induce RNase H cleavage of a complementary strand, albeit weakly [33, 147, 148].

Expanding the ring still further, to seven atoms, gives an oxepane ring with increased flexibility (Figure 2.3). The oxepane nucleic acids based on this sugar show some RNase H recognition but decreased binding affinity [148]. They are highly stable to nucleases [148].

2.3.2 Phosphate Modifications

2.3.2.1 Phosphorothioate

The PS linkage (Figures 2.4 and 2.5) is the most common and important chemical modification in the toolbox of therapeutic ASOs [149, 150]. In this conservative modification, one of the nonbridging oxygens is replaced by sulfur. Most of the sugar modifications above are used in combination with the PS backbone.

The first work on PS-modified ASOs was published in the late 1980s [151–154]. The most obvious advantage of PS oligonucleotides at the time was their very high nuclease resistance [152], which is particularly impressive in light of the conservative nature of the modification.

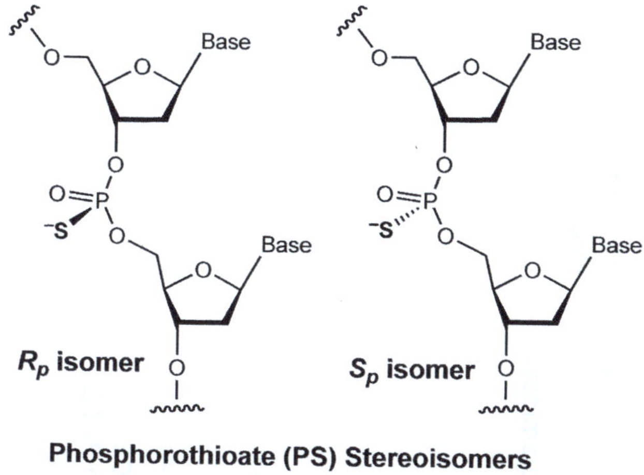

Figure 2.4 The R_p and S_p stereoisomers of the phosphorothioate linkage.

However, over time it has become clear that an equally important advantage of the PS backbone is its effect on serum stability and cell uptake. PS oligonucleotides bind plasma proteins, leading to a longer serum half-life for PS oligonucleotides than for other classes of compounds [155]. This is a relatively low-affinity, reversible interaction and allows the opportunity for the oligonucleotides to contact cell surface proteins. PS oligonucleotides have 100–1000-fold higher affinity for cell surface proteins compared with PO oligonucleotides [156]. Crucially, this includes trafficking proteins such as AP2M1 [157] and nucleolin [158]. PS-ASOs also demonstrate such extremely high affinity for collagen [159] as to cause problems for delivery to the pancreas and other cell types with a collagen-rich extracellular matrix. Thus there is an optimum affinity for cell surface proteins – high enough to interact with cell surfaces, but not so high as to prevent internalization. Moreover, this increased affinity for proteins can cause problems of its own, in the form of toxic interactions with undesired proteins [160].

The PS backbone reduces binding affinity to complementary nucleic acids, which partly explains the low potency of "first-generation" ASOs based on PS-DNA alone. A gapmer design, in which high-affinity wings can compensate for the relatively low affinity of the PS backbone, can mitigate this drawback of PS-ASOs.

PS linkages are chiral at phosphorus, and PS oligonucleotides are generally synthesized as a mixture of diastereomers at each phosphorus. This means that a 20-mer stereorandom PS-ASO is more correctly a mixture of 2^{19} (524 288) diastereomeric ASOs. The R_P and S_P stereoisomers have different properties in terms of nuclease stability, hydrophobicity, binding affinity, and immunorecognition [161–164]. For example, the S_P linkage is more stable to nucleases than

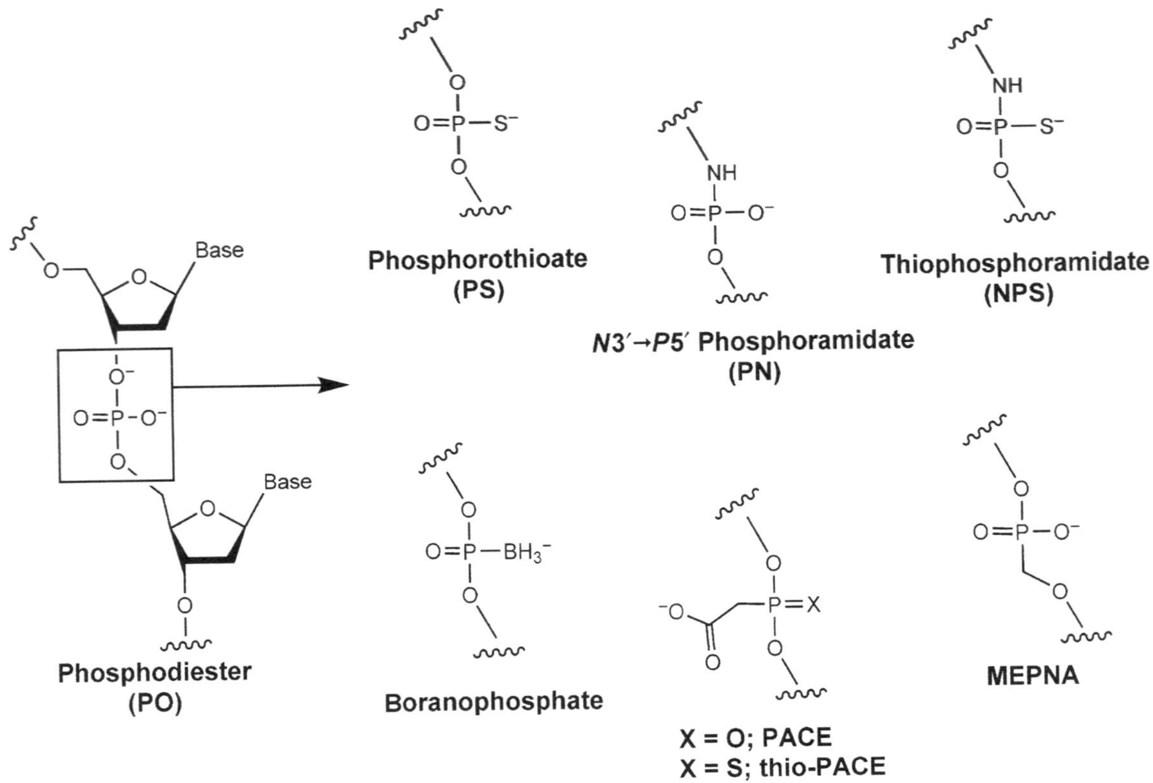

Figure 2.5 Charged mimics of the internucleotide phosphate linkage.

the R_P congener [162]. The [all-R_P]-PS oligonucleotides bind complementary RNA with higher affinity than the [all-S_P]-PS congeners [161, 162], perhaps related to a disruption of duplex hydration patterns by the S_P-PS linkages [161]. The T_m of the mixed stereochemistry PS oligomers is intermediate between the [all-R_P] and [all-S_P] oligomers.

Building on the foundational synthetic work on stereoselective PS synthesis by the Stec lab [162, 165, 166] and others [167–171], increasingly robust chemical methods to synthesize oligonucleotides with stereopure PS linkages have recently been developed by the Wada lab [164, 172–175] and WAVE Life Sciences [176]. This has allowed *in vitro* and *in vivo* activity comparison of ASOs with different patterns of backbone stereochemistry [176]. A crucial insight is that oligonucleotides with certain defined patterns of R_p and S_p linkages outperform both stereorandom ASOs and those with [all-R_p] or [all-S_p] linkages. One such promising pattern is S_p S_p R_p in the section of the oligomer that elicits RNase H cleavage [176]. More globally, however, there is a wide variation of activity between different diastereomeric sequences – some being highly potent and others inactive [176]. This suggests that the use of ASOs with stereopure PS linkages may increase the potency of an ASO relative to the stereorandom approach. Also consistent with this, Bohr et al. have recently shown that the electronic structure of different stereoisomeric PS-ASOs can differ dramatically [6]. Finally, besides showing optimal nuclease stability and RNase H activity, these stereopure ASOs may also augment allele selectivity, and WAVE's first clinical candidate is expected to be an allele-selective ASO for Huntington's disease [177].

PS appears to be the preeminent backbone modification in therapeutic oligonucleotides for the foreseeable future, particularly if its therapeutic index improves through the use of stereoselective synthesis. However, for improving specificity and especially when targeting tissues other than the liver and kidney, it may be helpful to find other options for backbone modification. The advantages of PS linkages can, theoretically at least, be attained in other ways – for example, it is conceivable that conjugates may be developed that allow more specific interactions with appropriate serum and cell surface proteins, without so many nonspecific interactions that can lead to toxicity. PEGylation is another known route to increase serum half-life. And a large number of modifications give excellent nuclease stability, so that ought not to be the main reason for continuing to use PS modification.

Another way forward to less toxic therapeutic ASOs might be to reduce the PS content of an oligonucleotide without eliminating PS linkages altogether. Indeed, in work by Topigen/Pharmaxis, ASOs containing partial PS backbones also showed increased potency over fully modified PS oligomers [178]. Ionis pharmaceuticals has also described examples of mixed PS/PO backbone ASOs for applications in the central nervous system [179, 180] and in GalNAc conjugates for targeting hepatocytes [181].

2.3.2.2 Other Charged Phosphate Analogues

Another important modification is the N3′→P5′ phosphoramidate (PN) linkage, in which the 3′ bridging oxygen is replaced with an NH group (Figure 2.5) [182–184]. In contrast to PS linkages, oligodeoxynucleotides joined by PN linkages are not able to elicit RNase H activity. However, also in contrast to PS linkages, the linkage is not chiral, and the binding affinity of PN oligomers is increased by 2.2–2.6 °C per modification.

Combining elements of the PN and PS chemistries produces the thiophosphoramidate (NPS) linkage (Figure 2.5), which is another phosphate mimic in clinical trials: Geron and Janssen Oncology are developing a telomerase inhibitor, imetelstat, based on NPS chemistry [185, 186] for myelofibrosis and myelodysplastic syndromes.

Boranophosphate linkages (Figure 2.5), like PS linkages, are chiral, nuclease resistant, and compatible with RNase H cleavage [187]. Boranophosphate oligonucleotides are more lipophilic than PS or PO oligomers. However, boranophosphate linkages lead to reduced binding affinity (even more so than the PS), and this may limit their therapeutic applicability to the antisense field [187]. Boranophosphates may be useful in siRNA development to a larger extent [188], as siRNAs can make use of modifications with differential binding affinities to favor appropriate strand loading and reduce off-target effects.

Phosphonoacetate (PACE) and thio-PACE modifications (Figure 2.5), which contain an acetic acid group in place of one of the nonbridging oxygens, retain the negative charge of phosphate groups (PACE $pK_a \sim 3.8$). Unsurprisingly the structural change leads to very high nuclease resistance [189, 190]. And importantly, RNase H cleavage of complementary RNA is retained by PACE modifications [189, 190]. Both PACE and thio-PACE linkages reduce binding affinity severely; this drawback can be mitigated by combining the modified linkages with 2′-*O*-Me sugar modifications at the cost of losing RNase H activity [191, 192]. Neutral, esterified PACE/thio-PACE moieties can serve as prodrugs of these modified oligonucleotides and show increased cellular uptake [189, 191, 193].

The recently described 5′-*O*-methylenephosphate (MEPNA) linkage (Figure 2.5) extends the DNA backbone by one atom, containing an extra methylene group between the 5′-oxygen and the phosphorus. Chimeric oligonucleotides containing combinations of MEPNA and PO linkages show significantly increased binding affinity compared with the parent PO oligonucleotides ($\Delta T_m \sim +3$ °C/ modified linkage) and enhanced rates of RNase H cleavage of complementary RNA [194].

2.3.2.3 Neutral Mimics of the Phosphate Linkage

The oligonucleotide backbone can also be modified to remove the charge on the phosphate group. By removing the charge repulsion associated with bringing two negative strands together, neutral linkages should allow for increased binding affinity. Two neutral analogues of current interest are amide and triazole linkages, both of which can be constructed in various ways (Figure 2.6).

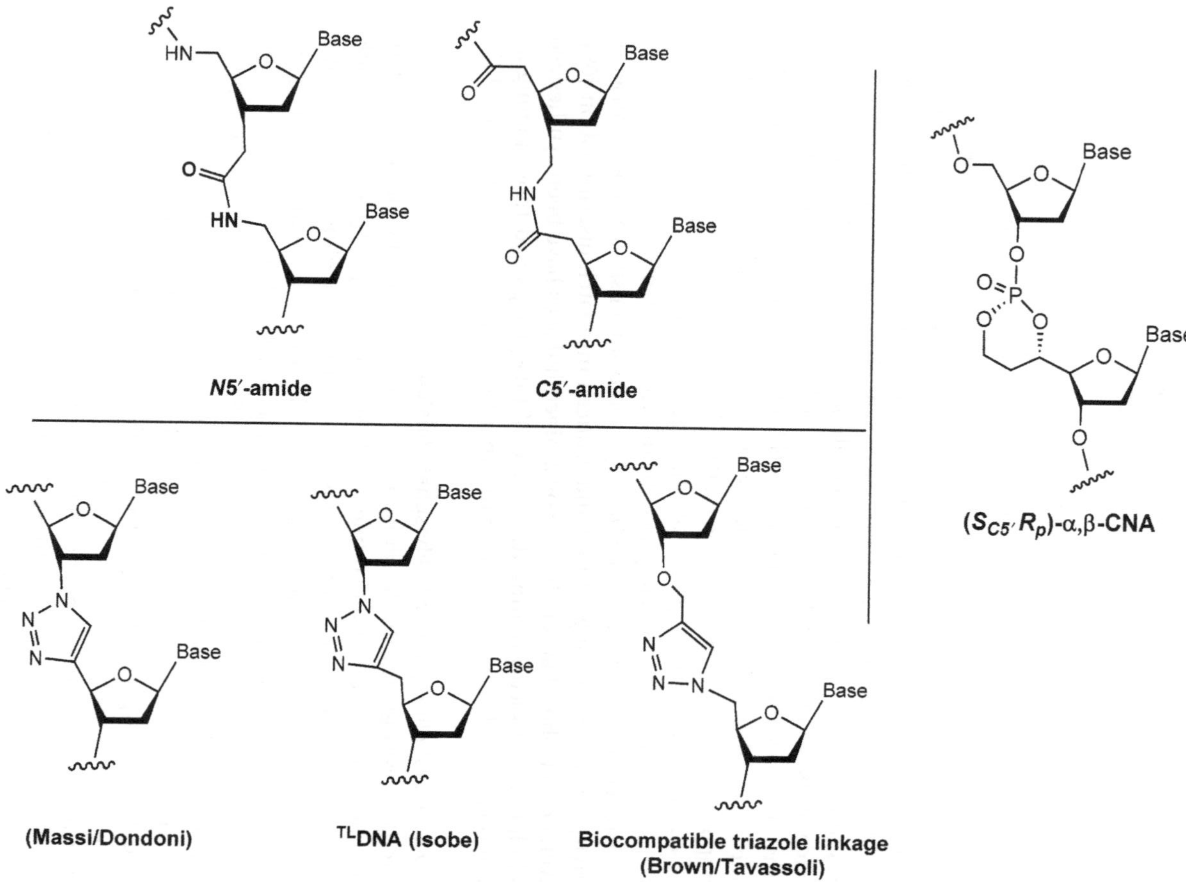

Figure 2.6 Neutral mimics of the phosphate linkage based on amides (top), triazoles (bottom), and cyclic phosphotriesters (right).

Freier and Altman reviewed a range of carbamate, urea, and amide linkages of various lengths and designs [28]. Of these, all of the three-atom or five-atom amides, along with the carbamates and ureas, were destabilizing. Two of the four-atom amide linkages showed neutral or improved binding affinity (Figure 2.6) [28]. These two amides have been extensively tested in both RNA and DNA oligonucleotides. The N5′-amide in particular is an excellent mimic of phosphate linkages. It produces a slight increase in binding affinity when included in DNA [195] and maintains binding affinity, structure, and hydration when included in RNA [196].

A number of triazole linkages can be included within oligonucleotides and provide high nuclease resistance (Figure 2.6) [28, 197–201]. A fully triazole-linked sequence showed high binding to complementary DNA [201], but a single triazole linkage within a phosphate-linked oligonucleotide reduces binding affinity. Triazole linkages have been combined with modified nucleobases (such as G-clamp [202]) to improve their binding affinity. The combination of two different triazole linkages with LNA sugars has also been tested but only leads to improved affinity in certain combinations (e.g. an LNA sugar on the 3′-side of the Brown triazole linkage) [203, 204]. The triazole linkage is unique in that it allows assembly of oligonucleotides by the copper-catalyzed alkyne–azide cycloaddition. This allows access to long and complex oligonucleotides for applications in synthetic biology [205, 206].

A constrained, neutral phosphate linkage (($S_{C5}{}'$ R_p)-α,β-CNA; Figure 2.6) [207] was recently shown to improve the discrimination of RNase H cleavage of mismatched transcripts [69]. The ($R_{C5}{}'$ S_p) stereoisomer showed good antisense activity but lower binding affinity and reduced mismatch discrimination [69]. Several other phosphacycle-based constrained linkages have been synthesized including those incorporating the 2′ and 3′ OH groups and those joining an additional 4′-hydroxymethyl group to either the 3′ or 5′ OH group [208–212].

2.3.2.4 Metabolically Stable 5′-Phosphate Analogues

Argonaute recognizes the 5′-phosphate of the guide strand of an siRNA or other oligonucleotide. While duplex siRNAs can be 5′-phosphorylated inside cells by the kinase Clp-1 [213], this kinase does not phosphorylate single strands. Thus single-stranded oligonucleotides must be chemically phosphorylated in order to be taken up by Argonaute [39, 40]. The metabolically stable phosphate analogue (*E*)-vinylphosphonate (EVP) (Figure 2.7) is the best phosphate mimic for this purpose [45], likely due to its ability to adopt an appropriate structure for Argonaute binding [214, 215]. A significant investment in structure–activity relationships of stable phosphate analogues revealed no new analogues with properties superior to those of EVP [45]. However, the saturated analogue 5′-methylenephosphonate (Figure 2.7) also shows some activity [45]. It is interesting that 5′-phosphonate modification results in increased nuclease stability as well as potent recruitment of Argonaute proteins [17].

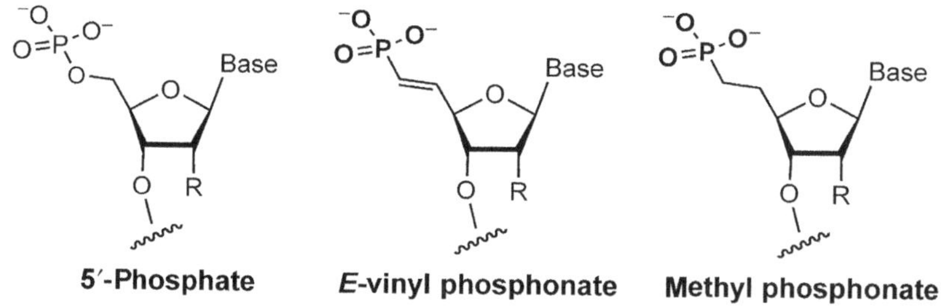

Figure 2.7 Terminal phosphonates used as phosphatase-resistant 5′-phosphate analogues.

Chemical modification of the 5′-nucleotide of duplex RNAs may impair phosphorylation by Clp-1, so the use of a metabolically stable 5′-phosphate analogue such as EVP on the guide strand can also improve the potency of duplex siRNAs, particularly when these are delivered subcutaneously as unformulated conjugates [216].

2.3.3 Total Replacement of the Sugar-Phosphate Backbone

Two families of modified ASOs are based on structures that diverge dramatically from the natural sugar-phosphate backbone. Both of them show increased binding affinity, due in part to their neutral character.

Phosphorodiamidate morpholino oligomers (PMOs) (Figure 2.8) are based on a morpholine ring joined by neutral phosphorodiamidate linkages [217]. PMOs are steric blockers and have been used for many years for translational inhibition as research tools (particularly in models such as zebrafish, where they can be delivered to the embryos by microinjection) [218, 219]. More recently, PMOs have also found significant application as splice-switching oligomers. Sarepta's drug eteplirsen is a splice-switching PMO that was FDA approved for Duchenne muscular dystrophy (DMD) in August 2016 [220].

Modifying the PMO backbone to include piperazine units in the phosphorodiamidate backbone gives the PMO positive charge at physiological pH (Figure 2.8) [221]. Sarepta refers to this chemistry as "PMO*plus*." It was the basis for three candidates in phase 1 clinical trials to treat Ebola, Marburg, and influenza viruses, but those drugs have now been withdrawn from development. A second type of cationic PMO derivative, marketed as "vivo-morpholino," consists of an octaguanidine dendrimer conjugate of PMO. It has shown good activity for exon skipping applications in particular, including in animal studies [222, 223]. However, a more clinically relevant platform would appear to be peptide-conjugated PMOs (PPMOs): For example, specific peptides can lead to uptake by the muscle and heart, the tissues of relevance for DMD [224, 225].

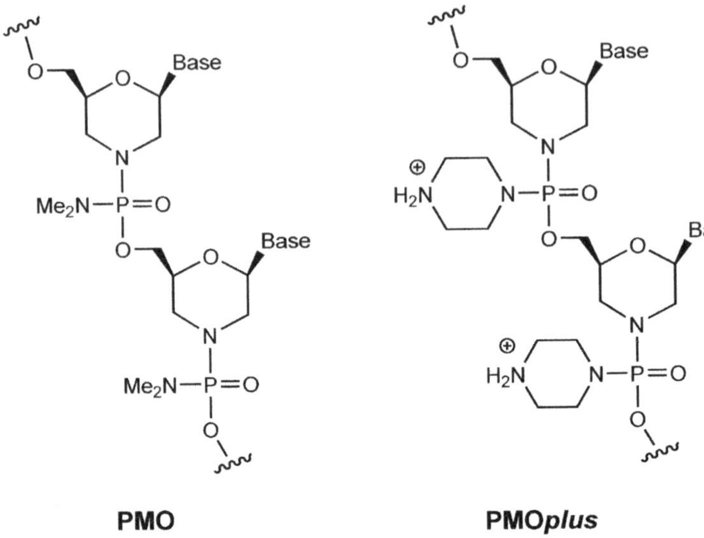

Figure 2.8 Phosphorodiamidate morpholino oligomers and a recent derivative termed PMO *plus*.

Peptide nucleic acid consists of nucleobases arranged on a neutral aminoethylglycine backbone (Figure 2.9) [228, 229]. PNA, though invented nearly 25 years ago, has not attained clinical trials, though it has shown promise in some animal models [230, 231]. The obstacle is its relatively poor solubility and bioavailability.

The therapeutic potential of PNA is likely to increase through improved chemistries developed in recent years. For example, α-substituted PNAs (Figure 2.9) were developed by Ugichem (Austria). Introduction of phosphonate esters into the side chains of these modified PNAs [82, 226, 232] led to localization in the mitochondria [82]. The company's name was derived from Ivar Ugi and his eponymous reaction, a four-component coupling that allows the facile preparation of the modified PNA monomers [233, 234].

Another approach involves substitution of the γ position of PNA (Figure 2.9). Even a methyl group at the γ-position can ensure that PNA preorganizes into the correct helical conformation, increasing the binding affinity and strand invasion capacity of the PNA [235–237]. Other modifications at this position, including lysine-derived amino side chains [238, 239] or diethylene glycol chains [240], confer additional favorable properties in addition to the conformational preorganization.

2.3.4 Nucleobase Modifications

Nucleobase modifications can lead to enhanced affinity and specificity of binding as well as reduced immunostimulation. In addition, fluorescent nucleobase

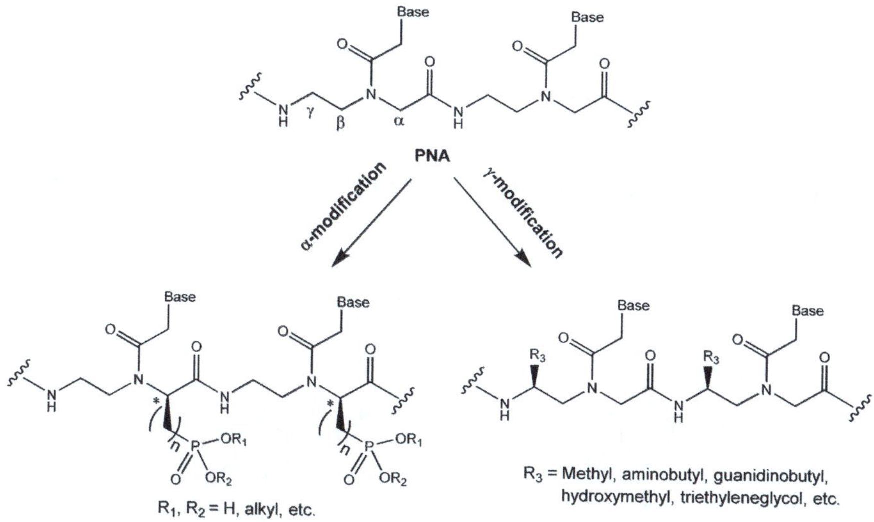

Figure 2.9 Peptide nucleic acid and recent derivatives. The stereochemistry of the α-modified analogues developed by Ugichem is not disclosed [226], but older literature [227] suggests that the oligomer is likely to bind with higher affinity when the stereocenters marked * are configured as drawn. Note that for both α and γ modification of PNA, favorable properties are obtained with partial modification – there is no need to modify every unit of the PNA backbone.

analogues allow for monitoring oligonucleotide uptake and even sensing of the base pairing environment in some cases. For the purposes of this review, however, we will discuss only three types of modified nucleobases.

2.3.4.1 Sulfur-Modified Nucleobases

A minimalist modification is the replacement of a carbonyl with a thiocarbonyl group. Both 4-thio and 2-thio analogues of thymine are well known, but the 2-thio analogue in particular (Figure 2.10) shows promise in the context of allele-selective ASOs. The sulfur atom, isoelectronic with oxygen but larger in size, maintains the Watson–Crick hydrogen bonding of thymine and allows cleavage of the Watson–Crick (fully matched) target but leads to reduced RNase H cleavage of a target containing a G • 2-thio-T wobble pair, which suggests utility in allele-selective inhibition [68, 70].

2.3.4.2 5-Modified Pyrimidines

The simplest nucleobase modification is the 5-methylpyrimidine (Figure 2.10). This modification is naturally occurring in T and 5-methyl-C. The presence of a methyl group increases the binding affinity ($\Delta T_m = +0.5–1.5\,°C$ per methyl); this has been explained as an increase in polarizability of the nucleobase, leading to increased base stacking [241, 242]. Methylation of cytosine at CpG

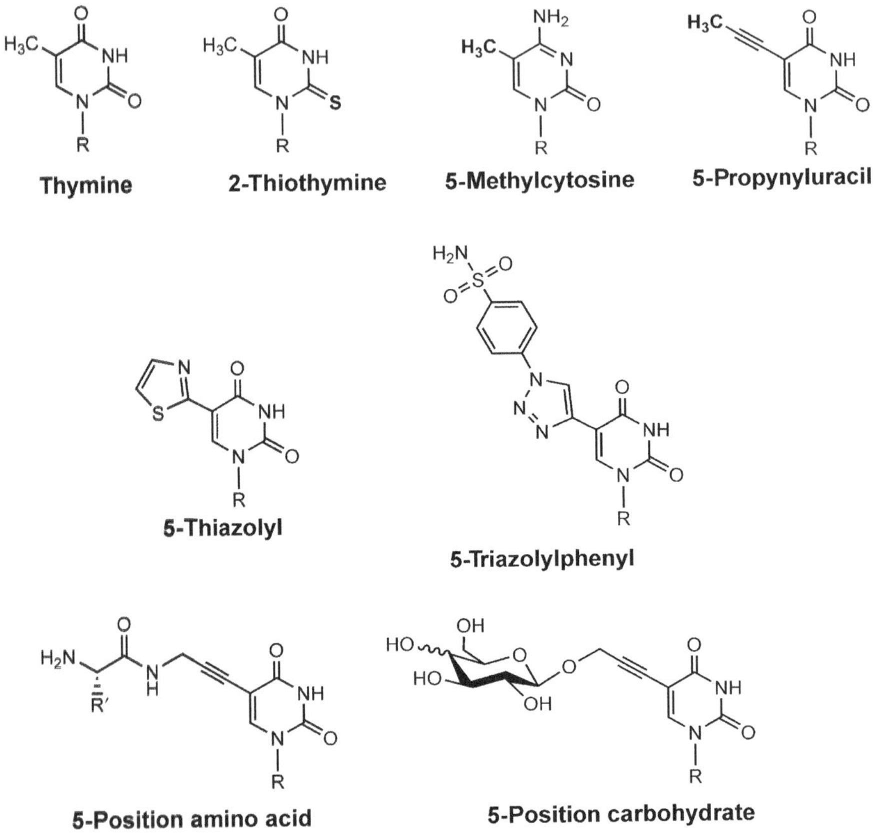

Figure 2.10 Simple nucleobase modifications include S-for-O substitutions and a large family of C5-modified pyrimidines.

(i.e. 5′-CG-3′) dinucleotides in DNA oligonucleotides has the additional advantage of reducing the TLR9-mediated immune response otherwise seen by these sequences [243, 244]. Importantly, the methyl modification is fully compatible with RNase H activity.

Some unnatural 5-modified pyrimidines provide an even larger advantage in terms of binding affinity. Of these, a 5-propynyl group (Figure 2.10) provides outstanding stabilization ($\Delta T_m = 1.5–2$°C per modification) [245–247]. Like for the methyl group, this is likely related to increased polarizability of the pyrimidine nucleobases. Once again, it also supports RNase H activity. Unfortunately, the propynyl analogues show substantial toxicity [248].

Heteroaromatic 5-membered rings can also be added to the 5-position, with thiazoles leading to higher binding affinity than thiophenes or imidazoles (Figure 2.10) [249]. They adopt a coplanar conformation with the pyrimidine ring and increase stacking with neighboring nucleobases. Triazoles including substituted triazoles are also suitable; a recent study showed that pyrimidines modified with a triazolyl-phenyl group at the 5-position (Figure 2.10) showed enhanced mismatch selectivity of RNase H cleavage [68].

The combination of a locked sugar with a 5-modified pyrimidine base has also proven fruitful. For example, locked nucleotides with amino acids [250] or carbohydrates [251] linked to the 5-position (Figure 2.10) provide further increases in binding affinity and nuclease resistance relative to the parent LNA.

2.3.4.3 Nucleobases with Expanded Hydrogen Bonding Networks

One notable modified nucleobase was rationally designed to increase the number of hydrogen bonding contacts to its complementary base. This is the G-clamp, a cytosine mimic built on the tricyclic phenoxazine framework (Figure 2.11) [252]. A pendant aminoethoxy group allows a fourth H-bond to be formed with the complementary guanine base. As such, a single G-clamp provides a large increase in affinity and specificity (ΔT_m up to +18 °C relative to 5-MeC).

Including a G-clamp in a previously optimized ASO design led to a 25-fold increase in potency *in vitro* [253]. RNase H activity was not impaired by inclusion of one or two G-clamps in the DNA window of gapmers [253, 254]. G-clamp has found much less use *in vivo* and when delivered unassisted to cultured cells; it may compromise the ability of oligonucleotides to be taken up efficiently by cells or released from endosomes [254].

Postsynthetic modification of the aminoethoxy tail of the G-clamp can be used to generate a guanidiniumethoxy tail that allows five H-bonds in total (Figure 2.11) [255]. Structural work carried out on the guanidinium G-clamp showed that the dramatic stabilization was related to a number of factors working together – including the two extra H-bonds to guanine, of course, but also additional stacking through the phenoxazine moiety, as well as a water-mediated network of H-bonds between the guanidinium group and the negatively charged phosphate oxygens of the complementary strand [256].

Without all of the additional factors (stacking, charge, etc.), simply adding hydrogen bonds yields a more subtle increase in binding affinity. For example, a much simpler nucleobase with expanded hydrogen bonding is diaminopurine (also known as 2-aminoadenine; Figure 2.11), which can form three H-bonds to thymine. Accordingly, replacing adenine with this base increases binding affinity, particularly toward RNA targets (ΔT_m = 1.3–2 °C per insert) [257, 258].

The N2-substituted guanines are likely to expand hydrogen bonding, but not to other nucleobases. Instead, adding an imidazolylpropyl [259] or aminopropyl [260] chain to the N2 of guanine is thought to create energetically favorable electrostatic interactions in the minor groove. This in turn increases the binding affinity (ΔT_m = 1–3.3 °C per insert).

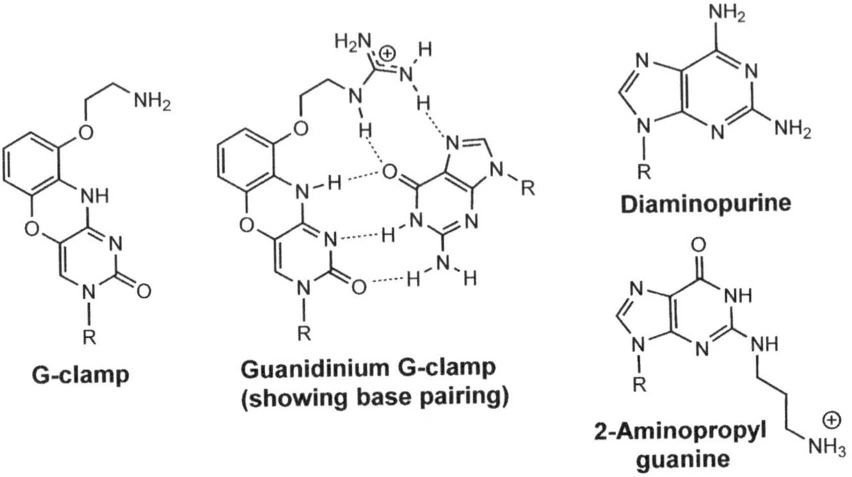

Figure 2.11 Nucleobase analogues with extended hydrogen bonding.

Other reviews include more detail on nucleobase modifications in ASOs [261] as well as modified nucleobases that can serve in diagnostic applications or as intracellular probes [262–264]. It is hoped that future work in the field will generate modifications that maintain large increases in binding affinity and specificity (like G-clamp) but allow such properties to be combined with efficient uptake.

2.3.5 Assembly of Oligonucleotides into Multimeric Structures

Multimeric oligonucleotide structures have been observed to show better cellular uptake and more effective silencing than the corresponding monomeric oligonucleotides. Some studies have used small clusters (e.g. two [265, 266], three [266], or nine [267] oligonucleotides), and others very large (e.g. nanoparticles modified with 80–100 oligonucleotides) [268]. One group observed that oligonucleotides with an inherent tendency to assemble into larger structures containing multiple oligomers were more active than oligomers that did not self-assemble [269].

The simplest possible multimer is a dimer. Linking two first-generation (PS-DNA) oligonucleotides by their 5′ ends (i.e. with their 3′ ends exposed) can increase their potency in both cells and animals while decreasing innate immunostimulation [265]. In contrast, when the oligonucleotides are linked by their 3′ ends, they show greater engagement with immune recognition proteins [65, 66, 270–272] and show no increase in silencing potency [265]. The mechanism of the increased silencing potency by the 5′-linked dimeric constructs is not clear.

Linear biocleavable dimeric and trimeric structures can be made by synthesizing PS oligonucleotides joined by short sequences of PO-DNA [266]. These constructs are stable in plasma but are cleaved to release the monomeric oligomers inside cells. The primary advantage of these linear multimeric designs is apparently their increased binding to plasma proteins and biodistribution to the liver, leading to approximately two fold greater efficacy and approximately four- to five fold potency *in vivo* [266].

Larger clusters of oligonucleotides can be developed by assembling alkanethiol-modified oligonucleotides around a metallic nanoparticle [273] or hydrophobic oligonucleotide conjugates around a liposome [274]. These so-called spherical nucleic acids [268] show increased cellular uptake through increased binding to proteins including scavenger receptors [275]. They can be designed to regulate gene expression [273] or to be highly potent immunomodulatory agents [67], in spite of earlier reports that SNAs were inherently less immunogenic than linear oligonucleotides [276].

Smaller branched structures have been made either by convergent [277] or divergent [267] dendrimerization approaches using branch-point phosphoramidites, and these too have demonstrated improved cellular uptake [267], though the field awaits data on their silencing activity.

Many different designs are possible for multimeric oligonucleotide constructs besides the examples given above – indeed, while still at very early stages, some DNA structures even show promise for delivering nonnucleic acid drugs [278–280].

2.4 Conclusion

Reading the antisense literature, particularly historical news pieces, is a little like riding a roller coaster. It seems the field of oligonucleotide therapeutics has been perpetually in a state of "gaining momentum at last" or being "finally about to realize its potential" for at least 20 years [281–286].

Beyond the fluctuations in optimism within the field, however, there has been real and steady progress in the underlying technologies required for effective antisense therapeutics. Most strikingly, some of the ASO chemistries in the early stages of clinical development today allow effective gene silencing using <1% of the doses necessary for first-generation oligonucleotides.

The field has picked up momentum from the FDA approvals of mipomersen in January 2013 and eteplirsen and nusinersen in late 2016. However, even these approvals were not a sure thing: An FDA advisory panel narrowly voted to support mipomersen's approval, 9–6, while the European Medicines Agency ultimately chose not to approve it. Similarly, the splice-switching activity of eteplirsen is marginal, and the approval was controversial.

A tremendous diversity of cellular nucleic acid targets are now considered clinically relevant. MiRNAs and pre-mRNA, which were not major targets 10 years ago, are now being actively developed by several companies, and there are multiple drugs in advanced clinical trials or even approved in the field of miRNAs [287] and splicing [288]. The medicinal chemist can take note of oligonucleotides chemically optimized for miRNA inhibition [86, 126, 289–298] and splice switching [53, 55, 84, 299, 300].

Similarly, as the biology of long noncoding RNAs has developed tremendously over the past decade, the idea of targeting long noncoding RNAs with ASOs has developed tremendously [301], and three companies have focused on this approach. Opko-Curna uses duplex RNAs and ASOs to activate gene expression by *cleaving and degrading* repressive noncoding transcripts, based on technology from Claus Wahlestedt's lab [302]. Their papers and patents make use of either duplex RNAs or gapmer ASOs. RaNA Therapeutics, in contrast, disclosed the superior performance of a steric block approach using short LNA mixmer oligonucleotides to modulate the interactions of long noncoding RNAs based on technology from Jeannie Lee's lab [303], but more recently they have switched their focus to mRNA therapeutics and renamed themselves Translate Bio. Finally, MiNA Therapeutics uses noncoding RNAs as docking sites for Argonaute-bound oligonucleotides, leading to gene activation without degrading their target RNAs [304].

Other new types of targets also continue to offer new opportunities for ASO therapeutics. ASOs have recently been proven capable of activating gene expression by disrupting R-loop formation [305] or inhibiting nonsense-mediated decay [306]. Finally, targeting an upstream open reading frame (uORF) or translation inhibitory element within an mRNA can activate protein production at the translational level [307, 308]. Medicinal chemists need to fully engage with these novel mechanisms since new target classes often have different chemical requirements for optimal activity.

It will be several years before we have data from large clinical trials of many of the recent, advanced ASO chemistries and approaches discussed in this review. Coupled with the post-approval safety and efficacy monitoring of recently approved drugs, the field stands to learn a tremendous amount in the next 5 years.

There are promising chemical innovations across the whole spectrum of development from newly discovered to newly FDA approved. The field is obtaining more clinical data than ever using a wider variety of ASO chemistries than ever before. This can only help in the process of determining whether some modifications are more advantageous for particular applications or tissues. Remaining obstacles to clinical success will also be seen more clearly in light of this data, and this knowledge can be fed back into the drug development process to find solutions. This in turn is very good for the field and ultimately, we trust, will be very good for patients.

References

1 Stephenson, M.L. and Zamecnik, P.C. (1978). Inhibition of Rous sarcoma viral RNA translation by a specific oligodeoxyribonucleotide. *Proc. Natl. Acad. Sci. U. S. A.* 75: 285–288.

2 Zamecnik, P.C. and Stephenson, M.L. (1978). Inhibition of Rous sarcoma virus replication and cell transformation by a specific oligodeoxynucleotide. *Proc. Natl. Acad. Sci. U. S. A.* 75: 280–284.

3 Khvorova, A. and Watts, J.K. (2017). The chemical evolution of oligonucleotide therapies of clinical utility. *Nat. Biotechnol.* 35: 238–248.

4 Cohen, J.S. (1991). Informational drugs: a new concept in pharmacology. *Antisense Res. Dev.* 1: 191–193.

5 Koch, T., Shim, I., Lindow, M. et al. (2014). Quantum mechanical studies of DNA and LNA. *Nucleic Acid Ther.* 24: 139–148.

6 Bohr, H.G., Shim, I., Stein, C. et al. (2017). Electronic structures of LNA phosphorothioate oligonucleotides. *Mol. Ther. Nucl. Acids* 8: 428–441.

7 Hagedorn, P.H., Yakimov, V., Ottosen, S. et al. (2013). Hepatotoxic potential of therapeutic oligonucleotides can be predicted from their sequence and modification pattern. *Nucleic Acid Ther.* 23: 302–310.

8 Burdick, A.D., Sciabola, S., Mantena, S.R. et al. (2014). Sequence motifs associated with hepatotoxicity of locked nucleic acid – modified antisense oligonucleotides. *Nucleic Acids Res.* 42: 4882–4891.

9 Shen, W., Liang, X.-h., Sun, H., and Crooke, S.T. (2015). 2′-Fluoro-modified phosphorothioate oligonucleotide can cause rapid degradation of P54nrb and PSF. *Nucleic Acids Res.* 43: 4569–4578.

10 Burel, S.A., Hart, C.E., Cauntay, P. et al. (2016). Hepatotoxicity of high affinity gapmer antisense oligonucleotides is mediated by RNase H1 dependent promiscuous reduction of very long pre-mRNA transcripts. *Nucleic Acids Res.* 44: 2093–2109.

11 Kamola, P.J., Kitson, J.D., Turner, G. et al. (2015). In silico and in vitro evaluation of exonic and intronic off-target effects form a critical element of therapeutic ASO gapmer optimization. *Nucleic Acids Res.* 43: 8638–8650.

12 Kasuya, T., Hori, S.-i., Watanabe, A. et al. (2016). Ribonuclease H1-dependent hepatotoxicity caused by locked nucleic acid-modified gapmer antisense oligonucleotides. *Sci. Rep.* 6: 30377.

13 Kamola, P.J., Maratou, K., Wilson, P.A. et al. (2017). Strategies for in vivo screening and mitigation of hepatotoxicity associated with antisense drugs. *Mol. Ther. Nucl. Acids* 8: 383–394.

14 Watts, J.K., Deleavey, G.F., and Damha, M.J. (2008). Chemically modified siRNA: tools and applications. *Drug Discov. Today* 13: 842–855.

15 Deleavey, G.F. and Damha, M.J. (2012). Designing chemically modified oligonucleotides for targeted gene silencing. *Chem. Biol.* 19: 937–954.

16 de Fougerolles, A., Vornlocher, H.-P., Maraganore, J., and Lieberman, J. (2007). Interfering with disease: a progress report on siRNA-based therapeutics. *Nat. Rev. Drug Discov.* 6: 443–453.

17 Haraszti, R.A., Roux, L., Coles, A.H. et al. (2017). 5-Vinylphosphonate improves tissue accumulation and efficacy of conjugated siRNAs in vivo. *Nucleic Acids Res.* 45: 7581–7592.

18 Nair, J.K., Willoughby, J.L., Chan, A. et al. (2014). Multivalent *N*-acetylgalactosamine-conjugated siRNA localizes in hepatocytes and elicits robust RNAi-mediated gene silencing. *J. Am. Chem. Soc.* 136: 16958–16961.

19 Fitzgerald, K., White, S., Borodovsky, A. et al. (2017). A highly durable RNAi therapeutic inhibitor of PCSK9. *N. Engl. J. Med.* 376: 41–51.

20 Khvorova, A. (2017). Oligonucleotide therapeutics – a new class of cholesterol-lowering drugs. *N. Engl. J. Med.* 376: 4–7.

21 Straarup, E.M., Fisker, N., Hedtjärn, M. et al. (2010). Short locked nucleic acid antisense oligonucleotides potently reduce apolipoprotein B mRNA and serum cholesterol in mice and non-human primates. *Nucleic Acids Res.* 38: 7100–7111.

22 Yamamoto, T., Yasuhara, H., Wada, F. et al. (2012). Superior silencing by 2′,4′-BNA(NC)-based short antisense oligonucleotides compared to 2′,4′-BNA/LNA-based apolipoprotein B antisense inhibitors. *J. Nucleic Acids* 2012: 707323.

23 Pedersen, L., Hagedorn, P.H., Lindholm, M.W., and Lindow, M. (2014). A kinetic model explains why shorter and less affine enzyme-recruiting oligonucleotides can be more potent. *Mol. Ther. Nucl. Acids* 3: e149.

24 Yamamoto, T., Fujii, N., Yasuhara, H. et al. (2014). Evaluation of multiple-turnover capability of locked nucleic acid antisense oligonucleotides in cell-free RNase H-mediated antisense reaction and in mice. *Nucleic Acid Ther.* 24: 283–290.

25 Hagedorn, P.H., Hansen, B.R., Koch, T., and Lindow, M. (2017). Managing the sequence-specificity of antisense oligonucleotides in drug discovery. *Nucleic Acids Res.* 45: 2262–2282.

26 Shimo, T., Tachibana, K., Saito, K. et al. (2014). Design and evaluation of locked nucleic acid-based splice-switching oligonucleotides in vitro. *Nucleic Acids Res.* 42: 8174–8187.

27 Aartsma-Rus, A., van Vliet, L., Hirschi, M. et al. (2009). Guidelines for antisense oligonucleotide design and insight into splice-modulating mechanisms. *Mol. Ther.* 17: 548–553.

28 Freier, S.M. and Altmann, K.-H. (1997). The ups and downs of nucleic acid duplex stability: structure-stability studies on chemically-modified DNA:RNA duplexes. *Nucleic Acids Res.* 25: 4429–4443.

29 Haeuptle, M.T., Frank, R., and Dobberstein, B. (1986). Translation arrest by oligodeoxynucleotides complementary to mRNA coding sequences yields polypeptides of predetermined length. *Nucleic Acids Res.* 14: 1427–1448.

30 Minshull, J. and Hunt, T. (1986). The use of single-stranded DNA and RNase H to promote quantitative "hybrid arrest of translation" of mRNA/DNA hybrids in reticulocyte lysate cell-free translations. *Nucleic Acids Res.* 14: 6433–6451.

31 Walder, R.Y. and Walder, J.A. (1988). Role of RNase H in hybrid-arrested translation by antisense oligonucleotides. *Proc. Natl. Acad. Sci. U. S. A.* 85: 5011–5015.

32 Damha, M.J., Wilds, C.J., Noronha, A. et al. (1998). Hybrids of RNA and arabinonucleic acids (ANA and 2′F-ANA) are substrates of ribonuclease H. *J. Am. Chem. Soc.* 120: 12976–12977.

33 Verbeure, B., Lescrinier, E., Wang, J., and Herdewijn, P. (2001). RNase H mediated cleavage of RNA by cyclohexene nucleic acid (CeNA). *Nucleic Acids Res.* 29: 4941–4947.

34 Shibahara, S., Mukai, S., Nishihara, T. et al. (1987). Site-directed cleavage of RNA. *Nucleic Acids Res.* 15: 4403–4415.

35 Monia, B.P., Lesnik, E.A., Gonzalez, C. et al. (1993). Evaluation of 2′-modified oligonucleotides containing 2′-deoxy gaps as antisense inhibitors of gene expression. *J. Biol. Chem.* 268: 14514–14522.

36 Liu, J., Carmell, M.A., Rivas, F.V. et al. (2004). Argonaute2 is the catalytic engine of mammalian RNAi. *Science* 305: 1437–1441.

37 Song, J.-J., Smith, S.K., Hannon, G.J., and Joshua-Tor, L. (2004). Crystal structure of argonaute and its implications for RISC slicer activity. *Science* 305: 1434–1437.

38 Tomari, Y. and Zamore, P.D. (2005). Perspective: machines for RNAi. *Genes Dev.* 19: 517–529.

39 Lima, W.F., Prakash, T.P., Murray, H.M. et al. (2012). Single-stranded siRNAs activate RNAi in animals. *Cell* 150: 883–894.

40 Yu, D., Pendergraff, H., Liu, J. et al. (2012). Single-stranded RNAs use RNAi to potently and allele-selectively inhibit mutant huntingtin expression. *Cell* 150: 895–908.

41 Prakash, T.P., Lima, W.F., Murray, H.M. et al. (2013). Lipid nanoparticles improve activity of single-stranded siRNA and gapmer antisense oligonucleotides in animals. *ACS Chem. Biol.* 8: 1402–1406.

42 Liu, J., Yu, D., Aiba, Y. et al. (2013). ss-siRNAs allele selectively inhibit ataxin-3 expression: multiple mechanisms for an alternative gene silencing strategy. *Nucleic Acids Res.* 41: 9570–9583.

43 Hu, J., Liu, J., Narayanannair, K.J. et al. (2014). Allele-selective inhibition of mutant atrophin-1 expression by duplex and single-stranded RNAs. *Biochemistry* 53: 4510–4518.

44 Hu, J., Liu, J., Yu, D. et al. (2014). Exploring the effect of sequence length and composition on allele-selective inhibition of human huntingtin expression by single-stranded silencing RNAs. *Nucleic Acid Ther.* 24: 199–209.

45 Prakash, T.P., Lima, W.F., Murray, H.M. et al. (2015). Identification of metabolically stable 5′-phosphate analogs that support single-stranded siRNA activity. *Nucleic Acids Res.* 43: 2993–3011.
46 Pendergraff, H.M., Debacker, A.J., and Watts, J.K. (2016). Single-stranded silencing RNAs: hit rate and chemical modification. *Nucleic Acid Ther.* 26: 216–222.
47 Byrne, M., Tzekov, R., Wang, Y. et al. (2013). Novel hydrophobically modified asymmetric RNAi compounds (sd-rxRNA) demonstrate robust efficacy in the eye. *J. Ocul. Pharmacol. Ther.* 29: 855–864.
48 Nikan, M., Osborn, M.F., Coles, A.H. et al. (2017). Synthesis and evaluation of parenchymal retention and efficacy of a metabolically stable *O*-phosphocholine-*N*-docosahexaenoyl-l-serine siRNA conjugate in mouse brain. *Bioconjug. Chem.* 28: 1758–1766.
49 Ly, S., Navaroli, D.M., Didiot, M.C. et al. (2017). Visualization of self-delivering hydrophobically modified siRNA cellular internalization. *Nucleic Acids Res.* 45: 15–25.
50 Alterman, J.F., Hall, L.M., Coles, A.H. et al. (2015). Hydrophobically modified siRNAs silence huntingtin mRNA in primary neurons and mouse brain. *Mol. Ther. Nucl. Acids* 4: e266.
51 Nikan, M., Osborn, M.F., Coles, A.H. et al. (2016). Docosahexaenoic acid conjugation enhances distribution and safety of siRNA upon local administration in mouse brain. *Mol. Ther. Nucl. Acids* 5: e344.
52 Castanotto, D., Lin, M., Kowolik, C. et al. (2015). A cytoplasmic pathway for gapmer antisense oligonucleotide-mediated gene silencing in mammalian cells. *Nucleic Acids Res.* 43: 9350–9361.
53 Skordis, L.A., Dunckley, M.G., Yue, B. et al. (2003). Bifunctional antisense oligonucleotides provide a trans-acting splicing enhancer that stimulates SMN2 gene expression in patient fibroblasts. *Proc. Natl. Acad. Sci. U. S. A.* 100: 4114–4119.
54 Liu, B., Han, Y., Ferdous, A. et al. (2003). Transcription activation by a PNA-peptide chimera in a mammalian cell extract. *Chem. Biol.* 10: 909–916.
55 Rigo, F., Hua, Y., Chun, S.J. et al. (2012). Synthetic oligonucleotides recruit ILF2/3 to RNA transcripts to modulate splicing. *Nat. Chem. Biol.* 8: 555–561.
56 Janas, M.M., Jiang, Y., Schlegel, M.K. et al. (2016). Impact of oligonucleotide structure, chemistry, and delivery method on in vitro cytotoxicity. *Nucleic Acid Ther.* 27: 11–22.
57 Crooke, S.T., Wang, S., Vickers, T.A. et al. (2017). Cellular uptake and trafficking of antisense oligonucleotides. *Nat. Biotechnol.* 35: 230–237.
58 Barchet, W., Wimmenauer, V., Schlee, M., and Hartmann, G. (2008). Accessing the therapeutic potential of immunostimulatory nucleic acids. *Curr. Opin. Immunol.* 20: 389–395.
59 Takeda, K. and Akira, S. (2005). Toll-like receptors in innate immunity. *Int. Immunol.* 17: 1–14.

60 Cornélie, S., Hoebeke, J., Schacht, A.M. et al. (2004). Direct evidence that Toll-like Receptor 9 (TLR9) functionally binds plasmid DNA by specific cytosine-phosphate-guanine motif recognition. *J. Biol. Chem.* 279: 15124–15129.

61 Robbins, M., Judge, A., Liang, L. et al. (2007). 2′-*O*-methyl-modified RNAs act as TLR7 antagonists. *Mol. Ther.* 15: 1663–1669.

62 Hamm, S., Latz, E., Hangel, D. et al. (2010). Alternating 2′-*O*-ribose methylation is a universal approach for generating non-stimulatory siRNA by acting as TLR7 antagonist. *Immunobiology* 215: 559–569.

63 Bodera, P., Stankiewicz, W., and Kocik, J. (2012). Synthetic immunostimulatory oligonucleotides in experimental and clinical practice. *Pharmacological reports* 64: 1003–1010.

64 Krieg, A.M. (2012). CpG still rocks! Update on an accidental drug. *Nucleic Acid Ther.* 22: 77–89.

65 Agrawal, S. and Kandimalla, E.R. (2007). Synthetic agonists of Toll-like receptors 7, 8 and 9. *Biochem. Soc. Trans.* 35: 1461–1467.

66 Kandimalla, E.R., Bhagat, L., Li, Y. et al. (2005). Immunomodulatory oligonucleotides containing a cytosine-phosphate-2′-deoxy-7-deazaguanosine motif as potent toll-like receptor 9 agonists. *Proc. Natl. Acad. Sci. U. S. A.* 102: 6925–6930.

67 Radovic-Moreno, A.F., Chernyak, N., Mader, C.C. et al. (2015). Immunomodulatory spherical nucleic acids. *Proc. Natl. Acad. Sci. U. S. A.* 112: 3892–3897.

68 Ostergaard, M.E., Kumar, P., Nichols, J. et al. (2015). Allele-selective inhibition of mutant huntingtin with 2-thio- and C5- triazolylphenyl-deoxythymidine-modified antisense oligonucleotides. *Nucleic Acid Ther.* 25: 266–274.

69 Ostergaard, M.E., Gerland, B., Escudier, J.M. et al. (2014). Differential effects on allele selective silencing of mutant huntingtin by two stereoisomers of alpha, beta-constrained nucleic acid. *ACS Chem. Biol.* 9: 1975–1979.

70 Ostergaard, M.E., Southwell, A.L., Kordasiewicz, H. et al. (2013). Rational design of antisense oligonucleotides targeting single nucleotide polymorphisms for potent and allele selective suppression of mutant huntingtin in the CNS. *Nucleic Acids Res.* 41: 9634–9650.

71 Southwell, A.L., Skotte, N.H., Kordasiewicz, H.B. et al. (2014). In vivo evaluation of candidate allele-specific mutant huntingtin gene silencing antisense oligonucleotides. *Mol. Ther.* 22: 2093–2106.

72 Juliano, R.L., Ming, X., and Nakagawa, O. (2012). The chemistry and biology of oligonucleotide conjugates. *Acc. Chem. Res.* 45: 1067–1076.

73 Juliano, R.L. (2016). The delivery of therapeutic oligonucleotides. *Nucleic Acids Res.* 44: 6518–6548.

74 Biessen, E.A.L., Vietsch, H., Rump, E.T. et al. (1999). Targeted delivery of oligodeoxynucleotides to parenchymal liver cells in vivo. *Biochem. J.* 340: 783–792.

75 Manoharan, M., Rajeev, K.G., Narayanannair, J.K. and Maier, M. (2008). Carbohydrate conjugates as delivery agents for oligonucleotides. US Patent 8, 106, 022, filed 4 December 2008 and issued 30 December 2009.

76 Matsuda, S., Keiser, K., Nair, J.K. et al. (2015). siRNA conjugates carrying sequentially assembled trivalent *N*-acetylgalactosamine linked through nucleosides elicit robust gene silencing in vivo in hepatocytes. *ACS Chem. Biol.* 10: 1181–1187.

77 Huang, X., Leroux, J.C., and Castagner, B. (2017). Well-defined multivalent ligands for hepatocytes targeting via asialoglycoprotein receptor. *Bioconjug. Chem.* 28: 283–295.

78 Rajeev, K.G., Nair, J.K., Jayaraman, M. et al. (2015). Hepatocyte-specific delivery of siRNAs conjugated to novel non-nucleosidic trivalent *N*-acetylgalactosamine elicits robust gene silencing in vivo. *Chem Bio Chem* 16: 903–908.

79 Prakash, T.P., Graham, M.J., Yu, J. et al. (2014). Targeted delivery of antisense oligonucleotides to hepatocytes using tri-antennary *N*-acetyl galactosamine improves potency 10-fold in mice. *Nucleic Acids Res.* 42: 8796–8807.

80 Østergaard, M.E., Yu, J., Kinberger, G.A. et al. (2015). Efficient synthesis and biological evaluation of 5′-GalNAc conjugated antisense oligonucleotides. *Bioconjug. Chem.* 26: 1451–1455.

81 Lehto, T., Ezzat, K., Wood, M.J.A., and El Andaloussi, S. (2016). Peptides for nucleic acid delivery. *Adv. Drug Deliv. Rev.* 106, *Part A*: 172–182.

82 Bock, H. and Lindhorst, T. (2007). Method for selective localization of active agents at and in mitochondria and corresponding active agents. US Patent application 12/446, 860, filed 23 October 2007 and issued 2 May 2008.

83 Ittig, D., Liu, S., Renneberg, D. et al. (2004). Nuclear antisense effects in cyclophilin a pre-mRNA splicing by oligonucleotides: a comparison of tricyclo-DNA with LNA. *Nucleic Acids Res.* 32: 346–353.

84 Goyenvalle, A., Griffith, G., Babbs, A. et al. (2015). Functional correction in mouse models of muscular dystrophy using exon-skipping tricyclo-DNA oligomers. *Nat. Med.* 21: 270–275.

85 Rettig, G.R. and Behlke, M.A. (2012). Progress toward in vivo use of siRNAs-II. *Mol. Ther.* 20: 483–512.

86 Lennox, K.A. and Behlke, M.A. (2010). A direct comparison of anti-micro RNA oligonucleotide potency. *Pharm. Res.* 27: 1788–1799.

87 Choung, S., Kim, Y.J., Kim, S. et al. (2006). Chemical modification of siRNAs to improve serum stability without loss of efficacy. *Biochem. Biophys. Res. Commun.* 342: 919–927.

88 Ng, E.W.M., Shima, D.T., Calias, P. et al. (2006). Pegaptanib, a targeted anti-VEGF aptamer for ocular vascular disease. *Nat. Rev. Drug Discov.* 5: 123–132.

89 Manoharan, M. (1999). 2′-Carbohydrate modifications in antisense oligonucleotide therapy: importance of conformation, configuration and conjugation. *Biochim. Biophys. Acta, Gene Struct. Expr.* 1489: 117–130.

90 Mangos, M.M. and Damha, M.J. (2002). Flexible and frozen sugar-modified nucleic acids – modulation of biological activity through furanose ring dynamics in the antisense strand. *Curr. Top. Med. Chem.* 2: 1147–1171.

91 Prakash, T.P. (2011). An overview of sugar-modified oligonucleotides for antisense therapeutics. *Chem. Biodivers.* 8: 1616–1641.

92 Martin, P. (1995). A new access to 2′-*O*-alkylated ribonucleosides and properties of 2′-*O*-alkylated oligoribonucleotides. *Helv. Chim. Acta* 78: 486–504.

93 Raal, F.J., Santos, R.D., Blom, D.J. et al. (2010). Mipomersen, an apolipoprotein B synthesis inhibitor, for lowering of LDL cholesterol concentrations in patients with homozygous familial hypercholesterolaemia: a randomised, double-blind, placebo-controlled trial. *Lancet* 375: 998–1006.

94 Chiriboga, C.A., Swoboda, K.J., Darras, B.T. et al. (2016). Results from a phase 1 study of nusinersen (ISIS-SMN(Rx)) in children with spinal muscular atrophy. *Neurology* 86: 890–897.

95 Egli, M., Minasov, G., Tereshko, V. et al. (2005). Probing the influence of stereoelectronic effects on the biophysical properties of oligonucleotides: comprehensive analysis of the RNA affinity, nuclease resistance, and crystal structure of ten 2′-*O*-ribonucleic acid modifications. *Biochemistry* 44: 9045–9057.

96 Kawasaki, A.M., Casper, M.D., Freier, S.M. et al. (1993). Uniformly modified 2′-deoxy-2′-fluoro-phosphorothioate oligonucleotides as nuclease-resistant antisense compounds with high affinity and specificity for RNA targets. *J. Med. Chem.* 36: 831–841.

97 Pallan, P.S., Greene, E.M., Jicman, P.A. et al. (2011). Unexpected origins of the enhanced pairing affinity of 2′-fluoro-modified RNA. *Nucleic Acids Res.* 39: 3482–3495.

98 Patra, A., Paolillo, M., Charisse, K. et al. (2012). 2′-fluoro RNA shows increased Watson-Crick H-bonding strength and stacking relative to RNA: evidence from NMR and thermodynamic data. *Angew. Chem. Int. Ed. Eng.* 51: 11863–11866.

99 Deleavey, G.F., Watts, J.K., Alain, T. et al. (2010). Synergistic effects between analogs of DNA and RNA improve the potency of siRNA-mediated gene silencing. *Nucleic Acids Res.* 38: 4547–4557.

100 Morrissey, D.V., Lockridge, J.A., Shaw, L. et al. (2005). Potent and persistent in vivo anti-HBV activity of chemically modified siRNAs. *Nat. Biotechnol.* 23: 1002–1007.

101 Wilds, C.J. and Damha, M.J. (2000). 2′-Deoxy-2′-fluoro-β-D-arabinonucleosides and oligonucleotides (2′F-ANA): synthesis and physicochemical studies. *Nucleic Acids Res.* 28: 3625–3635.

102 Watts, J.K. and Damha, M.J. (2008). 2′F-Arabinonucleic acids (2′F-ANA) – history, properties, and new frontiers. *Can. J. Chem.* 86: 641–656.

103 Noronha, A.M., Wilds, C.J., Lok, C.-N. et al. (2000). Synthesis and biophysical properties of arabinonucleic acids (ANA): circular dichroic spectra, melting temperatures, and ribonuclease H susceptibility of ANA. RNA hybrid duplexes. *Biochemistry* 39: 7050–7062.

104 Lok, C.-N., Viazovkina, E., Min, K.-L. et al. (2002). Potent gene-specific inhibitory properties of mixed-backbone antisense oligonucleotides comprised of 2′-deoxy-2′-fluoro-D-arabinose and 2′-deoxyribose nucleotides. *Biochemistry* 41: 3457–3467.

105 Mangos, M.M., Min, K.-L., Viazovkina, E. et al. (2003). Efficient RNase H-directed cleavage of RNA promoted by antisense DNA or 2′F-ANA constructs containing acyclic nucleotide inserts. *J. Am. Chem. Soc.* 125: 654–661.

106 Kalota, A., Karabon, L., Swider, C.R. et al. (2006). 2′-deoxy-2′-fluoro-beta-D-arabinonucleic acid (2′ F-ANA) modified oligonucleotides (ON) effect highly efficient, and persistent, gene silencing. *Nucleic Acids Res.* 34: 451–461.

107 Ferrari, N., Bergeron, D., Tedeschi, A.-L. et al. (2006). Characterization of antisense oligonucleotides comprising 2′-deoxy-2′-fluoro-beta-D-arabinonucleic acid (FANA): specificity, potency, and duration of activity. *Ann. N. Y. Acad. Sci.* 1082: 91–102.

108 Souleimanian, N., Deleavey, G.F., Soifer, H. et al. (2012). Antisense 2′-deoxy, 2′-fluoroarabino nucleic acid (2′F-ANA) oligonucleotides: in vitro gymnotic silencers of gene expression whose potency is enhanced by fatty acids. *Mol. Ther. Nucl. Acids* 1: e43.

109 Guimond, A., Viau, E., Aube, P. et al. (2008). Advantageous toxicity profile of inhaled antisense oligonucleotides following chronic dosing in non-human primates. *Pulm. Pharmacol. Ther.* 21: 845–854.

110 Watts, J.K., Katolik, A., Viladoms, J., and Damha, M.J. (2009). Studies on the hydrolytic stability of 2′-fluoroarabinonucleic acid (2′F-ANA). *Org. Biomol. Chem.* 7: 1904–1910.

111 Watts, J.K., Martin-Pintado, N., Gomez-Pinto, I. et al. (2010). Differential stability of 2′F-ANA:RNA and ANA:RNA hybrid duplexes: roles of structure, pseudohydrogen bonding, hydration, ion uptake and flexibility. *Nucleic Acids Res.* 38: 2498–2511.

112 Yahyaee Anzahaee, M., Watts, J.K., Alla, N.R. et al. (2011). Energetically important C-H...F-C pseudohydrogen bonding in water: evidence and application to rational design of oligonucleotides with high binding affinity. *J. Am. Chem. Soc.* 133: 728–731.

113 Martin-Pintado, N., Deleavey, G.F., Portella, G. et al. (2013). Backbone FC-H.O hydrogen bonds in 2′F-substituted nucleic acids. *Angew. Chem. Int. Ed. Eng.* 52: 12065–12068.

114 Martinez-Montero, S., Deleavey, G.F., Kulkarni, A. et al. (2014). Rigid 2′,4′-difluororibonucleosides: synthesis, conformational analysis, and incorporation into nascent RNA by HCV polymerase. *J. Org. Chem.* 79: 5627–5635.

115 Martinez-Montero, S., Deleavey, G.F., Martin-Pintado, N. et al. (2015). Locked 2′-deoxy-2′,4′-difluororibo modified nucleic acids: thermal stability, structural studies, and siRNA activity. *ACS Chem. Biol.* 10: 2016–2023.

116 Martinez-Montero, S., Deleavey, G.F., Dierker-Viik, A. et al. (2015). Synthesis and properties of 2′-deoxy-2′,4′-difluoroarabinose-modified nucleic acids. *J. Org. Chem.* 80: 3083–3091.

117 Kool, E.T. (1997). Preorganization of DNA: design principles for improving nucleic acid recognition by synthetic oligonucleotides. *Chem. Rev.* 97: 1473–1487.

118 Koshkin, A.A., Singh, S.K., Nielsen, P. et al. (1998). LNA (Locked Nucleic Acids): synthesis of the adenine, cytosine, guanine, 5-methylcytosine, thymine and uracil bicyclonucleoside monomers, oligomerisation, and unprecedented nucleic acid recognition. *Tetrahedron* 54: 3607–3630.

119 Singh, S.K., Nielsen, P., Koshkin, A.A., and Wengel, J. (1998). LNA (locked nucleic acids): synthesis and high-affinity nucleic acid recognition. *Chem. Commun.* 455–456.

120 Obika, S., Nanbu, D., Hari, Y. et al. (1998). Stability and structural features of the duplexes containing nucleoside analogues with a fixed N-type conformation, 2′-*O*,4′-*C*-methyleneribonucleosides. *Tetrahedron Lett.* 39: 5401–5404.

121 Koch, T. (2013). LNA antisense: a review. *Current Physical Chemistry* 3: 55–68.

122 Watts, J.K. (2013). Locked nucleic acid: tighter is different. *Chem. Commun.* 49: 5618–1520.

123 Vester, B. and Wengel, J. (2004). LNA (Locked nucleic acid): high-affinity targeting of complementary RNA and DNA. *Biochemistry* 43: 13233–13241.

124 Seth, P.P., Siwkowski, A., Allerson, C.R. et al. (2009). Short antisense oligonucleotides with novel 2′-4′ conformationaly restricted nucleoside analogues show improved potency without increased toxicity in animals. *J. Med. Chem.* 52: 10–13.

125 Murray, S., Ittig, D., Koller, E. et al. (2012). Tricyclo DNA-modified oligo-2′-deoxyribonucleotides reduce scavenger receptor B1 mRNA in hepatic and extra-hepatic tissues – a comparative study of oligonucleotide length, design and chemistry. *Nucleic Acids Res.* 40: 6135–6143.

126 Obad, S., Dos Santos, C.O., Petri, A. et al. (2011). Silencing of microRNA families by seed-targeting tiny LNAs. *Nat. Genet.* 43: 371–378.

127 Bartel, D.P. (2009). MicroRNAs: target recognition and regulatory functions. *Cell* 136: 215–233.

128 Morita, K., Hasegawa, C., Kaneko, M. et al. (2002). 2′-*O*,4′-*C*-ethylene-bridged nucleic acids (ENA): highly nuclease-resistant and thermodynamically stable oligonucleotides for antisense drug. *Bioorg. Med. Chem. Lett.* 12: 73–76.

129 Imanishi, T. and Obika, S. (2002). BNAs: novel nucleic acid analogs with a bridged sugar moiety. *Chem. Commun.* 1653–1659.

130 Srivastava, P., Barman, J., Pathmasiri, W. et al. (2007). Five- and six-membered conformationally locked 2′,4′-carbocyclic ribo-thymidines: synthesis, structure, and biochemical studies. *J. Am. Chem. Soc.* 129: 8362–8379.

131 Mitsuoka, Y., Kodama, T., Ohnishi, R. et al. (2009). A bridged nucleic acid, 2′,4′-BNA COC: synthesis of fully modified oligonucleotides bearing thymine, 5-methylcytosine, adenine and guanine 2′,4′-BNA COC monomers and RNA-selective nucleic-acid recognition. *Nucleic Acids Res.* 37: 1225–1238.

132 Seth, P.P., Allerson, C.R., Berdeja, A. et al. (2010). An exocyclic methylene group acts as a bioisostere of the 2′-oxygen atom in LNA. *J. Am. Chem. Soc.* 132: 14942–14950.

133 Rahman, S.M.A., Seki, S., Obika, S. et al. (2008). Design, synthesis and properties of 2′,4′-BNA(NC): a bridged nucleic acid analogue. *J. Am. Chem. Soc.* 130: 4886–4896.

134 Prakash, T.P., Siwkowski, A., Allerson, C.R. et al. (2010). Antisense oligonucleotides containing conformationally constrained 2′,4′-(N-methoxy) aminomethylene and 2′,4′-aminooxymethylene and 2′-*O*,4′-*C*-aminomethylene bridged nucleoside analogues show improved potency in animal models. *J. Med. Chem.* 53: 1636–1650.

135 Zhou, C., Liu, Y., Andaloussi, M. et al. (2009). Fine tuning of electrostatics around the internucleotidic phosphate through incorporation of modified 2′,4′-carbocyclic-LNAs and -ENAs leads to significant modulation of antisense properties. *J. Org. Chem.* 74: 118–134.

136 A number of different pendant groups and linkers have been tested. For one excellent example of this approach see Hrdlicka, P.J., Babu, B.R., Sorensen, M.D., and Wengel, J. (2004). Interstrand communication between 2′-*N*-(pyren-1-yl)methyl-2′-amino-LNA monomers in nucleic acid duplexes: directional control and signalling of full complementarity. *Chem. Commun.* 1478–1479.

137 Rajwanshi, V.K., Hakansson, A.E., Sorensen, M.D. et al. (2000). The eight stereoisomers of LNA (Locked Nucleic Acid): a remarkable family of strong RNA binding molecules. *Angew. Chem. Int. Ed. Eng.* 39: 1656–1659.

138 Nielsen, J.T., Stein, P.C., and Petersen, M. (2003). NMR structure of an a-L-LNA:RNA hybrid: structural implications for RNase H recognition. *Nucleic Acids Res.* 31: 5858–5867.

139 Frieden, M., Christensen, S.M., Mikkelsen, N.D. et al. (2003). Expanding the design horizon of antisense oligonucleotides with alpha-L-LNA. *Nucleic Acids Res.* 31: 6365–6372.

140 Giacometti, R.D., Salinas, J.C., Ostergaard, M.E. et al. (2016). Design, synthesis, and duplex-stabilizing properties of conformationally constrained tricyclic analogues of LNA. *Org. Biomol. Chem.* 14: 2034–2040.

141 Hanessian, S., Wagger, J., Merner, B.L. et al. (2013). A constrained tricyclic nucleic acid analogue of alpha-L-LNA: investigating the effects of dual conformational restriction on duplex thermal stability. *J. Org. Chem.* 78: 9064–9075.

142 Hanessian, S., Schroeder, B.R., Giacometti, R.D. et al. (2012). Structure-based design of a highly constrained nucleic acid analogue: improved duplex stabilization by restricting sugar pucker and torsion angle gamma. *Angew. Chem. Int. Ed. Eng.* 51: 11242–11245.

143 Herdewijn, P. (2010). Nucleic acids with a six-membered "Carbohydrate" mimic in the backbone. *Chem. Biodivers.* 7: 1–59.

144 Egli, M., Pallan, P.S., Allerson, C.R. et al. (2011). Synthesis, improved antisense activity and structural rationale for the divergent RNA affinities of 3′-fluoro hexitol nucleic acid (FHNA and Ara-FHNA) modified oligonucleotides. *J. Am. Chem. Soc.* 133: 16642–16649.

145 Herdewijn, P., Allart, B., De Bouvere, B. et al. (1999). Properties of oligonucleotides with six membered carbohydrate mimics and a 1,4-relationship between the base moiety and the hydroxymethyl group. *Nucleosides Nucleotides Nucl. Acids* 18: 1371–1376.

146 Van Aerschot, A., Meldgaard, M., Schepers, G. et al. (2001). Improved hybridisation potential of oligonucleotides comprising *O*-methylated anhydrohexitol nucleoside congeners. *Nucleic Acids Res.* 29: 4187–4194.

147 Wang, J., Verbeure, B., Luyten, I. et al. (2000). Cyclohexene nucleic acids (CeNA): serum stable oligonucleotides that activate RNase H and increase duplex stability with complementary RNA. *J. Am. Chem. Soc.* 122: 8595–8602.

148 Sabatino, D. and Damha, M.J. (2007). Oxepane nucleic acids: synthesis, characterization, and properties of oligonucleotides bearing a seven-membered carbohydrate ring. *J. Am. Chem. Soc.* 129: 8259–8270.

149 Eckstein, F. (2000). Phosphorothioate oligodeoxynucleotides: what is their origin and what is unique about them? *Antisense Nucleic Acid Drug Dev.* 10: 117–121.

150 Eckstein, F. (2014). Phosphorothioates, essential components of therapeutic oligonucleotides. *Nucleic Acid Ther.* 24: 374–387.

151 Matsukura, M., Shinozuka, K., Zon, G. et al. (1987). Phosphorothioate analogs of oligodeoxynucleotides: inhibitors of replication and cytopathic effects of human immunodeficiency virus. *Proc. Natl. Acad. Sci. U. S. A.* 84: 7706–7710.

152 Spitzer, S. and Eckstein, F. (1988). Inhibition of deoxyribonucleases by phosphorothioate groups in oligodeoxyribonucleotides. *Nucleic Acids Res.* 16: 11691–11704.

153 Matsukura, M., Zon, G., Shinozuka, K. et al. (1989). Regulation of viral expression of human immunodeficiency virus in vitro by an antisense phosphorothioate oligodeoxynucleotide against rev (art/trs) in chronically infected cells. *Proc. Natl. Acad. Sci. U. S. A.* 86: 4244–4248.

154 Agrawal, S., Goodchild, J., Civeira, M.P. et al. (1988). Oligodeoxynucleoside phosphoramidates and phosphorothioates as inhibitors of human immunodeficiency virus. *Proc. Natl. Acad. Sci. U. S. A.* 85: 7079–7083.

155 Geary, R.S. (2009). Antisense oligonucleotide pharmacokinetics and metabolism. *Expert Opin Drug Met* 5: 381–391.

156 Cy Stein (2013). Delivery of therapeutic oligonucleotides to cells: a retrospective of the past and some thoughts on moving forward. Presented at EuroTides 2013, Prague (13 November 2013).

157 Koller, E., Vincent, T.M., Chappell, A. et al. (2011). Mechanisms of single-stranded phosphorothioate modified antisense oligonucleotide accumulation in hepatocytes. *Nucleic Acids Res.* 39: 4795–4807.

158 Weidner, D.A., Valdez, B.C., Henning, D. et al. (1995). Phosphorothioate oligonucleotides bind in a non sequence-specific manner to the nucleolar protein C23/nucleolin. *FEBS Lett.* 366: 146–150.

159 Stein, C.A., Wu, S., Voskresenskiy, A.M. et al. (2009). G3139, an anti-Bcl-2 antisense oligomer that binds heparin-binding growth factors and collagen I, alters in vitro endothelial cell growth and tubular morphogenesis. *Clin. Cancer Res.* 15: 2797–2807.

160 Brown, D.A., Kang, S.H., Gryaznov, S.M. et al. (1994). Effect of phosphorothioate modification of oligodeoxynucleotides on specific protein binding. *J. Biol. Chem.* 269: 26801–26805.

161 Boczkowska, M., Guga, P., and Stec, W.J. (2002). Stereodefined phosphorothioate analogues of DNA: relative thermodynamic stability of the model PS-DNA/DNA and PS-DNA/RNA complexes. *Biochemistry* 41: 12483–12487.

162 Stec, W.J., Cierniewski, C.S., Okruszek, A. et al. (1997). Stereodependent inhibition of plasminogen activator inhibitor type 1 by phosphorothioate oligonucleotides: proof of sequence specificity in cell culture and in vivo rat experiments. *Antisense Nucleic Acid Drug Dev.* 7: 567–573.

163 Krieg, A.M., Guga, P., and Stec, W. (2003). P-chirality-dependent immune activation by phosphorothioate CpG oligodeoxynucleotides. *Oligonucleotides* 13: 491–499.

164 Oka, N. and Wada, T. (2011). Stereocontrolled synthesis of oligonucleotide analogs containing chiral internucleotidic phosphorus atoms. *Chem. Soc. Rev.* 40: 5829–5843.

165 Koziolkiewicz, M., Krakowiak, A., Kwinkowski, M. et al. (1995). Stereodifferentiation – the effect of P chirality of oligo (nucleoside phosphorothioates) on the activity of bacterial RNase H. *Nucleic Acids Res.* 23: 5000–5005.

166 Stec, W.J., Karwowski, B., Boczkowska, M. et al. (1998). Deoxyribonucleoside 3′-*O*-(2-Thio- and 2-Oxo-“spiro”-4, 4-pentamethylene-1, 3, 2-oxathiaphospholane)s: monomers for stereocontrolled synthesis of oligo (deoxyribonucleoside phosphorothioate)s and chimeric PS/PO oligonucleotides. *J. Am. Chem. Soc.* 120: 7156–7167.

167 Wang, J.-C. and Just, G. (1997). A stereoselective synthesis of dinucleotide phosphorothioates, using chiral indol-oxazaphosphorine intermediates. *Tetrahedron Lett.* 38: 3797–3800.
168 Jin, Y. and Just, G. (1998). Stereoselective synthesis of dithymidine phosphorothioates using xylose derivatives as chiral auxiliaries. *J. Org. Chem.* 63: 3647–3654.
169 Wang, J.-C. and Just, G. (1999). Indolyl-oxazaphosphorine precursors for stereoselective synthesis of phosphite triesters and dithymidinyl phosphorothioates. *J. Org. Chem.* 64: 8090–8097.
170 Lu, Y. and Just, G. (2000). Stereoselective synthesis of RP- and SP-dithymidine phosphorothioates via chiral indolooxazaphosphorine intermediates derived from tryptophan. *Angew. Chem. Int. Ed.* 39: 4521–4524.
171 Hayakawa, Y., Hirabayashi, Y., Hyodo, M. et al. (2006). A strategy for the stereoselective preparation of thymidine phosphorothioates with the (R) or the (S) configuration at the stereogenic phosphorus atom and their application to the synthesis of oligodeoxyribonucleotides with stereochemically pure phosphate/phosphorothioate chimeric backbones. *Eur. J. Org. Chem.* 2006: 3834–3844.
172 Nukaga, Y., Oka, N., and Wada, T. (2016). Stereocontrolled solid-phase synthesis of phosphate/phosphorothioate (PO/PS) chimeric oligodeoxyribonucleotides on an automated synthesizer using an oxazaphospholidine-phosphoramidite method. *J. Org. Chem.* 81: 2753–2762.
173 Nukaga, Y., Yamada, K., Ogata, T. et al. (2012). Stereocontrolled solid-phase synthesis of phosphorothioate oligoribonucleotides using 2′-*O*-(2-cyanoethoxymethyl)-nucleoside 3′-*O*-oxazaphospholidine monomers. *J. Org. Chem.* 77: 7913–7922.
174 Oka, N., Yamamoto, M., Sato, T., and Wada, T. (2008). Solid-phase synthesis of stereoregular oligodeoxyribonucleoside phosphorothioates using bicyclic oxazaphospholidine derivatives as monomer units. *J. Am. Chem. Soc.* 130: 16031–16037.
175 Iwamoto, N., Oka, N., Sato, T., and Wada, T. (2009). Stereocontrolled solid-phase synthesis of oligonucleoside H-phosphonates by an oxazaphospholidine approach. *Angew. Chem. Int. Ed. Eng.* 48: 496–499.
176 Iwamoto, N., Butler, D.C.D., Svrzikapa, N. et al. (2017). Control of phosphorothioate stereochemistry substantially alters the efficacy of antisense oligonucleotides. *Nat. Biotechnol.* 35: 845–851. doi: 10.1038/nbt.3948.
177 Pipeline. www.wavelifesciences.com (accessed 8 August2017).
178 Ferrari, N. and Seguin, R. (2012). Oligonucleotide inhibitors with chimeric backbone and 2-amino-2′-deoxyadenosine. PCT application PCT/CA2012/000169, filed 23 February 2012 and issued 30 August 2012.
179 Moore, L.R., Rajpal, G., Dillingham, I.T. et al. (2017). Evaluation of antisense oligonucleotides targeting ATXN3 in SCA3 mouse models. *Mol. Ther. Nucleic Acids* 7: 200–210.

180 Rigo, F. (2017). Compositions for Modulating C9orf72 Expression. US 9,605,263 B2.

181 Prakash, T.P., Yu, J., Migawa, M.T. et al. (2016). Comprehensive structure-activity relationship of triantennary N-acetylgalactosamine conjugated antisense oligonucleotides for targeted delivery to hepatocytes. *J. Med. Chem.* 59: 2718–2733.

182 Schultz, R. and Gryaznov, S. (1996). Oligo-2′-fluoro-2′-deoxynucleotide N3′-->P5′ phosphoramidates: synthesis and properties. *Nucl. Acids Res.* 24: 2966–2973.

183 Schultz, R.G. and Gryaznov, S.M. (2000). arabino-fluorooligonucleotide N3′-->P5′ phosphoramidates: synthesis and properties. *Tetrahedron Lett.* 41: 1895–1899.

184 Gryaznov, S.M., Lloyd, D.H., Chen, J.K. et al. (1995). Oligonucleotide N3′-->P5′ phosphoramidates. *Proc. Natl. Acad. Sci. U. S. A.* 92: 5798–5802.

185 Gryaznov, S., Asai, A., Oshima, Y. et al. (2003). Oligonucleotide N3′ --> P5′ thio-phosphoramidate telomerase template antagonists as potential anticancer agents. *Nucleos Nucleot Nucl* 22: 577–581.

186 Joseph, I., Tressler, R., Bassett, E. et al. (2010). The telomerase inhibitor imetelstat depletes cancer stem cells in breast and pancreatic cancer cell lines. *Cancer Res.* 70: 9494–9504.

187 Rait, V.K. and Shaw, B.R. (1999). Boranophosphates support the RNase H cleavage of polyribonucleotides. *Antisense Nucleic Acid Drug Dev.* 9: 53–60.

188 Hall, A.H.S., Wan, J., Shaughnessy, E.E. et al. (2004). RNA interference using boranophosphate siRNAs: structure-activity relationships. *Nucleic Acids Res.* 32: 5991–6000.

189 Sheehan, D., Lunstad, B., Yamada, C.M. et al. (2003). Biochemical properties of phosphonoacetate and thiophosphonoacetate oligodeoxyribonucleotides. *Nucleic Acids Res.* 31: 4109–4118.

190 Dellinger, D.J., Sheehan, D.M., Christensen, N.K. et al. (2003). Solid-phase chemical synthesis of phosphonoacetate and thiophosphonoacetate oligodeoxynucleotides. *J. Am. Chem. Soc.* 125: 940–950.

191 Threlfall, R.N., Torres, A.G., Krivenko, A. et al. (2012). Synthesis and biological activity of phosphonoacetate- and thiophosphonoacetate-modified 2′-*O*-methyl oligoribonucleotides. *Org. Biomol. Chem.* 10: 746–754.

192 Matsui, M., Threlfall, R.N., Caruthers, M.H., and Corey, D.R. (2014). Effect of 2′-*O*-methyl/thiophosphonoacetate-modified antisense oligonucleotides on huntingtin expression in patient-derived cells. *Artificial DNA, PNA & XNA* 5: e1146391.

193 Yamada, C.M., Dellinger, D.J., and Caruthers, M.H. (2007). Synthesis and biological activity of phosphonocarboxylate DNA. *Nucleos Nucleot Nucl* 26: 539–546.

194 Sipova, H., Springer, T., Rejman, D. et al. (2014). 5′-*O*-methylphosphonate nucleic acids – new modified DNAs that increase the Escherichia coli RNase H cleavage rate of hybrid duplexes. *Nucleic Acids Res.* 42: 5378–5389.

195 Nina, M., Fonne-Pfister, R., Beaudegnies, R. et al. (2005). Recognition of RNA by amide modified backbone nucleic acids: molecular dynamics simulations of DNA-RNA hybrids in aqueous solution. *J. Am. Chem. Soc.* 127: 6027–6038 and references therein.

196 Selvam, C., Thomas, S., Abbott, J. et al. (2011). Amides as excellent mimics of phosphate linkages in RNA. *Angew. Chem. Int. Ed.* 50: 2068–2070 and references therein.

197 El-Sagheer, A.H. and Brown, T. (2010). Click chemistry with DNA. *Chem. Soc. Rev.* 39: 1388–1405.

198 El-Sagheer, A.H., Sanzone, A.P., Gao, R. et al. (2011). Biocompatible artificial DNA linker that is read through by DNA polymerases and is functional in Escherichia coli. *Proc. Natl. Acad. Sci. U. S. A.* 108: 11338–11343.

199 Fujino, T., Yamazaki, N., and Isobe, H. (2009). Convergent synthesis of oligomers of triazole-linked DNA analogue ((TL)DNA) in solution phase. *Tetrahedron Lett.* 50: 4101–4103.

200 Nuzzi, A., Massi, A., and Dondoni, A. (2007). Model studies toward the synthesis of thymidine oligonucleotides with triazole internucleosidic linkages via iterative Cu(I)-Promoted azide-alkyne ligation chemistry. *QSAR Comb. Sci.* 26: 1191–1199.

201 Isobe, H., Fujino, T., Yamazaki, N. et al. (2008). Triazole-linked analogue of deoxyribonucleic acid ((TL)DNA): design, synthesis, and double-strand formation with natural DNA. *Org. Lett.* 10: 3729–3732.

202 El-Sagheer, A.H. and Brown, T. (2014). Combined nucleobase and backbone modifications enhance DNA duplex stability and preserve biocompatibility. *Chem. Sci.* 5: 253–259.

203 Kumar, P., El-Sagheer, A.H., Truong, L., and Brown, T. (2017). Locked nucleic acid (LNA) enhances binding affinity of triazole-linked DNA towards RNA. *Chem. Commun.* 53: 8910–8913.

204 Sharma, V.K., Singh, S.K., Krishnamurthy, P.M. et al. (2017). Synthesis and biological properties of triazole-linked locked nucleic acid. *Chem. Commun.* 53: 8906–8909.

205 El-Sagheer, A.H. and Brown, T. (2010). New strategy for the synthesis of chemically modified RNA constructs exemplified by hairpin and hammerhead ribozymes. *Proc. Natl. Acad. Sci. U. S. A.* 107: 15329–15334.

206 Qiu, J., El-Sagheer, A.H., and Brown, T. (2013). Solid phase click ligation for the synthesis of very long oligonucleotides. *Chem. Commun.* 49: 6959.

207 Dupouy, C., Iché-Tarrat, N., Durrieu, M.-P. et al. (2006). Watson–Crick base-pairing properties of nucleic acid analogues with stereocontrolled α and β torsion angles (α,β-D-CNAs). *Angew. Chem. Int. Ed.* 45: 3623–3627.

208 Le Clézio, I., Dupouy, C., Lavedan, P., and Escudier, J.-M. (2007). Synthesis and structure of an α,β-D-CNA featuring a noncanonical α/β torsion angle combination within a tetranucleotide. *Eur. J. Org. Chem.* 2007: 3894–3900.

209 Dupouy, C., Lavedan, P., and Escudier, J.-M. (2007). Synthesis and structure of dinucleotides featuring canonical and non-canonical A-type duplex α, β and δ torsion angle combinations (LNA/α, β-D-CNA). *Eur. J. Org. Chem.* 2007: 5256–5264.

210 Maturano, M., Catane, D.-A., Tarrat, N. et al. (2012). Synthesis and structural study of ribo-dioxaphosphorinane-constrained nucleic acid dinucleotides (ribo-α, β-D-CNA). *Eur. J. Org. Chem.* 2012: 721–730.

211 Dupouy, C., Lavedan, P., and Escudier, J.-M. (2008). Synthesis and structure of dinucleotides with S-type sugar puckering and noncanonical ϵ and ζ torsion angle combination (ν2,ϵ, ζ-D-CNA). *Eur. J. Org. Chem.* 2008: 1285–1294.

212 Boissonnet, A., Dupouy, C., Millard, P. et al. (2011). [α], [β]-D-CNA featuring canonical and noncanonical [α]/[β] torsional angles behaviours within oligonucleotides. *New J. Chem.* 35: 1528–1533.

213 Weitzer, S. and Martinez, J. (2007). The human RNA kinase hClp1 is active on 3′ transfer RNA exons and short interfering RNAs. *Nature* 447: 222–226.

214 Schirle, N.T., Kinberger, G.A., Murray, H.F. et al. (2016). Structural analysis of human argonaute-2 bound to a modified siRNA guide. *J. Am. Chem. Soc.* 138: 8694–8697.

215 Elkayam, E., Parmar, R., Brown, C.R. et al. (2017). siRNA carrying an (E)-vinylphosphonate moiety at the 5′ end of the guide strand augments gene silencing by enhanced binding to human argonaute-2. *Nucleic Acids Res.* 45: 3528–3536. doi: 10.1093/nar/gkw1171.

216 Parmar, R., Willoughby, J.L., Liu, J. et al. (2016). 5′-(E)-vinylphosphonate: a stable phosphate mimic can improve the RNAi activity of siRNA-GalNAc conjugates. *Chem Bio Chem* 17: 985–989.

217 Summerton, J. and Weller, D. (1997). Morpholino antisense oligomers: design, preparation, and properties. *Antisense Nucleic Acid Drug Dev.* 7: 187–195.

218 Corey, D.R. and Abrams, J.M. (2001). Morpholino antisense oligonucleotides: tools for investigating vertebrate development. *Genome Biol.* 2, REVIEWS: 1015.

219 Bill, B.R., Petzold, A.M., Clark, K.J. et al. (2009). A primer for morpholino use in zebrafish. *Zebrafish* 6: 69–77.

220 Mendell, J.R., Rodino-Klapac, L.R., Sahenk, Z. et al. (2013). Eteplirsen for the treatment of Duchenne muscular dystrophy. *Ann. Neurol.* 74: 637–647.

221 Weller, D.D., Hassinger, J.N. and Cai, B.Z. (2011). Oligonucleotide analogs having cationic intersubunit linkages. US Patent 7943762, filed 10 May 2007 and issued 17 May 2011.

222 Wu, B., Li, Y.F., Morcos, P.A. et al. (2009). Octa-guanidine morpholino restores dystrophin expression in cardiac and skeletal muscles and ameliorates pathology in dystrophic mdx mice. *Mol. Ther.* 17: 864–871.

223 Aoki, Y., Yokota, T., Nagata, T. et al. (2012). Bodywide skipping of exons 45–55 in dystrophic mdx52 mice by systemic antisense delivery. *Proc. Natl. Acad. Sci. U. S. A.* 109: 13763–13768.
224 Betts, C., Saleh, A.F., Arzumanov, A.A. et al. (2012). Pip6-PMO, a new generation of peptide-oligonucleotide conjugates with improved cardiac exon skipping activity for DMD treatment. *Mol. Ther. Nucl. Acids* 1: e38.
225 Moulton, H.M. and Iversen, P.L. (2009). Tissue specific peptide conjugates and methods. US Patent application 12/217, 040, filed 30 June 2008 and issued 28 May 2009.
226 Bock, H., Werner, B., and Lindhorst, T. (2012). Chiral compounds substituted with phosphonic acid ester functions or phosphonic acid functions. US Patent application 13/527, 217, filed 20 July 2007 and issued 24 February 2015.
227 Haaima, G., Lohse, A., Buchardt, O., and Nielsen, P.E. (1996). Peptide nucleic acids (PNAs) containing thymine monomers derived from chiral amino acids: hybridization and solubility properties of D-lysine PNA. *Angew. Chem. Int. Ed.* 35: 1939–1942.
228 Nielsen, P.E., Egholm, M., Berg, R.H., and Buchardt, O. (1991). Sequence-selective recognition of DNA by strand displacement with a thymine-substituted polyamide. *Science* 254: 1497–1500.
229 Egholm, M., Turner, B.J., Bruce, S. et al. (1993). PNA hybridizes to complementary oligonucleotides obeying the Watson-Crick hydrogen-bonding rules. *Nature* 365: 566–568.
230 Turner, B.J., Cheah, I.K., Macfarlane, K.J. et al. (2003). Antisense peptide nucleic acid-mediated knockdown of the p 75 neurotrophin receptor delays motor neuron disease in mutant SOD1 transgenic mice. *J. Neurochem.* 87: 752–763.
231 Rembach, A., Turner, B.J., Bruce, S. et al. (2004). Antisense peptide nucleic acid targeting GluR3 delays disease onset and progression in the SOD1 G93A mouse model of familial ALS. *J. Neurosci. Res.* 77: 573–582.
232 Bock, H. and Lindhorst, T. (2006). Oligomers substituted by phosphite acid ester, phosphonic acid or carbaborane functions and the corresponding PNA monomers. US Patent 7105648, filed 3 March 2000 and issued 12 September 2006.
233 Maison, W., Schlemminger, I., Westerhoff, O., and Martens, J. (1999). Modified PNAs: a simple method for the synthesis of monomeric building blocks. *Bioorg. Med. Chem. Lett.* 9: 581–584.
234 Domling, A., Richter, W., and Ugi, I. (1997). Combinatorial generation of nucleobase libraries by MCR. *Nucleosides Nucleotides* 16: 1753–1756.
235 Dragulescu-Andrasi, A., Rapireddy, S., Frezza, B.M. et al. (2006). A simple gamma-backbone modification preorganizes peptide nucleic acid into a helical structure. *J. Am. Chem. Soc.* 128: 10258–10267.
236 Rapireddy, S., He, G., Roy, S. et al. (2007). Strand invasion of mixed-sequence B-DNA by acridine-linked, gamma-peptide nucleic acid (gamma-PNA). *J. Am. Chem. Soc.* 129: 15596–15600.

237 He, G., Rapireddy, S., Bahal, R. et al. (2009). Strand invasion of extended, mixed-sequence B-DNA by gammaPNAs. *J. Am. Chem. Soc.* 131: 12088–12090.

238 Tedeschi, T., Sforza, S., Corradini, R., and Marchelli, R. (2005). Synthesis of new chiral PNAs bearing a dipeptide-mimic monomer with two lysine-derived stereogenic centres. *Tetrahedron Lett.* 46: 8395–8399.

239 Englund, E.A. and Appella, D.H. (2007). gamma-substituted peptide nucleic acids constructed from L-lysine are a versatile scaffold for multifunctional display. *Angew. Chem. Int. Ed.* 46: 1414–1418.

240 Sahu, B., Sacui, I., Rapireddy, S. et al. (2011). Synthesis and characterization of conformationally preorganized, (R)-diethylene glycol-containing gamma-peptide nucleic acids with superior hybridization properties and water solubility. *J. Org. Chem.* 76: 5614–5627.

241 Wang, S. and Kool, E.T. (1995). Origins of the large differences in stability of DNA and RNA helixes: C-5 methyl and 2′-hydroxyl effects. *Biochemistry* 34: 4125–4132.

242 Sowers, L.C., Shaw, B.R., and Sedwick, W.D. (1987). Base stacking and molecular polarizability – effect of a methyl-group in the 5-position of pyrimidines. *Biochem. Biophys. Res. Commun.* 148: 790–794.

243 Henry, S., Stecker, K., Brooks, D. et al. (2000). Chemically modified oligonucleotides exhibit decreased immune stimulation in mice. *J. Pharmacol. Exp. Ther.* 292: 468–479.

244 Krieg, A.M. (2002). CpG motifs in bacterial DNA and their immune effects. *Annu. Rev. Immunol.* 20: 709–760.

245 Froehler, B.C., Wadwani, S., Terhorst, T.J., and Gerrard, S.R. (1992). Oligodeoxynucleotides containing C-5 propyne analogs of 2′-deoxyuridine and 2′-deoxycytidine. *Tetrahedron Lett.* 33: 5307–5310.

246 Wagner, R.W., Matteucci, M.D., Lewis, J.G. et al. (1993). Antisense gene inhibition by oligonucleotides containing C-5 propyne pyrimidines. *Science* 260: 1510–1513.

247 Moulds, C., Lewis, J.G., Froehler, B.C. et al. (1995). Site and mechanism of antisense inhibition by C-5 propyne oligonucleotides. *Biochemistry* 34: 5044–5053.

248 Shen, L.J., Siwkowski, A., Wancewicz, E.V. et al. (2003). Evaluation of C-5 propynyl pyrimidine-containing oligonucleotides in vitro and in vivo. *Antisense Nucleic Acid Drug Dev.* 13: 129–142.

249 Gutierrez, A.J., Terhorst, T.J., Matteucci, M.D., and Froehler, B.C. (1994). 5-heteroaryl-2′-deoxyuridine analogs. Synthesis and incorporation into high-affinity oligonucleotides. *J. Am. Chem. Soc.* 116: 5540–5544.

250 Guenther, D.C., Kumar, P., Anderson, B.A., and Hrdlicka, P.J. (2014). C5-amino acid functionalized LNA: positively poised for antisense applications. *Chem. Commun. (Camb.)* 50: 9007–9009.

251 Kaura, M., Guenther, D.C., and Hrdlicka, P.J. (2014). Carbohydrate-functionalized locked nucleic acids: oligonucleotides with extraordinary

binding affinity, target specificity, and enzymatic stability. *Org. Lett.* 16: 3308–3311.
252 Lin, K.Y. and Matteucci, M.D. (1998). A cytosine analogue capable of clamp-like binding to a guanine in helical nucleic acids. *J. Am. Chem. Soc.* 120: 8531–8532.
253 Flanagan, W.M., Wolf, J.J., Olson, P. et al. (1999). A cytosine analog that confers enhanced potency to antisense oligonucleotides. *Proc. Natl. Acad. Sci. U. S. A.* 96: 3513–3518.
254 Stanton, R., Sciabola, S., Salatto, C. et al. (2012). Chemical modification study of antisense gapmers. *Nucleic Acid Ther.* 22: 344–359.
255 Wilds, C.J., Maier, M.A., Tereshko, V. et al. (2002). Direct observation of a cytosine analogue that forms five hydrogen bonds to guanosine: guanidino G-clamp. *Angew. Chem. Int. Ed. Eng.* 41: 115–117.
256 Wilds, C.J., Maier, M.A., Manoharan, M., and Egli, M. (2003). Structural basis for recognition of guanosine by a synthetic tricyclic cytosine analogue: guanidinium G-clamp. *Helv. Chim. Acta* 86: 966–978.
257 Lamm, G.M., Blencowe, B.J., Sproat, B.S. et al. (1991). Antisense probes containing 2-aminoadenosine allow efficient depletion of U5 Snrnp from hela splicing extracts. *Nucleic Acids Res.* 19: 3193–3198.
258 Gryaznov, S. and Schultz, R.G. (1994). Stabilization of DNA-DNA and DNA-RNA duplexes by substitution of 2′-deoxyadenosine with 2′-deoxy-2-aminoadenosine. *Tetrahedron Lett.* 35: 2489–2492.
259 Ramasamy, K.S., Zounes, M., Gonzalez, C. et al. (1994). Remarkable enhancement of binding-affinity of heterocycle-modified DNA to DNA and Rna – synthesis, characterization and biophysical evaluation of N-2-imidazolylpropylguanine and N-2-imidazolylpropyl-2-aminoadenine modified oligonucleotides. *Tetrahedron Lett.* 35: 215–218.
260 Manoharan, M., Ramasamy, K.S., Mohan, V., and Cook, P.D. (1996). Oligonucleotides bearing cationic groups: N-2-(3-aminopropyl) deoxyguanosine. Synthesis, enhanced binding properties and conjugation chemistry. *Tetrahedron Lett.* 37: 7675–7678.
261 Herdewijn, P. (2000). Heterocyclic modifications of oligonucleotides and antisense technology. *Antisense Nucleic Acid Drug Dev.* 10: 297–310.
262 Dodd, D.W. and Hudson, R.H.E. (2009). Intrinsically fluorescent base-discriminating nucleoside analogs. *Mini-Rev. Org. Chem.* 6: 378–391.
263 Wojciechowski, F. and Hudson, R.H.E. (2007). Nucleobase modifications in peptide nucleic acids. *Curr. Top. Med. Chem.* 7: 667–679.
264 Krueger, A.T., Lu, H., Lee, A.H., and Kool, E.T. (2007). Synthesis and properties of size-expanded DNAs: toward designed, functional genetic systems. *Acc. Chem. Res.* 40: 141–150.
265 Bhagat, L., Putta, M.R., Wang, D. et al. (2011). Novel oligonucleotides containing two 3′-ends complementary to target mRNA show optimal gene-silencing activity. *J. Med. Chem.* 54: 3027–3036.

266 Subramanian, R.R., Wysk, M.A., Ogilvie, K.M. et al. (2015). Enhancing antisense efficacy with multimers and multi-targeting oligonucleotides (MTOs) using cleavable linkers. *Nucleic Acids Res.* 43: 9123–9132.
267 Hussain, M., Shchepinov, M., Sohail, M. et al. (2004). A novel anionic dendrimer for improved cellular delivery of antisense oligonucleotides. *J. Control. Release* 99: 139–155.
268 Cutler, J.I., Auyeung, E., and Mirkin, C.A. (2012). Spherical nucleic acids. *J. Am. Chem. Soc.* 134: 1376–1391.
269 Ezzat, K., Aoki, Y., Koo, T. et al. (2015). Self-assembly into nanoparticles is essential for receptor mediated uptake of therapeutic antisense oligonucleotides. *Nano Lett.* 15: 4364–4373.
270 Yu, D., Kandimalla, E.R., Bhagat, L. et al. (2002). "Immunomers" – novel 3′-3′-linked CpG oligodeoxyribonucleotides as potent immunomodulatory agents. *Nucleic Acids Res.* 30: 4460–4469.
271 Yu, D., Zhao, Q., Kandimalla, E.R., and Agrawal, S. (2000). Accessible 5′-end of CpG-containing phosphorothioate oligodeoxynucleotides is essential for immunostimulatory activity. *Bioorg. Med. Chem. Lett.* 10: 2585–2588.
272 Putta, M.R., Bhagat, L., Wang, D. et al. (2013). Immune-stimulatory dinucleotide at the 5′-end of oligodeoxynucleotides is critical for TLR9-mediated immune responses. *ACS Med. Chem. Lett.* 4: 302–305.
273 Rosi, N.L., Giljohann, D.A., Thaxton, C.S. et al. (2006). Oligonucleotide-modified gold nanoparticles for intracellular gene regulation. *Science* 312: 1027–1030.
274 Banga, R.J., Chernyak, N., Narayan, S.P. et al. (2014). Liposomal spherical nucleic acids. *J. Am. Chem. Soc.* 136: 9866–9869.
275 Choi, C.H., Hao, L., Narayan, S.P. et al. (2013). Mechanism for the endocytosis of spherical nucleic acid nanoparticle conjugates. *Proc. Natl. Acad. Sci. U. S. A.* 110: 7625–7630.
276 Massich, M.D., Giljohann, D.A., Seferos, D.S. et al. (2009). Regulating immune response using polyvalent nucleic acid-gold nanoparticle conjugates. *Mol. Pharm.* 6: 1934–1940.
277 Hudson, R.H.E. and Damha, M.J. (1993). Nucleic-acid dendrimers – novel biopolymer structures. *J. Am. Chem. Soc.* 115: 2119–2124.
278 Hamblin, G.D., Carneiro, K.M., Fakhoury, J.F. et al. (2012). Rolling circle amplification-templated DNA nanotubes show increased stability and cell penetration ability. *J. Am. Chem. Soc.* 134: 2888–2891.
279 Zhao, Y.X., Shaw, A., Zeng, X.H. et al. (2012). DNA origami delivery system for cancer therapy with tunable release properties. *ACS Nano* 6: 8684–8691.
280 Bujold, K.E., Fakhoury, J., Edwardson, T.G.W. et al. (2014). Sequence-responsive unzipping DNA cubes with tunable cellular uptake profiles. *Chem. Sci.* 5: 2449.
281 Gura, T. (1995). Antisense has growing pains. *Science* 270: 575–577.
282 Roush, W. (1997). Antisense aims for a renaissance. *Science* 276: 1192–1193.

283 Frederickson, R.M. (2007). RNAi gets vote of confidence from big pharma. *Mol. Ther.* 15: 221–222.

284 Potera, C. (2007). Antisense – down, but not out. *Nat. Biotechnol.* 25: 497–499.

285 Schmidt, C. (2011). RNAi momentum fizzles as pharma shifts priorities. *Nat. Biotechnol.* 29: 93–94.

286 Garber, K. (2016). Big win possible for Ionis/Biogen antisense drug in muscular atrophy. *Nat. Biotechnol.* 34: 1002–1003.

287 miRNA mimics and/or inhibitors are in clinical development by companies including Miragen, Mirna and Regulus. Three miRNA mimics and three miRNA inhibitors are currently in clinical trials. Progress in miRNA therapeutics has been reviewed by van Rooij, E., Purcell, A.L., and Levin, A.A. (2012). Developing microRNA therapeutics. *Circ. Res.* 110: 496–507.

288 Oligonucleotides that modulate splicing are being developed by multiple companies. Sarepta's PMO drug eteplirsen was FDA approved in August 2016 for the treatment of Duchenne muscular dystrophy, while Ionis' MOE drug nusinersen has was approved for the treatment of spinal muscular atrophy in December 2016.

289 Fabani, M.M. and Gait, M.J. (2008). miR-122 targeting with LNA/2′-*O*-methyl oligonucleotide mixmers, peptide nucleic acids (PNA), and PNA-peptide conjugates. *RNA* 14: 336–346.

290 Torres, A.G., Fabani, M.M., Vigorito, E., and Gait, M.J. (2011). MicroRNA fate upon targeting with anti-miRNA oligonucleotides as revealed by an improved Northern-blot-based method for miRNA detection. *RNA* 17: 933–943.

291 Fabani, M.M., Abreu-Goodger, C., Williams, D. et al. (2010). Efficient inhibition of miR-155 function in vivo by peptide nucleic acids. *Nucleic Acids Res.* 38: 4466–4475.

292 Elmen, J., Lindow, M., Schutz, S. et al. (2008). LNA-mediated microRNA silencing in non-human primates. *Nature* 452: 896–899.

293 Orom, U.A., Kauppinen, S., and Lund, A.H. (2006). LNA-modified oligonucleotides mediate specific inhibition of microRNA function. *Gene* 372: 137–141.

294 Davis, S., Lollo, B., Freier, S., and Esau, C. (2006). Improved targeting of miRNA with antisense oligonucleotides. *Nucl. Acids Res.* 34: 2294–2304.

295 Krutzfeldt, J., Rajewsky, N., Braich, R. et al. (2005). Silencing of microRNAs in vivo with "antagomirs". *Nature* 438: 685–689.

296 Kloosterman, W.P., Lagendijk, A.K., Ketting, R.F. et al. (2007). Targeted inhibition of miRNA maturation with morpholinos reveals a role for miR-375 in pancreatic islet development. *PLoS Biol.* 5: e203.

297 Esau, C.C. (2008). Inhibition of microRNA with antisense oligonucleotides. *Methods* 44: 55–60.

298 Li, Z. and Rana, T.M. (2014). Therapeutic targeting of microRNAs: current status and future challenges. *Nat. Rev. Drug Discov.* 13: 622–638.

299 Jarver, P., O'Donovan, L., and Gait, M.J. (2014). A chemical view of oligonucleotides for exon skipping and related drug applications. *Nucleic Acid Ther.* 24: 37–47.

300 Bauman, J., Jearawiriyapaisarn, N., and Kole, R. (2009). Therapeutic potential of splice-switching oligonucleotides. *Oligonucleotides* 19: 1–13.

301 Matsui, M. and Corey, D.R. (2017). Non-coding RNAs as drug targets. *Nat. Rev. Drug Discov.* 16: 167–179.

302 Modarresi, F., Faghihi, M.A., Lopez-Toledano, M.A. et al. (2012). Inhibition of natural antisense transcripts in vivo results in gene-specific transcriptional upregulation. *Nat. Biotechnol.* 30: 453–459.

303 Sarma, K., Levasseur, P., Aristarkhov, A., and Lee, J.T. (2010). Locked nucleic acids (LNAs) reveal sequence requirements and kinetics of Xist RNA localization to the X chromosome. *Proc. Natl. Acad. Sci. U. S. A.* 107: 22196–22201.

304 Voutila, J., Reebye, V., Roberts, T.C. et al. (2017). Development and mechanism of small activating RNA targeting CEBPA, a novel therapeutic in clinical trials for liver cancer. *Mol. Ther.* 25: 2705–2714.

305 Li, L., Matsui, M., and Corey, D.R. (2016). Activating frataxin expression by repeat-targeted nucleic acids. *Nat. Commun.* 7: 10606.

306 Nomakuchi, T.T., Rigo, F., Aznarez, I., and Krainer, A.R. (2016). Antisense oligonucleotide-directed inhibition of nonsense-mediated mRNA decay. *Nat. Biotechnol.* 34: 164–166.

307 Liang, X.-H., Shen, W., Sun, H. et al. (2016). Translation efficiency of mRNAs is increased by antisense oligonucleotides targeting upstream open reading frames. *Nat. Biotechnol.* 34: 875–880.

308 Liang, X.-H., Sun, H., Shen, W. et al. (2017). Antisense oligonucleotides targeting translation inhibitory elements in 5′ UTRs can selectively increase protein levels. *Nucleic Acids Res.* 45: 9528–9546. doi: 10.1093/nar/gkx1632.

3

Cellular Pharmacology of Antisense Oligonucleotides

Xin Ming

Division of Molecular Pharmaceutics, UNC Eshelman School of Pharmacy, University of North Carolina, Chapel Hill, NC, USA

3.1 Introduction

RNAs have become important targets for therapeutic intervention in light of the diverse roles that they play in human physiology and pathology. Being the most direct therapeutic strategy to modulate RNA function, antisense oligonucleotides (ASOs) have been emerging as an important therapeutic entity. ASOs are defined as the oligonucleotides with 8 to 50 nucleotides in length that can bind to RNA through Watson–Crick base pairing and thereafter modulate its function [1]. Based on this definition, ASOs include classic ASOs as well as siRNA and microRNA (miRNA) oligonucleotides, which are double-stranded oligonucleotides.

ASOs can be designed to target and modulate any RNA, the pharmacological actions of antisense-based drugs are diverse, and so clinical application and testing of ASOs involve various diseases [1]. All of the antisense pharmacological actions involve three common processes: (i) access of ASO to their action sites in the cells, (ii) binding of their target RNA, and (iii) post-binding events, such as degradation of the RNA through endogenous enzymes. This chapter deals with these basic and common processes of ASOs, instead of the diverse pharmacological outcomes that ASOs can produce. Thus, we will first review the process of ASOs' binding to the target RNA and the variety of post-binding events. Then, we will focus on cellular processes in which synthetic oligonucleotides reach their RNA targets.

Oligonucleotide-Based Drugs and Therapeutics: Preclinical and Clinical Considerations for Development, First Edition. Edited by Nicolay Ferrari and Rosanne Seguin.

3.2 Molecular Mechanisms of Antisense Oligonucleotides

ASOs bind to RNA in a sequence-specific manner but have different mechanisms of actions and effects. Up to the involvement of enzyme in post-binding events, they can be classified into enzyme-dependent oligonucleotides and enzyme-independent oligonucleotides. Antisense and siRNA oligonucleotides downregulate gene expression by inducing enzyme-dependent degradation of target mRNA. Steric-blocking oligonucleotides prevent the access of cellular machinery to pre-mRNA or mRNA without causing enzymatic degradation of the RNA. Through this mechanism, steric-blocking oligonucleotides can redirect alternative splicing, repair defective RNA, restore protein production, or downregulate gene expression. In this chapter, we will discuss two types of enzyme-dependent oligonucleotides, antisense and siRNA oligonucleotides, and three types of enzyme-independent oligonucleotides including splice switching oligonucleotides, miRNA antagomirs, and long noncoding RNAs (lncRNAs) antagomirs.

3.2.1 Classic Antisense Oligonucleotides

Since the first application of a short fragment of unmodified DNA in cell culture as an ASO by Zamecnik and Stephenson in 1978 [2], there has been great advance in understanding antisense mechanism and application of ASOs for therapeutic purposes. Upon introduction into cells, ASOs can enter the nucleus and engage in specific Watson–Crick base pairing with complementary sequences in pre-mRNAs. Relatively short (≥16 bases) oligonucleotides that are readily prepared by convenient solid phase synthesis can theoretically provide unique recognition among the approximately three billion base pairs in the human genome. The formation of a DNA/RNA hybrid results in the recruitment of ribonuclease H (RNase H), a ribonuclease that degrades duplexed RNAs. This then triggers further degradation of the cleaved fragments by other enzymatic processes (Figure 3.1).

Native DNA, with phosphodiester internucleotide linkages, proved to be unstable in the cells and blood circulation. Thus, a large number of chemical modifications were developed to enhance the enzymatic stability and thereafter the biological effectiveness of antisense molecules. One of the most popular modifications is the substitution of sulfur for oxygen in the phosphate backbone leading to phosphorothioate (PS) oligonucleotides, which are considerably more stable than native DNA in the biological milieu. Another important set of modifications involves substitution of aliphatic residues at the 2′ position of the nucleoside sugar, thus creating 2′-*O*-methyl and similar derivatives. This

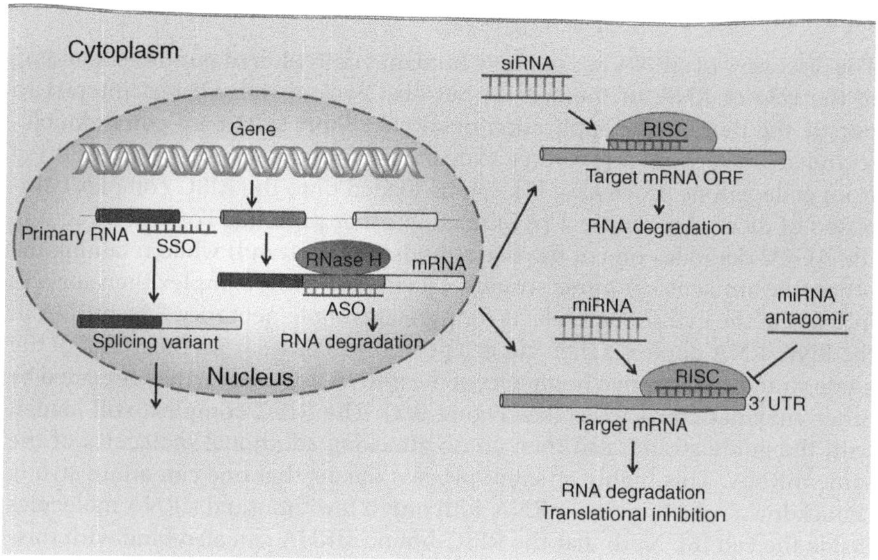

Figure 3.1 Oligonucleotide mechanisms of action. Four mechanisms are illustrated. Starting from the top: (i) classical ASO-mediated mRNA degradation via ribonuclease H; (ii) siRNA-mediated mRNA degradation via the Ago 2/RISC complex; (iii) removal of a mutant extron using splice switching oligonucleotides; and (iv) miRNA's modulation of mRNA function.

modification results in molecules that are more RNA-like and often bind with higher affinity to RNA. However, fully substituting this type of modification leads to loss of the ability to recruit RNase H. Therefore, the commonly used ASOs have a PS backbone with flanks that are additionally modified with 2′-*O*-methoxyethyl (2′-MOE) or 2′-*O*-methyl (2′-OMe) residues. Flank modifications increase the resistance of the ASO to degradation and enhance binding to targeted mRNA. The unmodified "gap" in a gapmer–mRNA duplex is recognized by RNase H. Another important modification is represented by locked nucleic acids (LNAs) where a 2′–4′ bridge results in oligonucleotides with very high binding affinity. Finally families of uncharged oligonucleotides have been created by radical modification of the backbone. Replacement of the phosphodiester backbone with peptide linkages gives rise to peptide nucleic acids (PNAs), while inclusion of nonionic phosphorodiamidate linkages leads to morpholino oligonucleotides. These fully modified oligonucleotides, as well as the 2′ modifications, do not recruit RNase H; however, they do retain excellent base-pairing specificity and the ability to block the access of cellular machinery to pre-mRNA and mRNA without degrading the RNA and thereby are used as steric-blocking oligonucleotides.

3.2.2 siRNA

The discovery of siRNA has not only fundamentally altered our understanding of the role of RNA in the cell [3] but also evoked the greatest interest to pursue the therapeutic application of siRNA. Short (~21 base pairs) double-stranded RNA segments, either exogenously administered [4] or resulting from endogenous processing [5], can be loaded onto the RISC complex composed of dicer, Argonaute 2 (AGO2), and other proteins. For siRNA action, the AGO2 degrades one of the two strands (sense strand) while retaining the other (the antisense or guide strand). The loaded RISC complex then surveys mRNAs in the cytosol and selects for perfect complementarity. The mRNA in the RNA/RNA duplex is then cut at a position 10 bases from the 5′ end of the guide strand, and the two fragments of the mRNA are then further degraded by other enzymatic processes (see Figure 3.1). The RISC complex, still loaded with the guide strand, can then go on attacking additional molecules of the same mRNA. This highly efficient process means that one can attain strong "knockdown" of a targeted mRNA with only a few thousand siRNA molecules inside the cell [6]. Note that the RISC-bound siRNA can also bind with mismatches to unintended mRNAs, leading to off-target effects. Although both classic antisense and siRNA oligonucleotides mediate target mRNA degradation, a key difference is that the siRNA oligonucleotide is bound to the cleavage enzyme prior to interacting with the target RNA for the *RNAi* activity. In contrast, ASOs bind to the target RNA prior to interaction with the RNase H enzyme. In the *RNAi* action, the oligonucleotide in the complex presumably has fewer nonproductive interactions with proteins and other nucleic acids, which may contribute to higher potency in mRNA degradation by siRNA than antisense in cell culture [1].

3.2.3 Splice Switching Oligonucleotides

As a type of steric-blocking oligonucleotides, splice switching antisense oligonucleotides (SSOs) have become an important tool to modulate gene expression and function and have moved to clinical testing. Although the human genome contains only about twenty thousand genes, it can direct the production of hundreds of thousands of proteins via alternative splicing of pre-mRNA. Actually, over 95% of all human genes produce splice variant proteins by alternative splicing; thus, modulation of alternative splicing may be applicable to multiple diseases [7]. Steric-blocking oligonucleotides can be designed to target splice sites or other sequences that are essential for splicing in pre-mRNA and thereby to prevent the proper assembly of the spliceosome on the exon and redirect splicing to another pathway to generate an alternatively spliced mRNA [8] (Figure 3.1). This splice variant may encode a novel protein with favorable properties, or it may restore translation if the skipped exon causes a frameshifting mutation; alternatively, it may change the balance of alternative splice variants. These outcomes cannot be accomplished by classic antisense and siRNA oligonucleotides, which degrade target mRNAs. The SSO must be designed so that it does not recruit RNase H, or it would result in degradation rather than altered splicing of the target RNAs. Fortunately this is easy to do using either oligonucleotides that are fully substituted at the 2′ position or those that have uncharged backbones as represented by PNAs or morpholinos. Clinical studies have showed encouraging results in applying SSOs for the systemic treatment of Duchenne muscular dystrophy [9, 10].

3.2.4 microRNA Antagomirs

miRNAs are an abundant class of small (~22 nt) endogenous noncoding RNAs that direct posttranscriptional regulation of gene expression. The actions of miRNAs are similar to siRNAs and are also mediated by the RISC complex, but in this case there is not full complementarity between the guide strand and the target mRNA. Rather the guide strand engages a partially complementary region, usually in the 3′ untranslated region of the message [11] (see Figure 3.1). This results in message sequestration into P-bodies [12], interruption of translation, and often in degradation. Since there is only partial complementarity involved, this means that a single miRNA may interact with multiple mRNAs and thus be involved in the regulation of multiple genes. While in some cases miRNAs themselves have been used to modify expression of disease-related genes [13], a more popular approach is the use of "antagomirs," ASOs complementary to the miRNA that can bind and inactivate it [14, 15]. Chemically modified antagomirs can sequester the mature miRNA in competition with cellular target mRNAs leading to functional inhibition of the miRNA and derepression of the direct targets. A phase II clinical study has shown that miRNA 122 antagomirs can reduce HCV RNA in patients with chronic HCV infection without viral resistance [16].

3.2.5 lncRNAs Antagomirs

Besides miRNAs, lncRNAs, another regulatory RNA, have begun to emerge as a potential important therapeutic target [17]. lncRNAs can positively or negatively regulate gene expression and chromatin architecture. Single-stranded oligonucleotides can be designed as inhibitors of the natural antisense transcript (NAT) class of lncRNAs and thereby lead to mRNA derepression [18]. The inhibition can be achieved by blocking the interactions of NATs with effector proteins and/or by causing RNase H-mediated degradation of the NATs [18].

3.3 Cellular Pharmacology of Antisense Oligonucleotides

Regardless of the molecular mechanisms, ASOs cannot access their pharmacological targets until they are present in their action sites in the cells, being either the cytoplasm or nucleus. Therefore, cellular processes through which the oligonucleotides can access their intracellular targets largely determine pharmacological activity of ASOs. Like other biological macromolecules that are large, polar, and sometimes charged, the oligonucleotides reach their pharmacological targets from extracellular space by some form of endocytosis. In general, endocytosis of oligonucleotides includes four major steps:

1) Initial entry of oligonucleotides into cells. After the oligonucleotides or their delivery systems bind to the receptor, the ligand–receptor complex enters vesicles that bud from the cytosolic face of the plasma membrane and then pinch off, often with the assistance of the dynamin GTPase [19]. The vesicles are formed under additional assistance of clathrin or caveolin, which then selectively sort cargos to their destination. The endocytosis pathways are classified into five major classes based on the proteins involved in the internalization step [20] (Figure 3.2): (i) the "classic" clathrin-coated pit pathway (also known as clathrin-mediated endocytosis [CME]); (ii) the caveolar pathway; (iii) multiple noncaveolar and clathrin-independent pathways (CLIC pathways); (iv) phagocytosis, which mainly takes place in "professional phagocytes" such as macrophages and granulocytes; and (v) macropinocytosis, in which the macromolecules are simply engulfed along with the ambient medium.
2) Initial uptake is followed by sequential intracellular trafficking into a variety of low pH endomembrane compartments, such as early endosomes and lysosomes. After internalization, many vesicles first traffic to early endosomes. In some instances, the receptors, which may be coupled with the cargos, are sorted into recycling endosomes and then head back to the cell surface, while in polarized cells they are transported across cells in a process known as "transcytosis." In other cases, the receptors enter late endosomes and further traffic to lysosomes, where they can be degraded. Vesicular fusion events are controlled by SNARE and SM proteins [22], whereas the complex flow of endomembrane traffic is guided by the Rab GTPases, which are localized in distinct membrane vesicles [23]. For example, transferrin receptor and epidermal growth factor receptors (EGFRs) converge in early endosomes after internalization; however, Rab11-containing vesicles deliver transferrin receptor to recycling endosomes and eventually back to the plasma membrane, whereas Rab7 guides the sorting of EGFRs to the lysosomes, where they are degraded.

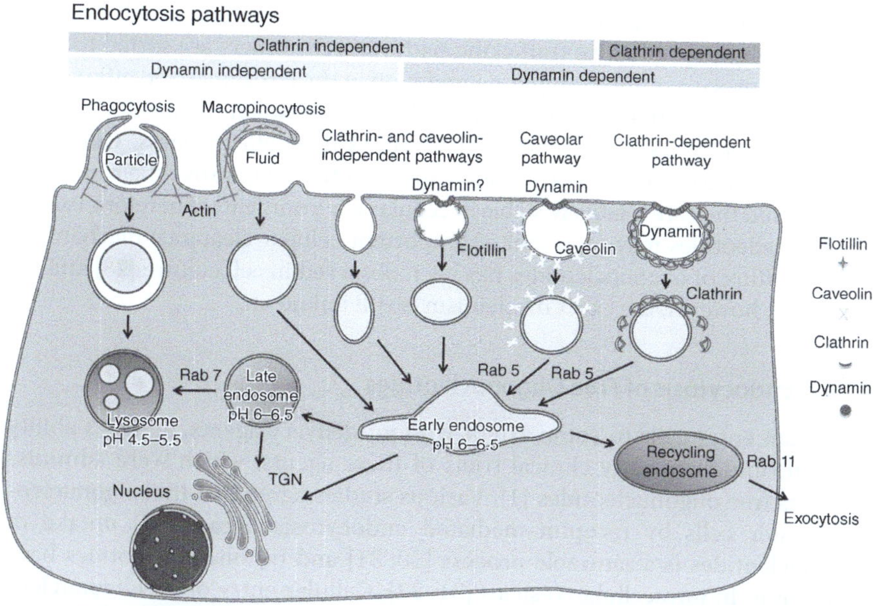

Figure 3.2 Pathways of endocytosis and intracellular trafficking. There are multiple pathways for cellular entry of macromolecules including oligonucleotides. In all cases the initial step of endocytosis involves initial binding to the plasma membrane and engulfment of cargo into intracellular vesicles. The second step often involves sorting of the cargo through endosomes, which is followed by the final stage during which the cargo is delivered to its final destination or recycled to extracellular milieu. Adapted from Ming [21].

3) Ultimately, the oligonucleotides must exit from the endosomal vesicles to reach the site of action in the cytoplasm or nucleus. Endosomal release is a mandated step for delivery of oligonucleotide via endocytosis [24]. Its importance has been highlighted by various examples in which the oligonucleotides are trapped in endosomal vesicles and treatment with endosomal disrupting agents improves the functional delivery of oligonucleotides without enhancing the overall cellular uptake [25, 26]. The innate activity of oligonucleotides taken up by cells is likely due to a modest amount of continuous leakage from endomembrane compartments that spontaneously occurs during intracellular trafficking, while much of the current technology for enhancing oligonucleotide effects is aimed at increasing the extent of endosomal release.
4) After endosomal release, nuclear entry may not be the rate-limiting step for monomeric oligonucleotides in light of the observation that oligonucleotides rapidly accumulate in the nucleus within minutes after being microinjected into the cytoplasm of mammalian cells [27]. When oligonucleotides are bound to a nanocarrier, nuclear entry may become rate limiting due to the significant increase in size.

In a typical delivery process, only some of the oligonucleotides reach their targets through the productive trafficking pathways; the others are sorted to nonproductive processes, for example, degradation in the lysosomes and efflux from the cells. The distribution of the oligonucleotides between the productive and nonproductive processes essentially determines their effectiveness. Currently, therapeutic oligonucleotides are chemically modified to improve their nuclease stability, and they are relatively stable in cellular environments. Therefore, export of oligonucleotides may represent an important cellular clearance mechanism. Cellular efflux of oligonucleotides has been observed in cell culture [28] and rat livers [29]; however, the exact mechanism is still unknown.

3.3.1 Endocytosis of Free Oligonucleotides

ASOs can enter cells to some extent without delivery agents, and this ability has contributed to early clinical trials of these agents, which were administrated as free oligonucleotides [1]. Various studies agree that the oligonucleotides enter cells by receptor-mediated endocytosis because (i) uptake of oligonucleotides is a saturable process [30, 31] and (ii) oligonucleotides have been found in intracellular vesicles [31, 32]. Cellular entry of PS oligonucleotides does not depend on actin and dynamin in cell culture [33], and they are sorted into the vesicles other than clathrin-coated vesicles [30]. Recent evidence supports that there are productive and nonproductive intracellular pathways for delivery of PS oligonucleotides to hepatocytes, with the nonproductive pathway probably trafficking to lysosomes [34]. Another study indicated that LNA PS gapmers traffic into P-bodies in the cytoplasm [35], where multiple proteins involving mRNA degradation are localized [12], to produce gene suppression. The receptors that are responsible for oligonucleotides' cellular entry have not been identified, and therefore, the precise trafficking pathways are yet to be clarified.

3.3.2 Endocytosis of Oligonucleotide Conjugates

Cell-penetrating peptides (CPPs) are the short basic amino acid-rich peptides, which not only enter cells themselves but also facilitate the transport of molecular cargos across the plasma membrane. Many types of CPPs have been utilized in oligonucleotides delivery, including prototypical TAT peptide [36], chimeric CPPs [37], and novel CPPs that are designed from structure–activity studies [38]. Conjugation with CPPs dramatically improved functional delivery of ASOs. For example, the direct conjugate of synthetic peptide $(R\text{-}Ahx\text{-}R)_4$ and SSO has been shown to restore dystrophin function in a Duchenne muscular dystrophy mouse model [39]. Despite the rapid progress in application of CPPs to oligonucleotides delivery, controversies still exist concerning their cellular entry mechanisms, first whether CPP internalization involves endocytosis, and if so, what is the endocytic pathway. Initial studies on prototypical CPPs such as TAT and ANT suggested a mechanism of receptor- and energy-independent translocation across the plasma membrane [40]; however, this observation was then considered as an artifact due to the high membrane binding of CPPs [41, 42]. By carefully removing the membrane-bound CPPs, the experimental data supported an active process of cellular entry of Tat CPP [43, 44]. However, using similar tools of pharmacological inhibitors and endocytosis markers, various endocytosis pathways have been proposed for Tat CPP internalization, such as CME [43], macropinocytosis [44, 45], and caveolar endocytosis [46]. Knocking out dynamin [44], clathrin, or caveolin [47] in the cultured cells, causing the molecular inhibition of corresponding endocytosis pathways, did not prevent the cellular entry of the Tat CPP, creating more uncertainties about the exact molecular mechanism of CPPs' endocytosis. The internalization mechanism of the CPPs coupled with oligonucleotides is more complicated and may be different from that of the CPPs themselves. In a study, while Tat CPP was localized in the nucleus in the cells, CPP–siRNA conjugates were localized to perinuclear sites, where siRNAs interact with RISC to cause *RNAi* [48]. It was suggested that siRNAs' interaction with RISC determines localization of the conjugates rather than the CPP peptide [48].

Cellular delivery of oligonucleotides can be enhanced by utilizing targeting ligand–oligonucleotide conjugates. For example, RGD peptide conjugates of SSOs or of siRNA have been used to deliver the oligonucleotides to tumor cells overexpressing αvβ3 integrin [33, 49, 50]. In addition, bombesin peptide (a ligand for the BB2 receptor) and anisamide (a high affinity small molecule ligand for the sigma receptor) have been conjugated to SSOs for delivery to cancer cells [51, 52]. Interestingly, although the enhancement in cellular uptake of RGD–SSO was modest, the functional activity caused by the RGD–SSO conjugate was disproportionately higher than that by "free" SSO [49]. The disconnection of the cellular uptake and functional activity was also observed for RGD–siRNA conjugates. In this study, luciferase siRNA was linked with bi-, tri-, and tetravalent cRGD peptides [50]. The bi-, tri-, and tetravalent cRGD conjugates were taken up by $M21^{+}$ cells at approximately the same rate. However, only the tri- and tetravalent versions produced *RNAi* activity, while the bivalent version had little effect [50]. One cause of this disaccord may be the distinct internalization pathways that the peptide–oligonucleotide conjugates undertake. A study was followed to resolve endocytosis pathways of RGD–SSO [33]. Using endocytosis inhibitors and dominant-negative form of dynamin GTPase, we have identified an actin-dependent, dynamin-dependent pathway for uptake of the RGD–SSO conjugate in A375 cells and an unusual actin-independent, dynamin-independent pathway for uptake of the unconjugated PS oligonucleotide, suggesting that the route of internalization could affect the magnitude of the biological response [33].

3.3.3 Uptake and Trafficking of Oligonucleotides Incorporated into Nanocarriers

A widely used strategy for promoting delivery of antisense, siRNA, or other types of oligonucleotides is to complex cationic lipids and polymers with negatively charged oligonucleotides, and then the resultant nanoparticles potentially can help to overcome biological barriers, increase cell uptake, and enhance escape from membrane compartments [53–56]. Lipid-based nanocarriers have proven to be very efficacious for the delivery of siRNA to the liver [57]. Based on the mechanism study on cellular transfection of liposomal plasmid DNA, the conventional view is that cationic lipid carriers functionally deliver oligonucleotides via endocytosis followed by escape from endosomes. However, recent studies indicated that lipoplexes of oligonucleotides enter the cells in a way that is different from those of plasmid DNA. Using both chemical and molecular inhibitors of endocytosis, Lu et al. found that only a minor component of the cell-associated siRNA that contributed to "knockdown" function and that this component probably came from fusion between the siRNA lipoplexes and the plasma membrane [58]. A latter study has compared the uptake and trafficking pathways of SSOs delivered via cationic lipids and via PEI, a cationic polymer [59]. Interestingly, in agreement with the study on siRNA delivery discussed above [58], functional delivery of antisense associated with lipoplexes was apparently due to fusion at the plasma membrane, while delivery via polyplexes took place through an unconventional form of endocytosis.

Two recent studies provided more detailed information on the cellular uptake and intracellular trafficking of siRNA [60, 61]. In both studies, the siRNA was incorporated into (two different) lipid nanoparticles that had shown excellent efficacy for siRNA delivery *in vivo*. Both studies used advanced imaging techniques and chemical and molecular tools to manipulate uptake and trafficking pathways. One study suggests the siRNA lipoplexes enter the cells initially by CME that then triggers a more robust uptake via macropinocytosis [60]. This study found a low level of siRNA escape (~2%) from endosomes via gradual release rather than a bursting process and that the most likely site of escape is from an early endosomal compartment. The other study found evidence for substantial recycling of siRNA to the external medium via exocytosis from a late endosomal/lysosomal compartment [61]. This study identified that the transmembrane glycoprotein NPC1, which is present on multivesicular late endosomes and is known to be involved in trafficking of lipids, plays a role in recycling of the lipid/siRNA complex via a Rab27a-dependent pathway. The two studies differ in many details, but they both reveal the importance of intracellular processing in the functional delivery of siRNA. These different mechanisms revealed by these studies indicate that endocytosis mechanisms are so sophisticated that one cannot assume a single type of trafficking pathway for one type of nanoparticles.

3.4 Conclusion

Our understanding of antisense mechanisms and application of ASOs for therapeutic purposes have advanced dramatically over the past 35 years since DNA antisense molecules were first described. In contrast, we lack an equivalent depth of knowledge about how these molecules access their molecular targets in the cells. Previous studies have shown that cellular delivery of oligonucleotides requires functional synergy of cellular uptake, vesicular trafficking, and endosomal release. Thus, these steps govern the magnitude and duration of oligonucleotide exposure to the targets and ultimately determine its effectiveness. Depending on their relative efficiency in different delivery modes, each step may become rate-limiting and control the cellular response of oligonucleotides. Therefore, mechanistic studies that can identify the rate-limiting step can help design a specific strategy to overcome this barrier. This suggests that it will be very productive to continue to investigate uptake and trafficking issues for antisense, siRNA, and SSOs and that the knowledge gleaned will be very helpful in therapeutic development of these types of molecules.

References

1 Bennett, C.F. and Swayze, E.E. (2010). RNA targeting therapeutics: molecular mechanisms of antisense oligonucleotides as a therapeutic platform. *Annu. Rev. Pharmacol. Toxicol.* 50: 259–293.

2 Zamecnik, P.C. and Stephenson, M.L. (1978). Inhibition of *Rous sarcoma* virus replication and cell transformation by a specific oligodeoxynucleotide. *Proc. Natl. Acad. Sci. U. S. A.* 75: 280–284.

3 Moazed, D. (2009). Small RNAs in transcriptional gene silencing and genome defence. *Nature* 457: 413–420.

4 Elbashir, S.M., Harborth, J., Lendeckel, W. et al. (2001). Duplexes of 21-nucleotide RNAs mediate RNA interference in cultured mammalian cells. *Nature* 411: 494–498.

5 Pratt, A.J. and MacRae, I.J. (2009). The RNA-induced silencing complex: a versatile gene-silencing machine. *J. Biol. Chem.* 284: 17897–17901.

6 Overhoff, M., Wunsche, W., and Sczakiel, G. (2004). Quantitative detection of siRNA and single-stranded oligonucleotides: relationship between uptake and biological activity of siRNA. *Nucleic Acids Res.* 32: e170.

7 Ozsolak, F. and Milos, P.M. (2011). RNA sequencing: advances, challenges and opportunities. *Nat. Rev. Genet.* 12: 87–98.

8 Sazani, P. and Kole, R. (2003). Therapeutic potential of antisense oligonucleotides as modulators of alternative splicing. *J. Clin. Invest.* 112: 481–486.

9 Goemans, N.M., Tulinius, M., van den Akker, J.T. et al. (2011). Systemic administration of PRO051 in Duchenne's muscular dystrophy. *N. Engl. J. Med.* 364: 1513–1522.

10 Cirak, S., Arechavala-Gomeza, V., Guglieri, M. et al. (2011). Exon skipping and dystrophin restoration in patients with Duchenne muscular dystrophy after systemic phosphorodiamidate morpholino oligomer treatment: an open-label, phase 2, dose-escalation study. *Lancet* 378: 595–605.
11 Djuranovic, S., Nahvi, A., and Green, R. (2011). A parsimonious model for gene regulation by miRNAs. *Science* 331: 550–553.
12 Eulalio, A., Behm-Ansmant, I., and Izaurralde, E. (2007). P bodies: at the crossroads of post-transcriptional pathways. *Nat. Rev. Mol. Cell Biol.* 8: 9–22.
13 Rossbach, M. (2010). Small non-coding RNAs as novel therapeutics. *Curr. Mol. Med.* 10: 361–368.
14 Krutzfeldt, J., Rajewsky, N., Braich, R. et al. (2005). Silencing of microRNAs in vivo with "antagomirs". *Nature* 438: 685–689.
15 Mattes, J., Yang, M., and Foster, P.S. (2007). Regulation of microRNA by antagomirs: a new class of pharmacological antagonists for the specific regulation of gene function? *Am. J. Respir. Cell Mol. Biol.* 36: 8–12.
16 Janssen, H.L., Reesink, H.W., Lawitz, E.J. et al. (2013). Treatment of HCV infection by targeting microRNA. *N. Engl. J. Med.* 368: 1685–1694.
17 Gutschner, T. and Diederichs, S. (2012). The hallmarks of cancer: a long non-coding RNA point of view. *RNA Biol.* 9: 703–719.
18 Wahlestedt, C. (2013). Targeting long non-coding RNA to therapeutically upregulate gene expression. *Nat. Rev. Drug Discov.* 12: 433–446.
19 Damke, H. (1996). Dynamin and receptor-mediated endocytosis. *FEBS Lett.* 389: 48–51.
20 Conner, S.D. and Schmid, S.L. (2003). Regulated portals of entry into the cell. *Nature* 422: 37–44.
21 Ming, X. (2011). Cellular delivery of siRNA and antisense oligonucleotides via receptor-mediated endocytosis. *Expert Opin. Drug Deliv.* 8: 435–449.
22 Sudhof, T.C. and Rothman, J.E. (2009). Membrane fusion: grappling with SNARE and SM proteins. *Science* 323: 474–477.
23 Stenmark, H. (2009). Rab GTPases as coordinators of vesicle traffic. *Nat. Rev. Mol. Cell Biol.* 10: 513–525.
24 Dominska, M. and Dykxhoorn, D.M. (2010). Breaking down the barriers: siRNA delivery and endosome escape. *J. Cell Sci.* 123: 1183–1189.
25 Kwon, E.J., Bergen, J.M., and Pun, S.H. (2008). Application of an HIV gp41-derived peptide for enhanced intracellular trafficking of synthetic gene and siRNA delivery vehicles. *Bioconjug. Chem.* 19: 920–927.
26 Detzer, A., Overhoff, M., Wunsche, W. et al. (2009). Increased RNAi is related to intracellular release of siRNA via a covalently attached signal peptide. *RNA* 15: 627–636.
27 Fisher, T.L., Terhorst, T., Cao, X., and Wagner, R.W. (1993). Intracellular disposition and metabolism of fluorescently-labeled unmodified and modified oligonucleotides microinjected into mammalian cells. *Nucleic Acids Res.* 21: 3857–3865.

28 Iversen, P.L., Zhu, S., Meyer, A., and Zon, G. (1992). Cellular uptake and subcellular distribution of phosphorothioate oligonucleotides into cultured cells. *Antisense Res. Dev.* 2: 211–222.
29 Takakura, Y., Mahato, R.I., Yoshida, M. et al. (1996). Uptake characteristics of oligonucleotides in the isolated rat liver perfusion system. *Antisense Nucleic Acid Drug Dev.* 6: 177–183.
30 de Diesbach, P., N'Kuli, F., Berens, C. et al. (2002). Receptor-mediated endocytosis of phosphodiester oligonucleotides in the HepG2 cell line: evidence for non-conventional intracellular trafficking. *Nucleic Acids Res.* 30: 1512–1521.
31 Beltinger, C., Saragovi, H.U., Smith, R.M. et al. (1995). Binding, uptake, and intracellular trafficking of phosphorothioate-modified oligodeoxynucleotides. *J. Clin. Invest.* 95: 1814–1823.
32 Vaishnaw, A.K., Gollob, J., Gamba-Vitalo, C. et al. (2010). A status report on RNAi therapeutics. *Silence* 1: 14.
33 Alam, M.R., Ming, X., Dixit, V. et al. (2010). The biological effect of an antisense oligonucleotide depends on its route of endocytosis and trafficking. *Oligonucleotides* 20: 103–109.
34 Koller, E., Vincent, T.M., Chappell, A. et al. (2011). Mechanisms of single-stranded phosphorothioate modified antisense oligonucleotide accumulation in hepatocytes. *Nucleic Acids Res.* 39: 4795–4807.
35 Stein, C.A., Hansen, J.B., Lai, J. et al. (2010). Efficient gene silencing by delivery of locked nucleic acid antisense oligonucleotides, unassisted by transfection reagents. *Nucleic Acids Res.* 38: e3.
36 Astriab-Fisher, A., Sergueev, D.S., Fisher, M. et al. (2000). Antisense inhibition of P-glycoprotein expression using peptide-oligonucleotide conjugates. *Biochem. Pharmacol.* 60: 83–90.
37 Simeoni, F., Morris, M.C., Heitz, F., and Divita, G. (2003). Insight into the mechanism of the peptide-based gene delivery system MPG: implications for delivery of siRNA into mammalian cells. *Nucleic Acids Res.* 31: 2717–2724.
38 Abes, S., Moulton, H.M., Clair, P. et al. (2006). Vectorization of morpholino oligomers by the (R-Ahx-R)4 peptide allows efficient splicing correction in the absence of endosomolytic agents. *J. Control. Release* 116: 304–313.
39 Wu, B., Moulton, H.M., Iversen, P.L. et al. (2008). Effective rescue of dystrophin improves cardiac function in dystrophin-deficient mice by a modified morpholino oligomer. *Proc. Natl. Acad. Sci. U. S. A.* 105: 14814–14819.
40 Derossi, D., Calvet, S., Trembleau, A. et al. (1996). Cell internalization of the third helix of the Antennapedia homeodomain is receptor-independent. *J. Biol. Chem.* 271: 18188–18193.
41 Drin, G., Mazel, M., Clair, P. et al. (2001). Physico-chemical requirements for cellular uptake of pAntp peptide. Role of lipid-binding affinity. *Eur. J. Biochem.* 268: 1304–1314.

42 Richard, J.P., Melikov, K., Vives, E. et al. (2003). Cell-penetrating peptides. A reevaluation of the mechanism of cellular uptake. *J. Biol. Chem.* 278: 585–590.
43 Richard, J.P., Melikov, K., Brooks, H. et al. (2005). Cellular uptake of unconjugated TAT peptide involves clathrin-dependent endocytosis and heparan sulfate receptors. *J. Biol. Chem.* 280: 15300–15306.
44 Wadia, J.S., Stan, R.V., and Dowdy, S.F. (2004). Transducible TAT-HA fusogenic peptide enhances escape of TAT-fusion proteins after lipid raft macropinocytosis. *Nat. Med.* 10: 310–315.
45 Kaplan, I.M., Wadia, J.S., and Dowdy, S.F. (2005). Cationic TAT peptide transduction domain enters cells by macropinocytosis. *J. Control. Release* 102: 247–253.
46 Fittipaldi, A., Ferrari, A., Zoppe, M. et al. (2003). Cell membrane lipid rafts mediate caveolar endocytosis of HIV-1 Tat fusion proteins. *J. Biol. Chem.* 278: 34141–34149.
47 Ter-Avetisyan, G., Tunnemann, G., Nowak, D. et al. (2009). Cell entry of arginine-rich peptides is independent of endocytosis. *J. Biol. Chem.* 284: 3370–3378.
48 Chiu, Y.L., Ali, A., Chu, C.Y. et al. (2004). Visualizing a correlation between siRNA localization, cellular uptake, and RNAi in living cells. *Chem. Biol.* 11: 1165–1175.
49 Alam, M.R., Dixit, V., Kang, H. et al. (2008). Intracellular delivery of an anionic antisense oligonucleotide via receptor-mediated endocytosis. *Nucleic Acids Res.* 36: 2764–2776.
50 Alam, M.R., Ming, X., Fisher, M. et al. (2011). Multivalent cyclic RGD conjugates for targeted delivery of small interfering RNA. *Bioconjug. Chem.* 22: 1673–1681.
51 Ming, X., Alam, M.R., Fisher, M. et al. (2010). Intracellular delivery of an antisense oligonucleotide via endocytosis of a G protein-coupled receptor. *Nucleic Acids Res.* 38: 6567–6576.
52 Nakagawa, O., Ming, X., Huang, L., and Juliano, R.L. (2010). Targeted intracellular delivery of antisense oligonucleotides via conjugation with small-molecule ligands. *J. Am. Chem. Soc.* 132: 8848–8849.
53 Juliano, R., Bauman, J., Kang, H., and Ming, X. (2009). Biological barriers to therapy with antisense and siRNA oligonucleotides. *Mol. Pharm.* 6: 686–695.
54 Whitehead, K.A., Langer, R., and Anderson, D.G. (2009). Knocking down barriers: advances in siRNA delivery. *Nat. Rev. Drug Discov.* 8: 129–138.
55 Akhtar, S. and Benter, I.F. (2007). Nonviral delivery of synthetic siRNAs in vivo. *J. Clin. Invest.* 117: 3623–3632.
56 Tamura, A. and Nagasaki, Y. (2010). Smart siRNA delivery systems based on polymeric nanoassemblies and nanoparticles. *Nanomedicine (London)* 5: 1089–1102.

57 Love, K.T., Mahon, K.P., Levins, C.G. et al. (2010). Lipid-like materials for low-dose, in vivo gene silencing. *Proc. Natl. Acad. Sci. U. S. A.* 107: 1864–1869.
58 JJ, L., Langer, R., and Chen, J. (2009). A novel mechanism is involved in cationic lipid-mediated functional siRNA delivery. *Mol. Pharm.* 6: 763–771.
59 Ming, X., Sato, K., and Juliano, R.L. (2011). Unconventional internalization mechanisms underlying functional delivery of antisense oligonucleotides via cationic lipoplexes and polyplexes. *J. Control. Release* 153: 83–92.
60 Gilleron, J., Querbes, W., Zeigerer, A. et al. (2013). Image-based analysis of lipid nanoparticle-mediated siRNA delivery, intracellular trafficking and endosomal escape. *Nat. Biotechnol.* 31: 638–646.
61 Sahay, G., Querbes, W., Alabi, C. et al. (2013). Efficiency of siRNA delivery by lipid nanoparticles is limited by endocytic recycling. *Nat. Biotechnol.* 31: 653–658.

4

Pharmacokinetics and Pharmacodynamics of Antisense Oligonucleotides

Helen Lightfoot[1,*], *Anneliese Schneider*[2,*], *and Jonathan Hall*[1]

[1] *ETH-Zurich, Zurich, Switzerland*
[2] *Preclinical Services & Consulting, Munich, Germany*

4.1 Introduction

A major hurdle to the breakthrough of first-generation phosphorothioate antisense oligonucleotide (PTO) drugs has been their inability to reach target tissues, for example, tumors, in sufficient amounts for a clinical response within the safety range. Consequently, key drivers for the development of new oligonucleotide chemistries were increased affinity for the target mRNA and higher drug concentrations at their sites of action. Such developments, in turn, have provided considerable insight into the factors that determine the pharmacokinetic properties of this compound class.

Insight into the specific profile of an oligonucleotide in terms of its plasma or tissue concentration–time course (pharmacokinetics [PK]) and effect (pharmacodynamics [PD]) is essential for efficient antisense oligonucleotide (ASO) drug development. With respect to its PK, the assessment of parameters in the plasma as well as the concentration and half-life of the ASO in the tissue expressing the target mRNA is of high importance. In clinical investigations biopsies are often not available and this information is predominately obtained from animal studies. Surrogate systems such as the plasma trough concentration can also be useful. Depending on the mode of action of an ASO, PD parameters are assessed at the level of its target mRNA, the protein, or a downstream biomarker. Any physiological or therapeutic effect is generally assumed to be reflected by a PD parameter, but if the target is not clinically validated, then this also has to be established.

---

*Contributed equally

Oligonucleotide-Based Drugs and Therapeutics: Preclinical and Clinical Considerations for Development, First Edition. Edited by Nicolay Ferrari and Rosanne Seguin.

For the development of an oligonucleotide drug, inappropriate target selection from either the PK or the PD perspective will inevitably lead to failure. The majority of ASOs present in clinical development are phosphorothioate-modified second-generation ASOs. Their PK profiles are well characterized and comparable to a large extent. Hence, the rational selection of mRNA targets based on this general PK profile can be an advantage of this compound class. On the contrary, the PD aspect of target selection is predominantly dependent on the nature of the target. Here, the target's validation status, the existence of clinical biomarkers, and, in the majority of the cases, its translation rate/protein stability are factors to be considered.

The purpose of gathering PK information on an ASO compound is to guide the selection of doses and treatment schedules for clinical development. However, first, the relationship between the PK and PD of the ASO compound has to be established. Factors that should be considered include the identification of:

1) The concentration of drug needed at the target site to observe an activity that translates into a therapeutic effect.
2) The appropriate dose, route of administration, and dosing frequency to obtain and maintain this concentration.
3) Biomarkers to assess the PD effect in clinical trials.
4) The relationship of target inhibition or biomarker effect and clinical response.

Implementation of the PK/PD concepts and exposure–response analyses provide an important rational tool for efficient decision-making during clinical development, and several reports describing integrated PK/PD approaches have been published. To date, modeling and simulation has rarely been applied and is in its infancy for oligonucleotides.

This chapter will mainly cover the PK, PD, and PK/PD relationships of single-stranded ASOs, with a focus on phosphorothioated oligonucleotides because of their wide-scale use and the large amount of publicly available information.

4.2 Pharmacokinetic Properties of Antisense Oligonucleotides

Knowledge of the PK profile of an ASO in preclinical models and patients is a prerequisite for the selection and optimization of dosing regimens in clinical investigations. As a guide, first- and second-generation single-stranded phosphorothiated ASOs generally share many of their PK properties, including:

- Peak and area-under-the-plasma-concentration-versus-time-curve (AUC) levels increase proportionally at doses typical for clinical administration.
- High plasma protein binding across the species with 85–95% of the ASO bound.
- High bioavailability after subcutaneous (s.c.) administration.

- Short distribution half-lives in the range of minutes to 1.5 h.
- Rapid and wide tissue distribution with a similar pattern, i.e. highest concentrations in the kidney and liver.
- Relatively long terminal elimination half-lives from 1 to 2 days for first-generation PTOs and weeks for second-generation ASOs.
- Clearance driven by nuclease metabolism.
- Renal excretion of metabolites mainly as a result of their lower protein binding abilities relative to full-length ASOs.
- PK plasma parameters comparable across animal species and scalable to human application based on body weight rather than surface.

The section below provides an overview of the PK properties of ASOs.

4.2.1 Protein Binding

The PK properties of oligonucleotides, like other molecules, are largely determined by their physicochemical properties. Chemical modifications of oligonucleotides can result in marked differences in their distribution patterns, which is at least partially a result of their altered protein binding capacities [1, 2].

Phosphorothioate-modified backbones enhance hydrophobicity and increase protein binding. As a result, phosphorothioated ASOs bind unspecifically to plasma and cellular proteins, albeit with low affinity. In the plasma, PTOs bind to hydrophilic sites of albumin and to a lesser extent to alpha2-macroglobulin, with more than 90–95% of the ASO sequestered [3]. Protein binding is dependent on the length of the oligonucleotide, and truncated/fragmented metabolites have significantly lower binding capacities, leading to a higher fraction of free metabolites compared with full-length ASO. Protein binding seems to be independent of the nucleotide sequence.

Binding to plasma and cellular proteins is essential for the persistence, tissue distribution, and cellular uptake of ASOs. In addition, it protects the ASO from glomerular filtration. Full-length ASOs are excreted in the urine to a small extent, reflecting the low proportion of the unbound form. Consequently, in the urine, mainly unbound truncated or fragmented metabolites are found. Oligonucleotide chemistries without the phosphorothioate backbone display lower protein binding and as a result are subject to rapid renal clearance. The protein binding capacity of ASOs is reported to be similar across species, with mouse having a notably lower capacity compared with rat, monkey, and human [1, 4].

The effect of the phosphorothioate backbone on the protein binding properties of second-generation ASOs was investigated through the comparison of the protein binding capacity of 2′-*O*-methoxyethyl (2′-MOE) oligoribonucleotides comprising different contents of phosphorothioates, as well as different formats of the 2′-MOE modifications [1]. It was observed that the high binding and tissue distribution profiles of the second-generation molecules resembled those of the first-generation oligonucleotides. Hence, newer oligonucleotide modifications, e.g. 2′-MOE or locked nucleic acid (LNA), in clinical development also employ phosphorothioate backbone linkages to reduce renal clearance of the active compound and maintain tissue distribution. In additional studies, plasma protein binding of second-generation ASOs with phosphorothioate linkages was in the same range as PTO with approximately 95% bound across most species, except mouse (85%) [5, 6]. Similarly, protein binding of the second-generation ASOs was dependent on the length of the oligonucleotide with shorter metabolites binding to plasma proteins to a lower extent [7].

4.2.2 Dose Dependency of ASO Pharmacokinetics

The maximum concentration (C_{max}) and the AUCs are reported to increase concentration dependently at doses used in clinical therapy. However, peak levels do not increase with repeated administrations because ASOs distribute widely to tissues [6, 8]. On the other hand, plasma trough levels reflecting concentrations in tissue increased after weekly injections of the second-generation ASOs mipomersen and drisapersen. This is because tissue half-lives of the second-generation oligonucleotides are longer than 1 week; hence the ASOs accumulate in tissues after weekly injections [9, 10]. In animal studies, there are reports that plasma concentrations of ASO increase over proportionally with dose above certain levels, indicating the involvement of saturable factors in ASO tissue distribution, with differences noted among species [11].

4.2.3 Absorption

Oligonucleotides cannot be delivered by oral administration because their gastrointestinal absorption is low as a consequence of their high molecular weight, their hydrophilicity, and their charge. Therefore, ASOs are generally administered parenterally. PTOs are mostly applied intravenously. They are rarely administered s.c. because of an insufficient stability toward degrading nucleases and their propensity to cause inflammatory reactions at the injection site after continuous or repeated dosing. Since the nuclease stability of 2′-MOE and LNA-modified ASO compounds is greatly improved and their inflammatory potential reduced, likely a consequence of reduced injection frequency, s.c. administration in patients is feasible and tolerable. Other parenteral routes that have been employed in ASO clinical testing are intramuscular (i.m.), pulmonary, intratumoral (glioma), intrathecal, and intravitreal delivery [10, 12–15].

After s.c. injection ASOs are generally well absorbed from the injection site. The bioavailability reported for s.c. administered PTOs as well as for 2′-MOE and LNA oligonucleotides ranged from 26 to 100% of the exposure after i.v. delivery [6, 16, 17]. For example, the bioavailability of a PTO injected s.c. into monkeys ranged from 26 to 55%. Interestingly, bioavailability was found to be dependent on the concentration of the dosing solution but not on the dose [16]. The bioavailability of a 2′-MOE-modified ASO ranged from 80 to 100% in the same species [11]. In healthy volunteers, the bioavailability for the same 2′-MOE-modified ASO was 82% [18], confirming the results in the animal model.

The C_{max} observed in plasma was lower after s.c. than after i.v. administration because of slow absorption from the injection site. In addition, the time to reach maximum concentration (Tmax) was prolonged after s.c. administration: C_{max} was observed 3–4 h after s.c. dosing for mipomersen in patients and was thus similar to the Tmax of 3 h observed for a PTO after s.c. administration [17, 18].

Following ASO administration the mechanisms by which the ASOs cross biological membranes to reach their intracellular target are not fully understood. Due to the highly charged and hydrophilic nature of ASOs, passive diffusion is not expected to play a major role in the process. Rather, it is assumed that transporters mediate an active transport or a facilitated diffusion mechanism. In fact, the existence of a saturable component in the absorption process has been suggested [11].

4.2.4 Distribution

The tissue distribution properties of second-generation ASOs are generally similar to that observed for PTOs. First, ASOs are readily and almost completely distributed from the plasma to the tissues. The plasma concentration versus time profile of ASOs has been shown in many studies to be multiphasic with a fast decline in ASO concentration in the first 24 h after administration and an initial half-life reflecting tissue distribution following i.v. delivery in the range of 30–90 min or even shorter depending on the chemical modification and the specific ASO [11, 19]. For example, for ISIS 104838, a 2′-MOE-ASO, the distribution half-life after i.v. injection was 15–45 min in mouse, rat, dog, monkey, and human. After s.c. delivery however the distribution half-life is longer due to prolonged absorption from the injection site [11]. For example, for the 2′-MOE mipomersen, the distribution half-life was in the range of 2–5 h after s.c. injection [17].

The volume of distribution of ASO is often more than two orders of magnitude larger than the blood volume. For example, the volume of distribution was reported to be 1000 L for LY212181308 in patients [19], indicating a wide distribution and a localization in peripheral tissues, whereas the total body clearance was low [6, 20]. ASOs share a general tissue distribution pattern, with typically the highest concentration and accumulation in the kidney, liver, spleen, lymph nodes, adipocytes, and bone marrow [11, 21]. The organs to which ASOs distribute appear to depend on blood flow/perfusion but is clearly also an effect of the protein binding properties of the ASOs. For example, the use of different formats of the same sequence of an LNA-modified ASO targeting miR-122 demonstrated that different formats of a particular chemical modification could influence the relative concentrations of the ASO in a specific tissue. For example, a fully LNA-modified sequence produced the highest concentrations in the liver and kidney, whereas ASOs in a gapmer design with 2–5 LNA on either end of a 16-mer ASO distributed to the liver, the target tissue, in up to 10-fold lower concentrations [22]. There are also reports indicating that the mode of ASO administration can also affect the ASO concentration in the kidney and liver tissue. For example, i.v. administration of a 2′-*O*-methyl PTO, compared with s.c. or i.p. injections, leads to higher concentrations in the kidney and liver [23].

Cellular uptake of ASOs in organs can be cell type specific. Tissue macrophages or cells with a high potential for endocytosis typically show the highest uptake. In the kidney and liver, the organs with the highest tissue concentrations, renal proximal tubular epithelial cells, Kupffer cells, and sinusoidal endothelial cells contain significantly higher levels of oligonucleotides in comparison with other cells in the kidney or liver such as cells of the glomerular and distal tubules or hepatocytes [24].

4.2.5 Metabolism and Excretion

Oligonucleotides are metabolized by nucleases ubiquitously expressed by cells in most tissues. PTOs lacking ribose modifications at their 3′ and 5′ ends are primarily degraded by exonucleases generating 3′ or 5′ shortened fragments but also by endonucleases in tissues. These fragmented oligonucleotides may still possess antisense activity. On the contrary, second-generation ASOs protected at the 3′ and 5′ from exonuclease degradation by chemical modifications are initially metabolized by endonucleases in tissues, leading to short fragments, which may be further degraded by exonucleases. In contrast to PTOs, the metabolites of second-generation ASOs resulting from the initial endonuclease cleavage are too short to still possess antisense activity [24].

The elimination half-life of oligonucleotides in plasma reflects their metabolism in tissues, the equilibration of full-length ASOs and metabolites between tissues and blood, and their excretion by the kidneys [11, 24]. In general, the terminal half-life of 2′-MOE and LNA ASOs in tissues range from several days to weeks. This is significantly increased compared with PTOs, which displayed half-lives of 1–2 days, and is largely due to the enhanced nuclease stability of the additional chemical modifications at the 3′ and 5′ ends of the ASOs. The tissue half-life is specific for a given ASO but similar among the various tissues. Interestingly, the tissue half-life can vary among different species, which may be explained by differences in exo- or endonuclease activity. In mice, the half-life of an LNA-modified ASO in the liver ranged from 4 to 10 days, whereas it was 2–4 weeks in monkeys and up to 5 weeks in humans [6]. The prolonged half-lives of the second-generation ASOs are reflected by their increased persistence in tissues. Increased tissue persistence is a prerequisite for a less frequent dosing regime, and thus many of the second-generation ASO compounds are administered once weekly [9]. This is a major improvement to PTOs with which continuous infusions up to several days are required to attain the drug levels within therapeutic range.

As mentioned earlier in this chapter, phosphorothioated 2′-MOE and LNA-modified ASOs are mostly bound to plasma proteins, similar to PTOs. This limits their glomerular filtration to a small free fraction. As a result, full-length ASO with a low free fraction are rarely detectable in urine, for example, below 1% of the dose was detected after a single i.v. infusion of 2′-MOE modified ISIS 104838 in monkey. Furthermore, due to their high nuclease resistance as a result of the additional chemical modifications, the overall excretion was generally slow with 80% of the dose excreted mainly in the urine after 90 days' post-dosing. In contrast, the shorter metabolites bind to plasma proteins to a lower extent, and unbound ASOs are readily excreted via the kidney [11].

To predict PK parameters and dosing in humans, scaling of PK parameters among animal species to human was shown to be reasonably well directed by body weight and not by surface area for PTO and 2′-MOE and also for LNA ASOs [6, 11, 25].

There are several analytical methods available to estimate the concentrations of ASOs in plasma and tissues (Table 4.1). Method selection depends on the objectives in PK assessment and also the properties of the ASO.

4.3 Pharmacodynamic Properties of Antisense Oligonucleotides

The association of drugs such as ASOs with their biological targets and the mechanisms by which the drug affects the target to initiate a physiological response can collectively be referred to as pharmacodynamics (PD). Hence, preclinical and clinical studies of ASO PD have focused primarily on the effect of the ASO on the target mRNA and/or target protein levels, downstream effects that result from target modulation, and therapeutic effects within a given disease indication [5]. In this context, the process of target selection and validation, the time until onset of action and the duration of activity needed to achieve a therapeutic effect, and the existence of clinical biomarkers to monitor ASO function are all important parameters.

Table 4.1 Overview of bioanalytical methods for oligonucleotide quantification in biological matrices such as plasma and tissue.

Method	Principle	Sensitivity in plasma (ng/ml)	Selectivity (resolution of metabolites)	Reference
Total radioactivity	Scintillation counting or combustion	pg/ml range	None	—
Capillary gel electrophoreses	Electrophoretic separation, UV detection	ng/ml range	High	[26]
HPLC (anion exchange, ion pair reversed phase)	Chromatographic separation, UV or fluorescence detection	ng/ml range	Poor	[27]
Hybridization assay (various formats)	Sequence-specific binding to capture/detection probe	pg/ml range	Poor	[28]
LC/MS-MS	Chromatographic separation, MS detection	pg-ng/ml range	High	[29, 30]

4.3.1 ASO Target Selection and Validation

With greater than 50% of late-stage clinical attritions resulting from a lack of compound efficacy [31, 32] and a large fraction of these failures likely a result of an inappropriately chosen target, target selection has become a critical stage in the discovery of new drug candidates. For ASO compounds, this is of particular importance as the PK profiles of phosphorothiated chemistries are well characterized and very similar; therefore the general PK profile of ASOs should direct target selection.

The pharmaceutical industry considers a drug target for a defined therapeutic indication ideal if it adheres to the following criteria [33, 34]:

1) Has an established role in the relevant disease pathophysiology.
2) Is disease modifying by the chosen therapeutic modality.
3) Does not have an essential role in nondisease-related pathways and/or has restricted expression to the disease tissue.
4) Allows the selection of a well-defined patient population for clinical trials and treatment.
5) Has assayable PD endpoints and early biomarker(s) to monitor dosing and activity.
6) Has robust efficacy endpoints that can be monitored in patient studies.
7) Holds a favorable competitive situation.

Connecting a gene product to a disease is crucial for selecting a target. This connection is generally made through a combination of several approaches, including human genetics, pathway analysis, preclinical studies in model organisms, and target expression perturbation in the disease tissue [34]. However the strongest support for a targets' role in a specific disease pathology is ultimately achieved through target modulation in patients and is referred to as clinical validation [35]. For obvious reasons, the success rate of drug development involving validated targets is higher than for novel targets working by an as-yet untested mechanism of action [36–38]. On the other hand, utilization of an unvalidated target, defined as a molecule that conveys a new mechanism of action that has not previously been shown to have therapeutic utility in patients for a particular indication, has the potential to lead to breakthrough treatments and also benefits from a favorable competitive situation.

Many historically successful small molecule drugs serendipitously display therapeutic polypharmacology [39]. They are capable of bypassing cellular redundancy mechanisms and achieve efficacy through their partial selectivity with a number of targets concurrently. ASOs are designed in a selective target-focused manner based on the probability that in the human genome, any sequence longer than a minimal number of nucleotides (13 nt for RNA) occurs only once [40]. Hence, it is preferable for ASOs to be directed against clinically validated targets for a monogenic disease. Furthermore, it is favorable to not directly compete with small molecule drugs for the same drug target – a direction that is achieved by working outside of the druggable genome on novel mechanisms (unvalidated targets). Particularly favored are targets that are genetically validated but classified as undruggable. Gene products that are impaired or unfunctional as a result of alternations in their DNA sequence (e.g. mutations, translocations) or are misexpressed due to an altered epigenetic profile are genetically modified targets. If such modifications are proven to cause or be a high risk factor for a selection of patients in a particular indication, the target can then be referred to as a genetically validated target for that indication. Here it is crucial that the gene or gene product is the main contributor to the disease pathology and contributes to the disease process at the time of treatment. The Online Mendelian Inheritance in Man (OMIM) catalog of human genes and genetic disorders and trait is a rich source of such genetic mutation–disease associations. To put this into the context of present ASO candidate drugs, we found that a substantial number of ASOs in late-stage clinical trials for which target inhibition in man has been shown with targeted gene products, which are outside of the classical druggable genome, i.e. enzymes, transporters, and receptors [41]. Therefore, prior to the onset of the trials, the majority of these targets were clinically unvalidated. Notably, only three of the twenty ASO targets examined were genetically validated [41].

In contrast to monogenic diseases, for treatment of multifactorial diseases, the use of an ASO as a monotherapy targeting a single gene is unlikely to be successful. Consequently, for the treatment of these complex diseases, an ASO can be used in combination with either established therapies (i.e. small molecules and biologics) or additional ASOs targeting alternative pathology-relevant genes. For example, in cancer treatment, the use of ASOs as monotherapies targeting single genes has shown little success. In a number of examples, target suppression was observed; however patient responses were not [41]. As a consequence, clinical stage ASOs are now being used in combination with conventional therapies (chemotherapy, radiotherapy). Conventional cancer treatment response can be limited as many of the strategies used to kill cancer cells induce stress responses that activate survival pathways, resulting in the development of a treatment-resistant phenotype. Such resistance has been attributed to alterations in the expression of genes that regulate multiple cellular pathways involving apoptosis, inflammation, angiogenesis, and drug efflux pumps [42]. In advanced prostate cancer, for example, treatment resistance manifested by progression to castration-resistant prostate cancer (CRPC) is attributed to the reactivation of alternative mitogenic growth factor pathways, stress-induced prosurvival genes, and cytoprotective chaperone networks, to name a few [43]. ASOs targeting key members of the cancer treatment resistance cohort have the potential to partially eradicate redundancy mechanisms that cancer cells have evolved during treatment, enhancing the therapeutic response. Several ASOs in combination with chemotherapeutic agents are present in mid-stage clinical trials. Custirsen (OGX-011) and OGX-427, which target the stress-activated cytoprotective chaperone proteins, clusterin (CLU) and heat shock protein 27 (HSP-27), respectively, are examples of such ASOs. CLU and HSP-27 are both known to be associated with treatment resistance in a variety of cancers, protecting cells from numerous therapeutic stressors that induce apoptosis, including androgen or estrogen withdrawal, radiation, cytotoxic chemotherapy, and biologic agents [44]. However even as part of a combination treatment, ASOs have shown a general lack of robust effects on disease outcomes [45, 46]. This may be partially due to insufficient ASO delivery, resulting in incomplete target knockdown and/or inappropriate targets. The continuous application of functional genomics and proteomics to the field of cancer biology will facilitate the identification of new targets for tumor cell sensitization to standard therapy, as well as hopefully gain mechanistic insight into parameters influencing tumor-type specific uptake of ASOs.

An example of a clinical stage, solely ASO-based combination therapy is TPI ASM8. The TPI ASM8 investigational medicinal product comprises two first-generation 19-mer phosphorothioate oligodeoxynucleotides (adenines in the sequence are replaced with 20-amino-20-deoxyadenine) that are directed against the C–C chemokine receptor type 3 (CCR3) and beta chain as a treatment for asthma [12, 47]. TPI ASM8 is designed to attenuate the allergic inflammation response by targeting CCR3 and the beta chain of IL-3, IL-5, and granulocyte–macrophage colony-stimulating factor receptors, factors that possibly play redundant roles in the disease pathology of asthma.

4.3.2 Mechanisms of Action

Binding of an ASO to a target RNA in the nuclear and/or cytoplasmic compartments of a cell can affect the target function through a variety of mechanisms. These include induced cleavage of mRNAs via RNase H, modulation of splicing, and blockage of target RNA function through sequestration. The mechanism of action of an ASO is predominately directed by the chemical format of the ASO as well as its site of hybridization on the target (the criterion for which is extensive and is covered elsewhere [48]). ASOs also exert biological effects that are independent of their binding to complementary target RNAs [39, 46]. For example, ASOs containing G-quartets can interact with proteins. CpG-containing ASOs can activate immune cells, acting as potent TLR9 agonists, and phosphorothioates in general can bind nonspecifically to numerous proteins involved in the clotting and complement system and protein phosphorylation. In addition, oligonucleotides are metabolized to nucleotides, nucleosides, and bases, all of which can potentially display pharmacological activity [40, 49].

The time till onset of ASO action is defined as the time from ASO administration until a specific response can be observed. It is governed by both the PK and the PD properties of the ASO [50]. When considering the PK aspects of the onset of action, one must not only incorporate the kinetics of ASO distribution from the blood to the tissue but also other factors important for its transit to the target molecule such as target tissue composition, cell surface protein and extracellular matrix binding, endosomal capture and release, and transport from the cytoplasmic to the nuclear compartments [50]. The dependence on some of these processes likely depends on the target tissue and, in the case of cancer, the type of tumor. Upon reaching its target, the time taken for the transition between ASO–target interaction and downstream effects becomes important. In terms of mRNA target modulation, ASOs do not regulate the half-life of an expressed protein; instead they modulate the expression of the protein at the mRNA level. Hence the onset of action of these ASOs is highly dependent on the protein half-life and/or rate of translation of the target protein. For example, in terms of mRNA target inhibition, a rapid onset of action is expected when the mRNA is slowly translated and when the translated protein is rapidly degraded. On the contrary, the kinetics of the onset of action is limited by the protein half-life when the protein is rapidly translated and/or slowly degraded [51, 52]. In the case of the latter, an effect at the protein level may not be observed at all, or it may result in a significant delay between the oligonucleotide reaching the target mRNA and a detectable pharmacological effect. The turnover kinetics of many proteins is unknown – as a result the onset of action of an ASO may be variable or unpredictable.

A detailed understanding of the time until onset of action of an ASO prior to its use in patients is particularly important for appropriate sampling in clinical PD/biomarker assessments (see Section 4.3.3). Direct relationships between the ASO concentration in tissue and target inhibition have been reported [25, 53] and used to estimate protein half-lives [25]. For example, the peak tissue concentration of LY2181308 and maximum target inhibition of survivin were both detected at 24 h after a single dose. Consequently, for survivin, turnover was predicted to be rapid, and the protein half-life was estimated to be 2 h [25]. As discussed earlier in this chapter, modified ASOs to increase ASO serum stability, loading doses to reach efficacious doses faster, and multiple dosing regimens to maintain the drug at the site of action are possible measures to facilitate optimal onset of target modulation and downstream effects.

4.3.3 Biomarkers and PD Endpoints

For proof of concept of a novel therapeutic modality or an unvalidated target, as well as to establish the PK/PD relationship, it is vital that the ASO can also be shown to reach and alter the target it is directed against in the patients. Therefore, the availability of markers to assess activity and therapeutic response is an important aspect in target selection.

In patient samples, the effect of an ASO upon its respective target is analyzed by various methods including quantitative real-time PCR for the analysis of target mRNA and immunohistochemistry (IHC), immunofluorescence, Western blotting, and flow cytometry-based methods for the corresponding protein [41, 54]. In cases in which the target is secreted into body fluids (i.e. the blood), monitoring the effects of the ASO on the target is relatively straightforward and can facilitate evaluations of the PD effect. However, in a number of cases, particularly in cancer, target modulation and downstream effects in the relevant cell types in patients can only be detected through invasive interference (biopsies), the feasibility of which is highly dependent on the location of the target. Additional limitations for the use of biopsies (limited scope for repeated sampling, sample heterogeneity, etc.) have led to the increasing usage of normal surrogate tissues as replacements [55]. For example, in a phase I oncology trial of custirsen, an ASO targeting CLU, peripheral blood mononuclear cells (PBMCs) were used as surrogate tissues for PD and PK investigations alongside standard tumor biopsies [56]. Clearly there are limitations for the use of surrogate tissues, for example, differences in drug concentrations between the surrogate and the diseased tissue, as well as the lack of the relevant molecular pathology [55]. As a result a direct correlation between the effect observed in the target tissue and that of the surrogate tissue and/or the ability to link changes in these markers to clinical responses cannot be expected [55].

However, to aid identification of the most appropriate surrogate tissue, attempts should be made to establish that the surrogate tissue reflects the ASO tissue concentration, the target half-life, the molecular PD response, and the therapeutic response of the target tissue.

4.4 PD and PK Results and Strategies of ASOs in Clinical Development

ASOs are known to exert biological effects that are independent from the desired antisense mechanism of action, potentially leading to off-target effects. Therefore it is vital that not only an ASO treatment results in the desired alternation in the disease state in patients but also can be shown to reach its target and act to alter the factor it is directed against in patients. We found that over half of the ASOs in late-stage clinical trials have been reported to suppress their targets at the mRNAs and/or the protein level in the relevant cell types in patients (Table 4.2, [41]), including those present in the liver, muscle fibers, various solid tumors, lymph nodes, bone marrow, intestinal mucosa, and eosinophil progenitors and blood cells [41]. In the majority of the cases, the drug was delivered i.v. or s.c. Here, it can be safely assumed that the ASOs are being delivered to their cellular targets and are functioning via an antisense mechanism; however it cannot be assumed that the target is valid or that the extent and duration of target downregulation is sufficient to mediate the desired therapeutic effect.

Linking pharmacokinetic data for a specific ASO compound to its pharmacodynamic data is important in guiding the first dose into man and the dose and schedule optimization in later-stage clinical development. For an assessment of this PK/PD relationship knowledge of the ASO concentration in plasma and target tissue, tissue half-life and concentration needed to achieve relevant target inhibition, the level of target inhibition to achieve therapeutic effect and ideally in the case of target degrading mechanisms also the turnover rate or half-life of the target protein or mRNA is required. Although PK/PD modeling and simulation is in its early stages and has not been applied widely for oligonucleotides, there are several reports on the use of integrated assessments of PK and PD in preclinical and/or clinical stages of development [9, 17, 19, 23, 25, 50, 56].

Below we will summarize strategies for and results of selected PK and PD investigations of ASOs, with a focus on clinical stage, second-generation ASOs (miravirsen, drisapersen, LY2181308, mipomersen, and custirsen). Where applicable (and where reports are available), integrated assessments of PK/PD relationships used to aid dosing and dosing schedules are also described.

Table 4.2 ASOs recently under active development in phase II clinical trials or above demonstrating target alternations in patients.

Entry	Target	Drug (indication)	References to target inhibition in man
1	Apolipoprotein B-100 (ApoB): Nonsense or frameshift mutations in the LDL receptor binding domain of ApoB cause hypercholesterolemia (OMIM 107730)	Mipomersen (preregistered: FH): 2′-MOE ASO administered i.v. or s.c. reduces target protein and LDL cholesterol in serum	[57–60]
2	Dystrophin (DMD): Mutations induce a frameshift or nonsense residue and produce dysfunctional protein (OMIM 300377)	Eteplirsen (phase II: DMD): Morpholino ASO administered i.v. causes exon skipping and restores dystrophin levels in muscle biopsies	[61]
3	Dystrophin (DMD): Mutations induce a frameshift or nonsense residue and produce dysfunctional protein	Drisapersen (phase III: DMD): 2′-OMe ASO administered s.c. causes exon skipping and restores dystrophin levels in muscle biopsies of patients	[10, 62]
4	miR-122 (MIR122): MicroRNA-122 is necessary for the translation of HCV genome	Miravirsen (phase II: hepatitis C): LNA AMO administered s.c. reduces viral RNA in the blood of HCV patients	[63, 64]
5	Clusterin (CLU): Secreted stress-induced cytoprotective protein, associated with chemotherapy or radiotherapy resistance	Custirsen (phase III: cancers): 2′-MOE ASO administered i.v. suppresses target in prostate cancer tissue and in lymph nodes, increases apoptotic index, and improves survival	[56, 65, 66]
6	Survivin (BIRC5): Inhibitor of cancer cell apoptosis, highly expressed in tumors associated with chemotherapy or radiotherapy resistance	ISIS-23722 (phase II: cancer): 2′-MOE ASO administered i.v. suppresses target in tumor tissue, restores apoptosis in tumor cells	[19]
7	B-cell lymphoma-2 (BCL2): Inhibitor of cancer cell apoptosis associated with chemotherapy or radiotherapy resistance	Oblimersen (phase III: cancer) PTO administered i.v. continuous infusion inhibits target levels in PBMCs and bone marrow cells	[67–69]

8	Heat shock 27 kDa protein 1 (HSPB1): Chaperone protein associated with chemotherapy or radiotherapy resistance	OGX-427 (phase II: cancer): 2′-MOE ASO administered i.v. induces changes in tumor markers, measurable disease, and circulating tumor counts	[44]
9	Eukaryotic translation initiation factor 4E (EIF4E): Protein translation factor	ISIS-EIF4ERx (phase II: cancer) 2′-MOE ASO administered i.v. inhibits eIF-4E mRNA and protein expression, as well as downstream markers in tumor biopsies	[45]
10	Ribonucleoside-diphosphate reductase M2 chain (RRM2): Essential protein for synthesis of deoxyribonucleotides	GTI-2040 (phase II: cancer): PTO administered i.v. with cytarabine in AML patients: reduces RRM2 protein in bone marrow cells, in complete responders and not in nonresponders	[70, 71]
11	CCR3, beta chain (CCR3, CSF2RB): Chemokine signaling proteins mediate eosinophil trafficking in asthma	TPI ASM8 (phase II: allergic asthma): Combination of two PTOs targeting IL5 and the beta chain suppress target mRNAs when delivered via nebulizer	[12, 47]
12	ICAM1 (ICAM1): Glycoprotein involved in cell trafficking in inflammatory bowel pathophysiology	Alicaforsen (phase III: pouchitis): PTO administered i.v. inhibits ICAM-1 protein expression in mucosal biopsies	[72]
13	C-reactive protein (CRP): Elevated levels of CRP are a biomarker inflammation	ISIS-CRPRx (phase II: rheumatoid arthritis): 2′-MOE ASO administered s.c. reduces levels of CRP by 76.7% in patients with active rheumatoid arthritis	[73]

Data from Lightfoot and Hall [41].
OMIM: "Online Mendelian Inheritance in Man" catalog of human genes and genetic disorders and trait

4.4.1 Genetic Diseases

4.4.1.1 Mipomersen, Apolipoprotein B-100, and Hypercholesterolemia

Mipomersen, a subcutaneously delivered 20-mer 2′-MOE ASO targeting apolipoprotein B-100 (ApoB) mRNA, has recently been approved as a treatment for hypercholesterolemia. The liver-derived ApoB protein plays a fundamental role in cholesterol homeostasis as a structural component of very low-density lipoprotein (VLDL) and as a ligand for LDL receptor-mediated endocytosis of LDL by liver cells. Its overproduction and decreased clearance are seen in cardiovascular-associated diseases, as well as inherited diseases such as familial hypercholesterolemia (FH) (OMIM 107730). Familial defective ApoB hypercholesterolemia is a result of nonsense or frameshift mutations within the LDL receptor binding domain of ApoB and causes severe hypercholesterolemia and premature cardiovascular disease. ApoB is an ideal target for an mRNA-targeting oligonucleotide in FH, as it is not only a genetically validated target but it is also synthesized in the liver, one of the predominant target tissues of ASOs, where high ASO concentrations can be achieved. Furthermore, the target proteins, as well as associated factors such as LDL-C and triglycerides, are secreted into the blood in real time, allowing their use as biomarkers for PD analysis and as a result facilitating integrated PK/PD assessments of drugs targeting ApoB [74].

As with all ASOs, ApoB-targeting ASO concentrations cannot be easily measured in patient livers. As a substitute the relationship between the plasma trough concentrations and the liver concentrations of ApoB-targeting ASOs in preclinical models was utilized to estimate mipomersen liver concentrations in patients and subsequently facilitate patient PK/PD analysis. Preclinical studies demonstrated that the partition ratios of the plasma trough concentrations and the liver concentrations were similar in both mouse and monkey models [9]. Assuming that a similar partition ratio may apply for humans, the relationship between the plasma trough levels of mipomersen and the serum ApoB levels in healthy volunteers was determined and the trough plasma concentration reflecting 50% of the maximum effect (EC_{50}) estimated (employing a sigmoidal inhibitory effect Emax model) [17]. The clinical and preclinical data were used to estimate the ASO liver concentrations in man and subsequently the liver EC_{50}.

This first-in-man clinical trial demonstrated a rapid, dose-dependent long-term reduction of circulating ApoB and LDL-C of up to 50 and 35%, respectively, in volunteers upon treatment with mipomersen (50–400 mg) once weekly for 3 weeks following a loading dose. A significant reduction in serum ApoB, with a well-correlated reduction in LDL-C, was observed in healthy volunteers by the end of the second week. Maximal effects were observed at or just following the last dose consistent with the highest Cmin. Mipomersen displayed an elimination half-life of 31 days in the plasma. The prolonged pharmacodynamic effects were consistent with the long tissue half-life, for example, significant reductions of serum ApoB levels from baseline for the 200-mg treatment cohort were observed up to 75 days after the last dose.

The estimated trough plasma EC_{50} was 14 ng/ml, which was comparable with that obtained in mice (18 ng/ml). This equilibrium plasma concentration corresponded to an estimated liver exposure of approximately 80 µg/g in man based on the ratio between plasma and liver concentrations in monkeys and mice. Overall, the PK/PD data from this phase I study was found to be well predicted by the preclinical data.

In this example, the pharmacokinetic and pharmacodynamic data from both preclinical studies and first-in-man clinical trial has been further utilized to investigate the PK/PD relationship of mipomersen. The strategy and model applied accurately predicted the plasma trough concentrations of mipomersen reflecting sufficient levels in the liver to achieve an effect in man and were used to support dose selection in clinical development. Numerous additional clinical trials followed where mipomersen was tested as a single agent and in combination with statins. All demonstrated robust reductions in ApoB and when tested LDL-C, Lp(a), and triglycerides [57–60]. At present mipomersen is the only systemically delivered ASO on the market.

4.4.1.2 Drisapersen, Dystrophin, and Duchenne Muscular Dystrophy (DMD)

Drisapersen is a 2′-*O*-methyl-phosphorothioate RNA designed to hybridize to a 20 nt sequence in exon 51 of dystrophin pre-mRNA as a treatment for Duchenne muscular dystrophy (DMD). Dystrophin is essential for the normal function of skeletal and cardiac muscle. Its gene is prone to deletions that disrupt the reading frame and produce a dysfunctional protein, resulting in muscular dystrophy (OMIM: 300377). For example, an out-of-frame transcript and a premature stop codon within exon 51 of dystrophin mRNA are generated on deletion of exon 50 and the subsequent splicing of exon 49 to exon 51. Consequently, a nonfunctional protein is produced, which is rapidly degraded. The binding of drisapersen to dystrophin pre-mRNA masks the exon 51-inclusion signal, resulting in exon 51 skipping and accumulation of a shortened but nevertheless partially functional protein. This concept has been validated in numerous *in vivo* contexts [75].

The first clinical investigation of drisapersen focused on determining the feasibility of this novel therapeutic intervention in patients. Four DMD patients bearing genetic deletions correctable by exon 51 skipping received a single dose of drisapersen via local administration into the tibialis anterior muscle [62]. Biopsies of treated muscles from each patient were assessed for mRNA splicing and dystrophin expression. Each patient biopsy demonstrated specific skipping of exon 51. Dystrophin restoration in 64–97% of myofibers and to levels ranging from 17 to 35% of that of the control samples was observed, validating the rationale in the clinic.

To aid systemic delivery of the dystrophin-targeting ASO, more information on the PK/PD profile of this compound was needed. The PK/PD relationship was assessed in several studies in the *mdx* mouse model of DMD and in healthy mice with a mouse-specific ASO [23]. Here, following i.v., s.c., i.p., and i.m. ASO administration, the plasma PK, levels and half-lives of the ASO in muscle tissues, and the PK/PD relationship were compared. The PD parameters analyzed were exon 23 skipping, dystrophin protein expression, and, as a surrogate marker, the levels of plasma creatine kinase (CK). A functional analysis was also conducted with the *rotarod* assay.

ASO concentrations and tissue half-lives were estimated in dystrophin-deficient skeletal and heart muscle fibers of the *mdx* mouse model and in the same tissues in healthy mice, as well as in the liver and kidney of both models. Interestingly the tissue concentration of ASO was found to be different in the various muscle fibers analyzed. In dystrophin-deficient muscle fibers, the ASO concentrations were considerably higher than in the corresponding fiber tissue from healthy mice (~10 μg/g and 4 and 8 times higher in the gastrocnemius and triceps). This was reflected by higher levels of exon skipping in the disease model. This suggested that the pathomechanism of DMD increased ASO biodistribution to the affected muscle tissues, possibly due to an increased permeability of the diseased tissues. The ASO tissue half-lives differed between the different types of muscle tissues with values ranging from 10 days for the gastrocnemius and quadriceps, 33 days for triceps, and 46 days for cardiac muscle. Muscle tissue exposure was comparable between the different routes of delivery. However higher ASO exposure levels were observed in the liver and the kidney after i.v. administration (approximately twofold higher compared with s.c.). Surprisingly, the greatest PD effect in most muscle tissues was observed after i.v. administration. Dose-corrected plasma PK data indicated no accumulation at C_{max} after 2 weekly doses for 4–8 weeks; in contrast, trough levels increased with corresponding increases in tissue concentrations and PD effects (exon skipping and dystrophin levels). The PD effect was echoed by decreased CK levels and also on the functional level through improvements in the *rotarod* assay.

The second larger clinical phase I/IIa study [10] was designed to assess the safety, PK, the PD effects at the molecular level, and also the clinical outcomes following s.c. administration in 12 patients. This trial involved an initial dose-escalation stage consisting of weekly abdominal s.c. delivery of drisapersen for 5 weeks at doses ranging from 0.5 to 6 mg/kg. This was followed by a 12-week extension phase in which all patients received a dose of 6 mg/kg per week. Muscle biopsies were analyzed for mRNA splicing and dystrophin expression at the end of the dose-escalation phase. PK results indicated increasing plasma trough levels with repeated dosing, suggesting an increase in drisapersen tissue levels. The terminal plasma half-life (19–56 days) was in the range of other ASOs. Exon skipping was observed in all patients in the two highest dose cohorts, 4 and 6 mg/kg, and in one patient in the 2 mg/kg dose cohort. New dystrophin expression was observed in approximately 60–100% of muscle fibers in 10 of the 12 patients, and dystrophin expression restoration up to 15.6% of that found in healthy muscle was also detected. Furthermore, an improvement in the 6-min walk test performed after the 12-week extension phase provided a first proof of concept. However, in a later phase III trial, the primary endpoint for the 6-min walk test was not met [76].

4.4.2 Infectious Diseases

4.4.2.1 Miravirsen, miR-122, and Hepatitis C Virus (HCV)

Miravirsen is a 15-mer LNA-modified ASO that is designed to bind to the 5′ region of miR-122 and inactivate the host liver-specific miRNA miR-122 by sequestration for anti-hepatitis C virus (HCV) intervention. In addition to its role as a regulator of cholesterol and fatty acid metabolism, miR-122 interacts at two closely spaced target sites in the 5′ terminus of the HCV RNA, preventing nucleolytic degradation of the HCV genome, as well as inhibiting innate immune system activation. Overall this results in an increased abundance of viral RNA [77–80].

The link between anti-miR-122 effects and therapeutic response (HCV load decline) was assessed as a proof of concept in a primate model of chronic HPV injection [8]. The animals received either 1 or 5 mg/kg of i.v. administrated miravirsen weekly for 12 weeks. Both the presence of miravirsen and its effect on miR-122 were monitored in the liver by Northern blotting. The effect of miravirsen on HCV RNA levels in serum and in liver biopsies was also tested alongside cholesterol levels and interferon-responsive gene expression. In addition, the plasma PK was also assessed. The presence of miravirsen and its interaction with miR-122 in the target cells were confirmed, and, as expected, miravirsen–miR-122 complex formation resulted in undetectable miR-122 levels. Furthermore, a decrease in cholesterol levels, reflecting the sequestration of miR-122, was observed in the high-dose group, affirming cholesterol as a downstream marker for miR-122 inhibition. Viral load was significantly decreased at the highest dose in both serum and the liver. The plasma PK data reported was in the range of data obtained from reports of other phosphorothioated ASO, e.g. the terminal elimination half-life was approximately 21 days.

A phase I study of miravirsen performed in healthy male volunteers with doses of up to 12 mg/kg demonstrated dose-dependent pharmacology based on the level of cholesterol in the plasma [81]. This was followed by a phase IIa study to assess the antiviral activity and safety of miravirsen in treatment-naive HCV patients [64]. Miravirsen at doses of up to 7 mg/kg was administered weekly over a 29-day period to patients with chronic HCV (genotype 1) infection by s.c. injection. The plasma HCV RNA levels were monitored, and long-lasting dose-dependent reductions in HCV RNA, with a mean decrease of 2–3 logs from baseline HCV RNA levels ($\log_{10}$ IU/ml), were observed upon miravirsen treatment. In four out of nine patients in the high-dose cohort, HCV RNA became undetectable. Escape mutations were not detected in the miR-122 binding site of HCV RNA, indicating that therapy resistance might not occur easily. This clinical study demonstrated that blocking miR-122 through once-weekly injections of a complementary LNA ASO is a feasible approach for the treatment of chronic HCV infections and therefore clinically validates the target.

4.4.3 Cancer

4.4.3.1 Custirsen, Clusterin, and Cancer

Custirsen (OGX-011), a 21-mer 2′-MOE ASO targeting the CLU mRNA translation initiation site, is presently in phase III clinical studies for cancer. Its target CLU is a stress-induced cytoprotective chaperone protein that is expressed in most tissues. In cancer CLU inhibits apoptosis through its suppression of the proapoptotic factor BAX and activates the PI3K/AKT cell survival pathway. CLU levels are increased in a variety of cancers displaying resistance to radiation and chemotherapy. Furthermore, inhibition of CLU enhances apoptosis in xenograft models of cancer. The target is not yet clinically validated [43].

The first-in-man study of custirsen was performed in 25 patients with localized prostate cancer [56]. Subjects were treated with custirsen and androgen ablation therapy prior to radical prostatectomy. The aim of this study was to determine PK parameters and the correlating PD effects and in turn identify the optimal biological dose of the ASO for clinical phase II investigations. Custirsen was administered as a 2-h i.v. infusion with initial loading doses followed by weekly dosing of 40–640 mg for 4 weeks. The starting dose used was 40 mg, as this was one-sixth of the "no observed effect level" dose in monkeys. Notably, xenograft mouse models predicted that efficacy in terms of tumor growth reduction could be detected at doses of 320 mg and higher [82]. Radical prostatectomy was performed within 7 days of the last dose, and custirsen plasma and prostate tumor tissue concentrations were analyzed. CLU expression in the prostate was assayed before and after drug treatment (prostatectomy), as well as in the lymph nodes and PBMCs (surrogate tissues). Plasma pharmacokinetics were similar to that observed with other antisense molecules. At the time of prostatectomy, mean tissue concentrations of ASO in the 320-, 480-, and 640-mg dose cohorts were 1.67-, 2.29-, and 4.82 μg/g prostate tissue, respectively. Q-PCR showed efficient dose-dependent inhibition of CLU mRNA, resulting in greater than 90% inhibition at the maximum dose of 640 mg. Furthermore, IHC demonstrated dose-dependent inhibition of CLU protein with complete suppression in 51.2 and 56.7% of cells in the prostate tumor samples at doses of 480 and 640 mg, respectably. However, it is important to note that as androgen ablation therapy increases CLU expression, the effect of custirsen on the expression levels of CLU is likely underestimated in this study. Regardless, the clinical exposure–response relationship is in good agreement with preclinical model data where tumor tissue concentrations of approximately 4–5 μg/g produced marked reductions of CLU expression by approximately 80% in a xenograft tumor [50]. In the surrogate tissues, dose-dependent decreases in CLU expression, similar to the prostatectomy specimens, were noted, but due to high variations, dose–effect correlations were not possible. Finally, the apoptotic index, a parameter used as an indicator for increased apoptosis, was also increased at the 640-mg dose in the prostatectomy specimens. As a result of this study, an optimal biologic dose of 640 mg was determined for custirsen and carried forward into alternate phase I as well as phase II clinical trials for various cancers in combination with other chemotherapies [65, 66, 83–85]. These trials have shown promise for patients in terms of median survival or time to disease progression and support further evaluation in clinical phase III trials.

4.4.3.2 LY2181308 (ISIS-23722), Survivin, and Cancer

LY2181308 (ISIS-23722), an 18-mer 2′-MOE-ASO targeting survivin, was recently investigated in clinical trials for cancer therapy. Survivin is a member of the inhibitors of apoptosis (IAP) family. It functions as a stress-induced cytoprotective chaperone that inhibits apoptosis through its direct interaction with the second mitochondrial activator of caspases (SMAC). It is highly expressed in many human cancer types; however it is not expressed in differentiated tissues [86]. The target is not yet clinically validated.

For LY2181308 an integrated assessment of preclinical pharmacodynamic and pharmacokinetic data to guide the design of clinical treatment has been reported. As part of the development strategy, a PK/PD model was developed to predict the dose and dosing schedule for a first-in-human trial. The model was verified with the results from this phase I study and modified accordingly for its use in LY2181308 dosing simulation, facilitating clinical development [19, 25].

To predict doses that can achieve ASO concentrations in tumors that will lead to appropriate survivin downregulation, animal models were employed. For an estimation of the half maximal effective concentration for target inhibition (EC_{50}), the resulting tumor growth inhibition as well as an estimate of ASO concentrations in tumor tissue, a xenograft mouse model was utilized. For the PK simulation, which required tissue exposure, the tissue concentrations detected in monkey tissues were used. Here, the authors distinguished between low uptake tissues (i.e. lung and muscle) and high uptake tissues (i.e. kidney and liver), and as ASOs are known to be poorly distributed to tumors, ASO concentrations observed in the low uptake tissues were used as tumor surrogates. Human plasma PK parameters were predicted based on PK data derived from the studies in animals applying allometric scaling across the species based on body weight. The prediction capacity of this approach for determining ASO concentrations in human tumor tissue was tested by measuring ASO concentrations in clinical tumor biopsy samples. Based on these results an appropriate dosing regimen could be selected.

The plasma PK parameters and ASO tumor concentrations in patients were well predicted by the PK/PD model. For example, the desired target tumor concentrations in patients were achieved (13.9–52.8 µg/g, $n = 4$) on administration of the highest tolerable dose (750 mg) once weekly. However, these tumor concentrations, which were predicted to result in 50–90% target inhibition at an EC_{50} of 20 µg/g, only lead to reductions of 20 and 23% in survivin mRNA and protein, respectively, in this trial.

The underlying causes for the overestimation of the ASO PD effect in cancer patients remain unclear. This discrepancy may partially be explained by the fact that the simulation model was based on PD data from a single study, applying a single dose in a xenograft mouse model. Differences in cellular uptake and intracellular transport mechanisms among species and among different cell types may be partially responsible. For improved PK/PD models and predictions of clinical dosing, a more detailed insight into these processes will be required.

4.5 Summary and Conclusions

With the recent approval of mipomersen as the first in class of systemically delivered ASOs, oligonucleotide drugs have demonstrated their potential as a compound class. This has revitalized the discovery and development of ASOs representing a new paradigm in the field. The development strategy of mipomersen has also paved the way for other ASO compounds to a certain extent, for example, its investigation and application of the PK/PD relationship.

Emerging data, not only from mipomersen, has confirmed that ASOs can be delivered to many organs/tissues at concentrations sufficient to inhibit their RNA targets and to observe therapeutic effects in man. With development of the second-generation chemistries, several examples have proven that it is sufficient to administer ASO doses weekly, reflecting their greater stability, which, together with their added potency, is a tribute to extensive research on the modification of the ribose by medicinal chemists. Regardless, the failure rate of ASOs is still high. Although detailed information concerning the reasons for failure is often not publicly available, insufficient target inhibition, inappropriate patient selection, and the selection of unvalidated targets will likely be responsible for some cases.

The selection of the target tissue and target/indication pair is make or break for the success of an ASO drug. ASOs are rapidly and widely distributed to tissues, and the highest concentrations were always detected in the kidney and liver. Due to this selective intrinsic biodistribution pattern of ASOs, it is favorable for ASOs to be designed to target factors that are relevant to the tissues in which ASOs accumulate, e.g. targets in liver such as ApoB-100 and miRNA-122. Surprisingly, diseased muscle in DMD has also been identified as a tissue in which sufficient ASO levels can be achieved. This is likely due to the increased permeability of the affected tissues.

With regard to the targets, in principle, oligonucleotides can be targeted to any gene of the genome. However, in reality, a number of factors come into consideration when selecting the best targets for ASOs. The validation status of the target for the selected indication is important. ASOs will likely be most successful when targeting genes that are already clinically or genetically validated for an indication. Here the target has already been proven to be a driver in the disease pathology, and in the case of clinical validation, clinical responses on target modulation have been established. The use of genetically validated targets in later-stage clinical trials is uncommon. As competition from small molecules and alternate biologics is detrimental to ASO success, ASOs have been preferably designed against clinically nonvalidated targets and hence can be defined as high risk. In an alternate direction, ASOs have also shown promise in combination with chemotherapeutics in cancer therapy, e.g. as tumor chemosensitizers, to increase efficacy and/or to treat patients who have become resistant to treatment. An additional important factor that should be considered when selecting a target/indication pair is the accessibility of tissue biopsies for PD assessment or the existence of biomarkers in the blood.

The strategies reported for assessments of the PK, the PD, and the integrated PK/PD relationship are dependent on the target, the target tissue, and the availability of biopsies and markers. The most straightforward approach is to analyze the extent of mRNA alteration, effects on related downstream markers, and the corresponding ASO tissue concentration in animal models. Patient biopsies can be performed if feasible from an ethical point of view; however in many cases, this is impossible and alternate surrogate patient tissues/measures are employed. For example, for mipomersen the plasma trough level was applied as a surrogate measure to estimate the ASO liver exposure at steady state, removing the need for patient biopsies. In preclinical models trough ASO levels were correlated with liver ASO levels, PD effects on ApoB-100 mRNA, and downstream effects. An integrated PK/PD analysis facilitated the prediction and optimization of clinical dosing, as well as the confirmation of the mode of action in patients. This approach has set the ground for the development of other ASOs, with particular relevance to those ASOs that address liver targets.

For cancer indications PK/PD estimation and assessment in patients is a major challenge. There is still not sufficient information on the extent which animal models of cancer, either xenograft or orthotopic, reflect patient exposure. In addition, analysis of the PD effect can be quite limited. In studies of LY2181308, cancer tissue exposure in patients was simulated on the basis of concentrations in nontumor tissues of animals. The authors distinguished between low and high uptake tissues and used the levels in the low uptake tissues as a surrogate for tumors. The tumor concentration was predicted very well; however the degree of target downregulation was significantly lower than predicted, indicating that the model and information obtained did not accurately reflect the human context.

Few reports are available on the use of PK/PD modeling to predict human exposure. To establish more reliable PK/PD models, more research is required to identify the factors dictating ASO tissue/cellular uptake and intracellular trafficking in the models applied in preclinical research and in humans.

References

1 Geary, R.S., Watanabe, T.A., Truong, L. et al. (2001). Pharmacokinetic properties of 2′-*O*-(2-methoxyethyl)-modified oligonucleotide analogs in rats. *J. Pharmacol. Exp. Ther.* 296 (3): 890–897.

2 Dirin, M. and Winkler, J. (2013). Influence of diverse chemical modifications on the ADME characteristics and toxicology of antisense oligonucleotides. *Expert. Opin. Biol. Ther.* 13 (6): 875–888.

3 Watanabe, T.A., Geary, R.S., and Levin, A.A. (2006). Plasma protein binding of an antisense oligonucleotide targeting human ICAM-1 (ISIS 2302). *Oligonucleotides* 16 (2): 169–180.

4 Levin, A.A., Yu, R.Z., and Geary, R.S. (2007). Basic principles of the pharmacokinetics of antisense oligonucleotide drugs. In: *Antisense Drug Technology: Principles, Strategies and Applications*, 2e (ed. S.T. Crooke), 183–215. Boca Raton, FL: CRC Press Inc.

5 Yu, R.Z., Geary, R.S., and Levin, A.A. (2006). Pharmacokinetics and pharmacodynamics of antisense oligonucleotides. In: *Pharmacokinetics and Pharmacodynamics of Biotech Drugs: Principles and Case Studies in Drug Development* (ed. B. Meibohm), 93–119. Weinheim: Wiley-VCH Verlag, GmbH & Co. KGaA.

6 Lundin, K.E., Hojland, T., Hansen, B.R. et al. (2013). Biological activity and biotechnological aspects of locked nucleic acids. *Adv. Genet.* 82: 47–107.

7 Yu, R.Z., Kim, T.W., Hong, A. et al. (2007). Cross-species pharmacokinetic comparison from mouse to man of a second-generation antisense oligonucleotide, ISIS 301012, targeting human apolipoprotein B-100. *Drug Metab. Dispos.* 35 (3): 460–468.

8 Lanford, R.E., Hildebrandt-Eriksen, E.S., Petri, A. et al. (2010). Therapeutic silencing of microRNA-122 in primates with chronic hepatitis C virus infection. *Science* 327 (5962): 198–201.
9 Yu, R.Z., Lemonidis, K.M., Graham, M.J. et al. (2009). Cross-species comparison of *in vivo* PK/PD relationships for second-generation antisense oligonucleotides targeting apolipoprotein B-100. *Biochem. Pharmacol.* 77 (5): 910–919.
10 Goemans, N.M., Tulinius, M., van den Akker, J.T. et al. (2011). Systemic administration of PRO051 in Duchenne's muscular dystrophy. *N. Engl. J. Med.* 364 (16): 1513–1522.
11 Geary, R.S., Yu, R.Z., Watanabe, T. et al. (2003). Pharmacokinetics of a tumor necrosis factor-alpha phosphorothioate 2′-*O*-(2-methoxyethyl) modified antisense oligonucleotide: comparison across species. *Drug Metab. Dispos.* 31 (11): 1419–1428.
12 Imaoka, H., Campbell, H., Babirad, I. et al. (2011). TPI ASM8 reduces eosinophil progenitors in sputum after allergen challenge. *Clin. Exp. Allergy* 41 (12): 1740–1746.
13 Bogdahn, U., Hau, P., Stockhammer, G. et al. (2011). Targeted therapy for high-grade glioma with the TGF-beta2 inhibitor trabedersen: results of a randomized and controlled phase IIb study. *Neuro-Oncology* 13 (1): 132–142.
14 Miller, T.M., Pestronk, A., David, W. et al. (2013). An antisense oligonucleotide against SOD1 delivered intrathecally for patients with SOD1 familial amyotrophic lateral sclerosis: a phase 1, randomised, first-in-man study. *Lancet Neurol.* 12 (5): 435–442.
15 de Smet, M.D., Meenken, C.J., and van den Horn, G.J. (1999). Fomivirsen – a phosphorothioate oligonucleotide for the treatment of CMV retinitis. *Ocul. Immunol. Inflamm.* 7 (3–4): 189–198.
16 Leeds, J.M., Henry, S.P., Geary, R. et al. (2000). Comparison of the pharmacokinetics of subcutaneous and intravenous administration of a phosphorothioate oligodeoxynucleotide in cynomolgus monkeys. *Antisense Nucleic Acid Drug Dev.* 10 (6): 435–441.
17 Crooke, S.T. and Geary, R.S. (2013). Clinical pharmacological properties of mipomersen (Kynamro), a second generation antisense inhibitor of apolipoprotein B. *Br. J. Clin. Pharmacol.* 76 (2): 269–276.
18 Sewell, K.L., Geary, R.S., Baker, B.F. et al. (2002). Phase I trial of ISIS 104838, a 2′-methoxyethyl modified antisense oligonucleotide targeting tumor necrosis factor-alpha. *J. Pharmacol. Exp. Ther.* 303 (3): 1334–1343.
19 Talbot, D.C., Ranson, M., Davies, J. et al. (2010). Tumor survivin is downregulated by the antisense oligonucleotide LY2181308: a proof-of-concept, first-in-human dose study. *Clin. Cancer Res.* 16 (24): 6150–6158.
20 Hildebrandt-Eriksen, E.S., Aarup, V., Persson, R. et al. (2012). A locked nucleic acid oligonucleotide targeting microRNA 122 is well-tolerated in cynomolgus monkeys. *Nucleic Acid Ther.* 22 (3): 152–161.

21 Straarup, E.M., Fisker, N., Hedtjarn, M. et al. (2010). Short locked nucleic acid antisense oligonucleotides potently reduce apolipoprotein B mRNA and serum cholesterol in mice and non-human primates. *Nucleic Acids Res.* 38 (20): 7100–7111.

22 Laxton, C., Brady, K., Moschos, S. et al. (2011). Selection, optimization, and pharmacokinetic properties of a novel, potent antiviral locked nucleic acid-based antisense oligomer targeting hepatitis C virus internal ribosome entry site. *Antimicrob. Agents Chemother.* 55 (7): 3105–3114.

23 Heemskerk, H., de Winter, C., van Kuik, P. et al. (2010). Preclinical PK and PD studies on 2′-*O*-methyl-phosphorothioate RNA antisense oligonucleotides in the mdx mouse model. *Mol. Ther.* 18 (6): 1210–1217.

24 Geary, R.S. (2009). Antisense oligonucleotide pharmacokinetics and metabolism. *Expert Opin. Drug Metab. Toxicol.* 5 (4): 381–391.

25 Callies, S., Andre, V., Patel, B. et al. (2011). Integrated analysis of preclinical data to support the design of the first in man study of LY2181308, a second generation antisense oligonucleotide. *Br. J. Clin. Pharmacol.* 71 (3): 416–428.

26 Leeds, J.M., Graham, M.J., Truong, L., and Cummins, L.L. (1996). Quantitation of phosphorothioate oligonucleotides in human plasma. *Anal. Biochem.* 235 (1): 36–43.

27 Bigelow, J.C., Chrin, L.R., Mathews, L.A., and McCormick, J.J. (1990). High-performance liquid chromatographic analysis of phosphorothioate analogues of oligodeoxynucleotides in biological fluids. *J. Chromatogr.* 533: 133–140.

28 Yu, R.Z., Baker, B., Chappell, A. et al. (2002). Development of an ultrasensitive noncompetitive hybridization-ligation enzyme-linked immunosorbent assay for the determination of phosphorothioate oligodeoxynucleotide in plasma. *Anal. Biochem.* 304 (1): 19–25.

29 Hemsley, M., Ewles, M., and Goodwin, L. (2012). Development of a bioanalytical method for quantification of a 15-mer oligonucleotide at sub-ng/ml concentrations using LC-MS/MS. *Bioanalysis* 4 (12): 1457–1469.

30 Ewles, M., Goodwin, L., Schneider, A., and Rothhammer-Hampl, T. (2014). Quantification of oligonucleotides by LC-MS/MS: the challenges of quantifying a phosphorothioate oligonucleotide and multiple metabolites. *Bioanalysis* 6 (4): 447–464.

31 Arrowsmith, J. (2011). Trial watch: phase III and submission failures: 2007–2010. *Nat. Rev. Drug Discov.* 10 (2): 87.

32 Arrowsmith, J. (2011). Trial watch: phase II failures: 2008–2010. *Nat. Rev. Drug Discov.* 10 (5): 328–329.

33 Gashaw, I., Ellinghaus, P., Sommer, A., and Asadullah, K. (2011). What makes a good drug target? *Drug Discov. Today* 16 (23–24): 1037–1043.

34 Albright, C.F. (2010). Target selection in drug discovery. In: *Pharmaceutical Sciences Encyclopedia: Drug Discovery, Development, and Manufacturing*, 1–9. New York: John Wiley & Sons, Inc.

35 Knowles, J. and Gromo, G. (2003). A guide to drug discovery: target selection in drug discovery. *Nat. Rev. Drug Discov.* 2 (1): 63–69.

36 Lindsay, M.A. (2003). Target discovery. *Nat. Rev. Drug Discov.* 2 (10): 831–838.

37 Betz, U.A., Farquhar, R., and Ziegelbauer, K. (2005). Genomics: success or failure to deliver drug targets? *Curr. Opin. Chem. Biol.* 9 (4): 387–391.

38 Berggren, R., Moller, M., Moss, R. et al. (2012). Outlook for the next 5 years in drug innovation. *Nat. Rev. Drug Discov.* 11 (6): 435–436.

39 Jalencas, X. and Mestres, J. (2013). On the origins of drug polypharmacology. *Med. Chem. Commun.* 4: 80–87.

40 Mansoor, M. and Melendez, A.J. (2008). Advances in antisense oligonucleotide development for target identification, validation, and as novel therapeutics. *Gene Regul. Syst. Biol.* 2: 275–295.

41 Lightfoot, H.L. and Hall, J. (2012). Target mRNA inhibition by oligonucleotide drugs in man. *Nucleic Acids Res.* 40 (21): 10585–10595.

42 Mahon, K.L., Henshall, S.M., Sutherland, R.L., and Horvath, L.G. (2011). Pathways of chemotherapy resistance in castration-resistant prostate cancer. *Endocr. Relat. Cancer* 18 (4): R103–R123.

43 Zoubeidi, A., Chi, K., and Gleave, M. (2010). Targeting the cytoprotective chaperone, clusterin, for treatment of advanced cancer. *Clin. Cancer Res.* 16 (4): 1088–1093.

44 Zoubeidi, A. and Gleave, M. (2012). Small heat shock proteins in cancer therapy and prognosis. *Int. J. Biochem. Cell Biol.* 44 (10): 1646–1656.

45 Hong, D.S., Kurzrock, R., Oh, Y. et al. (2011). A phase 1 dose escalation, pharmacokinetic, and pharmacodynamic evaluation of eIF-4E antisense oligonucleotide LY2275796 in patients with advanced cancer. *Clin. Cancer Res.* 17 (20): 6582–6591.

46 Stadler, W.M., Desai, A.A., Quinn, D.I. et al. (2008). A phase I/II study of GTI-2040 and capecitabine in patients with renal cell carcinoma. *Cancer Chemother. Pharmacol.* 61 (4): 689–694.

47 Gauvreau, G.M., Pageau, R., Seguin, R. et al. (2011). Dose-response effects of TPI ASM8 in asthmatics after allergen. *Allergy* 66 (9): 1242–1248.

48 Chan, J.H., Lim, S., and Wong, W.S. (2006). Antisense oligonucleotides: from design to therapeutic application. *Clin. Exp. Pharmacol. Physiol.* 33 (5–6): 533–540.

49 Juliano, R.L., Alahari, S., Yoo, H. et al. (1999). Antisense pharmacodynamics: critical issues in the transport and delivery of antisense oligonucleotides. *Pharm. Res.* 16 (4): 494–502.

50 Siwkowski, A.A. (2008). Pharmacokinetic/pharmacodynamic properties of phosphorothiolate 2′-*O*-(2-methoxyethyl)-modified antisense oligonucleotides in animals and man. In: *Antisense Drug Technology: Principles, Strategies and Applications*, 2e (ed. S.T. Crooke), 305–321. Boca Raton, FL: CRC Press/Taylor & Francis Group.

51 Ramanathan, M., MacGregor, R.D., and Hunt, C.A. (1993). Predictions of effect for intracellular antisense oligodeoxyribonucleotides from a kinetic model. *Antisense Res. Dev.* 3 (1): 3–18.
52 Spiller, D.G., Giles, R.V., Broughton, C.M. et al. (1998). The influence of target protein half-life on the effectiveness of antisense oligonucleotide analog-mediated biologic responses. *Antisense Nucleic Acid Drug Dev.* 8 (4): 281–293.
53 Yu, R.Z., Zhang, H., Geary, R.S. et al. (2001). Pharmacokinetics and pharmacodynamics of an antisense phosphorothioate oligonucleotide targeting Fas mRNA in mice. *J. Pharmacol. Exp. Ther.* 296 (2): 388–395.
54 Olsen, A.L., Davies, J.M., Medley, L. et al. (2012). Quantitative analysis of survivin protein expression and its therapeutic depletion by an antisense oligonucleotide in human lung tumors. *Mol. Ther.–Nucleic Acids* 1: e30.
55 Ang, J.E., Kaye, S., and Banerji, U. (2012). Tissue-based approaches to study pharmacodynamic endpoints in early phase oncology clinical trials. *Curr. Drug Targets* 13 (12): 1525–1534.
56 Chi, K.N., Eisenhauer, E., Fazli, L. et al. (2005). A phase I pharmacokinetic and pharmacodynamic study of OGX-011, a 2′-methoxyethyl antisense oligonucleotide to clusterin, in patients with localized prostate cancer. *J. Natl. Cancer Inst.* 97 (17): 1287–1296.
57 Akdim, F., Stroes, E.S., Sijbrands, E.J. et al. (2010). Efficacy and safety of mipomersen, an antisense inhibitor of apolipoprotein B, in hypercholesterolemic subjects receiving stable statin therapy. *J. Am. Coll. Cardiol.* 55 (15): 1611–1618.
58 Akdim, F., Tribble, D.L., Flaim, J.D. et al. (2011). Efficacy of apolipoprotein B synthesis inhibition in subjects with mild-to-moderate hyperlipidaemia. *Eur. Heart J.* 32 (21): 2650–2659.
59 Visser, M.E., Wagener, G., Baker, B.F. et al. (2012). Mipomersen, an apolipoprotein B synthesis inhibitor, lowers low-density lipoprotein cholesterol in high-risk statin-intolerant patients: a randomized, double-blind, placebo-controlled trial. *Eur. Heart J.* 33 (9): 1142–1149.
60 Kastelein, J.J., Wedel, M.K., Baker, B.F. et al. (2006). Potent reduction of apolipoprotein B and low-density lipoprotein cholesterol by short-term administration of an antisense inhibitor of apolipoprotein B. *Circulation* 114 (16): 1729–1735.
61 Cirak, S., Arechavala-Gomeza, V., Guglieri, M. et al. (2011). Exon skipping and dystrophin restoration in patients with Duchenne muscular dystrophy after systemic phosphorodiamidate morpholino oligomer treatment: an open-label, phase 2, dose-escalation study. *Lancet* 378 (9791): 595–605.
62 van Deutekom, J.C., Janson, A.A., Ginjaar, I.B. et al. (2007). Local dystrophin restoration with antisense oligonucleotide PRO051. *N. Engl. J. Med.* 357 (26): 2677–2686.
63 Janssen, H.L., Reesink, H.W., Zeuzem, S. et al. (2012). Randomized, double-blind, placebo-controlled safety, anti-viral, proof of concept study of

miravirsen, an oligonucleotide targeting miR-122, in treatment-naive patients with genotype 1 chronic HCV infection. *Hepatology* 56: S26–S26.

64 Janssen, H.L., Reesink, H.W., Lawitz, E.J. et al. (2013). Treatment of HCV infection by targeting microRNA. *N. Engl. J. Med.* 368 (18): 1685–1694.

65 Chi, K.N., Siu, L.L., Hirte, H. et al. (2008). A phase I study of OGX-011, a 2′-methoxyethyl phosphorothioate antisense to clusterin, in combination with docetaxel in patients with advanced cancer. *Clin. Cancer Res.* 14 (3): 833–839.

66 Laskin, J.J., Nicholas, G., Lee, C. et al. (2012). Phase I/II trial of custirsen (OGX-011), an inhibitor of clusterin, in combination with a gemcitabine and platinum regimen in patients with previously untreated advanced non-small cell lung cancer. *J. Thorac. Oncol.* 7 (3): 579–586.

67 Marcucci, G., Byrd, J.C., Dai, G. et al. (2003). Phase 1 and pharmacodynamic studies of G3139, a Bcl-2 antisense oligonucleotide, in combination with chemotherapy in refractory or relapsed acute leukemia. *Blood* 101 (2): 425–432.

68 Marcucci, G., Stock, W., Dai, G. et al. (2005). Phase I study of oblimersen sodium, an antisense to Bcl-2, in untreated older patients with acute myeloid leukemia: pharmacokinetics, pharmacodynamics, and clinical activity. *J. Clin. Oncol.* 23 (15): 3404–3411.

69 Rheingold, S.R., Hogarty, M.D., Blaney, S.M. et al. (2007). Phase I Trial of G3139, a bcl-2 antisense oligonucleotide, combined with doxorubicin and cyclophosphamide in children with relapsed solid tumors: a Children's Oncology Group Study. *J. Clin. Oncol.* 25 (12): 1512–1518.

70 Klisovic, R.B., Blum, W., Wei, X. et al. (2008). Phase I study of GTI-2040, an antisense to ribonucleotide reductase, in combination with high-dose cytarabine in patients with acute myeloid leukemia. *Clin. Cancer Res.* 14 (12): 3889–3895.

71 Juhasz, A., Vassilakos, A., Chew, H.K. et al. (2006). Analysis of ribonucleotide reductase M2 mRNA levels in patient samples after GTI-2040 antisense drug treatment. *Oncol. Rep.* 15 (5): 1299–1304.

72 Yacyshyn, B.R., Bowen-Yacyshyn, M.B., Jewell, L. et al. (1998). A placebo-controlled trial of ICAM-1 antisense oligonucleotide in the treatment of Crohn's disease. *Gastroenterology* 114 (6): 1133–1142.

73 Warren, M.S., Hughes, S.G., Singleton, W. et al. (2015). Results of a proof of concept, double-blind, randomized trial of a second generation antisense oligonucleotide targeting high-sensitivity C-reactive protein (hs-CRP) in rheumatoid arthritis. *Arthritis Res. Ther.* 17 (1): 80–87.

74 Crooke, R.M. and Graham, M.J. (2011). Therapeutic potential of antisense oligonucleotides for the management of dyslipidemia. *Clin. Lipidol.* 6 (6): 675–692.

75 Dominski, Z. and Kole, R. (1993). Restoration of correct splicing in thalassemic pre-mRNA by antisense oligonucleotides. *Proc. Natl. Acad. Sci. U. S. A.* 90 (18): 8673–8677.

76 GlaxoSmithKline (2013). GSK and Prosensa announce primary endpoint not met in phase III study of drisapersen in patients with Duchenne Muscular Dystrophy. http://www.prosensa.eu/press-release/results-prosensa%E2%80%99s-extended-phase-iii-exon-skipping-trial-duchenne-muscular-dystrophy- (accessed 20 September 2013).

77 Machlin, E.S., Sarnow, P., and Sagan, S.M. (2011). Masking the 5′ terminal nucleotides of the hepatitis C virus genome by an unconventional microRNA-target RNA complex. *Proc. Natl. Acad. Sci. U. S. A.* 108 (8): 3193–3198.

78 Mortimer, S.A. and Doudna, J.A. (2013). Unconventional miR-122 binding stabilizes the HCV genome by forming a trimolecular RNA structure. *Nucleic Acids Res.* 41 (7): 4230–4240.

79 Jopling, C.L., Yi, M., Lancaster, A.M. et al. (2005). Modulation of hepatitis C virus RNA abundance by a liver-specific MicroRNA. *Science* 309 (5740): 1577–1581.

80 Li, Y., Masaki, T., and Lemon, S.M. (2013). miR-122 and the Hepatitis C RNA genome: more than just stability. *RNA Biol.* 10 (6): 919–924.

81 Hildebrandt-Eriksen, E.S., Bagger, Y.Z., Knudsen, T.B. et al. (2009). A unique therapy for hcv inhibits microrna-122 in humans and results in HCV RNA suppression in chronically infected chimpanzees: results from primate and first-in-human studies. *Hepatology* 50: 228A.

82 Zellweger, T., Miyake, H., Cooper, S. et al. (2001). Antitumor activity of antisense clusterin oligonucleotides is improved *in vitro* and *in vivo* by incorporation of 2′-*O*-(2-methoxy)ethyl chemistry. *J. Pharmacol. Exp. Ther.* 298 (3): 934–940.

83 Chia, S., Dent, S., Ellard, S. et al. (2009). Phase II trial of OGX-011 in combination with docetaxel in metastatic breast cancer. *Clin. Cancer Res.* 15 (2): 708–713.

84 Chi, K.N., Hotte, S.J., Yu, E.Y. et al. (2010). Randomized phase II study of docetaxel and prednisone with or without OGX-011 in patients with metastatic castration-resistant prostate cancer. *J. Clin. Oncol.* 28 (27): 4247–4254.

85 Saad, F., Hotte, S., North, S. et al. (2011). Randomized phase II trial of Custirsen (OGX-011) in combination with docetaxel or mitoxantrone as second-line therapy in patients with metastatic castrate-resistant prostate cancer progressing after first-line docetaxel: CUOG trial P-06c. *Clin. Cancer Res.* 17 (17): 5765–5773.

86 Church, D.N. and Talbot, D.C. (2012). Survivin in solid tumors: rationale for development of inhibitors. *Curr. Oncol. Rep.* 14 (2): 120–128.

5

Tissue Distribution, Metabolism, and Clearance

Mehrdad Dirin[1] *and Johannes Winkler*[1,2]

[1] *Department of Pharmaceutical Chemistry, University of Vienna, Vienna, Austria*
[2] *Department of Cardiology, Medical University of Vienna, Vienna, Austria*

5.1 Introduction

The main challenges for a more widespread therapeutic use of antisense (ASOs) and splice-switching oligonucleotides (SSOs) comprise good target affinity and specificity, adequate pharmacokinetic (PK) properties and target tissue distribution, sufficient stability against degradation, and efficient cellular uptake. The target affinity, e.g. the hybridization to the corresponding nucleic acid counter strand, is mediated by the oligonucleotide sequence, but the other issues are predominantly and essentially influenced by the chemical modification pattern and only marginally by a particular sequence. Wild-type nucleic acids are not therapeutically active because of rapid degradation through nucleases. A large number of different chemical derivatizations have been developed for antisense and splice-switching applications and tested in preclinical models, but only a few of them have been further developed for clinical evaluations. Those include mainly first-generation antisense agents, comprising a phosphorothioate (PS) backbone, second-generation compounds, which exhibit additional 2′-*O*-modifications (2′-*O*-MOE and others), and phosphoroamidate morpholino oligomers (PMOs) that are predominantly used for splice switching. In particular for neutral PMOs, the introduction of positively charged moieties, either through conjugation with a peptide or backbone modification, has enhanced their potencies [1–3]. All of these classes have been investigated clinically and preclinically for their PK behavior, including absorption after different application routes, biodistribution and cellular uptake, and metabolism and enzymatic degradation. Important differences in protein binding, nuclease degradation, and cellular uptake characteristics

Oligonucleotide-Based Drugs and Therapeutics: Preclinical and Clinical Considerations for Development, First Edition. Edited by Nicolay Ferrari and Rosanne Seguin.

between these groups are reflected by divergent distribution properties. This chapter focuses on tissue distribution and metabolism and the influence of the physicochemical and structural properties of oligonucleotide agents in addition to application routes.

5.2 Tissue Distribution

5.2.1 Dermal Delivery

Hair follicles and sebaceous glands may contribute significantly to topical and transdermal delivery of ASOs. Compared with other tissues, nucleases are less abundant in the skin. However, the high molecular weight and (if applicable) the negative charge of ASOs limit their effective systemic delivery through the skin. ASOs can distribute into the dermis via hair follicles only when formulated in lipophilic creams, but not with aqueous saline formulations. It has been suggested that sebum, which flows into hair follicle from the sebaceous duct, may necessitate a certain degree of lipophilicity for compounds to enter the follicles. Topical application of a PMO conjugated with a cell-penetrating peptide (PPMO) dissolved in 95% propylene glycol/5% linoleic acid reduced skin inflammation in sensitized adult male rats. In comparison, the PPMO dissolved in water was not effective [4]. Additionally, physical enhancement techniques (electroporation, low frequency ultrasound, laser ablation, and microneedles) and cationic polymers/liposomes have been investigated for the *in vitro* and *in vivo* delivery of ASOs through the skin [4, 5]. As opposed to intradermal injection, topical delivery can result in more uniform distribution of ASOs complexed with dendrimer carriers in the epidermis. Furthermore, intradermal injection can also result in systemic uptake by the dermal blood vessels [6]. In order to allow a PS (with two C-propyne cytidines) to achieve systemic effects in a midazolam-induced sleeping rat model after transdermal absorption, an iontophoretic current was applied [7, 8]. In analogy to intraperitoneal (IP) injection, full-length ASOs were found predominantly in the liver, kidney, and spleen, demonstrating the possibility of transdermal delivery for systemic effects. In another *in vivo* study, iontophoresis resulted in localization of about 30% of a PS-ODN in the whole epidermis and upper dermis of conventional NC/Nga mice, a human atopic dermatitis model [9].

Coapplication of iontophoresis and microinjection patch technology has been shown to increase oligonucleotide availability and resulted in therapeutically relevant amounts of a PS delivered both locally and systemically in hairless guinea pigs [10]. Compared with intravenous (IV) administration, topical delivery of a PS targeted to ICAM-1 showed a three orders of magnitude higher amount of the compound in the epidermis and a two orders of magnitude higher abundance in the dermis [4]. Subcutaneous (SC), IP, and intradermal administration of PSs resulted in a lower peak plasma concentration (C_{max}) than the IV route [11]. The SC route of administration of a PS led to a slower absorption into plasma and a lower C_{max}, suggestive of a depot effect or possibly lymphatic uptake with this route of administration [12–14]. The transdermal flux of uncharged PMOs *in vitro* is considerably higher than that of PSs. A few studies have proven the efficacy of transdermally delivered PMOs *in vivo* without the use of penetration enhancement techniques [15, 16]. These results are consistent with another report in which modifying an uncharged methylphosphonate oligonucleotide by adding a phosphate (PO) linkage and thus a single negative charge reduced dermal penetration by a factor of 10 [15, 17–19]. The results demonstrate the decisive influence of inherent oligonucleotide charge for transdermal delivery.

5.2.2 Ocular Delivery

First- and second-generation oligonucleotides are distributed sequence independently to most major ocular tissues. High concentrations of oligonucleotides were observed in the outer and inner plexiform layers, outer limiting membrane, ganglion cells, ciliary body, retinal pigment epithelium, and optic nerve. Lesser amounts of oligonucleotide appeared to be distributed in the inner and outer nuclear layers [20]. The most notable difference between first- and second-generation ASOs is the change of their half-lives in the retina. ISIS 13312 (2′-MOE ASO) was cleared very slowly from the retina, compared with ISIS 2922 (fomivirsen, a first-generation PS-ASO) in the monkey retina ($t_{1/2} = 2$ months vs. $t_{1/2} = 3$ days) [21]. The differences in clearance are caused by slower enzymatic degradation of 2′-modified compounds. In the case of fomivirsen (the first ASO approved for clinical use), the vitreal clearance was mediated by a combination of tissue distribution and metabolism [10, 22–24].

As a strategy for treating *toxoplasma gondii* infections, conjugates of transductive peptides with PMOs at the 3′ end have been shown to enter intracellular tachyzoites and encysted bradyzoites in the retina, following topical application to the eyes [25]. The peptides also carried cargos into the retina when applied topically to the eyes. Additional approaches including encapsulation of oligonucleotides in liposomes, PLGA matrix, or other carriers can sustain drug release, reduce degradation, and increase intravitreal retention and half-lives [26].

5.2.3 Oral Administration

Owing to high patient compliance and given the large surface area and the high vascularization of the gut epithelium and unwanted effects of SC or IV administration such as inflammation at injection sites, the oral route is of particular interest [27]. However, the physicochemical properties of ASOs are clearly unfavorable for oral delivery. Nucleic acids are prone to denaturation and depurination in acidic gastric media. In addition, nucleases present in the gastrointestinal tract together with pancreatic secretions degrade nucleic acids prior to cellular entry. Nevertheless, initial studies on PSs have shown reasonable stability in stomach contents, but they have demonstrated susceptibility to extensive nuclease degradation (primarily by 3′-exonuclease) in the small and large intestine [28].

Doses of a partially 2′-MOE-modified PS for the oral route required to achieve comparable biological effects were 48 and 20–28 times higher than that of the IV route (in dogs) and SC route (in humans), respectively. In both studies, full-length ASOs were taken up in the gastrointestinal tract and were effective in target organs/cells [28]. Compared with PSs, 2′-MOE POs showed significantly higher permeability in various segments of the rat intestine. The improved uptake characteristics in gut tissues of 2′-MOE POs relative to PSs were attributed to their different backbone polarities and were consistent with the extensive hydration of the 2′-*O*-substituents [29, 30].

Charge reduction (as in chimeric ASOs containing PS and methylphosphonate linkages or charge-masking formulations) or oral dosage forms formulated with intestinal membrane permeation enhancers are among various approaches that are being developed to further increase the oral absorption of ASOs [12, 27, 28, 31, 32]. Agents such as sodium dodecanoate (sodium caprate) that increase paracellular transport of biomacromolecules by opening tricellular tight junctions were investigated in animal models for enhancing oral bioavailability of oligonucleotides [28, 33]. Increased systemic bioavailability had been demonstrated, but safety concerns resulting from the opening of the intestinal barrier for larger molecules and xenobiotics need to be eliminated before advancing to a clinical setting. In dogs receiving three enteric-coated tablets, each containing ISIS104838 and sodium caprate, or IV injections of ISIS104838 once daily for a week, plasma bioavailability after oral application reached 1.4% of the IV dose. The tissue distribution profile and tissue ASO levels were similar for both routes, while tissue bioavailability ranged from 2 to 4.3% of what was achieved after IV injection. Plasma ASO levels were lower upon intestinal infusion, which may be due to the slow and incomplete absorption through the intestinal wall and resulted in lower tissue uptake (mainly by well-perfused organs such as the diaphragm and the heart). The relatively longer circulation time compared with injections may support this observation [28, 34, 35].

Pharmacologic effects of ASOs after oral administration are generally confined to gut-associated tissues and are of interest in a number of disorders such as inflammatory bowel disease. Topical administration is ostensibly an effective way of administering antisense agents to target tissue in inflammatory bowel disease. Retention enemas are an effective way of achieving therapeutically relevant concentrations within the mucosa of the distal and left colon

[35]. Alicaforsen enema in patients with mild to moderate ulcerative colitis caused a significant dose-dependent reduction in disease severity without any safety issues. A modified-release formulation of mongersen (formerly GED0301), a PS-ASO, was not absorbed following oral administration in patients with Crohn's disease [36, 37]. Mongersen is targeted at SMAD7, which inhibits TGF-β1-mediated immunosuppression by binding to the TGF-β1 receptor and is up regulated in the gut of Morbus Crohn patients. In a phase II clinical trial, mongersen resulted in significantly higher remission rates in a dose-dependent manner [36]. ASOs against the intercellular adhesion molecule 1 (ICAM-1) administered intravenously have been tested clinically in patients with Crohn's disease. In contrast to local application, they failed to result in significant treatment successes. However, population PK suggests that the ASOs may be effective after IV application when given in adequate doses [35, 36, 38].

PMOs have significant advantages over other ASO analogs from the perspective of oral delivery. Orally administered PMOs (AVI-4126) showed higher oral bioavailability (up to 78.8% over a period of 24 h). This could possibly be due to efflux of PMO that accumulates in the liver following oral delivery. PMOs display favorable plasma PK following oral administration compared with PSs, which show much lower oral bioavailability [39].

5.2.4 Intrathecal Delivery

Their high charge, large molecule size, and high hydrophilicity impede ASOs from crossing the intact blood–brain barrier. However, direct delivery of the ASO into the cerebral spinal fluid (CSF) resulted in broad distribution in the spinal cord and brain tissue. After intrathecal infusion of a 2′-MOE PS in SOD1 familial ALS patients, maximum concentrations were measured in CSF and plasma at the end of the infusion [40]. The exposure of the drug in the blood after direct administration into CSF was multiple orders of magnitude lower than exposures observed with direct administration by SC or IV administration, reflective of the lower total dose administered and the partial transfer from the CNS to the systemic circulation. Comparatively, bolus injection into the CSF results in better distribution in the CNS than slow infusion. Distribution kinetics from the CSF into CNS tissues is rapid, with a distribution half-life of less than 1 h that is likely due to the combination of uptake into CNS tissues and transfer to systemic circulation [41]. Similar results were recently reported for CSF injection of nusinersen, a 2′-*O*-methoxyethyl SSO in children suffering from spinal muscular atrophy (SMA) [42]. After the initial distribution of a substantial part of the dose into plasma with the peak plasma level reached around 4 h after application, concentrations in the circulation declined to less than 1% after 1 week. The oligonucleotide dose that remained in the CSF and brain tissue showed a long half-life of 4–6 months, consistent with preclinical results.

These findings indicate that although only a fraction of the injected dose remains in the CSF, the high stability allows for infrequent administration.

5.2.5 Intravesical Administration

The urinary bladder is relatively impermeable to toxic substances and easily accessible for drug application through a catheter. Systemic absorption of chemotherapeutic drugs through the bladder wall is thought to be unlikely for agents with a molecular weight over 300 g/mol. Transport across the urothelial barrier is determined not only by molecular size but also by the degree of lipophilicity of the compound [43]. However, uptake of ASO was postulated to be limited by the glycosaminoglycan layer of the inner bladder surface. Additionally, anionic charge and the inherent size of the effector molecules may inhibit the intracellular passage through the cellular lipid bilayer [44]. Few reports of local administration of oligonucleotide agents to the bladder are available.

Intravesical administration of apatorsen (OGX-427), a second-generation 2′-MOE PS gapmer, in a porcine model with and without liposomal formulations, demonstrated acceptable intracellular uptake across the entire depth of the urothelium [45, 46]. The observed systemic uptake in mice may also be mediated in part by ureteral reflux and absorption at the kidney level [46].

However, the uptake of PSs into normal urothelium and tumor cells did not require lipid carrier molecules. This finding is especially important in the treatment of diffuse urothelial cancer, as the targeted therapy applied intravesically has to penetrate the urothelium [44].

5.2.6 Pulmonary Administration

The lung represents a unique target organ for therapy, as it is readily accessible by direct noninvasive delivery to the site of action while minimizing systemic exposure and virtually eliminating the risk for adverse effects on internal organs as well as the known blood-related toxicities. The translocation of ASOs across the pulmonary mucosal epithelia is hampered by the physical barriers of the overlying mucus layer, the tight packing of epithelial cells, and the sweeping movement of the apical cilia [47–50]. ASM8 is an ASO with a full-length PS backbone, with all adenosines replaced with 2-amino-2′-deoxyadenosine, formulated for inhalation. It consists of two PS-ASOs (TOP004 and TOP005). TOP004 inhibits the β-subunit of IL-3, IL-5, and GM-CSF (granulocyte–macrophage colony-stimulating factor) receptors. TOP005 is targeted against the chemokine receptor CCR3. Both interfere in two distinct but overlapping cellular pathways involved in the initiation, maturation, and persistence of allergic inflammation and asthma [51]. In monkeys, after inhalation for 14 consecutive days, less than 2% of the amount measured in the lung was found in plasma, kidney, and liver. At low doses, the TPI ASM8 was effectively confined to the lung [14]. Initial clinical trials demonstrated significant reduction of inflammatory markers. While no significant depletion of target protein levels could be proven, the PS agents were not found in plasma, indicative of effective restriction to the lung and lack of uptake via airways of unformulated PSs [10]. However, downregulation of individual Th2 cytokines in individuals with asthma has failed to produce useful effects. These outcomes may be related to functional pleiotropy among Th2 cytokine family members or may be due to lack of adequate distribution or stability of the drug compounds within the lung.

Aerosol administration of a 2′-MOE PS against IL-4Rα in allergen-challenged mice has been shown to improve antisense potency, stability, and tolerability in animals compared with POs [52]. Inhaled ASOs of this chemical class were also well tolerated in a non-human primate study [52].

In another study, the potential of delivering Cy5-labeled PO-LNA or PS-LNA to the lungs of mice was investigated. Interestingly, none of these ASOs led to the knockdown of the target mRNAs or protein in any of the cell types investigated. However, the PS-LNA accumulated in the liver and kidney at therapeutically relevant concentrations, comparable with the level when administered intravenously, suggesting that pulmonary delivery is a possible noninvasive route for administration of ASOs to the liver and kidney [47]. Following direct administration to the lung, PO-LNAs were not readily taken up by lung cells but were instead rapidly absorbed and cleared renally. In contrast, PS-LNA was retained within macrophages and to a lesser extent in the alveolar and bronchial epithelia and after 24 h also in submucosal membranes and endothelial cells, but surprisingly no knockdown of the target mRNA or protein could be detected in any of the cell types [53]. The use of a PS backbone in LNA ASO resulted in notable lung macrophage accumulation but failed to show demonstrable activity. Despite the ubiquitous tissue penetration of the lung, rapid transport mechanisms might account for systemic access [53].

5.2.7 Distribution to Muscular Tissue

Since over 30% of the human body consists of muscle, systemic treatment is necessary in neuromuscular diseases such as Duchenne muscular dystrophy (DMD) and myotonic dystrophy. To counter the expression of nonfunctional proteins caused by missplicing such as dystrophin, exon-skipping oligonucleotides are being developed as a novel treatment option. The two major chemical modifications used for exon skipping drugs are 2′-OMe PSs and PMOs [54]. For the treatment of muscular dystrophies, the ASOs must be able to penetrate muscles of various types. However, most of the injected ASOs ended inherently up in the liver and/or kidney [3]. The uptake of ASOs appears to benefit from the pathophysiological conditions of the dystrophic muscle tissues. Both

SC and IV routes resulted in significant muscle uptake and exon skipping, but IV administration resulted in higher distribution to the liver and kidneys, while ASO levels in muscle were similar to those after SC treatment [55]. Naked morpholino ASOs lacking any cell penetration-enhancing modification induced an unexpectedly high degree of exon skipping *in vivo* although this same preparation had previously appeared ineffective *in vitro* [56]. While significant uptake and effects have been demonstrated for several antisense agents in skeletal muscle tissues, profound effects in heart muscle are much more difficult to achieve. Despite a longer half-life of ASOs in the heart (~65 days) compared with that in the skeletal muscle, liver, and kidney (~35 days) [55, 57], the lack of effects is mainly due to lower uptake into cardiomyocytes and entrapment in the interstitial space [56, 58]. In animal models, the effect of ASOs in the heart is significantly lower (2′-OMe PS) or minimal (PMO) compared with the skeletal muscle [57]. Notably, even direct intracardiac injections of naked ASOs resulted in very low exon skipping rates [3, 54, 59].

5.2.8 Intravenous Administration

Independent of the route of administration, ASOs are subject to very similar kinetics after reaching the plasma. Plasma antisense levels are highest after IV injection. Binding to plasma proteins prevents quick renal filtration and elimination. Upon IP and SC administration of a 2′-OMe PS in mice, the plasma ASO levels were lower due to a gradual uptake from the IP cavity and SC depot, respectively, allowing a relatively higher amount of ASO to bind serum proteins and a lower clearance of unbound ASOs [28, 60]. POs as well as PS- and 2′-OMe-ASOs accumulate predominantly in the liver. 2′-MOE partially modified ASOs are highly bound (>90%) to plasma proteins, most likely through nonspecific electrostatic interactions, though not as extensively as their PS counterparts [22]. PSs show a high propensity to bind proteins nonspecifically, which is simultaneously beneficial and detrimental [11]. The benefit of nonspecific binding is manifested in the rapid absorption into the plasma and extensive distribution to tissues, with very little loss by excretion [61, 62]. Plasma clearance of PSs is attributed to tissue distribution and to a lesser extent to plasma metabolism [27]. In plasma, the binding affinity of ASO to α_2-macroglobulin was found to be greater than to albumin. However, albumin appears to have a much greater nonspecific binding capacity for oligonucleotides. Together with the high abundance of albumin, it is mainly responsible for ASO binding [22, 63, 64]. Plasma protein binding is not saturated at clinically relevant doses due to the large capacity of proteins in plasma. However, charge-neutral ASOs or ASOs that are less extensively or more weakly bound to plasma proteins, such as PMOs, exhibit significantly more rapid clearance from the blood, primarily due to either metabolism or excretion in the urine, resulting in much lower or negligible tissue uptake. Therefore, the kidney is the primary organ of distribution for these compounds but with very low retained concentrations because of quick loss in the urine. Thus, appropriate and balanced plasma protein binding (or nanoparticle packaging) is required for broad and significant delivery to tissues and cells systemically. For all animal species evaluated, plasma protein bound ASOs distribute broadly into most tissues with the exception of the central nervous system after systemic administration. The major systemic tissues of distribution include the liver, kidney, bone marrow, adipocytes (cell body), and lymph nodes [41].

PSs are widely distributed among tissues in a satiable manner, with the kidney cortex containing the highest concentration and the liver the highest fraction of the total dose [11, 12]. Scavenger receptors in the liver, bone marrow, and kidney are suggested to contribute to the elimination of PSs and POs from the blood. Physiologically, these receptors are involved in fatty acid transport. However, due to their broad polyanionic ligand specificity, polynucleotides can also bind to them and become internalized [65–68]. Scavenger receptor knockout mouse models exhibit excellent cellular uptake and sequence-specific antisense effects. This has also been the case for specific nucleic acid transporters that appear to function in multiple tissues *in vivo* or *in situ*, but are not expressed *in vitro* [69]. PSs, including those that carry additional sugar modifications, appear to be retained in tissues over extended periods [12, 70]. The *in vivo* disposition of 2′-MOE partially modified PSs is primarily determined by their backbone chemistry. Thus, similar to first-generation PSs, the PK properties of these ASOs are remarkably sequence independent.

In LNAs, the more hydrophobic cyclic structure results in differential compartmentalization compared with other classes of ASOs, particularly within the liver tissue [71]. Distribution of LNA mixmer ODNs resembled that of PSs; however, LNA mixmers exhibit much less protein binding and are present in plasma mainly as free molecules being available for rapid tissue uptake [72, 73]. Accordingly, their accumulation in the kidney is less compared with PSs, whereas uptake in the spleen, liver, and adrenal gland seems to be higher [72]. Furthermore, varied positioning of the LNA modifications in the ASO strand alters the distribution pattern. EZN-2968, a 16-mer PS carrying LNAs at positions 1–3 and 13–15 administered intravenously, was broadly distributed to tissues, with the major organs of accumulation being the kidney, liver, intestine, bone marrow, uterus, lymph nodes, and spleen [71, 74].

Contrary to the backbone-specific profile of PSs, the PK profile of PMOs shows a higher degree of sequence dependence. This can be a result of variations in the three-dimensional structure, their protein binding characteristics or their hybridization affinity [75]. Passive diffusion through cell membrane is not evident for uncharged PMOs. PMOs undergo rapid distribution in a dose-dependent manner and produce significant tissue concentrations within 24 h. They accumulate primarily in the kidney and liver and to a lesser extent in the spleen, lung, heart, and skeletal muscle and thymus and at minor concentrations in the brain [58, 76, 77]. As explained above, PMO levels in the kidney are due to the fact that uncharged PMOs show a low degree of plasma protein binding and subsequently higher concentrations in the kidneys and lower concentrations in the muscle and liver [18, 58, 78]. Conjugation of PMOs to cell penetration-enhancing peptides (CPPs) increases their tissue uptake and retention and results in a broad tissue distribution with the kidney and liver as the primary sites of accumulation. Relatively lower concentrations are detected in the spleen, lungs, heart, and brain. Electrostatic interaction between the positively charged conjugate and cell surface proteoglycans is likely to account for increased tissue uptake of CPP–PMO compared with PMO [2, 77, 79].

PMO*plus*, displaying positively charged piperazinyl phosphorodiamidate linkages, have been shown to be efficacious by a number of delivery routes and possesses favorable PK properties [80, 81]. A consistent trend of increasing renal clearance with increasing dose is likely to be due to low-affinity interactions between PMO*plus* agents and plasma proteins [82]. AVI-6002 and AVI-6003 are two combination ASOs under evaluation for postexposure prophylaxis of Ebola virus and Marburg virus, respectively. Both drugs were developed in 1 : 1 combinations of two distinct oligonucleotides, but in the case of the Ebola drug, one sequence (AVI-7537) was later selected for single-drug component. Both drugs consist of PMOs that have positive charges in the form of piperazine residues at defined locations. In healthy volunteers, all PMO*plus* components displayed similar plasma PK properties, including similar time–concentration curves and absolute values of C_{max} and AUC at higher dose levels [81, 82].

5.3 Cellular Uptake

ASOs are observed to be associated with cell surface proteins and extracellular matrix 1 h after parenteral administration [22]. Regardless of having anionic charge or being charge neutral, ASOs possess very little ability to diffuse across cell membranes [22, 83]. ASOs appear in the cytoplasm of different cell types within a very short time after administration, including Kupffer cells and sinusoidal endothelial cells in the liver and renal proximal epithelial cells in the kidney. The rapid cellular uptake in these cells *in vivo* appears to be a class effect and independent of sequence. The intracellular disposition of oligonucleotides is believed to be a satiable and nonlinear process of protein binding interactions from high capacity low-affinity proteins in plasma to higher affinity binding in organs of disposition [69]. The liver, as a primary target site for ASO accumulation and cellular uptake, shows a complicated suborgan distribution pattern characterized with the lowest concentrations per cell in hepatocytes while highest concentrations in Kupffer and sinusoidal endothelial cells.

Kinetic observations further illustrate that the cells in the liver that take up the most ASOs are also the cells that take up ASOs first [69].

PS modification of ASOs has been shown to improve cellular uptake, proving that the backbone chemistry of the ASO determines the kinetic behavior and nuclease sensitivity of the ASO *in vivo* [84, 85]. Interestingly, minimally modified ASOs (with only a limited number of PS bonds) show a higher cellular uptake than those with a full PS backbone [86–88]. PSs accumulate within certain cell populations in certain tissues [84]. Recent research has identified several cell type-specific nucleic acid-binding surface proteins [89, 90], and the presence and abundance of these are essential for cellular uptake [61, 91]. Multiantennary GalNAc ASO conjugates have been developed for specific targeting of the asialoglycoprotein receptor in hepatocytes and have increased *in vivo* potency for respective hepatic targets several-fold [92]. Moreover, compared with PSs that appear to be largely sequestered in the nucleus, PMOs are seen to distribute more evenly throughout the cell [2, 93]. Not only the hybridization dynamics of ASOs but also their affinities for subcellular structures may enhance or attenuate their biological effects [27, 94]. Up to now, 32 proteins have been characterized for effects on ASO activity and localization [95]. They exert their effect through different mechanisms. Altering ASO subcellular distribution or competition with RNase H1 for binding to ASO/RNA duplex counts among the mechanisms proposed for moderate augmentation or inhibition/attenuation of ASO activity [96, 97].

Depending on the delivery approach, PSs distribute differently within cells. PSs localize predominately in the cytoplasm, mainly in punctate structures and/or perinuclear structures corresponding to endosomes or lysosomes. PS-ASOs undergo nucleocytoplasmic shuttling [98, 99]. PSs generally differ from their corresponding POs in showing substantially stronger unspecific reactivity with cellular proteins and, hence, higher cellular retention potentials. However, different degrees of PS accumulation in cells could be possibly a consequence of differential binding to intracellular substructures and not so much of their rates of cellular internalization per se. Such a view of the cellular internalization and compartmentalization may also provide the basis for the interpretation of the cell-type specificity of these processes [100–102].

Partial modification with 2′-MOE renders much longer tissue residence times of ASOs compared with first-generation oligonucleotides. The initial rapid distribution to tissues yet remains the dominant mechanism of plasma clearance [22]. The overall plasma clearance of POs is nearly 10-fold more rapid than that of their PS counterparts [94, 100]. The use of lipid carriers or lipoproteins as delivery vehicles for ASOs (LErafAON-ETU) has been demonstrated to direct ASOs mainly to the liver and spleen, rather than to the kidney [103, 104]. In contrast to free PS oligonucleotides, which are localized within endosomes or lysosome-like structures, liposome-complexed PSs are present in cytosolic and nuclear compartments, thereby reducing the efflux of oligonucleotides [105–109].

5.4 Metabolism and Clearance

5.4.1 Phosphorothioates Including 2′-Modifications

Metabolism of oligonucleotides is afforded nearly exclusively through nuclease digestion and cleavage at PO linkages. In the blood and tissues, endonucleases and 3′- and 5′-exonucleases are present. Unmodified oligonucleotides are degraded within minutes in the blood, serum, or tissue [110, 111]. The predominant nuclease degradation is thereby exonucleolytic, and the rate and extent of cleavage is generally sequence independent. As a result, shorter metabolites are generated by gradual loss of one nucleotide after the other. Both *in vitro* and *in vivo* 3′-exonucleases are more active in oligonucleotide metabolism than 5′-exonucleases [110, 112]. However, stabilization at the 3′-terminus is not sufficient for adequate protection because of 5′-exo- and endonuclease activity, which steadily degrades oligonucleotides upon inhibition of 3′-exonuclease activity. PSs show reduced enzymatic degradation, but 2′-alkylated oligonucleotides are even more stable against digestion.

The tissue clearance of oligonucleotides is mainly due to exonuclease degradation [113] with subsequent elimination of the degraded products. Consequently, tissue half-lives are primarily dependent on enzymatic resistance of the oligonucleotides. Wild-type ODNs suffer from insufficient plasma protein binding and are thus rapidly eliminated before undergoing enzymatic cleavage as primarily a full-length product by glomerular filtration. Full PS backbone modifications significantly increase enzymatic stability and minimize renal elimination. Plasma clearance is approximately 10-fold slower for PS- than for PO-ODNs. In liver homogenate, enzymatic degradation is more than 20-fold slower for PSs compared with phosphodiesters [111, 114]. From a physicochemical point of view, the substitution of sulfur for oxygen generates a chiral center at the PO. Interestingly, Sp PSs are significantly more stable against nuclease degradation than Rp diastereomers [115]. Although it is technically possible to prepare PSs in a stereospecific manner, it is much more challenging than nonstereospecific synthesis. Therefore, oligonucleotides are generally used as a diastereomeric mixture of Rp and Sp counterparts. In liver homogenate, a 3′-Sp-PS-modified oligonucleotide is barely metabolized, while the Rp counterpart is only slightly more stable than a phosphodiester oligonucleotide [111]. Biological and pharmacological effects, including sequence-dependent hybridization, are very similar regardless of the stereochemistry [85].

Only a small fraction of full-length PS-ODN is excreted in urine, but shorter metabolites resulting from exonuclease digestion are ultimately eliminated via the kidneys due to their lower plasma protein binding [116]. The principal serum degradation products result from 3′-exonucleolytic cleavage, but intracellular PS-ASON degradation appears to be more variable and is cell-type dependent with both 3′- and 5′-degradation profiles [117]. Within minutes after *in vivo* application, shorter, base-depleted PS oligonucleotide metabolites can be found [114].

The PS backbone not only confers resistance against nuclease digestion but also inhibits the enzyme. Preincubation of liver homogenate with a PS reduced subsequent enzymatic degradation of a phosphodiester oligonucleotide [111]. Besides chemical modification, sequence length also influences the rate of metabolism. A minimum of three internucleotide bonds is required for nuclease substrates, and shorter sequences are digested more slowly [111]. This fact implies that initial degradation products from endonuclease digestion compete with full-length sequences and contribute to their stability. Due to exonuclease digestion being the major metabolizing mechanism, oligonucleotides further stabilized by terminal 2′-*O*-modifications or LNA nucleosides have longer serum and tissue half-lives. Partially 2′-*O*-methoxyethyl-modified antisense agents show markedly decreased metabolism and a tissue half-life of 2–4 weeks [118]. These stability issues have resulted in the development of gapmer oligonucleotides, which consist of three to five 2′-modified nucleotides at each of the two termini and an internal stretch of 2′-deoxynucleotides. This inclusion of at least seven adjacent deoxynucleotides ensures activation of RNase H, a nuclease that recognizes hybrid duplexes of DNA/RNA and cleaves the RNA portion. Thus, RNase H-activating antisense agents have catalytic activity, while nonRNase H substrates work via a stoichiometric translation block mechanism. Phosphorothioate backbone modifications are usually incorporated to induce a high degree of plasma protein binding, which is necessary to prevent renal elimination. Due to slow metabolism in tissues, 2′-*O*-MOE and LNA-modified gapmer PS oligonucleotides demonstrate a significantly longer duration of downregulation effects [119, 120]. Both the 2′-MOE- and LNA-phosphodiester oligonucleotides are rapidly excreted as full-length sequences in the urine [74] due to low plasma protein binding, which prevents efficient tissue distribution.

The liver acts as the primary organ for drug metabolism in the body. For oligonucleotides, nuclease metabolism is by far the most abundant form of degradation. Although exo- and endonucleases are dispersed widely throughout all tissues and organs, the liver is the major site of oligonucleotide metabolism because of its biodistribution characteristics. No metabolism by the cytochrome P450 system has been found [121], indicating a low probability of drug–drug interaction with substrates of the cytochrome P450 system.

5.4.2 Phosphorodiamidate Morpholino Oligonucleotides

PMO-based gene knockdown has been extensively used in zebrafish due to the relatively short effective lifetime of morpholino oligonucleotides, which, although not degraded, are excreted to a variable extent just as any normal cellular metabolite [122, 123]. PMOs are stable against enzymatic degradation, and the parent compound can be detected in tissues even after six half-lives [77]. No metabolites have been found after incubation in plasma or cultured cells [124] or the urine after *in vivo* application [77]. Thus, PMOs are regarded as substrates neither to nucleases nor to other degrading enzymes, including proteases and hydrolases [125]. Moreover, no induction of enzymes of the cytochrome P450 systems was found with biologically relevant concentrations *in vitro* [126]. Generally, full-length PMOs are eliminated primarily in the urine, but for some sequences, significant excretion was found in fecal matter [127].

Evaluation of eteplirsen in cynomolgus monkeys established that there was no marked drug accumulation in plasma upon weekly dosing. This is most likely due to the relatively short plasma half-life of PMOs, which results in complete clearance prior to subsequent dosing. Accumulation of eteplirsen in renal cells reflects high renal clearance rates [128]. The plasma elimination half-life following IV administration of AVI-6003 (PMOplus) in rats has been demonstrated to be similar to that of a neutrally charged *c-myc* PMO but somewhat shorter than that of an arginine-rich peptide conjugated version of the same *c-myc* PMO. The primary route of excretion of PMO*plus* components administered via the IV route in rats seems to be renal [129].

5.5 Conclusion

Substantial differences in distribution and metabolism between the major chemical modifications of ASOs and SSOs exist. The PS backbone enhances enzymatic stability, plasma protein binding, and thus biodistribution, as well as cellular uptake. Alkylation at the 2′-position is useful for resistance against nuclease degradation but on its own has only minimal influence on protein binding and cellular uptake. The combination of terminal 2′-modified nucleotides with the PS backbone results in high stability together with adequate biodistribution profiles. However, while high accumulation in some organs such as the liver is promising for corresponding disease targets, insufficient amounts in other tissues and organs, including the brain and heart muscles, compromise a more universal application of the antisense technology for therapy.

PMOs have higher stability against degradation but suffer from low plasma protein binding and consequently quick elimination. In addition, unassisted cellular uptake is low, at least in healthy tissues. Thus, their therapeutic use so far is restricted to muscular dystrophy, with corresponding cellular uptake into compromised tissues. The attachment of positively charged functionalities to PMOs increases uptake and may increase the therapeutic potential of this class, if toxicity can be limited.

So far, only local and SC or IV systemic applications have been used in clinical evaluations. Recent preclinical studies of absorption over the gastrointestinal tract not only show that a patient-friendly oral application may be possible in the future but also that significant improvements regarding bioavailability and comprehensive safety evaluations are necessary.

Besides further optimization of chemical modifications, the use of nanoparticle formulations and the preparation of conjugates with specific targeting properties may give a boost to the field of antisense therapeutics in order to resolve the major issues associated with absorption, distribution, and cellular uptake [130–133]. Important basic research and progress has been achieved during the last years, and the necessary improvements for a more widespread pharmaceutical use can be expected in the near future.

References

1 Swenson, D.L., Warfield, K.L., Warren, T.K. et al. (2009). Chemical modifications of antisense morpholino oligomers enhance their efficacy against Ebola virus infection. *Antimicrob. Agents Chemother.* 53 (5): 2089–2099.
2 Summerton, J. (1999). Morpholino antisense oligomers: the case for an RNase H-independent structural type. *Biochim. Biophys. Acta* 1489 (1): 141–158.
3 Jirka, S.M.G., Heemskerk, H., Tanganyika-de Winter, C.L. et al. (2013). Peptide conjugation of 2′-*O*-methyl phosphorothioate antisense oligonucleotides enhances cardiac uptake and exon skipping in MDX mice. *Nucleic Acid Ther.* 24 (1): 25–36.
4 Mourich, D.V., Jendrzejewski, J.L., Marshall, N.B. et al. (2009). Antisense targeting of cFLIP sensitizes activated T cells to undergo apoptosis and desensitizes responses to contact dermatitis. *J. Invest. Dermatol.* 129 (8): 1945–1953.
5 Dokka, S., Cooper, S.R., Kelly, S. et al. (2005). Dermal delivery of topically applied oligonucleotides via follicular transport in mouse skin. *J. Invest. Dermatol.* 124 (5): 971–975.
6 Venuganti, V.V.K., Saraswathy, M., Dwivedi, C. et al. (2015). Topical gene silencing by iontophoretic delivery of an antisense oligonucleotide-dendrimer nanocomplex: the proof of concept in a skin cancer mouse model. *Nanoscale* 7 (9): 3903–3914.
7 Brand, R.M. and Iversen, P.L. (2000). Transdermal delivery of antisense compounds. *Adv. Drug Deliv. Rev.* 44 (1): 51–57.
8 Brand, R.M., Hannah, T.L., Norris, J., and Iversen, P.L. (2001). Transdermal delivery of antisense oligonucleotides can induce changes in gene expression in vivo. *Antisense Nucleic Acid Drug Dev.* 11 (1): 1–6.
9 Sakamoto, T., Miyazaki, E., Aramaki, Y. et al. (2004). Improvement of dermatitis by iontophoretically delivered antisense oligonucleotides for interleukin-10 in NC//Nga mice. *Gene Ther.* 11 (3): 317–324.

10 Dirin, M. and Winkler, J. (2013). Influence of diverse chemical modifications on the ADME characteristics and toxicology of antisense oligonucleotides. *Expert Opin. Biol. Ther.* 13 (6): 875–888.
11 Agrawal, S. (1996). Antisense oligonucleotides: towards clinical trials. *Trends Biotechnol.* 14 (10): 376–387.
12 Raoof, A.A., Chiu, P., Ramtoola, Z. et al. (2004). Oral bioavailability and multiple dose tolerability of an antisense oligonucleotide tablet formulated with sodium caprate. *J. Pharm. Sci.* 93 (6): 1431–1439.
13 Fidias, P., Pennell, N.A., Boral, A.L. et al. (2009). Phase I study of the c-raf-1 antisense oligonucleotide ISIS 5132 in combination with carboplatin and paclitaxel in patients with previously untreated, advanced non-small cell lung cancer. *J. Thorac. Oncol.* 4 (9): 1156–1162.
14 Guimond, A., Viau, E., Aubé, P. et al. (2008). Advantageous toxicity profile of inhaled antisense oligonucleotides following chronic dosing in non-human primates. *Pulm. Pharmacol. Ther.* 21 (6): 845–854.
15 Arora, V., Hannah, T.L., Iversen, P.L., and Brand, R.M. (2002). Transdermal use of phosphorodiamidate morpholino oligomer AVI-4472 inhibits cytochrome P450 3A2 activity in male rats. *Pharm. Res.* 19 (10): 1465–1470.
16 Pannier, A.K., Arora, V., Iversen, P.L., and Brand, R.M. (2004). Transdermal delivery of phosphorodiamidate Morpholino oligomers across hairless mouse skin. *Int. J. Pharm.* 275 (1–2): 217–226.
17 Kinali, M., Arechavala-Gomeza, V., Feng, L. et al. (2009). Local restoration of dystrophin expression with the morpholino oligomer AVI-4658 in Duchenne muscular dystrophy: a single-blind, placebo-controlled, dose-escalation, proof-of-concept study. *Lancet Neurol.* 8 (10): 918–928.
18 Freier, S.M. and Altmann, K.-H. (1997). The ups and downs of nucleic acid duplex stability: structure-stability studies on chemically-modified DNA:RNA duplexes. *Nucleic Acids Res.* 25 (22): 4429–4443.
19 Summerton, J. and Weller, D. (1997). Morpholino antisense oligomers: design, preparation, and properties. *Antisense Nucleic Acid Drug Dev.* 7 (3): 187–195.
20 Fattal, E. and Bochot, A. (2006). Ocular delivery of nucleic acids: antisense oligonucleotides, aptamers and siRNA. *Adv. Drug Deliv. Rev.* 58 (11): 1203–1223.
21 Henry, S.P., Miner, R.C., Drew, W.L. et al. (2001). Antiviral activity and ocular kinetics of antisense oligonucleotides designed to inhibit CMV replication. *Invest. Ophthalmol. Vis. Sci.* 42 (11): 2646–2651.
22 Meibohm, B. (2006). *Pharmacokinetics and Pharmacodynamics of Biotech Drugs: Principles and Case Studies in Drug Development*, 426. Weinheim: Wiley.
23 Yu, R.Z., Geary, R.S., Monteith, D.K. et al. (2004). Tissue disposition of 2′-*O*-(2-methoxy) ethyl modified antisense oligonucleotides in monkeys. *J. Pharm. Sci.* 93 (1): 48–59.
24 Short, B.G. (2008). Safety evaluation of ocular drug delivery formulations: techniques and practical considerations. *Toxicol. Pathol.* 36 (1): 49–62.

25 Lai, B.-S., Witola, W.H., El Bissati, K. et al. (2012). Molecular target validation, antimicrobial delivery, and potential treatment of toxoplasma gondii infections. *Proc. Natl. Acad. Sci. U. S. A.* 109 (35): 14182–14187.
26 Gomes Dos Santos, A.L., Bochot, A., and Fattal, E. (2005). Intraocular delivery of oligonucleotides. *Curr. Pharm. Biotechnol.* 6 (1): 7–15.
27 Tillman, L.G., Geary, R.S., and Hardee, G.E. (2008). Oral delivery of antisense oligonucleotides in man. *J. Pharm. Sci.* 97 (1): 225–236.
28 van Putten, M., Young, C., van den Berg, S. et al. (2014). Preclinical studies on intestinal administration of antisense oligonucleotides as a model for oral delivery for treatment of Duchenne muscular dystrophy. *Mol. Ther.-Nucleic Acids* 3: e211.
29 Murtola, M., Zaramella, S., Yeheskiely, E., and Strömberg, R. (2010). Cationic peptides that increase the thermal stabilities of 2′-O-MeRNA/RNA duplexes but do not affect DNA/DNA melting. *ChemBioChem* 11 (18): 2606–2612.
30 Manoharan, M. (1999). 2′-Carbohydrate modifications in antisense oligonucleotide therapy: importance of conformation, configuration and conjugation. *Biochim. Biophys. Acta* 1489 (1): 117–130.
31 Glick, B.R., Pasternak, J.J., and Patten, C.L. (2010). *Molecular Biotechnology: Principles and Applications of Recombinant DNA*, 4e, 1018. Washington, DC: ASM Press.
32 O'Neill, M.J., Bourre, L., Melgar, S., and O'Driscoll, C.M. (2011). Intestinal delivery of non-viral gene therapeutics: physiological barriers and preclinical models. *Drug Discov. Today* 16 (5–6): 203–218.
33 Krug, S.M., Amasheh, M., Dittmann, I. et al. (2013). Sodium caprate as an enhancer of macromolecule permeation across tricellular tight junctions of intestinal cells. *Biomaterials* 34 (1): 275–282.
34 Schreiber, S., Nikolaus, S., Malchow, H. et al. (2001). Absence of efficacy of subcutaneous antisense ICAM-1 treatment of chronic active Crohn's disease. *Gastroenterology* 120 (6): 1339–1346.
35 van Deventer, S.J.H., Tami, J.A., Wedel, M.K., and G. European Colitis Study (2004). A randomised, controlled, double blind, escalating dose study of alicaforsen enema in active ulcerative colitis. *Gut* 53 (11): 1646–1651.
36 Monteleone, G., Neurath, M.F., Ardizzone, S. et al. (2015). Mongersen, an oral SMAD7 antisense oligonucleotide, and Crohn's disease. *N. Engl. J. Med.* 372 (12): 1104–1113.
37 Monteleone, G., Fantini, M.C., Onali, S. et al. (2012). Phase I clinical trial of Smad7 knockdown using antisense oligonucleotide in patients with active Crohn's disease. *Mol. Ther.* 20 (4): 870–876.
38 Yacyshyn, B.R., Chey, W.Y., Goff, J. et al. (2002). Double blind, placebo controlled trial of the remission inducing and steroid sparing properties of an ICAM-1 antisense oligodeoxynucleotide, alicaforsen (ISIS 2302), in active steroid dependent Crohn's disease. *Gut* 51 (1): 30–36.

39 Arora, V., Knapp, D.C., Reddy, M.T. et al. (2002). Bioavailability and efficacy of antisense morpholino oligomers targeted to c-myc and cytochrome P-450 3A2 following oral administration in rats. *J. Pharm. Sci.* 91 (4): 1009–1018.
40 Miller, T., Pestronk, A., David, W. et al. (2013). A phase I, randomised, first-in-human study of an antisense oligonucleotide directed against SOD1 delivered intrathecally in SOD1-familial ALS patients. *Lancet Neurol.* 12 (5): 435–442.
41 Geary, R.S., Norris, D., Yu, R., and Bennett, C.F. (2015). Pharmacokinetics, biodistribution and cell uptake of antisense oligonucleotides. *Adv. Drug Deliv. Rev.* 87: 46–51.
42 Chiriboga, C.A., Swoboda, K.J., Darras, B.T. et al. (2016). Results from a phase 1 study of nusinersen (ISIS-SMNRx) in children with spinal muscular atrophy. *Neurology* 86 (10): 890–897.
43 Tyagi, P., Kashyap, M.P., Kawamorita, N. et al. (2014). Intravesical liposome and antisense treatment for detrusor overactivity and interstitial cystitis/ painful bladder syndrome. *ISRN Pharmacol.* 2014: Article ID 601653.
44 Blietz, C.E., Thode, B., Hauses, M. et al. (2009). *In vivo* studies on the availability and toxicity of antisense oligonucleotides in bladder cancer. *In Vivo* 23 (1): 13–19.
45 Hadchity, E., Aloy, M.-T., Paulin, C. et al. (2009). Heat shock protein 27 as a new therapeutic target for radiation sensitization of head and neck squamous cell carcinoma. *Mol. Ther.* 17 (8): 1387–1394.
46 Hadaschik, B.A., Jackson, J., Fazli, L. et al. (2008). Intravesically administered antisense oligonucleotides targeting heat-shock protein-27 inhibit the growth of non-muscle-invasive bladder cancer. *BJU Int.* 102 (5): 610–616.
47 Kjems, J. and Howard, K.A. (2012). Oligonucleotide delivery to the lung: waiting to inhale. *Mol. Ther.-Nucleic Acids* 1: e1.
48 Juliano, R.L., Alahari, S., Yoo, H. et al. (1999). Antisense pharmacodynamics: critical issues in the transport and delivery of antisense oligonucleotides. *Pharm. Res.* 16 (4): 494–502.
49 Ali, S., Leonard, S.A., Kukoly, C.A. et al. (2001). Absorption, distribution, metabolism, and excretion of a respirable antisense oligonucleotide for asthma. *Am. J. Respir. Crit. Care Med.* 163 (4): 989–993.
50 Holmlund, J.T. (2003). Applying antisense technology. *Ann. N. Y. Acad. Sci.* 1002 (1): 244–251.
51 Gauvreau, G.M., Pageau, R., Séguin, R. et al. (2011). Dose–response effects of TPI ASM8 in asthmatics after allergen. *Allergy* 66 (9): 1242–1248.
52 Karras, J.G., Crosby, J.R., Guha, M. et al. (2007). Anti-inflammatory activity of inhaled IL-4 receptor-α antisense oligonucleotide in mice. *Am. J. Respir. Cell Mol. Biol.* 36 (3): 276–285.
53 Moschos, S.A., Frick, M., Taylor, B. et al. (2011). Uptake, efficacy, and systemic distribution of naked, inhaled short interfering RNA (siRNA) and locked nucleic acid (LNA) antisense. *Mol. Ther.* 19 (12): 2163–2168.

54 Shabanpoor, F., McClorey, G., Saleh, A.F. et al. (2015). Bi-specific splice-switching PMO oligonucleotides conjugated via a single peptide active in a mouse model of Duchenne muscular dystrophy. *Nucleic Acids Res.* 43 (1): 29–39.
55 Verhaart, I.E.C., van Vliet-van den Dool, L., Sipkens, J.A. et al. (2014). The dynamics of compound, transcript, and protein effects after treatment with 2OMePS antisense oligonucleotides in MDX mice. *Mol. Ther.-Nucleic Acids* 3: e148.
56 Fletcher, S., Honeyman, K., Fall, A.M. et al. (2006). Dystrophin expression in the MDX mouse after localised and systemic administration of a morpholino antisense oligonucleotide. *J. Gene Med.* 8 (2): 207–216.
57 Järver, P., O'Donovan, L., and Gait, M.J. (2014). A chemical view of oligonucleotides for exon skipping and related drug applications. *Nucleic Acid Ther.* 24 (1): 37–47.
58 Heemskerk, H.A., de Winter, C.L., de Kimpe, S.J. et al. (2009). *In vivo* comparison of 2′-*O*-methyl phosphorothioate and morpholino antisense oligonucleotides for Duchenne muscular dystrophy exon skipping. *J. Gene Med.* 11 (3): 257–266.
59 Betts, C., Saleh, A.F., Arzumanov, A.A. et al. (2012). Pip6-PMO, a new generation of peptide-oligonucleotide conjugates with improved cardiac exon skipping activity for DMD treatment. *Mol. Ther.-Nucleic Acids* 1: e38.
60 Yu, R.Z., Lemonidis, K.M., Graham, M.J. et al. (2009). Cross-species comparison of *in vivo* PK/PD relationships for second-generation antisense oligonucleotides targeting apolipoprotein B-100. *Biochem. Pharmacol.* 77 (5): 910–919.
61 Wang, L., Prakash, R.K., Stein, C.A. et al. (2003). Progress in the delivery of therapeutic oligonucleotides: organ/cellular distribution and targeted delivery of oligonucleotides *in vivo*. *Antisense Nucleic Acid Drug Dev.* 13 (3): 169–189.
62 Frey, P. and Sammons, R. (1985). Bond order and charge localization in nucleoside phosphorothioates. *Science* 228 (4699): 541–545.
63 Wartlick, H., Spänkuch-Schmitt, B., Strebhardt, K. et al. (2004). Tumour cell delivery of antisense oligonuclceotides by human serum albumin nanoparticles. *J. Control. Release* 96 (3): 483–495.
64 Srinivasan, S.K., Tewary, H.K., and Iversen, P.L. (1995). Characterization of binding sites, extent of binding, and drug interactions of oligonucleotides with albumin. *Antisense Res. Dev.* 5 (2): 131–139.
65 Egli, M. and Pallan, P.S. (2010). Crystallographic studies of chemically modified nucleic acids: a backward glance. *Chem. Biodivers.* 7 (1): 60–89.
66 Canton, J., Neculai, D., and Grinstein, S. (2013). Scavenger receptors in homeostasis and immunity. *Nat. Rev. Immunol.* 13 (9): 621–634.
67 Järver, P., Zaghloul, E.M., Arzumanov, A.A. et al. (2015). Peptide nanoparticle delivery of charge-neutral splice-switching morpholino oligonucleotides. *Nucleic Acid Ther.* 25 (2): 65–77.

68 Greer, K.L., Lochmuller, H., Flanigan, K. et al. (2014). Targeted exon skipping to correct exon duplications in the dystrophin gene. *Mol. Ther.-Nucleic Acids* 3: e155.
69 Geary, R.S., Wancewicz, E., Matson, J. et al. (2009). Effect of dose and plasma concentration on liver uptake and pharmacologic activity of a 2′-methoxyethyl modified chimeric antisense oligonucleotide targeting PTEN. *Biochem. Pharmacol.* 78 (3): 284–291.
70 Coughlin, J.E., Pandey, R.K., Padmanabhan, S. et al. (2012). Metabolism, pharmacokinetics, tissue distribution, and stability studies of the prodrug analog of an anti-hepatitis B virus dinucleoside phosphorothioate. *Drug Metab. Dispos.* 40 (5): 970–981.
71 Greenberger, L.M., Horak, I.D., Filpula, D. et al. (2008). A RNA antagonist of hypoxia-inducible factor-1α, EZN-2968, inhibits tumor cell growth. *Mol. Cancer Ther.* 7 (11): 3598–3608.
72 Corey, D.R. (2007). Chemical modification: the key to clinical application of RNA interference? *J. Clin. Invest.* 117 (12): 3615–3622.
73 Moulds, C., Lewis, J.G., Froehler, B.C. et al. (1995). Site and mechanism of antisense inhibition by C-5 propyne oligonucleotides. *Biochemistry* 34 (15): 5044–5053.
74 Fluiter, K., ten Asbroek, A.L.M.A., de Wissel, M.B. et al. (2003). *In vivo* tumor growth inhibition and biodistribution studies of locked nucleic acid (LNA) antisense oligonucleotides. *Nucleic Acids Res.* 31 (3): 953–962.
75 Ghosh, C. and Iversen, P.L. (2000). Intracellular delivery strategies for antisense phosphorodiamidate morpholino oligomers. *Antisense Nucleic Acid Drug Dev.* 10 (4): 263–274.
76 Schnell, F.J., Crumley, S.L., Mourich, D.V., and Iversen, P.L. (2013). Development of novel bioanalytical methods to determine the effective concentrations of phosphorodiamidate morpholino oligomers in tissues and cells. *BioRes. Open Access* 2 (1): 61–66.
77 Amantana, A. and Iversen, P.L. (2005). Pharmacokinetics and biodistribution of phosphorodiamidate morpholino antisense oligomers. *Curr. Opin. Pharmacol.* 5 (5): 550–555.
78 Wu, B., Benrashid, E., Lu, P. et al. (2011). Targeted skipping of human dystrophin exons in transgenic mouse model systemically for antisense drug development. *PLoS One* 6 (5): e19906.
79 Amantana, A., Moulton, H.M., Cate, M.L. et al. (2007). Pharmacokinetics, biodistribution, stability and toxicity of a cell-penetrating peptide–morpholino oligomer conjugate. *Bioconjug. Chem.* 18 (4): 1325–1331.
80 Warren, T.K., Warfield, K.L., Wells, J. et al. (2010). Advanced antisense therapies for postexposure protection against lethal filovirus infections. *Nat. Med.* 16 (9): 991–994.
81 Warren, T.K., Whitehouse, C.A., Wells, J. et al. (2015). A single phosphorodiamidate morpholino oligomer targeting VP24 protects rhesus monkeys against lethal Ebola virus infection. *MBio* 6 (1): e02344–e02314.

82 Heald, A.E., Iversen, P.L., Saoud, J.B. et al. (2014). Safety and pharmacokinetic profiles of phosphorodiamidate morpholino oligomers with activity against Ebola virus and Marburg virus: results of two single-ascending-dose studies. *Antimicrob. Agents Chemother.* 58 (11): 6639–6647.

83 Sewell, K.L., Geary, R.S., Baker, B.F. et al. (2002). Phase I trial of ISIS 104838, a 2′-methoxyethyl modified antisense oligonucleotide targeting tumor necrosis factor-α. *J. Pharmacol. Exp. Ther.* 303 (3): 1334–1343.

84 Jason, T.L.H., Koropatnick, J., and Berg, R.W. (2004). Toxicology of antisense therapeutics. *Toxicol. Appl. Pharmacol.* 201 (1): 66–83.

85 Eckstein, F. (2014). Phosphorothioates, essential components of therapeutic oligonucleotides. *Nucleic Acid Ther.* 24 (6): 374–387.

86 Djavanbakht Samani, T., Jolles, B., and Laigle, A. (2001). Best minimally modified antisense oligonucleotides according to cell nuclease activity. *Antisense Nucleic Acid Drug Dev.* 11 (3): 129–136.

87 Agrawal, S., Iadarola, P.L., Temsamani, J. et al. (1996). Effect of G-rich sequences on the synthesis, purification, hybridization, cell uptake, and hemolytic activity of oligonucleotides. *Bioorg. Med. Chem. Lett.* 6 (18): 2219–2224.

88 Agrawal, S., Tan, W., Cai, Q. et al. (1997). *In vivo* pharmacokinetics of phosphorothioate oligonucleotides containing contiguous guanosines. *Antisense Nucleic Acid Drug Dev.* 7 (3): 245–249.

89 Schmidt, K., Prakash, T.P., Donner, A.J. et al. (2017). Characterizing the effect of GalNAc and phosphorothioate backbone on binding of antisense oligonucleotides to the asialoglycoprotein receptor. *Nucleic Acids Res.* 45 (5): 2294–2306.

90 Miller, C.M., Donner, A.J., Blank, E.E. et al. (2016). Stabilin-1 and stabilin-2 are specific receptors for the cellular internalization of phosphorothioate-modified antisense oligonucleotides (ASOs) in the liver. *Nucleic Acids Res.* 44 (6): 2782–2794.

91 Koller, E., Vincent, T.M., Chappell, A. et al. (2011). Mechanisms of single-stranded phosphorothioate modified antisense oligonucleotide accumulation in hepatocytes. *Nucleic Acids Res.* 39 (11): 4795–4807.

92 Prakash, T.P., Yu, J., Migawa, M.T. et al. (2016). Comprehensive structure–activity relationship of triantennary *N*-acetylgalactosamine conjugated antisense oligonucleotides for targeted delivery to hepatocytes. *J. Med. Chem.* 59 (6): 2718–2733.

93 Iversen, P.L., Aird, K.M., Wu, R. et al. (2009). Cellular uptake of neutral phosphorodiamidate morpholino oligomers. *Curr. Pharm. Biotechnol.* 10 (6): 579–588.

94 Shoeman, R.L., Hartig, R., Huang, Y. et al. (1997). Fluorescence microscopic comparison of the binding of phosphodiester and phosphorothioate (Antisense) oligodeoxyribonucleotides to subcellular structures, including intermediate filaments, the endoplasmic reticulum, and the nuclear interior. *Antisense Nucleic Acid Drug Dev.* 7 (4): 291–308.

95 Liang, X.-H., Sun, H., Shen, W., and Crooke, S.T. (2015). Identification and characterization of intracellular proteins that bind oligonucleotides with phosphorothioate linkages. *Nucleic Acids Res.* 43 (5): 2927–2945.
96 Straarup, E.M., Fisker, N., Hedtjärn, M. et al. (2010). Short locked nucleic acid antisense oligonucleotides potently reduce apolipoprotein B mRNA and serum cholesterol in mice and non-human primates. *Nucleic Acids Res.* 38 (20): 7100–7111.
97 Weers, J.G., Bell, J., Chan, H.-K. et al. (2010). Pulmonary formulations: what remains to be donze? *J. Aerosol Med.* 23 (S2): S-5–S-23.
98 Liang, X.-H., Shen, W., Sun, H. et al. (2014). TCP1 complex proteins interact with phosphorothioate oligonucleotides and can co-localize in oligonucleotide-induced nuclear bodies in mammalian cells. *Nucleic Acids Res.* 42 (12): 7819–7832.
99 Lima, W.F., Vickers, T.A., Nichols, J. et al. (2014). Defining the factors that contribute to on-target specificity of antisense oligonucleotides. *PLoS One* 9 (7): e101752.
100 Ho, S.P., Scully, M.S., Krauthauser, C.M. et al. (2002). Antisense oligonucleotides selectively regulate aspartyl β-hydroxylase and its truncated protein isoform *in vitro* but distribute poorly into A549 tumors *in vivo*. *J. Pharmacol. Exp. Ther.* 302 (2): 795–803.
101 Bijsterbosch, M.K., Rump, E.T., Vrueh, R.L.A.D. et al. (2000). Modulation of plasma protein binding and *in vivo* liver cell uptake of phosphorothioate oligodeoxynucleotides by cholesterol conjugation. *Nucleic Acids Res.* 28 (14): 2717–2725.
102 The, F.O., De Jonge, W.J., Bennink, R.J. et al. (2005). The ICAM-1 antisense oligonucleotide ISIS-3082 prevents the development of postoperative ileus in mice. *Br. J. Pharmacol.* 146 (2): 252–258.
103 Dritschilo, A., Huang, C.H., Rudin, C.M. et al. (2006). Phase I study of liposome-encapsulated c-raf antisense oligodeoxyribonucleotide infusion in combination with radiation therapy in patients with advanced malignancies. *Clin. Cancer Res.* 12 (4): 1251–1259.
104 Gutiérrez-Puente, Y., Tari, A.M., Stephens, C. et al. (1999). Safety, pharmacokinetics, and tissue distribution of liposomal P-ethoxy antisense oligonucleotides targeted to Bcl-2. *J. Pharmacol. Exp. Ther.* 291 (2): 865–869.
105 Tari, A.M., Gutierrez-Puente, Y., Monaco, G. et al. (2007). Liposome-incorporated Grb2 antisense oligodeoxynucleotide increases the survival of mice bearing bcr-abl-positive leukemia xenografts. *Int. J. Oncol.* 31 (5): 1243–1250.
106 Rudin, C.M., Marshall, J.L., Huang, C.H. et al. (2004). Delivery of a liposomal c-raf-1 antisense oligonucleotide by weekly bolus dosing in patients with advanced solid tumors. *Clin. Cancer Res.* 10 (21): 7244–7251.
107 Gokhale, P.C., Zhang, C., Newsome, J.T. et al. (2002). Pharmacokinetics, toxicity, and efficacy of ends-modified raf antisense

oligodeoxyribonucleotide encapsulated in a novel cationic liposome. *Clin. Cancer Res.* 8 (11): 3611–3621.

108 Cortes, J., Kantarjian, H., Ball, E.D. et al. (2011). Phase 2 randomized study of p53 antisense oligonucleotide (cenersen) plus idarubicin with or without cytarabine in refractory and relapsed acute myeloid leukemia. *Cancer* 118 (2): 418–427.

109 Dean, E., Jodrell, D., Connolly, K. et al. (2009). Phase I trial of AEG35156 administered as a 7-day and 3-day continuous intravenous infusion in patients with advanced refractory cancer. *J. Clin. Oncol.* 27 (10): 1660–1666.

110 Graham, M.J., Crooke, S.T., Monteith, D.K. et al. (1998). *In vivo* distribution and metabolism of a phosphorothioate oligonucleotide within rat liver after intravenous administration. *J. Pharmacol. Exp. Ther.* 286 (1): 447–458.

111 Crooke, R.M., Graham, M.J., Martin, M.J. et al. (2000). Metabolism of antisense oligonucleotides in rat liver homogenates. *J. Pharmacol. Exp. Ther.* 292 (1): 140–149.

112 Shaw, J.-P., Kent, K., Bird, J. et al. (1991). Modified deoxyoligonucleotides stable to exonuclease degradation in serum. *Nucleic Acids Res.* 19 (4): 747–750.

113 White, P.J., Anastasopoulos, F., Pouton, C.W., and Boyd, B.J. (2009). Overcoming biological barriers to *in vivo* efficacy of antisense oligonucleotides. *Expert Rev. Mol. Med.* 11: e10.

114 Gilar, M., Belenky, A., Smisek, D.L. et al. (1997). Kinetics of phosphorothioate oligonucleotide metabolism in biological fluids. *Nucleic Acids Res.* 25 (18): 3615–3620.

115 Yu, D., Kandimalla, E.R., Roskey, A. et al. (2000). Stereo-enriched phosphorothioate oligodeoxynucleotides: synthesis, biophysical and biological properties. *Bioorg. Med. Chem.* 8 (1): 275–284.

116 Yu, R.Z., Geary, R.S., and Levin, A.A. (2006). Pharmacokinetics and pharmacodynamics of antisense oligonucleotides. In: *Pharmacokinetics and Pharmacodynamics of Biotech Drugs: Principles and Case Studies in Drug Development* (ed. B. Meibohm), 93–146. Weinheim: Wiley-VCH.

117 Levin, A.A. (1999). A review of the issues in the pharmacokinetics and toxicology of phosphorothioate antisense oligonucleotides. *Biochim. Biophys. Acta* 1489 (1): 69–84.

118 Baek, M.-S., Yu, R.Z., Gaus, H. et al. (2010). *In vitro* metabolic stabilities and metabolism of 2′-*O*-(methoxyethyl) partially modified phosphorothioate antisense oligonucleotides in preincubated rat or human whole liver homogenates. *Oligonucleotides* 20 (6): 309–316.

119 Gupta, N., Fisker, N., Asselin, M.-C. et al. (2010). A locked nucleic acid antisense oligonucleotide (LNA) silences PCSK9 and enhances LDLR expression *in vitro* and *in vivo*. *PLoS One* 5 (5): e10682.

120 Hildebrandt-Eriksen, E.S., Aarup, V., Persson, R. et al. (2012). A locked nucleic acid oligonucleotide targeting microRNA 122 is well-tolerated in cynomolgus monkeys. *Nucleic Acid Ther.* 22 (3): 152–161.

121 Levin, A.A., Yu, R.Z., and Geary, R.S. (2007). Basic principles of the pharmacokinetics of antisense oligonucleotide drugs. In: *Antisense Drug Technology: Principles, Strategies, and Applications*, 2e (ed. S.T. Crooke), 183–216. Boca Raton, FL: CRC Press.

122 Zon, L.I. and Peterson, R.T. (2005). *In vivo* drug discovery in the zebrafish. *Nat. Rev. Drug Discov.* 4 (1): 35–44.

123 Bill, B.R., Petzold, A.M., Clark, K.J. et al. (2009). A primer for morpholino use in zebrafish. *Zebrafish* 6 (1): 69–77.

124 Youngblood, D.S., Hatlevig, S.A., Hassinger, J.N. et al. (2007). Stability of cell-penetrating peptide–morpholino oligomer conjugates in human serum and in cells. *Bioconjug. Chem.* 18 (1): 50–60.

125 Hudziak, R.M., Barofsky, E., Barofsky, D.F. et al. (1996). Resistance of morpholino phosphorodiamidate oligomers to enzymatic degradation. *Antisense Nucleic Acid Drug Dev.* 6 (4): 267–272.

126 Sazani, P., Magee, T., Charleston, J. et al. (2015). *In vitro* pharmacokinetic evaluation of eteplirsen, SRP-4045, and SRP-4053; three phosphorodiamidate morpholino oligomers (PMO) for the treatment of patients with Duchenne muscular dystrophy (DMD) (P5.061). *Neurology* 84 (14 Supplement): P5061.

127 Arora, V., Devi, G.R., and Iversen, P.L. (2004). Neutrally charged phosphorodiamidate morpholino antisense oligomers: uptake, efficacy and pharmacokinetics. *Curr. Pharm. Biotechnol.* 5 (5): 431–439.

128 Sazani, P., Van Ness, K.P., Weller, D.L. et al. (2011). Repeat-dose toxicology evaluation in cynomolgus monkeys of AVI-4658, a phosphorodiamidate morpholino oligomer (PMO) drug for the treatment of Duchenne muscular dystrophy. *Int. J. Toxicol.* 30 (3): 313–321.

129 Warren, T.K., Shurtleff, A.C., and Bavari, S. (2012). Advanced morpholino oligomers: a novel approach to antiviral therapy. *Antiviral Res.* 94 (1): 80–88.

130 Winkler, J. (2013). Oligonucleotide conjugates for therapeutic applications. *Ther. Deliv.* 4 (7): 791–809.

131 Lorenzer, C., Dirin, M., Winkler, A.-M. et al. (2015). Going beyond the liver: progress and challenges of targeted delivery of siRNA therapeutics. *J. Control. Release* 203: 1–15.

132 Dirin, M., Urban, E., Lachmann, B. et al. (2015). Concise postsynthetic preparation of oligonucleotide-oligopeptide conjugates through facile disulfide bond formation. *Future Med. Chem.* 7 (13): 1657–1673.

133 Dirin, M., Urban, E., Noe, C.R., and Winkler, J. (2016). Fragment-based solid-phase assembly of oligonucleotide conjugates with peptide and polyethylene glycol ligands. *Eur. J. Med. Chem.* 121: 132–142.

6

Hybridization-Independent Effects: Principles and Specific Considerations for Oligonucleotide Drugs

Nicolay Ferrari

Centre de recherche du CHUM – Tour Viger, Montreal, Quebec, Canada

6.1 Background

The development of oligonucleotides (ONs) as therapeutic modalities is maturing rapidly. The recent approvals, three ONs approved in 2016, are integral to the progress made for understanding the molecular mechanisms, pharmacology (*in vitro* and *in vivo*), pharmacokinetics, toxicology, manufacture, and clinical efficacy. Notably, the success or failure of a specific drug development program is determined by a range of different factors, including the efficacy of the therapeutic class, in this case ONs, as well as the clinical relevance of the target. As for any new drug, developers of ON therapeutics need to consider another important factor, the safety, questioning their toxicity potential or the risk of causing unwanted toxicities or side effects.

ON therapeutics come in a variety of structures, chemistries, and modes of action, principally divided on the basis of whether they target RNA or proteins. Single-stranded antisense ONs (ASOs) are the archetype RNA-targeting ONs. Two main mechanisms of toxicity have been described for ASO. The first and most commonly observed is independent of the hybridization of the ASO to RNA. The second mechanism is dependent of ASO hybridization to RNA, being its target RNA or other RNA sequences. Hybridization-independent toxicities are therefore unrelated to the intended mode of action – Watson and Crick base pairing to RNA – and are due to interactions between the ASO drug and proteins. Hybridization-independent toxicities entail that nucleotide sequence, chemistry, or composition of the delivery system accounts for this type of toxicity [1]. On the other hand, hybridization-dependent toxicities arise from hybridization of the ASO to cellular RNA using the normal base-pairing principles. Exaggerated pharmacology results from inhibition of the intended

Oligonucleotide-Based Drugs and Therapeutics: Preclinical and Clinical Considerations for Development, First Edition. Edited by Nicolay Ferrari and Rosanne Seguin.

target RNA to levels producing deleterious effects. Exaggerated pharmacology of ASO and the specific considerations for drug development are addressed separately in Chapter 9. Hybridization-dependent effects can also occur on unintended RNA targets that are completely or partially complementary to the ON sequence. This is referred to as "off-target" effects, and the potential for "off-target" effects and related considerations for ASO development are described in Chapter 7.

This chapter focuses on hybridization-independent toxicities of the most abundantly characterized ON class, ASO. Mechanisms, some specific concerns for the development of this therapeutic class, and an example of strategy to address hybridization-independent toxicities will be reviewed.

6.2 Mechanisms of Hybridization-independent Toxicities

6.2.1 Effects Related to Oligonucleotide Sequence

The specific sequence composition of ONs dictates its pharmacologic activity. ASO sequence can also produce unwanted pharmacologic effects that are independent of hybridization [2–6]. A growing body of evidence provides increasingly important information on the most common sequence motifs, potential mechanism of toxicity, and downstream effects.

6.2.1.1 Unmethylated CpG Motifs

Unmethylated CpG sequences occur more frequently in bacterial DNA than they do in eukaryotic DNA. These sequences are recognized by the immune system of many species as a defense mechanism. Synthetic ONs with unmethylated CpG dinucleotides [7] flanked with paired purines at the 5′ end and paired pyrimidines at the 3′ end (e.g. 5′-AACGTT), referred to as the CpG motif, can induce immune activation through activation of Toll-like receptor (TLR) 9 [8–10]. The nature of the immunostimulatory effect of CpG-containing ONs is complex and varies according to the composition of the CpG-flanking sequence [11] and is heavily dependent on the sugar backbone [12].

Many elements such as the CpG sequence, the cellular distribution of TLR9, and the fact that downstream effects following activation differ among species pose important challenges for developers when assessing potential TLR9-mediated unwanted effects [13]. The CpG sequence that is optimal for TLR9 activation differs between human, primates, and rodents, reflecting species differences in the TLR9 receptor itself. The presence of a core CpG sequence 5′-GTCGTT tends to produce a stronger TLR9 response in human, especially if one of these motifs is near the 5′ end of the ON. In mice, purine-purine-C-G-pyrimidine-pyrimidine such as 5′-GACGTT and 5′-AACGTT are stronger inducers [8]. Furthermore, although most CpG-containing ONs that modulate human TLR9 have been shown to also activate rodent TLR9, the reverse may not apply, i.e. CpG ON sequences that modulate TLR9 in rodents may not modulate human TLR9 [14]. In contrast, evidence suggests that TLR9 in human and nonhuman primates recognizes the same CpG ODN sequences [15]. This comparison suggests rodent-specific effects of TLR9 activation including splenomegaly and extramedullary hematopoiesis, multi-organ lymphoid infiltrates, lymphoid hyperplasia, or even lethal systemic inflammatory response syndrome.

Responses from TLR9 may also vary depending on species differences in terms of TLR9 localization. In humans, TLR9 has relatively limited distribution, and it is found exclusively in type I interferon-producing plasmacytoid dendritic cells and in B cells [16]. In mice, macrophages and myeloid dendritic cells also express high levels of TLR9 and respond to CpG stimulation [17, 18]. It is therefore proposed that species specificity of the CpG motif resides at the level of TLR9 [19, 20].

Avoiding CpG motifs does not completely eliminate the risk associated with immunostimulatory effects but does greatly mitigate the level of this activity. There are a number of models and biomarkers that can be used to evaluate whether a given ON triggers TLR9 activation as strongly as CpG-containing ONs do. *In vitro* models can be used to measure innate responses (for example, cytokine release by human peripheral blood lymphocytes or by rodent splenocytes), whereas *in vivo* models can evaluate innate responses (e.g. cytokine release) as well as responses that also reflect adaptive immune responses [21–24].

6.2.1.2 Poly-G Sequences

Some ON sequences form secondary structures such as duplexes and hairpins by Watson–Crick base pairing. These complex structures can comprise three, four, or more ON strands. In particular, ON sequences with two or more consecutive guanosines (poly-G) have the propensity to form tetraplexes that we may ascribe enhanced nonsequence-specific activity of both phosphodiester and phosphorothioate ONs, an important issue when developing ASO therapeutics [25, 26]. The nonsequence-specific activity of poly-G-based compounds stems from an aptamer-like interaction that has been shown to occur with various cellular proteins [27, 28]. For example, they have been shown to interfere with signal transducer and activator of transcription STAT-1, STAT-3, and STAT-4 and/or other downstream factors proximal to nuclear factor (NF)-κB activation involved in the signaling pathways of TLRs [29, 30]. Poly-G sequences containing 5′-GGTTGGTGTGGTTGG have been shown to have affinity for thrombin [31]. Nonspecific binding to thrombin may alter normal function leading to side effects associated with anticoagulation such as excessive bleeding. Other sequences investigated for potential antiviral activity, for example, containing 5′-TTGGGGTT, interfered with HIV envelop proteins resulting with a blockade of viral replication [32].

Poly-G sequences may also form other unusual structures depending on context. ONs containing repetitive 5′-TTAGGG motifs, which have been considered for the development of cancer immunotherapy, were shown to have multiple effects on immune activation, effects independent of CpG or TLR9 [33]. These ONs were capable of blocking the production of proinflammatory and Th1 cytokines (IL-2) induced not only with TLR9 ligands but also with a variety of polyclonal activators and antigens [33].

6.2.1.3 DNA Triplex-forming Oligonucleotides

Another cause for concern is the binding of ONs to the double-helix DNA forming a triple-helical DNA structure, or triplex DNA. Polypurine- or polypyrimidine-containing ON molecules can interact in the major groove of duplex DNA through Hoogsteen hydrogen bonds. Regardless of the composition of the ON, the binding seems restricted to DNA binding sites with runs of purines on one strand and pyrimidines on the other [34]. Triple-helix formation can induce a number of alterations at the genomic level. It has been linked with the induction of mutations and DNA recombination observed in mammalian cells treated with ONs [34, 35]. When the binding occurs in regulatory regions, it is believed that DNA triplex can also modulate normal cellular processes. For example, triplex DNA can compete with the binding of transcription factors preventing transcription initiation thus resulting with transcriptional repression [36].

6.2.1.4 Other Motifs

During the early discovery phase, typical screening for ASO against a given target allows to classify potential candidates based on activity and screen out less favorable candidates. In the latter category, some ON sequences, "bad players," are found to produce unexpected biological activity such as cellular cytotoxicity *in vitro* [37] or *in vivo* toxicity [38, 39]. Evidence from structure–toxicity analysis suggests that the observed toxicity is sequence dependent. In studies by Drygin et al., the only sequence motif that could be identified that contributed to cellular cytotoxicity was the presence of 5-methyl CT repeats [37]. The mechanism underlying this cytotoxicity is not fully understood, but study results suggest toxic effect may be mediated through proteases involved in cell proliferation. Similar efforts by Burdick et al. to understand sequence-dependent hepatotoxicity generated trinucleotide motifs (5′-TCC and 5′-TGC) that were present only in hepatotoxic sequences [39]. The ON sequences containing these motifs were more likely to bind liver proteins, interaction that triggered activation of stress pathways and increased hepatotoxicity.

6.2.2 Effects Related to Oligonucleotide Chemistry

Sequence-dependent effects unrelated to RNA hybridization, as described above, are currently somewhat predictable, as bioinformatics analysis to predict potential sequence-toxicity relationship are increasingly becoming part of the toolbox. There are other potential mechanisms of toxicity unrelated to the antisense mechanism or sequence composition. ASO therapeutic agents, like all other xenobiotics, have their own inherent toxicities that stem from the physicochemical characteristics of the compounds. Similarly to sequence-dependent effects, class-related toxicities of ONs are also mediated through unspecific ON–protein interactions. Most, if not all, toxicities observed in preclinical studies are toxicities relating to this class.

6.2.2.1 Phosphorothioate Oligonucleotides

ASOs are water-soluble, highly charged (polyanionic), and nonspecifically reactive molecules. For most therapeutic applications, they must withstand the rigors of transport via the bloodstream and uptake into target cells or penetrate through various other matrices such as the skin epithelium or lung alveolar epithelium, all of which are laden with nuclease enzymes capable of rapidly degrading natural nucleic acid structures. Besides metabolic instability, unmodified ONs are rapidly eliminated through renal glomerular clearance.

During the last three decades, the most common strategy for imparting drug-like properties to ASOs and other types of ON to overcome these initial drawbacks is to introduce chemical modifications that afford dramatically increased nuclease resistance, which effectively achieves greater stability in the blood and tissues. The most common chemical modification that is incorporated in the majority of antisense drugs is the substitution of a sulfur atom for one of the nonbonding oxygens in each internucleotide linkage of the phosphodiester backbone, i.e. the phosphorothioate chemistry. This modification (and other related backbone alterations) typically strengthens the polyanionic character of the molecule accentuating its reactivity. This, in addition to the superior tissue persistence, translates into more pronounced nonspecific effects, such that systemic administration of phosphorothioate ASOs results in various forms of toxicity that are largely class effects and unrelated to the mechanism of action, i.e. the effects are hybridization independent. Because of the extensive body of literature, we will focus on the class effects of systemically administered phosphorothioate ASOs. Class effects of locally administered phosphorothioate ONs will be described in a separate section.

Phosphorothioate ASO compounds are known to bind nonspecifically to proteins. Affinities of these interactions span the micromolar to the millimolar ranges [40–42]. The augmented affinity of phosphorothioate ONs for proteins is inherent to the increased lipophilicity of the phosphorothioate linkage over the phosphodiester linkage and stickiness from the additional negative charges of the thioate compared with the diester bond. The increased nonspecific protein binding of phosphorothioate ONs has been substantially documented and is the key mechanism underpinning the toxicity of this class. Examples of proteins with altered function following phosphorothioate ON binding include heparin-binding proteins such as factor H [43], basic fibroblast growth factor (bFGF), acidic fibroblast growth factor (aFGF), Kaposi's growth factor (FGF-4) [44], DNA polymerases [45], CD4 [46], gp120 [47], Mac-1 [48], RNase H [45], human immunodeficiency virus type 1 (HIV-1) reverse transcriptase [49], herpes simplex virus (HSV) type 2-induced DNA polymerase [50], HIV-1 integrase [51], C23/nucleolin, and other nuclear proteins [52].

Nonspecific binding of ASOs to proteins could interfere with normal protein function with the potential of causing severe toxicity. But the protein binding property is not unique to thioated ONs. Naturally occurring sulfated polyanions, such as the glycosaminoglycans heparan, dermatan, and chondroitin sulfates, are widely distributed in biological fluids, at the cell membrane and inside the cell, where they are implicated in several physiological and/or pathological processes such as infectious diseases, angiogenesis, and tumor growth [53]. These molecules have also been used in pharmaceutical approaches such as tumor-specific delivery and targeting vehicles for toxins and chemotherapeutics [54]. Preventing completely the nonspecific protein binding of ONs may not be desirable either; for example, the nonsequence-specific binding of phosphorothioate ONs to albumin contributes to their relatively slow rate of total body clearance [55].

Manifestations of toxicity that are most commonly observed following systemic delivery of phosphorothioate ASOs include thrombocytopenia, activation of the alternative complement pathway (and possible downstream sequelae, including hypotension), inhibition of the intrinsic coagulation cascade (reflected by prolongation of aPTT), and various changes in tissues (observed in animal studies), which are typically most prominent in the major organs of uptake such as the kidneys, liver, spleen for systemically administered ONs, or lung in the instance of pulmonary delivery via inhalation [56–58].

6.2.2.1.1 Thrombocytopenia

Platelets play crucial roles in pathophysiological conditions including wound healing, inflammation, cardiovascular diseases, diabetes, arthritis, Alzheimer's disease, angiogenesis, and metastasis [59]. Abnormally low amount of platelets leads to thrombocytopenia and bleeding disorders. There are different causes of thrombocytopenia, the most important ones being drug-induced immune-mediated thrombocytopenia or the downstream sequelae of splenomegaly [60]. Thrombocytopenia is an important side effect commonly described for drugs; more than 200 drugs reportedly causing immune thrombocytopenia including antibiotics, anti-inflammatory drugs, antineoplastics, antihypertensive drugs, and antidiabetic drugs [61].

Platelet reductions have been observed in rodent and monkey studies of phosphorothioate ASOs and are likely the result from the sequestration of the platelets secondary to the phosphorothioate-induced splenomegaly [62]. Thrombocytopenia has also been reported in man following parenteral administration of phosphorothioate ONs in cancer patients as well as in rheumatoid arthritis patients [63–72]. Studies suggest that thrombocytopenia is dependent of the infusion schedule and when is observed the platelet count generally normalizes and causes no bleeding episodes. Thrombocytopenia has not been observed, however, when phosphorothioate ONs were administered via the pulmonary route in line with the lower plasma concentrations that can be achieved with this route [58, 73–76].

6.2.2.1.2 Complement Activation

The complement system is increasingly recognized as important in the pathogenesis of tissue injury *in vivo* following immune, ischemic, or infectious insults. Within the complement system, three pathways are capable of initiating the processes that result in C3 activation: classical, alternative, and lectin [77]. Activation of complement alternative pathway is another key class-related toxicity observed with phosphorothioate ONs [78–81]. Activation of this pathway by systemically administered phosphorothioate ASOs has been more an issue in monkey toxicology studies and has been shown to occur primarily after rapid infusion of the ON drug [80]. In the monkey, the threshold plasma concentration of phosphorothioate ASO required for activation has been well defined, i.e. when plasma concentrations approach or exceed 40–50 μg/ml. In contrast to the monkey, complement activation has not been a concern in clinical trials as plasma concentrations can be avoided by controlling the dose and rate of administration underscoring the importance of the clinical dosing regimen [78, 82].

Complement activation is characterized by the reduction of serum hemolytic potential (CH_{50} analysis) and concomitant increases in the liberation of complement split products Bb, C3a, and C5a. The split product C5a is the most highly biologically active split product that mediates almost all of the possible adverse sequelae in monkeys. The absence of C4a increase is consistent with the activation of complement through the alternative pathway and not the classical complement pathway. Evidence suggests the mechanism underpinning complement activation is the result of interference of the ONs with factor H, the regulatory protein that limits alternative pathway activation. Disruption of factor H interaction with C3 convertase by the ON could promote activation of this pathway [78, 81]. In monkeys, complement activation is responsible for acute toxicities secondary to the physiologic responses to the formation and release of biologically active complement split products C3a and C5a. The acute toxicity is evidenced by marked clinical signs (lethargy and periorbital edema) accompanied with hemodynamic changes (heart rate and blood pressure reduction) and hematologic effects (fluctuations in neutrophils) [80]. Greater sensitivity of monkey to activation of the alternative complement pathway compared with humans is likely because of differences in factor H inhibitory capacity [43]. It has been reported that complement inhibition for monkey factor H is about threefold greater than that for human protein. The absence of ON-induced complement activation observed in dogs also supports the species differences of this effect of phosphorothioate ONs [81].

6.2.2.1.3 Effect on Clot Formation

There are two pathways involved in the formation of a fibrin clot: the intrinsic and extrinsic pathway. Although they are initiated by distinct mechanisms, the two converge on a common pathway that leads to clot formation. The intrinsic pathway (also called the contact activation pathway) is much less significant to hemostasis under normal physiological conditions than the extrinsic pathway, but under abnormal physiology such as hyperlipidemic states or bacterial infiltration, it is responsible for activation of thrombosis [83].

As for complement activation, the inhibition of the intrinsic coagulation cascade reflected by prolongation of activated partial thromboplastin times (aPTT) represents an unwanted effect relating to the phosphorothioate class occurring concomitantly with systemic administration. The prolongation of aPTT is transient, more pronounced than prothrombin times (PT), indicating the effect involves mainly the intrinsic pathway rather than the extrinsic pathway. Similarly to complement activation, prolongation of aPTT correlates directly with plasma concentrations of the phosphorothioate ASO with maximum prolongation occurring at C_{max}. Plasma concentrations associated with increased aPTT are, however, lower (~10–15 μg/ml) than those required for complement activation. Therefore, it is also possible to avoid increased aPTT by controlling the administration of the phosphorothioate ONs using dosing regimen in such a way that plasma concentrations remain below threshold [82, 84].

Prolongation of aPTT is common to many species and has been observed in mice, rat, and monkey [85–87]. Because of the lower plasma concentrations needed to increase prolongation of aPTT compared to complement activation, observations of prolongation of aPTT have been made in man. In these instances, however, there was no evidence of altered coagulation status nor signs of hemorrhage [71, 82, 88]. The relative insignificance of the clinical manifestations of these coagulation changes are most likely due to the relatively small magnitude of change and the transient nature of the aPTT, even following repeated dosing [71, 89, 90].

The mechanism by which phosphorothioate ASOs affect clotting times (prolongation of aPTT) in human plasma is the result of the selective interference with the tenase complex [91]. Intrinsic tenase complex contains the active factor IX (IXa), its cofactor factor VIII (VIIIa), and the substrate (factor X) and can be activated by negatively charged molecules such as heparin [92]. *In vitro* studies using human plasma suggest that phosphorothioate ASOs bind to an exosite on factor IXa, modulating catalytic activity of the intrinsic tenase complex, interfering with the activation of factor X by the factor IXa-phospholipid complex [93].

6.2.2.1.4 Effects Relating to Accumulation in Organs

In contrast to the class effects described above, occurring when plasma concentrations of phosphorothioate ONs surpass certain threshold concentrations and involving different mechanisms, class effects in various organs seem to relate principally to the proinflammatory potential of phosphorothioate ONs that leads to infiltrates of monocytes, lymphocytes, and fibroblasts [94, 95]. The proinflammatory effects of ONs will be reviewed in more details in Chapter 8.

For pharmacologic agents in general, there is a close relationship between drug exposure and target organ effect. Notwithstanding small variations based on sequence, the basis of the class effect in organs is the very well conserved pharmacokinetic properties of phosphorothioate ASOs. Following systemic administration, the ONs accumulate in the kidney and liver, with the kidney representing the site of the highest concentration. The kidney, therefore, is possibly the primary target organ for toxicity in animals [96]. It is not a coincidence that the vast majority of ONs currently in development are against hepatic or kidney targets as, in these organs, the relationship between tissue concentration and effect is well defined since it is the tissue concentration that is required for pharmacologic activity as well. For the same reasons, there are a number of potential toxicities or potential target organs that are consistently not affected following systemic treatment. For example, absorption of phosphorothioate ONs in the brain, lung, gastrointestinal tract, eye, skeletal muscle, or heart muscle is basically absent, and no or very limited toxicities in these particular organs have been reported with systemic administration of these compounds [97]. There are however exceptions where the exposure–effect relationship is not as strict. The spleen and lymph nodes are not among those with the highest concentration, yet they are susceptible to concentration-dependent proinflammatory effects of phosphorothioate ONs; effects are described separately (Chapter 5).

The histologic hallmark of ON exposure and accumulation in tissues is the presence of basophilic granulation that has been observed in various tissues of animals treated with systemic administration of phosphorothioate ONs [97] as well as with local delivery such as via the pulmonary route of administration [58, 73]. The organs exhibiting basophilic granulation are typically those that are major sites of deposition of the ON [57]. It is well recognized that this granular material reflects deposition and clearance of the ON in the cell type(s) in which this is observed. For systemically administered ONs, the granulation is commonly seen in resident macrophages of the spleen, lymph nodes, and liver, in addition to the proximal tubule cells in the kidney. The uptake of ON by resident macrophages is often accompanied by increases in the number of macrophages (histiocytosis), which results in an induced clearance process for the ON.

The kidney serves the body as natural filter to remove excess organic molecules from the blood. As a result of the physiologic handling of the ON drug at the glomerulus, the kidney is the organ that contains the highest concentration of ON following systemic administration and is particularly high in the proximal tubular epithelium [96, 98, 99]. Evidence demonstrates that the uptake of phosphorothioate ONs in the kidney occurs through interaction of the compound with the brush boarder epithelium. The ONs that are taken up in these cells by endocytosis and accumulation of ONs can be revealed by the presence of the basophilic granules [86, 100, 101]. The pattern and distribution of these ON-containing basophilic granules in the tubular epithelium parallels that for normal low molecular weight proteins. In the absence of cell morphology changes (e.g. increased cytoplasm, vacuolation), the presence of basophilic granules in the kidney is considered nonadverse.

Another change commonly observed in the kidney following systemic administration of ONs is the presence of cytoplasmic vacuoles in tubular epithelium. It is suggested that the presence of these vacuoles arises from osmotic imbalances induced by high concentrations of hydroscopic material (i.e. ONs) in the phagolysosomes.

Other changes have been observed at relatively high doses (higher than 80 mg/kg/week) and high tissue concentration (as high as 2.6 mg/g). Dose-dependent renal tubular atrophic and degenerative changes have been reported. These changes have been generally not accompanied by marked renal functional abnormalities as determined by the typical measurements of blood urea nitrogen, serum creatinine, creatinine clearance, or urine specific gravity. Nor were there changes in glomerular filtration rate or renal blood flow [102]. In reported instances where alterations of renal function were observed in monkey (increased urine protein/creatinine ratio), these were rapidly reversible [103]. The potential for effect on the kidney is of concern for the clinical development of phosphorothioate ONs and thus necessitates clinical monitoring. To date, however, while changes in renal clinical parameters have been observed [66], no evidence of renal dysfunction has been described in clinical trials of systemically administered phosphorothioate ONs [63, 65–71, 104–107]. The one notable exception of renal dysfunction in man is a recent report of one patient with acute renal injury following administration of a phosphorothioate ON with locked nucleic acid (LNA) modifications. The patient recovered fully, and the kidney function was normal after treatment was stopped [108].

The liver is the second main target organ in which we can find the highest accumulation of phosphorothioate ONs following systemic administration. In contrast to the kidney where ONs are mainly found in tubular epithelium, ONs can be detected in all liver cells [99, 109, 110]. But similarly to the kidney, the most commonly observed histopathologic change in the liver from animals treated with systemic administration of phosphorothioate ONs is the presence of cytoplasmic basophilic granules in the liver macrophages, the Kupffer cells. Again, the presence of the basophilic granules and hypertrophy of Kupffer cells reflects the normal function of Kupffer cells of clearing foreign material and is normally not associated with hepatic dysfunction.

Additional dose-dependent changes include lymphohistiocytic cell infiltrates that can be associated with multifocal hepatocellular degeneration or single-hepatocyte necrosis, consistent with the proinflammatory class effect of phosphorothioate ONs [111]. These changes are usually accompanied by increase in serum transaminases (ALT and AST), decreased levels of albumin and cholesterol that could be indicative of altered liver function [85, 112]. In clinical trials, elevation of serum transaminases (AST and ALT) is commonly reported following infusion of phosphorothioate ONs [63, 65–67, 71, 88, 104, 106, 113].

6.2.2.2 Effects of Other Chemical Modifications

Antisense technology has progressed beyond the point of using only phosphorothioate ONs as therapeutic agents. These antisense molecules contain additional chemical modifications intended to improve the overall therapeutic properties by increasing potency, optimizing pharmacokinetic properties, and improving the safety profile and represent the next generation of therapeutic agents (reviewed in Chapter 2).

The modification conferring the greatest increase in potency and that has reached clinical development is those with modification on the ribose sugar of nucleobases such as 2′-*O*-methoxyethyl-modified ONs (2′-MOE) [112] and LNA [114]. These ONs still possess phosphorothioate linkages throughout the molecule thus have toxicological properties that are similar in nature as described for phosphorothioates ONs [102, 103]. Differences observed between phosphorothioate DNA ONs and 2′-MOE or LNA are generally due to slight differences in pharmacokinetics profiles, protein binding, and prolonged stability [38, 39, 115]. This is also the case for more contemporary modifications such as constrained ethyl (cEt) modified phosphorothioate ASO [116].

6.3 Hybridization-independent Effects Following Local Delivery of Oligonucleotides

6.3.1 Pulmonary Toxicity of Inhaled Oligonucleotides

As described above, a major impediment to the development of antisense therapeutics is the difficulty of noninvasive and targeted delivery and in turn achieving effective tissue concentrations without eliciting adverse effects. Another practical limitation in developing ON therapeutics is that parenteral administration of ASOs is by its nature not targeted and therefore requires relatively high doses in order to achieve concentrations in target tissues that are sufficient for efficacy. The previous section described the manifestations of toxicity that are most commonly observed following systemic delivery of phosphorothioate ONs and various changes in tissues (observed in animal studies), which are typically most prominent in the major organs of uptake such as the kidney and liver.

In contrast, the lung represents a distinctive target organ for ON therapy, as it is readily accessible by direct and noninvasive delivery to the site of action via inhalation. Distribution of phosphorothioate ASOs to the lung using nebulized solutions [58, 73, 74, 117–119] is much more extensive and complete to both the large and small airways relative to parenteral administration [120, 121] and thus provides a good opportunity for antisense activity [73, 122–125]. Concentrations of phosphorothioate ONs achieved in the lungs are dose dependent, and while ONs can be measured in the kidney and liver following inhalation of high doses; these are significantly lower than in the lungs, thus minimizing systemic exposure and virtually eliminating the risk for adverse effects on the kidney and liver as well as the known hemodynamic side effects noted above [58, 73, 74, 117].

The effects associated with pulmonary delivery of phosphorothioate ASOs by inhalation are primarily dose-dependent class effects as seen in the various organs with high concentrations of ASOs. Because of the route of administration and inhalation, the changes are limited to the lung, the respiratory tract, and the upper airways. With the understanding that the type and/or severity of findings may differ among the tissues, the typical changes observed in the lung following pulmonary delivery resemble those observed in other tissues of major accumulation via other routes of administration.

One common toxicological preclinical finding of inhaled ASOs is the accumulation of alveolar macrophages within the lungs of animals. As for macrophages in the kidney and Kupffer cells in the liver, the alveolar macrophages and histiocytes of animals treated with ONs were enlarged with cytoplasmic vacuolization containing basophilic granular material [58, 73, 117]. These intracytoplasmic vacuoles contain large amounts of basophilic granular material consistent with the normal pulmonary mechanism associated with ON phagocytosis and most likely reflecting the clearance of the inhaled material. This is particularly observable histologically with the fact that ONs stain blue with hematoxylin and eosin [57].

Other changes include interstitial macrophages (histiocytes) and mononuclear cell infiltration and accumulation. These changes are typically more abundant in the lung parenchyma than the upper airway tissues or the transbronchial lymph nodes. Additional changes that have been observed are consistent with what is considered the downstream sequelae of tissue inflammation and consist of occasional observations of hemorrhage and fibroplasia and metaplasia in the lung or associated tissues (e.g. the trachea and lymph nodes), usually with relatively pronounced inflammation [58]. Finally, correlating with the relatively low levels of plasma concentrations of phosphorothioate ASOs following inhalation, no changes have been observed on hematology nor clinical chemistry. In addition, there have been no observations of meaningful effects on alternative pathway activation (for example, changes in split complement products such as Bb). Similarly, plasma analysis of coagulation parameters (aPTT and PT) following inhalation exposure has not provided indications of prolongation of coagulation times [58, 73, 117].

All the changes noted in animal studies were reversible upon termination of treatment. Alveolar macrophage accumulation and evidence of lung inflammation or of impaired lung function in normal subjects and patients following inhalation of phosphorothioate ONs have not been reported [75, 76, 126, 127]. Moreover, as the case in animal studies, hypersensitivity testing revealed no effect of ON drug administration.

6.3.1.1 Specific Considerations for Inhaled Oligonucleotides

One of the primary challenges in progressing these types of molecules into clinical trials for the treatment of respiratory diseases is the observation of the findings described above and whether the changes represent safety concerns for humans.

The macrophage accumulation and staining, with no or little evidence of activation nor other inflammatory cell accumulation, would usually be considered a nonadverse adaptive mechanism in an otherwise healthy animal lung [128, 129]. Whether this finding might be considered adverse, however, particularly in patients with established lung disease, is not well understood. In addition, although toxicity has not been reported in normal subjects to date, there may be increased susceptibility in patients with diseased lungs, for example, due to impaired epithelial barrier function. Most likely reflecting the adequate safety margins, relatively low-inhaled doses, and clinical trials of relatively short duration (i.e. <1 month), there have been no reports of increased lung inflammation in patients following inhalation of phosphorothioate ONs, based on the presence of inflammatory cells in sputum of changes in pulmonary function [75, 76, 126, 127]. The effects of prolonged exposure to ONs to human lungs remain undetermined.

The functional consequences that could result from this change are unclear as there is still controversy whether the vacuoles/inclusion bodies reflect normal clearance of the inhaled ONs or are early indicators of lung toxicity. While today there are no reports in the open literature describing the effects of inhaled ONs on alveolar macrophages, Reasor et al. [130] showed that the vacuolated (foamy) appearance of macrophages did not impair pulmonary host defense processes. The presence of foamy macrophages is not limited to a response to ONs as there is a growing body of evidence of inhaled and noninhaled small-molecule drugs, many of which are marketed, that have been associated with the presence and accumulation of vacuolated macrophages [130] including tobramycin [131] that is currently used as an inhaled antibiotic in both adults and pediatrics and has been shown to be safe and well tolerated in patients with cystic fibrosis [132]. As such, the changes observed with the inhaled ONs are considered relatively benign and not a distinct or isolated response to ON therapy.

Similarly, exposure of macrophages to phosphorothioate ASOs *in vitro* led to morphological changes of increased vacuolization; however, the hypervacuolization did not lead to functional consequences that were observed with another vacuolization-inducing drug (amiodarone). Together, these observations confirm that the *in vivo* phenotypic changes observed following inhalation of phosphorothioate ONs may be consistent with a clearance mechanism and not an activation or impairment of macrophages [128]. Indeed, cytoplasmic vacuolization of macrophages exposed to phosphorothioate ONs could be reproduced using an *in vitro* model using murine macrophage cells. We demonstrated that the cytological appearance of the treated macrophages *in vitro* was phenotypically similar (presence of cytoplasmic vacuoles) to the histologic appearance of macrophages accumulating in the alveolar lumens observed histologically in the lungs from animals exposed to phosphorothioate ONs (Figure 6.1). Further characterization using electron microscopy analysis and Nile red staining confirmed the vacuoles contained phospholipid-forming lamellar structures. The morphological changes did not correlate with functional changes such as cell proliferation, phagocytosis, and cell activation [128].

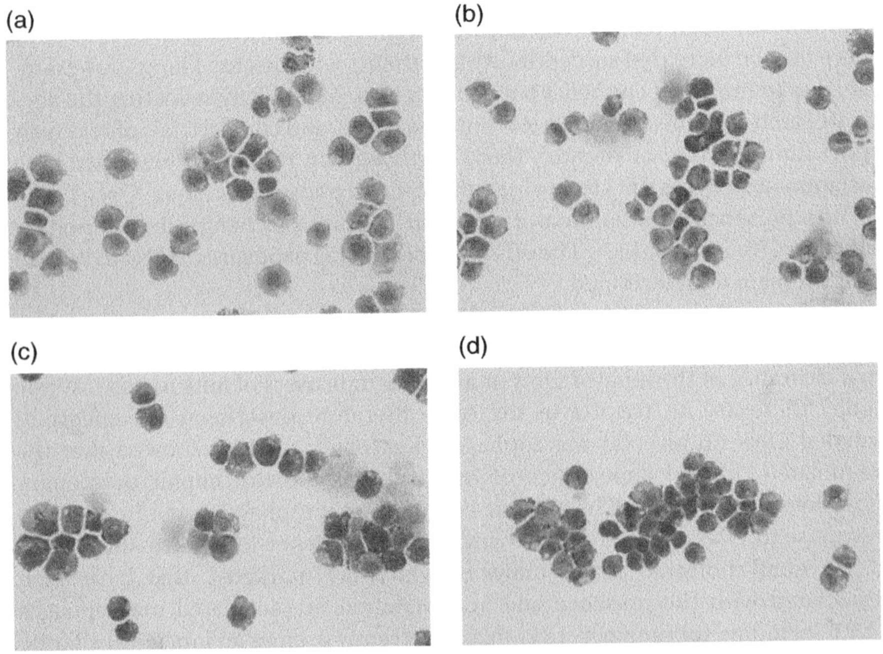

Figure 6.1 Morphological changes in macrophages following ASO exposure *in vitro*. Representative light micrographs (63×) of RAW 264.7 cells treated with vehicle (a) or a phosphorothioate ASO at 5 mM (b), 15 mM (c), or 30 mM (d) for 48 h. Incubation of cells with ASO resulted in minimal to mild vacuolation at the highest ASO concentration.

6.3.2 Approaches to Reduce Hybridization-independent Class Effects of Inhaled Oligonucleotides

Another shortcoming of phosphorothioate ASOs is their slightly reduced affinity toward complementary RNA molecules in comparison with their corresponding unmodified phosphodiester ASOs [133]. This weakness is, in part, compensated by an enhanced specificity of hybridization found for phosphorothioate ASOs compared with unmodified molecules [134].

6.3.2.1 Mixed Phosphorothioate/Phosphodiester Oligonucleotides

We found that reducing the phosphorothioate content of an ASO could potentially address the main challenge relating to the class effects of phosphorothioate ONs. However, such strategy would not represent a viable approach if reducing the phosphorothioate content resulted also in reduced pharmacologic activity of the ON. We have conducted structure–activity analysis of an ASO targeting phosphodiesterase (PDE) 7A with hybrid phosphorothioate/phosphodiester linkages (Table 6.1 and Figure 6.2). Reducing the phosphorothioate content of an ASO by selectively reintroducing phosphodiester internucleotide linkages produced molecules with increasing activity. We showed that it was not only the overall reduction of phosphorothioate linkages that modulated antisense activity but rather the location of the internucleotide phosphorothioate linkages within the ASO sequence (data not shown). In general, we found that alternating phosphorothioate/phosphodiester segments conferred the greatest antisense activity as long as the end linkages (5′- and

Table 6.1 List of antisense oligonucleotide sequences with mixed phosphorothioate/phosphodiester linkages.

Oligonucleotide	Sequence (5′-3′)	Phosphorothioate (%)
PS-DNA	$t_sc_sa_st_sg_sa_sg_st_sg_sg_sc_sa_sg_sc_st_sg_sc_sa_sa_st_st$	100
PO7	$t_sc_sa_ot_og_sa_sg_ot_og_sg_sc_sa_og_oc_st_sg_sc_oa_oa_st_st$	60
PO7-DAP	$t_sc_sX_ot_og_sX_sg_ot_og_sg_sc_sX_og_oc_st_sg_sc_oX_oX_st_st$	60
PO9	$t_sc_sa_ot_og_oa_og_st_sg_og_oc_oa_og_sc_st_og_oc_oa_oa_st_st$	40
PO9-DAP	$t_sc_sX_ot_og_oX_og_st_sg_og_oc_oX_og_sc_st_og_oc_oX_oX_st_st$	40
PO10	$t_sc_sa_ot_og_sa_sg_ot_og_og_sc_sa_og_oc_ot_sg_sc_oa_oa_st_st$	50
PO10-DAP	$t_sc_sX_ot_og_sX_sg_ot_og_og_sc_sX_og_oc_ot_sg_sc_oX_oX_st_st$	50
PS-FANA	$\mathbf{T}_s\mathbf{C}_sa_st_sg_sa_sg_st_sg_sg_sc_sa_sg_sc_st_sg_sc_s\mathbf{A}_s\mathbf{A}_s\mathbf{T}_s\mathbf{T}_s$	100
PS-LNA	$T_sC_sA_sT_sG_sg_st_sg_sg_sc_sa_sg_sc_st_sg_sC_sA_sA_sT_sT$	100

Lowercase letters = DNA; X = 2-amino-2′-deoxyadenosine; Bold letters = 2′F-ANA; underlined letters = LNA; lower script; S = phosphorothioate linkages; lower script O = phosphodiester linkages.

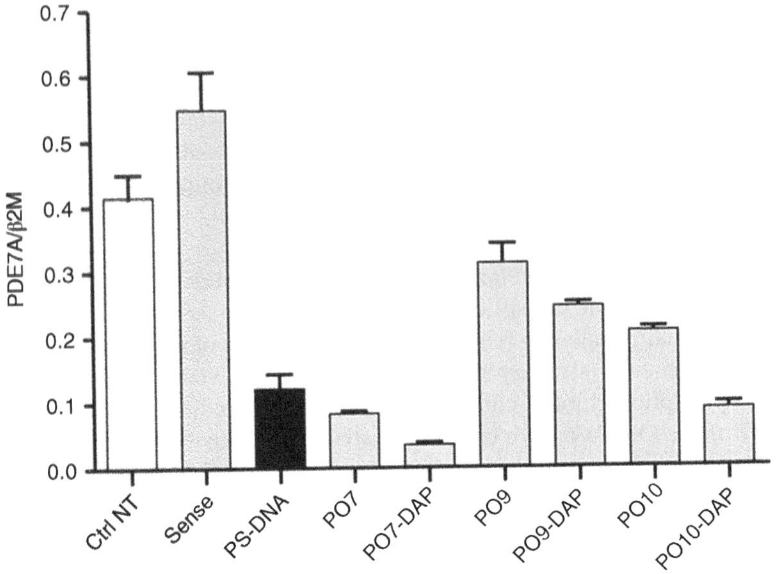

Figure 6.2 Comparison of target mRNA inhibition between selected mixed phosphorothioate/phosphodiester ASO sequences with (referred to as POX-DAP) or without (referred to as POX) substitution of adenosines with 2-amino-2′-deoxyadenosine (DAP). Flp-In T-REx 293 cells remained either untransfected (Ctrl NT) or transfected for 24 h with 236 nM of each ASO or a sense control sequence (sense). ASO sequences are listed in Table 6.1. The mRNA expression (mean±SEM) of PDE7A was quantified using the QuantiGene 2.0 assay with normalization to the expression levels of β2M.

3′-ends) were phosphorothioates (Figure 6.2). In addition, we found that substitution of adenosine with 2-amino-2′-deoxyadenosine (DAP) further increased the activity of the mixed phosphorothioate/phosphodiester ASOs. In the best instance, mixed phosphorothioate/phosphodiester ASOs with DAP were more potent than a phosphorothioate 2′F-ANA gapmer (Figure 6.3a) and comparable activity to a phosphorothioate LNA gapmer (Figure 6.3b).

Reducing the phosphorothioate by 40% was sufficient to abrogate the immunostimulatory effects of the ASOs *in vitro*. The immunostimulatory effect of a fully phosphorothioated ASOs was compared with the same sequence with mixed phosphorothioate/phosphodiester linkages in mouse macrophage cells (RAW264.7) and in human peripheral blood mononuclear cells (PBMC). Results showed no induction of secretion of TNF-α in mouse macrophage cells exposed to the mixed phosphorothioate/phosphodiester ASOs at concentrations up to 30 μM (Figure 6.4a). The same observation could be made in human PBMC (Figure 6.4b). Treating the PBMC with the mixed phosphorothioate/phosphodiester ASOs did not induce secretion of IFN-α in contrast to the fully

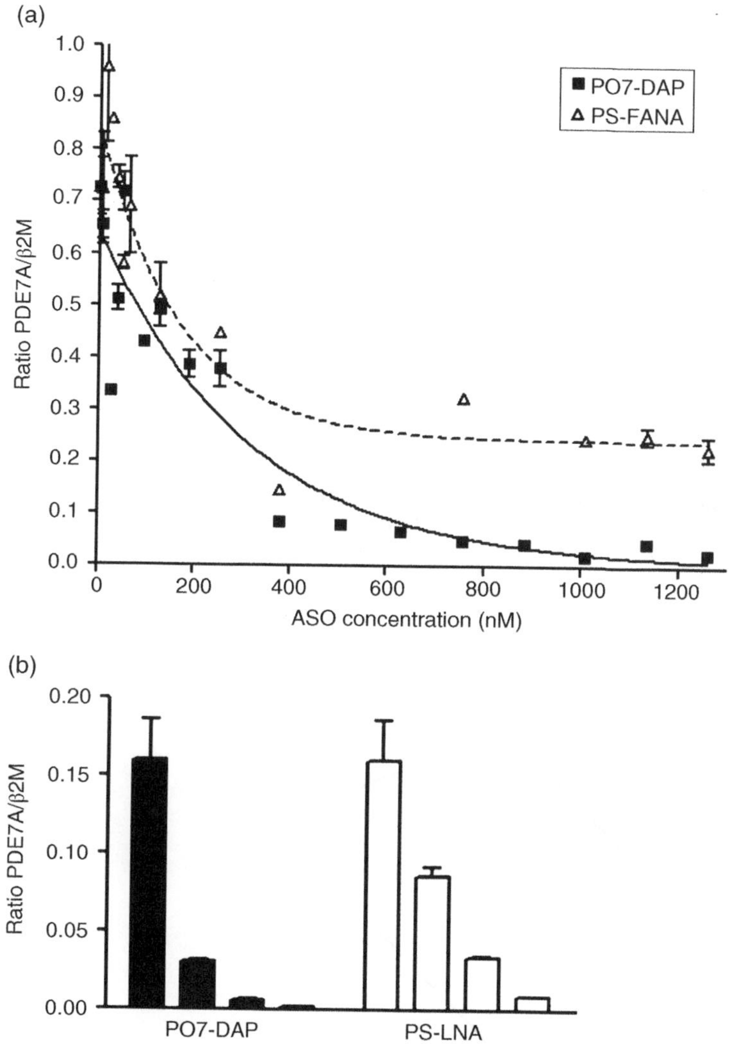

Figure 6.3 Comparison between a selected mixed phosphorothioate/phosphodiester ASO sequence with DAP (referred to as PO7-DAP) and (a) the same ASO sequence as a phosphorothioate 2′F-ANA gapmer (PS-FANA) (ASO concentrations up to 1250 nM), and (b) a phosphorothioate LNA gapmer (PS-LNA) (ASO concentrations up to 713 nM). ASO sequences are listed in Table 6.1. Flp-In T-REx 293 cells were transfected for 24 h. The mRNA expression (mean±SEM) of PDE7A was quantified using the QuantiGene 2.0 assay with normalization to the expression levels of β2M.

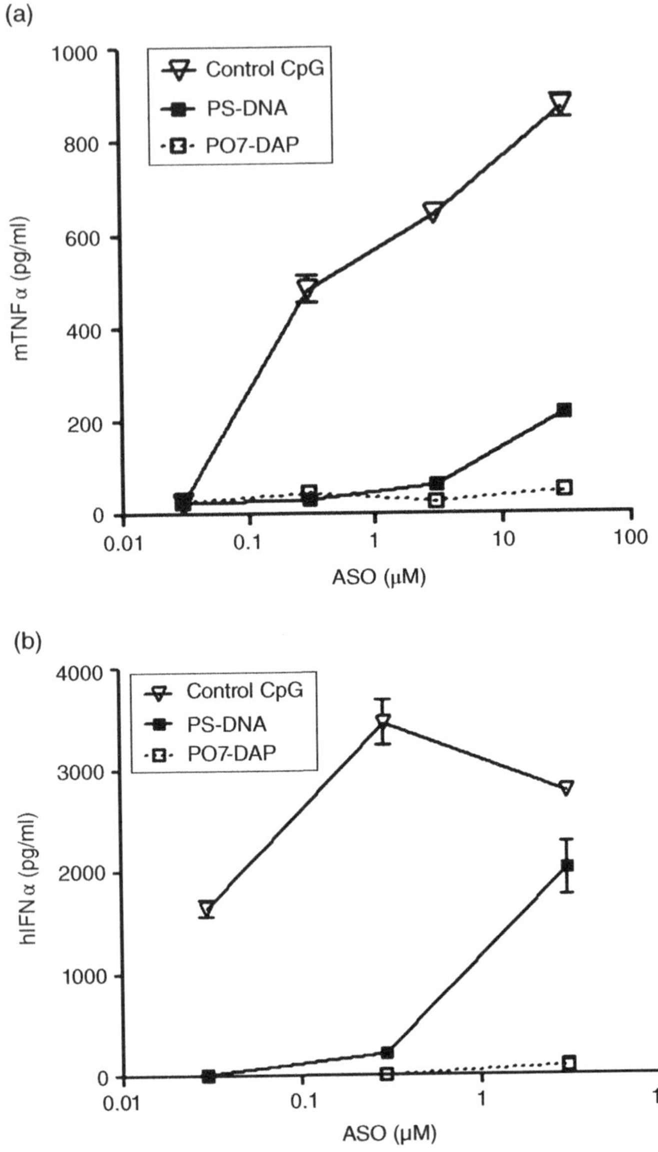

Figure 6.4 Comparison between a selected mixed phosphorothioate/phosphodiester ASO sequence (referred to as PO7-DAP) with the same sequence fully phosphorothioated (referred to as PS-DNA) on the immune response *in vitro*. (a) Mouse macrophage cells (RAW 264.7) or (b) human PBMC was exposed to different concentrations of ASO or a known immunostimulatory sequence CpG as positive control and levels of TNF-α and IFN-α in culture supernatants measured 24 h later (n = 3–5). ASO sequences are listed in Table 6.1.

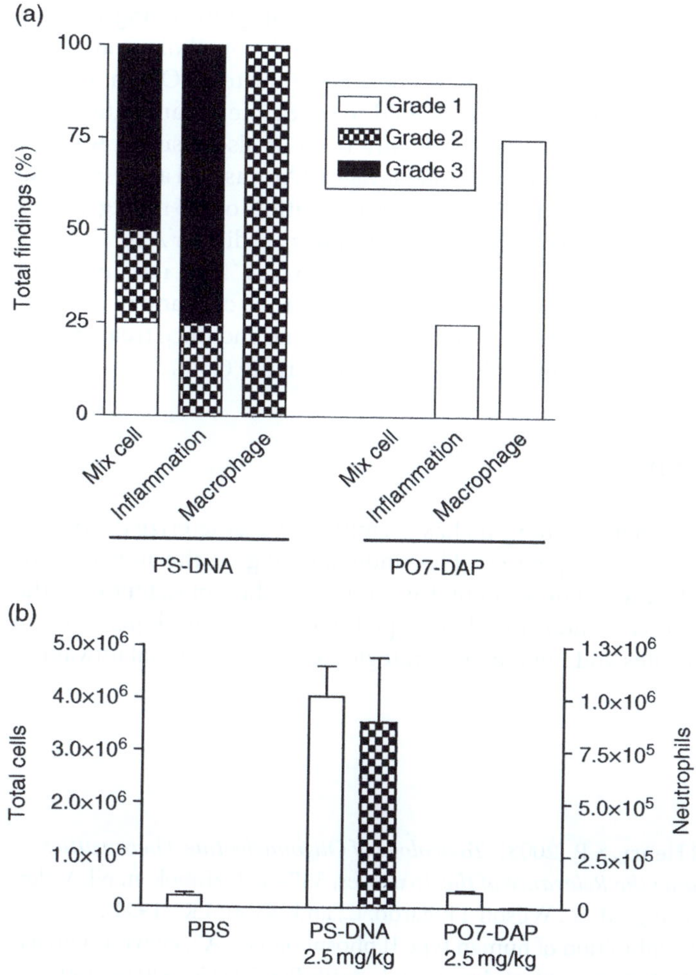

Figure 6.5 Comparison between a selected mixed phosphorothioate/phosphodiester ASO sequence (referred to as PO7-DAP) with the same sequence fully phosphorothioated (referred to as PS-DNA) on the inflammatory response in the lung of mice following repeated dosing. Mice ($n=4$ per group) received 10 doses (2.5 mg/kg) of the ASO over a 2-week period (intratracheal aerosol dosing using a MicroSprayer from PennCentury) and inflammatory response assessed by (a) lung tissue histopathology using a severity grade scoring (grade 1 = minimal; grade 2 = mild; grade 3 = moderate to severe) and in (b) bronchoalveolar lavage (total cells and neutrophils are shown). ASO sequences are listed in Table 6.1.

phosphorothioated molecule that induced TNF-α and IFN-α secretion (Figure 6.4a and b).

The reduced proinflammatory effect, or class effect, of the mixed phosphorothioate/phosphodiester ASO was also observed *in vivo*. As indicated above, administration of a phosphorothioate ASO via the pulmonary route of administration (intratracheal aerosol) induces lung findings including alveolar macrophage accumulation and mixed cell infiltration. Histopathological evaluation of the lungs of mice treated with a phosphorothioate ASO at a dose of 2.5 mg/kg for 10 consecutive days shows mild to moderate inflammation and accumulation of alveolar macrophages (Figure 6.5a). It also resulted with an increase in total cells in the bronchoalveolar lavage (BAL), as well as an increase in neutrophils (Figure 6.5b). In contrast administration of 2.5 mg/kg for 10 consecutive days of the mixed phosphorothioate/phosphodiester ASO induced only minimal accumulation of alveolar macrophages and inflammation (Figure 6.5a). Furthermore, there was no difference in total cells and neutrophil counts in BAL between control mice treated with saline and mice treated with the mixed phosphorothioate/phosphodiester ASO (Figure 6.5b).

6.4 Conclusion

The experience gained in recent years has permitted to characterize the potential toxicities of ASO therapeutics. The understanding of the mechanisms underlying hybridization-independent toxicities and the consistency of the properties of this class of molecules has helped inform ON developers for the design of the molecules and implement strategies to assess potential unwanted effects.

References

1 Levin, A.A. and Henry, S.P. (2008). *Toxicology of Oligonucleotide Therapeutics and Understanding the Relevance of the Toxicities*, 537–574. Hoboken, NJ: Wiley.

2 Bennett, F.C., Chiang, M.-Y., Wilson-Lingardo, L., and Wyatt, J.R. (1994). Sequence specific inhibition of human type II phospholipase A2, enzyme activity by phosphorothioate oligonucleotides. *Nucleic Acids Res.* 22 (15): 3202–3209.

3 Burgess, T.L., Fisher, E.F., Ross, S.L. et al. (1995). The antiproliferative activity of c-myb and c-myc antisense oligonucleotides in smooth muscle cells is caused by a nonantisense mechanism. *Proc. Natl. Acad. Sci. U. S. A.* 92 (9): 4051–4055.

4 Maltese, J.Y., Sharma, H.W., Vassilev, L., and Narayanan, R. (1995). Sequence context of antisense RelA/NF-kappa B phosphorothioates determines specificity. *Nucleic Acids Res.* 23 (7): 1146–1151.

5 Stein, C.A. (1996). Phosphorothioate antisense oligodeoxynucleotides: questions of specificity. *Trends Biotechnol.* 14 (5): 147–149.

6 Krieg, A.M. and Stein, C.A. (1995). Phosphorothioate oligodeoxynucleotides: antisense or anti-protein? *Antisense Res. Dev.* 5 (4): 241.

7 Takeda, K. and Akira, S. (2005). Toll-like receptors in innate immunity. *Int. Immunol.* 17 (1): 1–14.

8 Krieg, A.M., Yi, A.K., Matson, S. et al. (1995). CpG motifs in bacterial DNA trigger direct B-cell activation. *Nature* 374 (6522): 546–549.

9 Krieg, A.M. (2002). CpG motifs in bacterial DNA and their immune effects. *Annu. Rev. Immunol.* 20: 9–60.

10 Yamamoto, S., Yamamoto, T., Kataoka, T. et al. (1992). Unique palindromic sequences in synthetic oligonucleotides are required to induce IFN [correction of INF] and augment IFN-mediated [correction of INF] natural killer activity. *J. Immunol.* 148: 4072–4076.

11 Krieg, A.M. (2000). Immune effects and mechanisms of action of CpG motifs. *Vaccine* 19 (6): 618–622.

12 Haas, T., Metzger, J., Schmitz, F. et al. (2008). The DNA sugar backbone 2′ deoxyribose determines toll-like receptor 9 activation. *Immunity* 28 (3): 315–323.

13 Vollmer, J. and Krieg, A.M. (2009). Immunotherapeutic applications of CpG oligodeoxynucleotide TLR9 agonists. *Adv. Drug Deliv. Rev.* 61 (13): 195–204.

14 Verthelyi, D., Ishii, K.J., Gursel, M. et al. (2001). Human peripheral blood cells differentially recognize and respond to two distinct CPG motifs. *J. Immunol.* 166 (4): 2372–2377.

15 Verthelyi, D., Kenney, R.T., Seder, R.A. et al. (2002). CpG oligodeoxynucleotides as vaccine adjuvants in primates. *J. Immunol.* 168 (4): 1659–1663.

16 Hornung, V., Rothenfusser, S., Britsch, S. et al. (2002). Quantitative expression of toll-like receptor 1-10 mRNA in cellular subsets of human peripheral blood mononuclear cells and sensitivity to CpG oligodeoxynucleotides. *J. Immunol.* 168 (9): 4531–4537.

17 Roberts, T.L., Sweet, M.J., Hume, D.A., and Stacey, K.J. (2005). Cutting edge: species-specific TLR9-mediated recognition of CpG and non-CpG phosphorothioate-modified oligonucleotides. *J. Immunol.* 174 (2): 605–608.

18 Zarember, K.A. and Godowski, P.J. (2002). Tissue expression of human Toll-like receptors and differential regulation of Toll-like receptor mRNAs in leukocytes in response to microbes, their products, and cytokines. *J. Immunol.* 168 (2): 554–561.

19 Bauer, S., Kirschning, C.J., Häcker, H. et al. (2001). Human TLR9 confers responsiveness to bacterial DNA via species-specific CpG motif recognition. *Proc. Natl. Acad. Sci. U. S. A.* 98 (16): 9237–9242.

20 Hartmann, G., Weeratna, R.D., Ballas, Z.K. et al. (2000). Delineation of a CpG phosphorothioate oligodeoxynucleotide for activating primate immune responses *in vitro* and *in vivo*. *J. Immunol.* 164 (3): 617–624.

21 Hemmi, H., Takeuchi, O., Kawai, T. et al. (2000). A Toll-like receptor recognizes bacterial DNA. *Nature* 408 (6813): 740–745.

22 Puig, M., Tosh, K.W., Schramm, L.M. et al. (2012). TLR9 and TLR7 agonists mediate distinct type I IFN responses in humans and nonhuman primates *in vitro* and *in vivo*. *J. Leukoc. Biol.* 91 (1): 147–158.

23 Steinhagen, F., Meyer, C., Tross, D. et al. (2012). Activation of type I interferon-dependent genes characterizes the "core response" induced by CpG DNA. *J. Leukoc. Biol.* 92 (4): 775–785.
24 Verthelyi, D. and Klinman, D.M. (2003). Immunoregulatory activity of CpG oligonucleotides in humans and nonhuman primates. *Clin. Immunol.* 109 (1): 64–71.
25 Benimetskaya, L., Berton, M., Kolbanovsky, A. et al. (1997). Formation of a G-tetrad and higher order structures correlates with biological activity of the RelA (NF-kappaB p65) "antisense" oligodeoxynucleotide. *Nucleic Acids Res.* 25 (13): 2648–2656.
26 Wang, W., Chen, H.J., Sun, J. et al. (1998). A comparison of guanosine-quartet inhibitory effects versus cytidine homopolymer inhibitory effects on rat neointimal formation. *Antisense Nucleic Acid Drug Dev.* 8 (3): 227–236.
27 Agrawal, S., Tan, W., Cai, Q. et al. (1997). *In vivo* pharmacokinetics of phosphorothioate oligonucleotides containing contiguous guanosines. *Antisense Nucleic Acid Drug Dev.* 7 (3): 245–249.
28 Agrawal, S., Iadarolac, P.L., Temsamanic, J. et al. (1996). Effect of G-rich sequences on the synthesis, purification, hybridization, cell uptake, and hemolytic activity of oligonucleotides. *Bioorg. Med. Chem. Lett.* 6 (18): 2219–2224.
29 Lenert, P., Stunz, L., Yi, A.K. et al. (2001). CpG stimulation of primary mouse B cells is blocked by inhibitory oligodeoxyribonucleotides at a site proximal to NF-kappaB activation. *Antisense Nucleic Acid Drug Dev.* 11 (4): 247–256.
30 Shirota, H., Sano, K., Hirasawa, N. et al. (2001). Novel roles of CpG oligodeoxynucleotides as a leader for the sampling and presentation of CpG-tagged antigen by dendritic cells. *J. Immunol.* 167 (1): 66–74.
31 Griffin, L.C., Tidmarsh, G.F., Bock, L.C. et al. (1993). *In vivo* anticoagulant properties of a novel nucleotide-based thrombin inhibitor and demonstration of regional anticoagulation in extracorporeal circuits. *Blood* 81 (12): 3271–3276.
32 Stoddart, C.A., Rabin, L., Hincenbergs, M. et al. (1998). Inhibition of human immunodeficiency virus type 1 infection in SCID-hu Thy/Liv mice by the G-quartet-forming oligonucleotide, ISIS 5320. *Antimicrob. Agents Chemother.* 42 (8): 2113–2115.
33 Kobayashi, N., Hong, C., Klinman, D.M., and Shirota, H. (2013). Oligodeoxynucleotides expressing polyguanosine motifs promote antitumor activity through the upregulation of IL-2. *J. Immunol.* 190 (4): 1882–1889.
34 Knauert, M.P. and Glazer, P.M. (2001). Triplex forming oligonucleotides: sequence-specific tools for gene targeting. *Hum. Mol. Genet.* 10 (20): 2243–2251.
35 Wang, G., Seidman, M.M., and Glazer, P.M. (1997). Mutagenesis in mammalian cells induced by triple helix formation and transcription-coupled repair. *Science* 271 (5250): 802–805.

36 Jain, A.K. and Bhattacharya, S. (2010). Groove binding ligands for the interaction with parallel-stranded ps-duplex DNA and triplex DNA. *Bioconjug. Chem.* 21 (8): 1389–1403.
37 Drygin, D., Barone, S., and Bennett, C.F. (2004). Sequence-dependent cytotoxicity of second-generation oligonucleotides. *Nucleic Acids Res.* 32 (22): 6585–6594.
38 Hagedorn, P.H., Yakimov, V., Ottosen, S. et al. (2013). Hepatotoxic potential of therapeutic oligonucleotides can be predicted from their sequence and modification pattern. *Nucleic Acid Ther.* 23 (5): 302–310.
39 Burdick, A.D., Sciabola, S., Mantena, S.R. et al. (2014). Sequence motifs associated with hepatotoxicity of locked nucleic acid – modified antisense oligonucleotides. *Nucleic Acids Res.* 42 (8): 4882–4891.
40 Brown, D.A., Kang, S.H., Gryaznov, S.M. et al. (1994). Effect of phosphorothioate modification of oligodeoxynucleotides on specific protein binding. *J. Biol. Chem.* 269 (43): 26801–26805.
41 Srinivasan, S.K., Tewary, H.K., and Iversen, P.L. (1995). Characterization of binding sites, extent of binding, and drug interactions of oligonucleotides with albumin. *Antisense Res. Dev.* 5 (2): 131–139.
42 Krieg, A.M. and Stein, C.A. (1995). Phosphorothioate oligodeoxynucleotides: antisense or anti-protein? *Antisense Res. Dev.* 5 (4): 241.
43 Shen, L., Frazer-Abel, A., Reynolds, P.R. et al. (2014). Mechanistic understanding for the greater sensitivity of monkeys to antisense oligonucleotide-mediated complement activation compared with humans. *J. Pharmacol. Exp. Ther.* 351 (3): 709–717.
44 Guvakova, M.A., Yakubov, L.A., Vlodavsky, I. et al. (1995). Phosphorothioate oligodeoxynucleotides bind to basic fibroblast growth factor, inhibit its binding to cell surface receptors, and remove it from low affinity binding sites on extracellular matrix. *J. Biol. Chem.* 270 (6): 2620–2627.
45 Gao, W.Y., Han, F.S., Storm, C. et al. (1992). Phosphorothioate oligonucleotides are inhibitors of human DNA polymerases and RNase H: implications for antisense technology. *Mol. Pharmacol.* 41 (2): 223–229.
46 Yakubov, L.A., Khaled, Z., Zhang, L.M. et al. (1993). Oligodeoxynucleotides interact with recombinant CD4 at multiple sites. *J. Biol. Chem.* 268 (25): 18818–18823.
47 Stein, C.A., Pal, R., DeVico, A.L. et al. (1991). Mode of action of 5′-linked cholesteryl phosphorothioate oligodeoxynucleotides in inhibiting syncytia formation and infection by HIV-1 and HIV-2 *in vitro*. *Biochemistry* 30 (9): 2439–2444.
48 Benimetskaya, L., Loike, J.D., Khaled, Z. et al. (1997). Mac-1 (CD11b/CD18) is an oligodeoxynucleotide-binding protein. *Nat. Med.* 3 (4): 414–420.
49 Marshall, W.S., Beaton, G., Stein, C.A. et al. (1992). Inhibition of human immunodeficiency virus activity by phosphorodithioate oligodeoxycytidine. *Proc. Natl. Acad. Sci. U. S. A.* 89 (14): 6265–6269.

50 Gao, W.Y., Stein, C.A., Cohen, J.S. et al. (1989). Effect of phosphorothioate homo-oligodeoxynucleotides on herpes simplex virus type 2-induced DNA polymerase. *J. Biol. Chem.* 264 (19): 11521–11526.

51 Tramontano, E., Colla, P.L., and Cheng, Y.C. (1998). Biochemical characterization of the HIV-1 integrase 3′-processing activity and its inhibition by phosphorothioate oligonucleotides. *Biochemistry* 37 (20): 7237–7243.

52 Weidner, D.A., Valdez, B.C., Henning, D. et al. (1995). Phosphorothioate oligonucleotides bind in a non sequence-specific manner to the nucleolar protein C23/nucleolin. *FEBS Lett.* 366 (2–3): 146–150.

53 Urbinati, C., Chiodelli, P., and Rusnati, M. (2008). Polyanionic drugs and viral oncogenesis: a novel approach to control infection, tumor-associated inflammation and angiogenesis. *Molecules* 13 (11): 2758–2785.

54 Yip, G.W., Smollich, M., and Götte, M. (2006). Therapeutic value of glycosaminoglycans in cancer. *Mol. Cancer Ther.* 5 (9): 2139–2148.

55 Sands, H., Gorey-Feret, L.J., Cocuzza, A.J. et al. (1994). Biodistribution and metabolism of internally 3H-labeled oligonucleotides I. Comparison of a phosphodiester and a phosphorothioate. *Mol. Pharmacol.* 45 (5): 932–943.

56 Jason, T.L., Koropatnick, J., and Berg, R.W. (2004). Toxicology of antisense therapeutics. *Toxicol. Appl. Pharmacol.* 201 (11): 66–83.

57 Levin, A.A. (1999). A review of the issues in the pharmacokinetics and toxicology of phosphorothioate antisense oligonucleotides. *Biochim. Biophys. Acta* 1489 (1): 69–84.

58 Guimond, A., Viau, E., Aubé, P. et al. (2008). Advantageous toxicity profile of inhaled antisense oligonucleotides following chronic dosing in non-human primates. *Pulm. Pharmacol. Ther.* 21 (6): 845–854.

59 Thushara, R.M., Hemshekhar, M., Basappa et al. (2015). Biologicals, platelet apoptosis and human diseases: an outlook. *Crit. Rev. Oncol. Hematol.* 93 (3): 149–158.

60 Aster, R.H., Curtis, B.R., McFarland, J.G., and Bougie, D.W. (2009). Drug-induced immune thrombocytopenia: pathogenesis, diagnosis, and management. *J. Thromb. Haemost.* 7 (6): 911–918.

61 Thushara, R.M., Hemshekhar, M., Kemparaju, K. et al. (2014). Therapeutic drug-induced platelet apoptosis: an overlooked issue in pharmacotoxicology. *Arch. Toxicol.* 88 (2): 185–198.

62 Frazier, K.S., Sobry, C., Derr, V. et al. (2014). Species-specific inflammatory responses as a primary component for the development of glomerular lesions in mice and monkeys following chronic administration of a second-generation antisense oligonucleotide. *Toxicol. Pathol.* 42 (5): 923–935.

63 Grossman, S.A., Alavi, J.B., Supko, J.G. et al. (2005). Efficacy and toxicity of the antisense oligonucleotide aprinocarsen directed against protein kinase C-alpha delivered as a 21-day continuous intravenous infusion in patients with recurrent high-grade astrocytomas. *Neuro Oncol.* 7 (1): 32–40.

64 Nemunaitis, J., Holmlund, J.T., Kraynak, M. et al. (1999). Phase I evaluation of ISIS 3521, an antisense oligodeoxynucleotide to protein kinase C-alpha, in patients with advanced cancer. *J. Clin. Oncol.* 17 (11): 3586–3595.

65 Plummer, R., Vidal, L., Griffin, M. et al. (2009). Phase I study of MG98, an oligonucleotide antisense inhibitor of human DNA methyltransferase 1, given as a 7-day infusion in patients with advanced solid tumors. *Clin. Cancer Res.* 15 (9): 3177–3183.

66 Chi, K.N., Siu, L.L., Hirte, H. et al. (2008). A phase I study of OGX-011, a 2′-methoxyethyl phosphorothioate antisense to clusterin, in combination with docetaxel in patients with advanced cancer. *Clin. Cancer Res.* 14 (3): 833–839.

67 Paz-Ares, L., Douillard, J.Y., Koralewski, P. et al. (2006). Phase III study of gemcitabine and cisplatin with or without aprinocarsen, a protein kinase C-alpha antisense oligonucleotide, in patients with advanced-stage non-small-cell lung cancer. *J. Clin. Oncol.* 24 (9): 1428–1434.

68 Ritch, P., Rudin, C.M., Bitran, J.D. et al. (2006). Phase II study of PKC-alpha antisense oligonucleotide aprinocarsen in combination with gemcitabine and carboplatin in patients with advanced non-small cell lung cancer. *Lung Cancer* 52 (2): 173–180.

69 Tolcher, A.W., Chi, K., Kuhn, J. et al. (2005). A phase II, pharmacokinetic, and biological correlative study of oblimersen sodium and docetaxel in patients with hormone-refractory prostate cancer. *Clin. Cancer Res.* 11 (10): 3854–3861.

70 Vansteenkiste, J., Canon, J.L., Riska, H. et al. (2005). Randomized phase II evaluation of aprinocarsen in combination with gemcitabine and cisplatin for patients with advanced/metastatic non-small cell lung cancer. *Invest. New Drugs* 23 (3): 263–269.

71 Desai, A.A., Schilsky, R.L., Young, A. et al. (2005). A phase I study of antisense oligonucleotide GTI-2040 given by continuous intravenous infusion in patients with advanced solid tumors. *Ann. Oncol.* 16 (6): 958–965.

72 Rao, S., Watkins, D., Cunningham, D. et al. (2004). Phase II study of ISIS 3521, an antisense oligodeoxynucleotide to protein kinase C alpha, in patients with previously treated low-grade non-Hodgkin's lymphoma. *Ann. Oncol.* 15 (9): 1413–1418.

73 Templin, M.V., Levin, A.A., Graham, M.J. et al. (2000). Pharmacokinetic and toxicity profile of a phosphorothioate oligonucleotide following inhalation delivery to lung in mice. *Antisense Nucleic Acid Drug Dev.* 10 (5): 359–368.

74 Ali, S., Leonard, S.A., Kukoly, C.A. et al. (2001). Absorption, distribution, metabolism, and excretion of a respirable antisense oligonucleotide for asthma. *Am. J. Respir. Crit. Care Med.* 163 (4): 989–993.

75 Ball, H.A., Sandrasagra, A., Tang, L. et al. (2003). Clinical potential of respirable antisense oligonucleotides (RASONs) in asthma. *Am. J. Pharmacogenomics* 3 (2): 97–106.

76 Gauvreau, G.M., Boulet, L.P., Cockcroft, D.W. et al. (2008). Antisense therapy against CCR3 and the common beta chain attenuates allergen-induced eosinophilic responses. *Am. J. Respir. Crit. Care Med.* 177 (9): 952–958.

77 Thurman, J.M. and Holers, V.M. (2006). The central role of the alternative complement pathway in human disease. *J. Immunol.* 176 (3): 1305–1310.

78 Henry, S.P., Giclas, P.C., Leeds, J.M. et al. (1997). Activation of the alternative pathway of complement by a phosphorothioate oligonucleotide: potential mechanism of action. *J. Pharmacol. Exp. Ther.* 281 (2): 810–816.

79 Galbraith, W.M., Hobson, W.C., Giclas, P.C. et al. (1994). Complement activation and hemodynamic changes following intravenous administration of phosphorothioate oligonucleotides in the monkey. *Antisense Res. Dev.* 4 (5): 201–206.

80 Henry, S.P., Beattie, G., Yeh, G. et al. (2002). Complement activation is responsible for acute toxicities in rhesus monkeys treated with a phosphorothioate oligodeoxynucleotide. *Int. Immunopharmacol.* 2 (12): 1657–1666.

81 Henry, S.P., Jagels, M.A., Hugli, T.E. et al. (2014). Mechanism of alternative complement pathway dysregulation by a phosphorothioate oligonucleotide in monkey and human serum. *Nucleic Acid Ther.* 24 (5): 326–335.

82 Sewell, K.L., Geary, R.S., Baker, B.F. et al. (2002). Phase I trial of ISIS 104838, a 2′-methoxyethyl modified antisense oligonucleotide targeting tumor necrosis factor-alpha. *J. Pharmacol. Exp. Ther.* 303 (3): 1334–1343.

83 Chu, A.J. (2010). Blood coagulation as an intrinsic pathway for proinflammation: a mini review. *Inflamm. Allergy Drug Targets* 9 (1): 32–44.

84 Yacyshyn, B.R., Chey, W.Y., Goff, J. et al. (2002). Double blind, placebo controlled trial of the remission inducing and steroid sparing properties of an ICAM-1 antisense oligodeoxynucleotide, alicaforsen (ISIS 2302), in active steroid dependent Crohn's disease. *Gut* 51 (1): 30–36.

85 Henry, S.P., Bolte, H., Auletta, C., and Kornbrust, D.J. (1997). Evaluation of the toxicity of ISIS 2302, a phosphorothioate oligonucleotide, in a four-week study in cynomolgus monkeys. *Toxicology* 120 (2): 145–155.

86 Monteith, D.K., Geary, R.S., Leeds, J.M. et al. (1998). Preclinical evaluation of the effects of a novel antisense compound targeting C-raf kinase in mice and monkeys. *Toxicol. Sci.* 46 (2): 365–375.

87 Webb, M.S., Tortora, N., Cremese, M. et al. (2001). Toxicity and toxicokinetics of a phosphorothioate oligonucleotide against the c-myc oncogene in cynomolgus monkeys. *Antisense Nucleic Acid Drug Dev.* 11 (13): 155–163.

88 McHutchison, J.G., Patel, K., Pockros, P. et al. (2006). A phase I trial of an antisense inhibitor of hepatitis C virus (ISIS 14803), administered to chronic hepatitis C patients. *J. Hepatol.* 44 (1): 88–96.

89 Kandimalla, E.R., Shaw, D.R., and Agrawal, S. (1998). Effects of phosphorothioate oligodeoxyribonucleotide and oligoribonucleotides on human complement and coagulation. *Bioorg. Med. Chem. Lett.* 8 (16): 2103–2108.

90 Advani, R., Lum, B.L., Fisher, G.A. et al. (2005). A phase I trial of aprinocarsen (ISIS 3521/LY900003), an antisense inhibitor of protein kinase C-alpha administered as a 24-hour weekly infusion schedule in patients with advanced cancer. *Invest. New Drugs* 23 (5): 467–477.

91 Sheehan, J.P. and Lan, H.C. (1998). Phosphorothioate oligonucleotides inhibit the intrinsic tenase complex. *Blood* 92 (5): 1617–1625.

92 Barrow, R.T., Healey, J.F., and Lollar, P. (1994). Inhibition by heparin of thrombin-catalyzed activation of the factor VIII-von Willebrand factor complex. *J. Biol. Chem.* 269 (1): 593–598.

93 Sheehan, J.P. and Phan, T.M. (2001). Phosphorothioate oligonucleotides inhibit the intrinsic tenase complex by an allosteric mechanism. *Biochemistry* 40 (16): 4980–4989.

94 Monteith, D.K., Henry, S.P., Howard, R.B. et al. (1997). Immune stimulation – a class effect of phosphorothioate oligodeoxynucleotides in rodents. *Anticancer Drug Des.* 12 (5): 421–432.

95 Henry, S.P., Taylor, J., Midgley, L. et al. (1997). Evaluation of the toxicity of ISIS 2302, a phosphorothioate oligonucleotide, in a 4-week study in CD-1 mice. *Antisense Nucleic Acid Drug Dev.* 7 (5): 473–481.

96 Oberbauer, R., Schreiner, G.F., and Meyer, T.W. (1995). Renal uptake of an 18-mer phosphorothioate oligonucleotide. *Kidney Int.* 48 (4): 1226–1232.

97 Geary, R.S., Norris, D., Yu, R., and Bennett, C.F. (2015). Pharmacokinetics, biodistribution and cell uptake of antisense oligonucleotides. *Adv. Drug Deliv. Rev.* 87: 46–51.

98 Butler, M., Stecker, K., and Bennett, C.F. (1997). Cellular distribution of phosphorothioate oligodeoxynucleotides in normal rodent tissues. *Lab. Invest.* 77 (4): 379–388.

99 Ferdinandi, E.S., Vassilakos, A., Lee, Y. et al. (2011). Preclinical toxicity and toxicokinetics of GTI-2040, a phosphorothioate oligonucleotide targeting ribonucleotide reductase R2. *Cancer Chemother. Pharmacol.* 68 (1): 193–205.

100 Rappaport, J., Hanss, B., Kopp, J.B. et al. (1995). Transport of phosphorothioate oligonucleotides in kidney: implications for molecular therapy. *Kidney Int.* 47 (5): 1462–1469.

101 Monteith, D.K., Horner, M.J., Gillett, N.A. et al. (1999). Evaluation of the renal effects of an antisense phosphorothioate oligodeoxynucleotide in monkey. *Toxicol. Pathol.* 27 (3): 307–317.

102 Zanardi, T.A., Han, S.C., Jeong, E.J. et al. (2012). Pharmacodynamics and subchronic toxicity in mice and monkeys of ISIS 388626, a second-generation antisense oligonucleotide that targets human sodium glucose cotransporter. *J. Pharmacol. Exp. Ther.* 343 (2): 489–496.

103 Henry, S.P., Johnson, M., Zanardi, T.A. et al. (2012). Renal uptake and tolerability of a 2′-*O*-methoxyethyl modified antisense oligonucleotide (ISIS 113715) in monkey. *Toxicology* 301 (1–3): 13–20.

104 Bianchini, D., Omlin, A., Pezaro, C. et al. (2013). First-in-human phase I study of EZN-4176, a locked nucleic acid antisense oligonucleotide to exon 4 of the androgen receptor mRNA in patients with castration-resistant prostate cancer. *Br. J. Cancer* 109 (10): 2579–2586.

105 Fidias, P., Pennell, N.A., Boral, A.L. et al. (2009). Phase I study of the c-raf-1 antisense oligonucleotide ISIS 5132 in combination with carboplatin and paclitaxel in patients with previously untreated, advanced non-small cell lung cancer. *J. Thorac. Oncol.* 4 (9): 1156–1162.

106 Klisovic, R.B., Stock, W., Cataland, S. et al. (2008). A phase I biological study of MG98, an oligodeoxynucleotide antisense to DNA methyltransferase 1, in patients with high-risk myelodysplasia and acute myeloid leukemia. *Clin. Cancer Res.* 14 (8): 2444–2449.

107 van Deventer, S.J., Wedel, M.K., Baker, B.F. et al. (2006). A phase II dose ranging, double-blind, placebo-controlled study of alicaforsen enema in subjects with acute exacerbation of mild to moderate left-sided ulcerative colitis. *Aliment. Pharmacol. Ther.* 23 (10): 1415–1425.

108 van Poelgeest, E.P., Swart, R.M., Betjes, M.G. et al. (2013). Acute kidney injury during therapy with an antisense oligonucleotide directed against PCSK9. *Am. J. Kidney Dis.* 62 (4): 796–800.

109 Geary, R.S., Yu, R.Z., Watanabe, T. et al. (2003). Pharmacokinetics of a tumor necrosis factor-alpha phosphorothioate 2′-*O*-(2-methoxyethyl) modified antisense oligonucleotide: comparison across species. *Drug Metab. Dispos.* 31 (11): 1419–1428.

110 Graham, M.J., Crooke, S.T., Monteith, D.K. et al. (1998). *In vivo* distribution and metabolism of a phosphorothioate oligonucleotide within rat liver after intravenous administration. *J. Pharmacol. Exp. Ther.* 286 (1): 447–458.

111 Leeds, J.M., Henry, S.P., Geary, R. et al. (2000). Comparison of the pharmacokinetics of subcutaneous and intravenous administration of a phosphorothioate oligodeoxynucleotide in cynomolgus monkeys. *Antisense Nucleic Acid Drug Dev.* 10 (6): 435–441.

112 Henry, S.P., Geary, R.S., Yu, R., and Levin, A.A. (2001). Drug properties of second-generation antisense oligonucleotides: how do they measure up to their predecessors? *Curr. Opin. Investig. Drugs* 2 (10): 1444–1449.

113 Winquist, E., Knox, J., Ayoub, J.P. et al. (2006). Phase II trial of DNA methyltransferase 1 inhibition with the antisense oligonucleotide MG98 in patients with metastatic renal carcinoma: a National Cancer Institute of Canada Clinical Trials Group investigational new drug study. *Invest. New Drugs* 24 (2): 159–167.

114 Orum, H. and Wengel, J. (2001). Locked nucleic acids: a promising molecular family for gene-function analysis and antisense drug development. *Curr. Opin. Mol. Ther.* 3 (3): 239–243.

115 Swayze, E.E., Siwkowski, A.M., Wancewicz, E.V. et al. (2007). Antisense oligonucleotides containing locked nucleic acid improve potency but cause significant hepatotoxicity in animals. *Nucleic Acids Res.* 35 (2): 687–700.

116 Burel, S.A., Han, S.R., Lee, H.S. et al. (2013). Preclinical evaluation of the toxicological effects of a novel constrained ethyl modified antisense compound targeting signal transducer and activator of transcription 3 in mice and cynomolgus monkeys. *Nucleic Acid Ther.* 23 (3): 213–227.

117 Fey, R.A., Templin, M.V., McDonald, J.D. et al. (2014). Local and systemic tolerability of a 2′-*O*-methoxyethyl antisense oligonucleotide targeting interleukin-4 receptor-α delivery by inhalation in mouse and monkey. *Inhal. Toxicol.* 26 (8): 452–463.

118 Moschos, S.A., Frick, M., Taylor, B. et al. (2011). Uptake, efficacy, and systemic distribution of naked, inhaled short interfering RNA (siRNA) and locked nucleic acid (LNA) antisense. *Mol. Ther.* 19 (12): 2163–2168.

119 Garbuzenko, O.B., Saad, M., Betigeri, S. et al. (2009). Intratracheal versus intravenous liposomal delivery of siRNA, antisense oligonucleotides and anticancer drug. *Pharm. Res.* 26 (2): 382–394.

120 Nicklin, P.L., Bayley, D., Giddings, J. et al. (1998). Pulmonary bioavailability of a phosphorothioate oligonucleotide (CGP 64128A): comparison with other delivery routes. *Pharm. Res.* 15 (4): 583–591.

121 Holder, E., Griesenbach, U., Li, S. et al. (2006). Intravenously administered oligonucleotides can be delivered to conducting airway epithelium via the bronchial circulation. *Gene Ther.* 13 (23): 1628–1638.

122 Fortin, M., D'Anjou, H., Higgins, M.E. et al. (2009). A multi-target antisense approach against PDE4 and PDE7 reduces smoke-induced lung inflammation in mice. *Respir. Res.* 10: 39.

123 Fortin, M., Ferrari, N., Higgins, M.E. et al. (2006). Effects of antisense oligodeoxynucleotides targeting CCR3 on the airway response to antigen in rats. *Oligonucleotides* 16 (3): 203–212.

124 Allakhverdi, Z., Allam, M., Guimond, A. et al. (2006). Multitargeted approach using antisense oligonucleotides for the treatment of asthma. *Ann. N. Y. Acad. Sci.* 1082: 62–73.

125 Nyce, J.W. (1997). Respirable antisense oligonucleotides as novel therapeutic agents for asthma and other pulmonary diseases. *Expert Opin. Investig. Drugs* 6 (9): 1149–1156.

126 Gauvreau, G.M., Hessel, E.M., Boulet, L.P. et al. (2006). Immunostimulatory sequences regulate interferon-inducible genes but not allergic airway responses. *Am. J. Respir. Crit. Care Med.* 174 (1): 15–20.

127 Imaoka, H., Campbell, H., Babirad, I. et al. (2011). TPI ASM8 reduces eosinophil progenitors in sputum after allergen challenge. *Clin. Exp. Allergy* 41 (12): 1740–1746.

128 Brasey, A., Igue, R., Djemane, L. et al. (2011). The effect of *in vitro* exposure to antisense oligonucleotides on macrophage morphology and function. *J.Nucleic Acid Invest.* 2 (1): 77–84.

129 Mosser, D.M. and Edwards, J.P. (2008). Exploring the full spectrum of macrophage activation. *Nat. Rev. Immunol.* 8 (12): 958–969.

130 Reasor, M.J., McCloud, C.M., DiMatteo, M. et al. (1996). Effects of amiodarone-induced phospholipidosis on pulmonary host defense functions in rats. *Proc. Soc. Exp. Biol. Med.* 211 (4): 346–352.

131 De Broe, M.E., Paulus, G.J., Verpooten, G.A. et al. (1984). Early effects of gentamicin, tobramycin, and amikacin on the human kidney. *Kidney Int.* 25 (4): 643–652.

132 Lenoir, G., Antypkin, Y.G., Mian, O.A. et al. (2007). Efficacy, safety, and local pharmacokinetics of highly concentrated nebulized tobramycin in patients with cystic fibrosis colonized with Pseudomonas aeruginosa. *Paediatr. Drugs* 9 (Suppl. 1): 11–20.

133 Kurreck, J. (2003). Antisense technologies. Improvement through novel chemical modifications. *Eur. J. Biochem.* 270 (8): 1628–1644.

134 Crooke, S.T. (2000). Progress in antisense technology: the end of the beginning. *Methods Enzymol.* 313: 3–45.

7

Hybridization-Dependent Effects: The Prediction, Evaluation, and Consequences of Unintended Target Hybridization

Jeremy D. A. Kitson[1], *Piotr J. Kamola*[2,*], *and Lauren Kane*[1,**]

[1] *GlaxoSmithKline R&D, Stevenage, Hertfordshire, UK*
[2] *GlaxoSmithKline R&D, Ware, Hertfordshire, UK*

7.1 Introduction

The silencing of RNA expression achieved through antisense technology is based on the interaction between an oligonucleotide analogue and a complementary target sequence. The recognition and binding events are driven by Watson–Crick hydrogen bonding, which makes it relatively straightforward to create gene-selective reagent and therapeutics. Theoretically, the hybridization-mediated mechanism makes it possible to target the entire transcriptome, something that is not possible with conventional small molecules. Those unique features, coupled with a relatively short design and development cycle, made the RNase H1-mediated gene knockdown an attractive platform for therapeutic development.

Similar to other therapeutic modalities, several safety concerns were identified with therapeutic antisense oligonucleotides (ASOs), particularly following *in vivo* administration. For ASOs, the mechanism of toxicity can be subdivided into hybridization independent (structure-based interactions with proteins, covered in previous chapters) and hybridization dependent, driven by Watson–Crick base pairing. The hybridization-dependent effects can be subdivided into effects caused by exaggerated pharmacology, i.e. inhibition of the intended target to a degree that produces deleterious effects, and hybridization-dependent

*Current address: RIKEN Center for Integrative Medical Sciences, Yokohama, Japan
**Current address: MRC Human Genetics Unit, Institute of Genetics and Molecular Medicine, University of Edinburgh, Scotland

Oligonucleotide-Based Drugs and Therapeutics: Preclinical and Clinical Considerations for Development, First Edition. Edited by Nicolay Ferrari and Rosanne Seguin.

effects on unintended targets that happen to be completely or partially complementary to the oligonucleotide sequence used.

The importance of specificity for antisense molecules was recognized in the very first studies [1] and has been primarily addressed through the application of bioinformatics tools to avoid complementary or partially complementary sequences. The significance of hybridization-mediated toxicities has been considered to be only a modest concern for antisense molecules [2, 3], with few hybridization-dependent effects reported for gapmer–ASOs [2, 4]. However, these specificity studies have often been limited in scope, and more recent studies have revealed off-target activity to be a genuine concern [5, 6].

The phenomenon of off-target RNA effects can be considered as analogous to secondary pharmacology for small-molecule drugs, i.e. interacting with unintended proteins or receptors. Minimizing these liabilities is a routine part of a small-molecule drug development program and should also be considered an essential part of the lead optimization process for ASOs. However, as this activity is sequence related, an important consideration for ASO is that the activity will differ between species and will require species-specific bioinformatics screening. While recent findings have indicated that Watson–Crick hybridization-dependent off-target effects (OTEs) are a more significant issue than has previously been acknowledged, one of the unique opportunities of ASO-based therapeutics is that putative off-target interactions can be predicted using sequence alignment algorithms allowing the risk to be addressed.

7.1.1 Scope of this Review: RNase H1-dependent ASOs

This chapter will focus on the hybridization-dependent effects on unintended targets that happen to be completely or partially complementary to the oligonucleotide used. In this chapter "OTEs" will only refer to hybridization-dependent scission of unintended targeted RNAs. These have also been referred to as "irrelevant cleavage" or "inadvertent target" [7, 8]. The focus is solely on ASOs that utilize RNase H1 for activity, as these represent the vast majority of ASO molecules in development. This chapter will review studies on specificity and describe the factors that influence the probability of a predicted off-target being confirmed and implications of OTEs on the safety of the molecules. We will also cover the computational methods used to predict potential off-targets and the methodologies employed to investigate if these putative off-targets are observed.

7.2 Specificity Studies with ASOs

Computational approaches are utilized in the design phase of an ASO to select molecules with relatively lowest off-target potential (i.e. lowest number of unintended targets with perfect or partial homology to the oligonucleotide).

The likelihood of encountering genes with a high sequence homology is higher with shorter sequences and decreases with longer oligonucleotides [9]. With shorter ASOs, partial homology to unintended targets is largely unavoidable as it is difficult to identify sequences without at least one or two mismatches to unintended transcripts. Nevertheless, RNase H1-competent interaction can also occur for longer sequence, and efforts should be made to avoid them. A duplex with low affinity may not be antisense active, but with the addition of modified bases to the oligomer to increase affinity, a potent interaction may result. For a 16-mer ASO, analysis is generally limited to three mismatches, as a lower sequence similarity usually results in a complete loss of activity [9].

RNase H1-dependent ASO activity could also be demonstrated in cell-free experimental systems including wheat germ extract [10] and Xenopus oocytes [11]. In a number of early studies, concerns were raised over the low stringency requirements of RNase H1, as additional cleavage sites other than the intended one were also identified. For example, Cazenave et al. demonstrated that RNase H1 could cleave the beta-globin mRNA at a site that was only a 13-bp complement to a 17-mer ASO [11]. Further studies reported that as little as five contiguous base regions of homology were sufficient to elicit RNase H1 activity under certain circumstances [12]. In cell-based and *in vivo* studies, the low stringency of RNase H1 appeared to be of less concern. In a mouse *in vivo* model, Fas-mediated apoptosis was blocked by a 20-mer 2′MOE gapmer ASO-targeting Fas mRNA, but not by the variants of the ASO with two, four, six, or eight mismatches [4]. In another study by ISIS Pharmaceuticals (now Ionis Pharmaceuticals), C-raf kinase knockdown was reduced when one mismatch was introduced into the ASO [13, 14]. In a study comparing ASOs with siRNAs, it was shown that the position of the mismatches within the ASO can affect the level of knockdown observed [15]. Using an ASO against PTEN, two mismatches in the gap were not tolerated, while one in each wing (two mismatches in total) reduced knockdown [15]. A study using LNA gapmers showed that two mismatches in the wing resulted in greatly reduced activity against the intended target [16]. Allele-selective gapmer–ASOs have been developed that are able to distinguish between a single mismatch and cleave one variant more efficiently [17–21]. In one example, a fivefold selectivity has been observed for the mutant allele compared with the normal Ha-Ras allele [17]. Allele-specific suppression of mutant huntingtin has been achieved using gapmer–ASOs [20, 21]. These studies have led to the perception that ASO molecules are highly specific for their target sequence [2, 3, 22, 23].

However there are limitations to the published work on specificity. In most of these studies, the use of mismatched ASOs was to demonstrate that the observed target knockdown was due to the antisense effect and a comprehensive analysis of potential off-targets was not performed. Therefore, these studies did not demonstrate specificity in the context of the entire transcriptome. The studies were also limited in scope, with very few targets examined and mismatches being incorporated into only a small number of the possible positions in the ASO/target duplex. There are no published systematic studies defining the effect of mismatches in each position; frequently only mismatches in the middle of the gap region were reported [13–16]. However, the importance of gap length and mismatch position on specificity has been demonstrated in allele-specific studies [20, 21]. An additional limitation of these studies is that the mismatched ASOs have only been tested against a target region with the same secondary structure and did not address the possibility of off-target sites that are more accessible to the ASO.

7.3 Implications of the Nuclear Site of Action of RNase H1

Early studies identified RNase H1 as the enzyme responsible for the target cleavage activity of ASOs [24], and it was initially reported to be present throughout the cell [2, 25]. However RNaseH1 has recently been shown to be mainly active in the nucleus, where it is involved in maintaining genome stability, as well as present in the mitochondria, where it plays an essential role in mitochondrial DNA replication [26]. The function of RNase H1 in the nucleus is to mitigate the deleterious effects of the formation of R-loops during transcription [27]. R-loops are structures composed of an RNA/DNA hybrid, which is formed when the RNA hybridizes to a complementary DNA strand and leaves the non-transcribed DNA strand displaced [28]. R-loops can be beneficial or deleterious to cells, being required for important biological processes but can also promote DNA damage and genome instability. By implication, RNase H1 must be closely associated with the transcriptional complex to perform this function. This would suggest that RNase H1-dependent ASO activity is a cotranscriptional event and that the predominant target for ASO is the primary RNA transcript, containing both intron and exon sequences. It has been demonstrated in a number of studies that ASO molecules can enter the nucleus, target intron sequences, and degrade the transcript via the RNase H1 mechanism [2, 15]. ASOs have also been identified with targets that span an exon–exon junction and demonstrate target knockdown *in vivo* [14, 29]. This demonstrates that ASOs can also act on either the partially spliced transcript or the mature mRNA. This activity is most likely occurring in the nucleus after splicing but before mRNA export, although activity on the mRNA in the cytoplasm has not been formally excluded.

The major limitation to the published off-target studies is that the computational prediction and subsequent analysis of potential off-targets has been restricted to mature mRNA sequences [4, 13, 30–33]. The nuclear site of action of RNase H1 requires that the computational analysis methods used to predict potential off-target hybridization need to be expanded to include primary

RNA and other nuclear noncoding RNA species [5, 34]. The noncoding RNAs include long noncoding RNAs (lncRNAs), enhancer RNAs (eRNAs), microRNAs (miRNAs), small nuclear RNAs (snRNAs), small nucleolar RNAs (snoRNAs), and piwi-interacting RNAs (piRNAs) [35–40]. These RNAs mainly function in the regulation of gene expression and have important and complex roles during transcription and translation. Similar to protein-coding genes, complete or partial hybridization of ASO to a noncoding RNA could lead to a cleavage and consequent phenotypic effects.

An emerging class of noncoding RNAs are the lncRNAs. These RNAs are involved in many molecular functions within the cell, such as imprinting, development, and cell cycle control, and misregulation of lncRNAs has been implicated in a number of diseases [39, 40]. The expression of lncRNAs can regulate gene expression by directing chromatin remodeling, interacting with the transcriptional machinery and posttranscriptional processing. ASOs have been used to target lncRNAs and cause RNase H1-mediated cleavage of lncRNAs, which resulted in the upregulation of the corresponding mRNA by changing the local chromatin environment [41–43]. Many of these noncoding RNA species are normally found at lower levels in cells than coding RNA, potentially this makes a phenotypic change more likely following unintended ASO hybridization and the degradation of these RNA species. There are several studies where ASOs have been used to target nuclear RNA species via an RNase H1-dependent mechanism, including the Ube3a_ATS [44] and MALAT1 [45] lncRNAs. Furthermore, a class of lncRNAs expressed from enhancer elements called eRNAs are involved in regulation of enhancer activity, communication between enhancers and promoters, and initiation of gene expression. Sequence complementarity between ASO and eRNAs could result in disruption of the regulatory role eRNAs play during transcription of genes, as has been demonstrated by Schaukowitch et al. [46].

7.3.1 Confirmation of Unintended Targets within Introns

The inclusion of unspliced primary transcript sequences in specificity studies dramatically increased the number of potential off-targets. It is not untypical for several hundred hits, with one and two mismatches, to be identified for a 16 nucleotide-long (16-mer) ASO with the majority of these hits within introns as illustrated in Figure 7.1. A study to evaluate off-targets for a series of LNA-containing gapmer–ASOs designed against the BACH1 transcript has recently been published [5]. In these and additional studies at GSK, extensive analysis of many 14- and 16-mer LNA gapmers, which were designed against several different targets, has revealed a significant confirmation rate for predicted off-target genes, in particular off-targets that lie within intron regions. A wide range of different mismatch patterns were confirmed. These included single

ASOID	Sequence (5′-> 3′)	Total 0–2 MM	Protein coding genes–exons PM	Protein coding genes–exons 1 MM	Protein coding genes–exons 2 MM	Protein coding genes–introns PM	Protein coding genes–introns 1 MM	Protein coding genes–introns 2 MM
GSK2910546A	TCAGTTTAGCAGTGTA	328	0	1	16	0	18	293
GSK2910557A	AGTGTGATGAAAAGCA	806	0	0	46	2	53	705
GSK2910579A	GGTCATATGTGTGTAA	353	0	0	20	0	17	316
GSK2910584A	AGATGCAACACTACTA	346	0	0	15	1	8	322
GSK2910594A	ATCGTTTCCAATTTCC	549	0	2	25	0	8	514
GSK2910613A	GAGTTATTACTAGAGT	304	0	0	21	1	8	274
GSK2910632A	TGAGAAGATCCATAGC	313	0	0	20	0	16	277

Figure 7.1 A summary of unintended interactions predicted for six BACH1-targeting ASOs. As a part of the design process, ASOs with potential interactions with mature mRNA at the perfect or near-perfect level were screened out. "PM" stands for perfect match, "1MM" for one mismatch, and "2MM" for two mismatches.

mismatches or single insertions/deletions (indels) between the ASO/target alignment, two mismatches, and combinations of both indels and mismatches. The knockdown potency is a concern, with the activity against many off-targets on a par with, or greater than, the activity against the intended target. In an extreme example, with a 14-mer, the activity against an intron off-target (with a single mismatch) was 30-fold greater than the on-target activity. For some ASOs, the confirmation rate was greater than 30% of the predicted off-target hits, giving rise to an unacceptable number of potent off-target hits for a therapeutic entity. However, it was possible to identify ASOs with a much lower confirmation rate (<10%) with only a few potent off-target interactions of concern, these ASOs have the potential for further development.

7.4 Mechanism of OTE

There are numerous factors that influence the potency of ASOs, and in general these factors apply to both the intended and unintended targets. These include both sequence-dependent and nonsequence-dependent factors, as shown in Figures 7.2 and 7.3. The nonsequence-dependent factors concern both the nature of the transcript and determinants that effect the distribution of the ASO. Factors such as biostability, tissue distribution, cellular uptake, subcellular distribution, and protein interactions will influence the likelihood of an ASO gapmer to get to the tissue/cell where the target transcript is expressed and to the intracellular site of action, the nucleus. Other factors are specific to the transcript and include tissue expression, transcript accumulation, and nuclear residency. An example of the latter is the demonstration that targeting a transcript with long nuclear residency gave preferential knockdown, compared to a nonmutant allele without the CUG repeat [47]. If tissue distribution is similar, the influence of these nonsequence-dependent factors on both on-target and off-target hybridization activity will be similar. If tissue distribution is different, the relative on/off-target activity will depend on which transcript

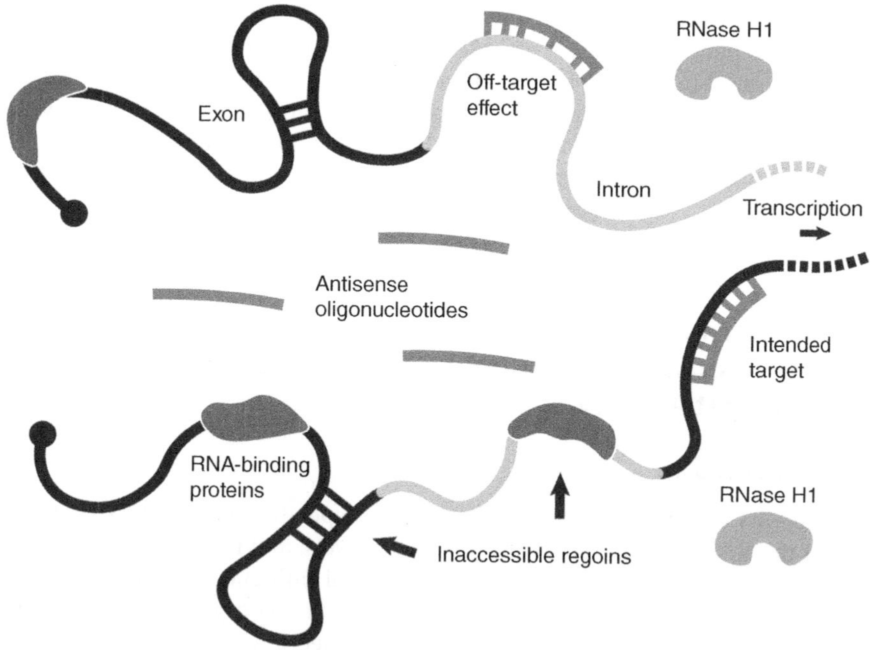

Figure 7.2 Graphical representation of ASO gapmer-mediated on- and off-target events. The ASO can bind to accessible regions within the intended target or off-target transcript and recruit RNase H1. Some regions of the transcripts are inaccessible to the ASO due to secondary or tertiary RNA structure or because they are masked by RNA-binding proteins.

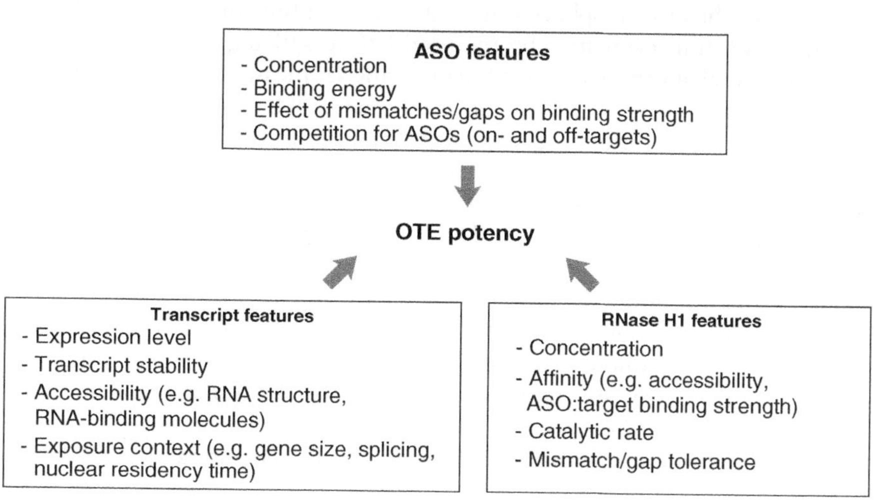

Figure 7.3 List of ASO-, transcript-, and RNase H1-based features that affect the potency of off-target knockdown.

is expressed more strongly in the tissues more amenable for delivery. The sequence-dependent factors include hybridization affinity of ASO/target interaction and target accessibility (masking of site due to RNA structure or protein binding). These factors are the same for on- and off-target activity, with the exception of whether the RNase H1 enzyme will tolerate mismatches in particular positions within the duplex and the impact of the mismatch on the binding affinity.

7.5 Determining the Extent that Accessibility, Affinity and, Mismatch Tolerance Contribute to Off-target Activity

7.5.1 Accessibility

Accessibility is dependent on a number of gene-specific factors. The structural fold of the RNA target, transcript expression levels, and the time that a target sequence is available to bind to the ASO will all influence the probability of unintended target knockdown being observed. The computational methods used to study the structural fold are covered in Section 7.7.1.5. The nature of the unspliced transcript and the position of the off-target site within the intron have a significant influence on temporal accessibility of the site to the ASO, and also whether the site is masked by RNA-binding proteins or tertiary fold of the RNA (Figure 7.2). A primary RNA transcript can be processed as it is transcribed, with the introns spliced out sequentially. At the other extreme, the full primary RNA transcript may be required before splicing occurs to create a single or several alternative-spliced mature mRNA messages. Introns can vary in length from a few hundred to tens of thousands of bases. The transcription rate varies between different genes but is typically in the order of 3 kb min^{-1} [48], and consequently target sequences will vary considerably in terms of nuclear residency. Increased nuclear residency has been associated with greater knockdown activity [19, 47], and this factor also applies to off-targets within transcripts with long synthesis times. A recent paper associates hepatotoxicity with the unintended knockdown of long transcripts [6]. The long transcripts are more vulnerable to OTEs because of the higher probability of a partially matched sequence, but also due to the extended nuclear residency, which gives rise to greater temporal accessibility. Furthermore, longer transcripts have a higher likelihood of more than one predicted interaction site. Very potent off-targets with more than one predicted interaction site have been reported [5], and it is possible that interactions at multiple sites work in an additive fashion.

Using a minigene system, Lima and colleagues [49] investigated the effect of secondary structure and protein binding on ASO activity. In this system, the higher-order RNA structure had a much greater effect on the activity of ASO than the RNA-binding protein interactions. Bearing in mind the caveat that the minigene is limited in size and may not fully represent what would happen with a complete transcript, they also examined off-target activity in this system and showed that off-target hybridization only occurred at highly accessible, unstructured sites on the mRNA. These findings are of potential relevance to the reported findings in human cells [5] where more potent activity against intron off-target sites was observed and may indicate that intron sequences are less structured than exon structures.

7.5.2 Affinity

Affinity can be characterized through melting temperature (T_m), which is the temperature at which half of the oligonucleotides are in a single-stranded form, while the remaining half are forming a duplex. A number of different methods are available to experimentally determine the T_m of oligonucleotide duplexes. These include ultraviolet (UV)–visible spectroscopy, which is low throughput and requires significant quantities of oligonucleotides [50]. Fluorescence-based assays employ 5′ fluorophore dye and 3′ quencher, although the high price of the labeled probes is often prohibitive [51–53]. Because of the limitations of these techniques, the experimental determination of T_m for most DNA/RNA modifications or mismatches is rare [51]. Instead, various thermodynamic calculations have been used to predict the T_m of ASOs [54]; as discussed in Section 7.7.1.6, there are limitations to these methods when predicting the T_m in the context of duplex containing modified bases and/or mismatches. As an alternative, a high throughput method was developed, which is based on an intercalating EvaGreen dye (Biotium), which fluoresce when bound to a double-stranded nucleic acid, e.g. ASO/RNA duplex [5]. Using only small quantities of ASO and target sequence, it has been possible to measure the effect mismatches at different positions in the context of a large number of different ASOs and targets/off-targets.

To explore the extent to which binding affinity and accessibility may influence off-target activity, 16-mer RNA oligonucleotides representing the target region of nine off-target genes were analyzed using a T_m assay [5]. Each RNA was duplexed to a target sequence that was fully complementary to GSK2910546A ASO. The aim of this experiment was to study the effect of various mismatches on binding affinity, and further examine the correlation between affinity and activity. The melting temperature decrease compared with the fully complementary ASO/RNA duplex varied between 4.5 and 11.5 °C. A correlation between hybridization affinity of the ASO and activity observed in cells or *in vivo* has been reported in numerous studies for on-target activity [12, 17, 55–57], and in general we observed a correlation between T_m and off-target activity. The most potent off-target activity was observed with those targets that retain the highest affinity for the ASO (Figure 7.4). However there are outliers, which demonstrate that some regions may be more accessible to the ASO than others. The knockdown activity of the ASO was similar to the on-target knockdown for several of the two mismatch off-targets examined, despite a 5 °C decrease in T_m. This suggests that the loss of binding affinity is compensated by increased accessibility to the off-target site.

7.5.3 The Interaction of RNase H1 with the RNA/ASO Duplex

Mammalian RNase H1 enzymes consist of the hybrid-binding domain connected to the catalytic domain by a "flexible" linker [58]. The three-dimensional structures of both the catalytic domain and HBD complexed with a RNA/DNA duplex have been solved [59, 60], but due to the flexible linker between the two domains, it has not been possible to obtain crystals of the whole enzyme. These structures can provide insights into how ASOs will interact with the enzyme and how mismatches might impact cleavage activity. The catalytic domain contains a DNA-binding pocket and a RNA-binding pocket separated by a ridge of conserved protein residues [59]. The active site is located in the RNA-binding groove, which interacts with the 2′-OH groups of four consecutive ribonucleotides, two on either site of the scissile phosphate. A phosphate-binding pocket that is specific for B-form nucleic acids mediates preferential binding of DNA/RNA hybrids, with the interaction occurring with the phosphate two base pairs from the scissile bond [59]. All the HBD protein–RNA/DNA interactions are with the backbones of the hybrid along the minor groove; this type of interaction is expected with nonsequence-specific recognition [60]. The specificity for DNA is mediated by the aromatic patch with an interface that is compatible with 2′-deoxyriboses only, not modified bases. The HBD contact region covers five base pairs [60], which is also the minimum number of 2′-deoxynucleotides in the central "gap" required to maintain the activity of a gapmer–ASO [12]. Due to the short length of the target RNA/ASO duplex, it is unlikely that both the HBD and catalytic domain would interact with the duplex simultaneously. A number of reports have demonstrated that although the enzyme has a preferred cleavage site, secondary cleavage sites are possible [24, 61, 62]. This indicates a degree of flexibility in the interaction between the enzyme and duplex. A more recent paper identified a number of unusual cleavage sites [49]. This may indicate that while the modified bases on the ASO prevent interaction with the DNA-binding pocket, the corresponding residues on the transcript are still able to interact with the RNA-binding region

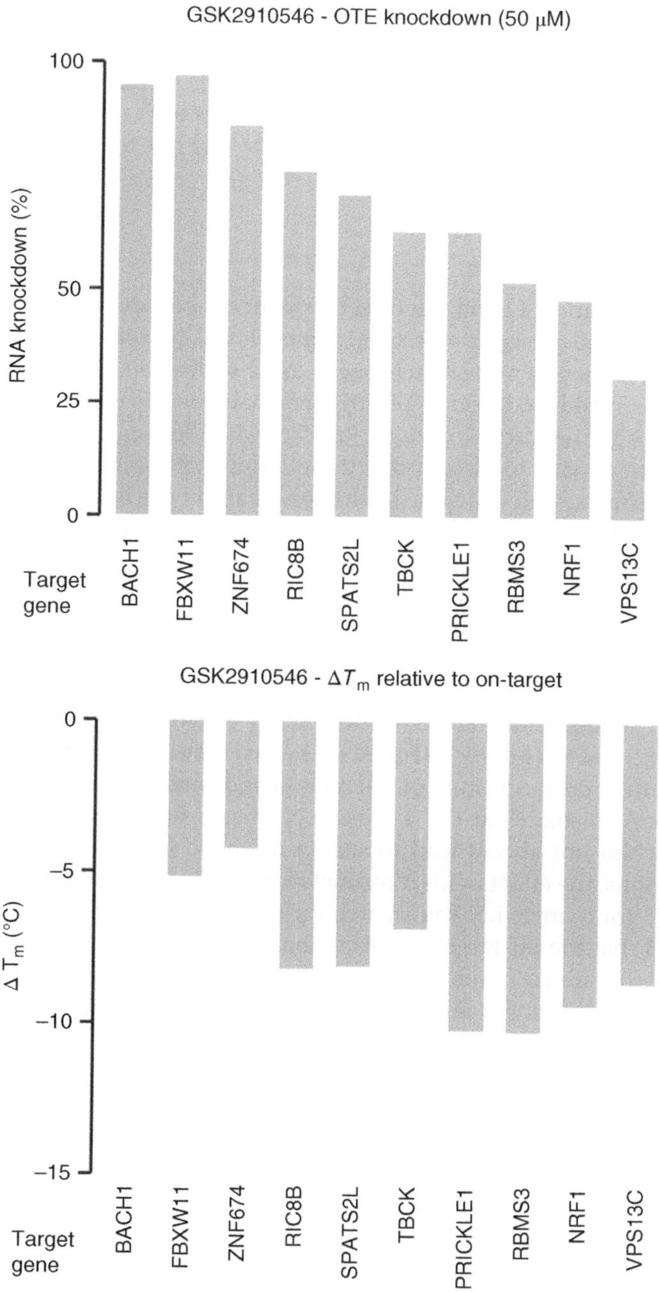

Figure 7.4 Comparison between the change of binding affinity (T_m) and activity for nine off-target hits. Top: Off-target activity measured 48 h following unassisted transfection of human NHBE cells with ASO GSK2910546A. Below: T_m measured for nine off-target RNA oligos duplexed with 16-mer ASO GSK2910546A. Change in T_m from the perfect matched duplex GSK2910546A/Bach1 RNA oligo is shown. Each off-target had two mismatches in the following positions: FBXW11 and ZNF674 1+16, RIC8B 13+14, SPATS2L 1+15, TBCK 13+16, PRICKLE1 12+16, RBMS3 2+16, NRF1 9+16, and VPS13C 12+14. Correlation between activity and change in T_m Pearson $r = 0.84$, P-value $= 0.0025$.

of the catalytic domain. In an *in vitro* cleavage assay, the effect of mismatches in positions at or near the junction between the gap and wings were examined, and in the majority of examples shown, the cleavage rate was not reduced by the mismatch [62].

7.5.4 Mismatch Tolerance

The interaction between an ASO and the target sequence is context dependent. Therefore, in order to measure which mismatches will be tolerated by RNase H1, the consequence of a mismatch needs to be studied in a defined context. To do this we introduced a series of single mismatches into each position within a 16-mer and measured on-target knockdown following transfection into cells. It is necessary to also measure the affinity of interaction to distinguish between effects, which are due to changes in affinity or interaction with the enzyme. The results demonstrated that mismatches could be tolerated in the majority of positions within the duplex; however, some key interactions were identified (see Figure 7.5). A mismatch at position 10 of a 16-mer ASO was detrimental for knockdown activity, but the effect on affinity was not greater than other mismatches where the activity was retained at high levels. This suggested that this position is of key importance for interaction with both the RNA- and DNA-binding domains of the RNase H1 enzyme [59]. A mismatch at the inner (5′) LNA is poorly tolerated in the majority of ASOs examined, with a major effect on the affinity of interaction and consequent lack of knockdown activity. A mismatch at the 3′ inner LNA has a similar effect on binding affinity, but did not always lead to such a dramatic loss of activity. Within the wing regions, the effect of LNA mismatches on affinity is not equal, and the change in T_m for a single LNA mismatch varied between 2 and 13 °C.

We have observed that the off-target confirmation rate differs between different ASOs despite the similar numbers of predicted off-targets being identified (Figure 7.1). A correlation was noted between how well a particular ASO will tolerate mismatches at different positions within the duplex and with the off-target confirmation rate. In the example shown in Figure 7.5, ASO GSK2910613A is sensitive to mismatches across the whole sequence, and in contrast ASO GSK2910546A is able to tolerate mismatches at the 3′ region of the ASO, despite a considerable effect on the measured T_m. Interestingly GSK2910613A has a low off-target confirmation rate with less than 10% of the predicted off-targets tested showing confirmed knockdown activity, compared with GSK2910546A, which has a much higher confirmation rate with over 30% of the predicted off-targets being active. The apparent correlation between tolerance for mismatches and confirmation rate for OTEs suggests that this is an inherent property of the ASO, as a consequence of the sequence and structural design, i.e. the sequence; the number of, and position of, high-affinity residues; oligo length, and gap size.

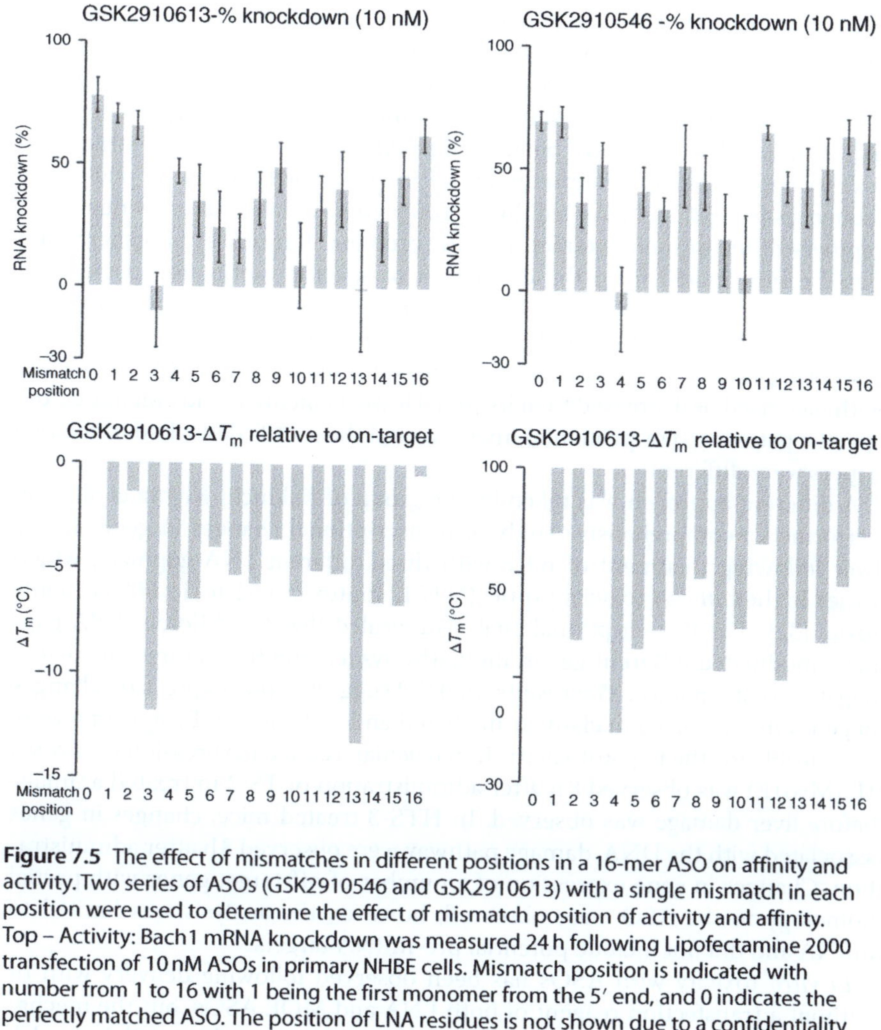

Figure 7.5 The effect of mismatches in different positions in a 16-mer ASO on affinity and activity. Two series of ASOs (GSK2910546 and GSK2910613) with a single mismatch in each position were used to determine the effect of mismatch position of activity and affinity. Top – Activity: Bach1 mRNA knockdown was measured 24 h following Lipofectamine 2000 transfection of 10 nM ASOs in primary NHBE cells. Mismatch position is indicated with number from 1 to 16 with 1 being the first monomer from the 5′ end, and 0 indicates the perfectly matched ASO. The position of LNA residues is not shown due to a confidentiality agreement. Below – Affinity: An RNA oligo of the appropriate Bach1 target sequences was annealed with the mismatch ASO and the T_m values of the duplexes were measured using the assay described in Ref. [5]. Graph is change in T_m from the perfectly matched duplex value.

7.6 Consequences of Unintended Transcript Knockdown: *In Vivo* and *In Vitro* Toxicity

In vivo toxicity issues with ASOs have long been recognized [63] and can be due to a number of reasons that are covered in other chapters. However, more recently hepatotoxicity has been observed following administration of

LNA-modified ASOs to mice, with between 20 and 40% of the ASOs used presenting hepatotoxicity issues [33, 64–68]. Further analysis has indicated that these toxicities are sequence dependent, as small changes in sequence or LNA spiking pattern could result in large differences in phenotype *in vivo* [65, 66]. Additional studies have been performed in order to validate structure–toxicity approaches for the prediction of *in vivo* toxicity based on nucleotide sequences [67, 68] and have led these groups to postulate that individual LNA gapmers differ in their interaction with proteins. Burdick et al. generated a classification model from the evaluation of specific motifs associated with hepatotoxicity and hypothesized that ASOs containing toxic sequences TGC and/or TCC are more likely to bind hepatocellular proteins in mice [68]. Using a similar process, Hagedorn et al. identified dinucleotide motifs associated with increased or decreased toxicity [67]. However, neither study identified the specific pathway(s) or protein(s) involved, and the motifs described in the two papers were different.

A different approach was taken by the group of Kakiuchi-Kiyota et al., who performed a comprehensive study of transcriptional changes in genes in the liver following treatment of mice with three different LNA gapmers, which varied in their *in vivo* toxicity profile (NTS1, nontoxic; TS2, toxic; HTS3, highly toxic) [32]. The transcriptional analysis revealed that the different LNA gapmers modulated different genes and pathways leading to nontarget-mediated hepatotoxicity in mice. The results for TS-2 suggested that expression changes in genes that involved in clathrin-mediated endocytosis (CME) signaling were responsible for the hepatotoxicity. In particular, reduced expression of myosin 1E (Myo1E) was observed 8 h after administration of TS-2 in treated animals, before liver damage was observed. In HTS-3 treated mice, changes in genes associated with the DNA-damage pathway were observed 8 h after administration. Changes in gene expression of a number of off-target genes with partial homology to the ASO were observed; however, the analysis was limited to mRNA and did not include potential pre-mRNA targets.

In vitro toxicity with ASOs has been observed following delivery with or without a transfection reagent of both LNA and MOE ASOs, but the mechanism underlying the adverse phenotype was not determined [66, 69]. In our lab we have observed toxicity with LNA–ASO following assisted delivery but noted that there are significant differences in the toxicity induced by different ASOs against the same target under the same experimental conditions (Figure 7.6a). This indicates that it is not due to knockdown of the intended target or due to the use of a transfection reagent. Subsequent studies demonstrated that the apoptosis observed is RNase H1 dependent, indicating that it is mechanism based, most probably due to Watson–Crick hybridization and knockdown of off-targets (Figure 7.6b). Changes in the LNA spiking pattern can alter the toxicity profile of an individual LNA–ASO (Figure 7.6e). Modifying the spiking pattern will affect the affinity for off-target sequences, thereby

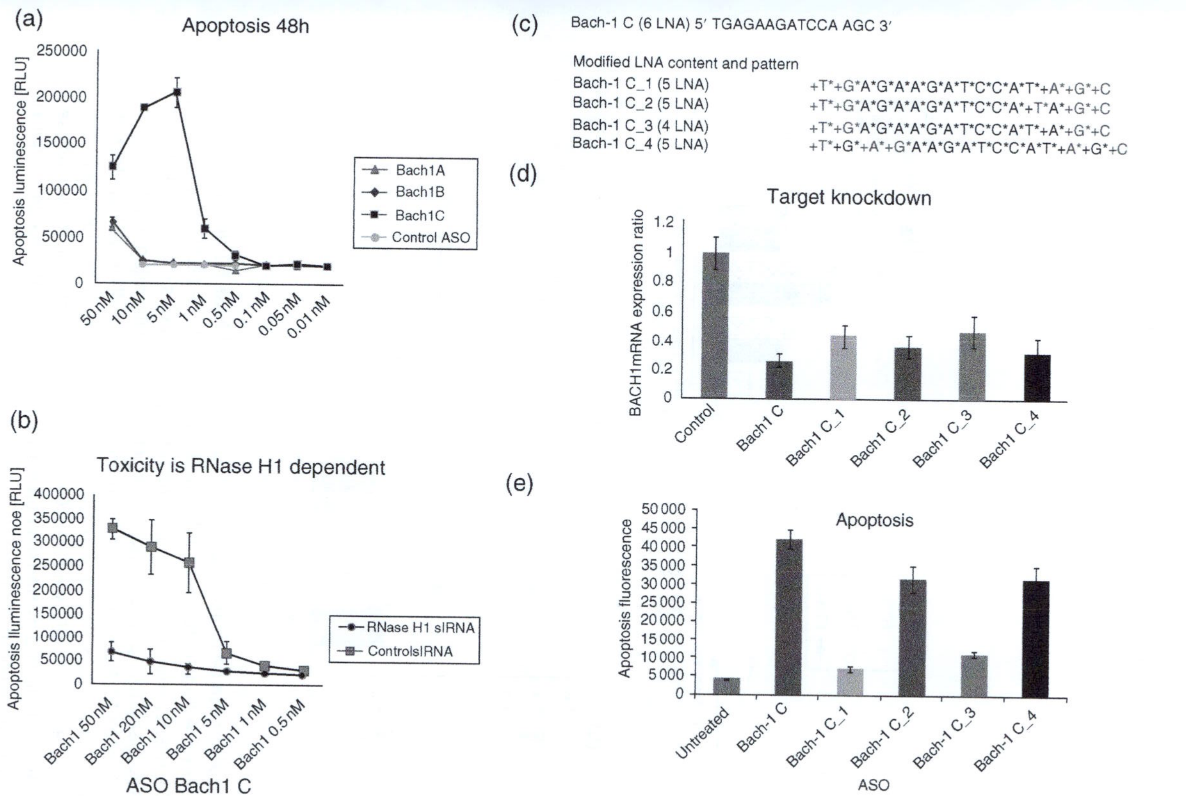

Figure 7.6 Apoptosis induced by an ASO *in vitro* is RNase H1 dependent and can be ablated by changing the LNA content and spiking pattern. (a) A549 cells were transfected with 0.01–50 nM of three different BACH1 targeting or control ASOs using Lipofectamine 2000. Forty-eight hours after transfection, apoptosis was measured using Promega ApoTox-Glo assay. Decline in signal at higher concentrations of ASO BACH1 C is due to cell loss at this time point. (b) A549 cells were transfected with 20 nM control siRNA or siRNA-targeting RNase H1 for 24 h, followed by transfection of ASO BACH1 C for 24 h. Apoptosis was measured using Promega ApoTox-Glo assay. (c) Variants of BACH1 C with modified LNA content and position: LNA residues are indicated in red, and "*" represents PS residue. The LNA spiking pattern cannot be shown for "BACH1 C" due to a confidentiality agreement. (d) BACH1 target mRNA knockdown following Lipofectamine 2000 transfection of A549 cells with 5 nM of ASO. (e) Induction of apoptosis following Lipofectamine 2000 transfection of A549 cells with 5 nM of ASO. Apoptosis was measured using Promega Apo-One Assay.

altering how a mismatch will be tolerated. This can result in a reduction in toxicity if knockdown of an essential gene is abolished because of reduced binding, or alternatively the ASO could be sequestered by increased binding to a different unintended target for which knockdown does not have such a deleterious effect. Such an effect could explain the observation that small changes to the ASO could have dramatic changes in phenotype *in vivo* [64–69]. We were interested to see if the sequence-dependent *in vivo* hepatotoxicity was due to OTEs. We extended the studies of Kakiuchi-Kiyota [32] and tested the ASO described therein in a cell-based *in vitro* assay. We were able to reproduce the toxicity and demonstrate that it requires RNase H1 expression (Figure 7.7a

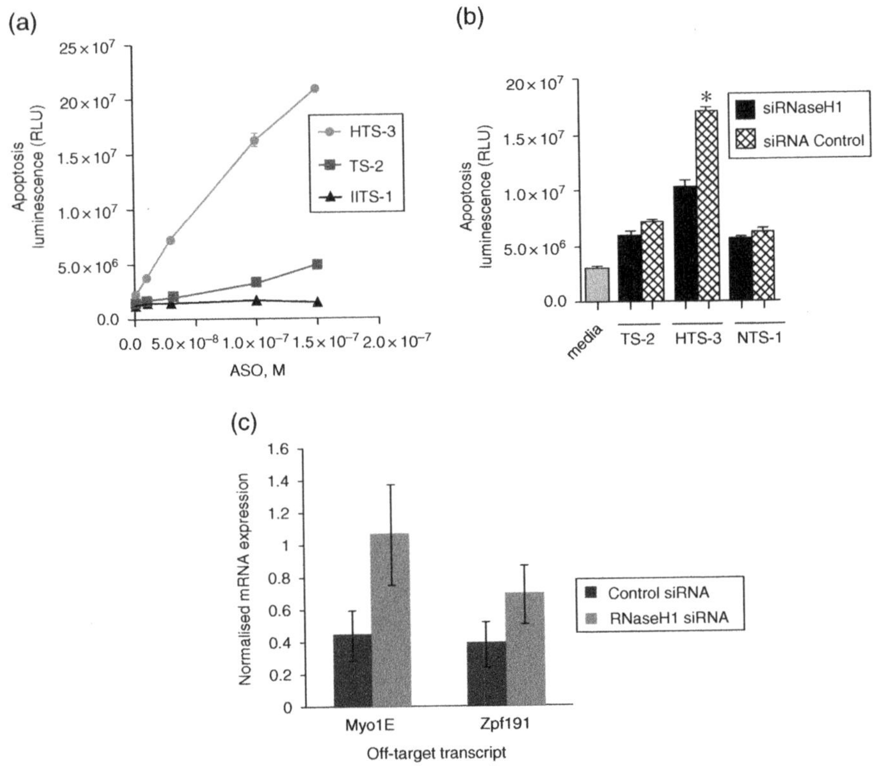

Figure 7.7 *In vivo* toxicity can be reproduced in cell culture and is RNase H1 dependent. (a) Mouse Hepa1-6 cells were transfected with different doses of NTS-1, TS-2 & HTS-3 [32] using Lipofectamine2000. Twenty-four hours after transfection, apoptosis was measured using Promega ApoTox-Glo assay. (b) Apoptosis induced by HTS-3 is partially mitigated by pretreatment with 25 nM siRNA designed against murine RNase H1. Cells were transfected with siRNA for 24 h, and then transfected with ASO for 24 h. (c) Myo1E knockdown is RNase H1 dependent. Murine Hepa1–6 cells were transfected with 25 nM control or mouse RNase H1-specific siRNA for 24 h followed by transfection with 100 nM of TS-2 ASO for 24 h. Zfp191 was previously reported as exon off-target [32].

and b). By including pre-mRNA sequences in our off-target analysis, we have demonstrated that the majority of the effects on gene expression observed are due to direct off-target RNA interaction mediated by the ASOs. For the TS-2 ASO, Myo1E was identified as an off-target within an intronic region (Figure 7.7c), and more than 70% of the downregulated genes reported for the highly toxic ASO HTS-3 were identified as off-target interactions.

An *in vivo* study by Ionis Pharmaceuticals [6] has demonstrated that hepatotoxicity of ASOs containing bicyclic modifications is a result of off-target activity. Whole transcriptome analysis revealed knockdown of a broad range of transcripts in the liver. Similar to the study of Kakiuchi-Kiyota, no common biological pathways were modulated by the hepatotoxic LNA–ASOs; however, a correlation between toxicity and large-scale downregulation of long transcripts was observed. The requirement for RNase H1 was demonstrated by pretreatment of mice with a MOE gapmer targeting RNase H1, with the hepatotoxicity in these animals greatly reduced following administration of "toxic" LNA–ASOs compared with animals pretreated with a control ASO. Furthermore, the studies were extended to demonstrate that other high-affinity modifications, including cEt, also display similar activity.

The results imply that off-target knockdown can contribute to the toxicity observed with some ASOs *in vivo* and *in vitro*. Although the doses of oligonucleotides used in these *in vivo* studies are high, they are typical of the doses used for safety testing where a therapeutic index between the highest tolerated dose and probable therapeutic dose needs to be established. ASOs can display a long half-life *in vivo* and show differential accumulation in certain tissues. Some ASOs have also shown higher activity against unintended targets compared with the intended target. The issue of Watson–Crick off-target hybridization is thus of central importance in the development of therapeutic ASOs.

7.7 Identification and Evaluation of Putative OTEs

It is now clear that Watson–Crick OTEs are more than a theoretical risk for therapeutic ASO, and that these effects can have serious consequences. Therefore, off-target profiling of a therapeutic ASO is an essential part of the development process. Fortunately, although time consuming, determining the off-target liability of an ASO is relatively straightforward and involves the computational identification of potential off-target hits followed by experimental verification, as illustrated in Figure 7.8.

7.7.1 Computational Prediction of Unintended Targeting

While the poor specificity of siRNA molecules is well recognized and addressed to some extent by both design [70] and software [71] solutions, at the time of

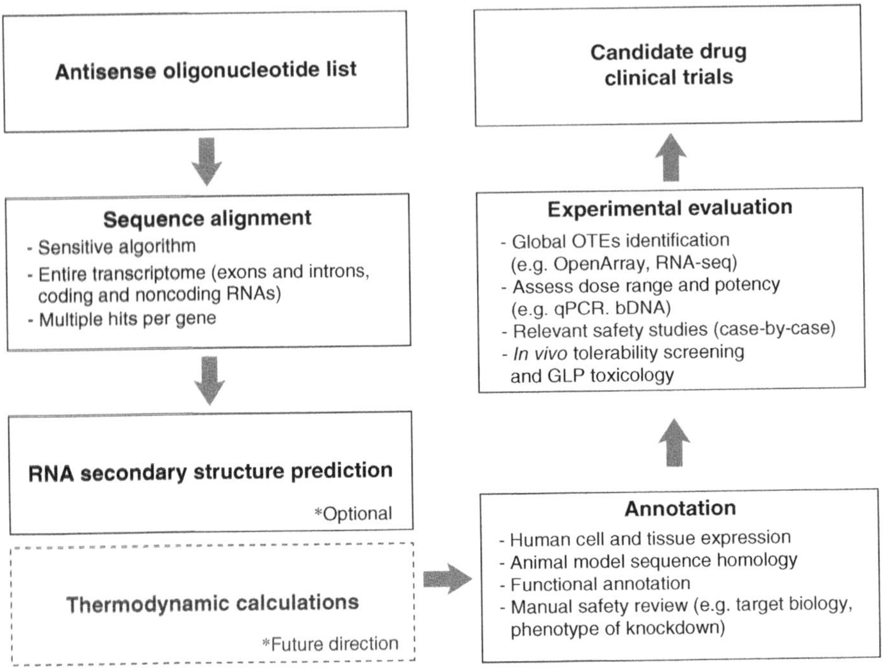

Figure 7.8 Graphical summary of computational and experimental steps used in ASO gapmer specificity evaluations. While RNA structure prediction and thermodynamic calculations are valuable and can bring substantial improvements to the prediction process in the future, they currently lack parameters and features that are necessary for comprehensive assessment of modified ASO-mediated interaction likelihood.

writing, there are no publicly available computational tools geared specifically toward assessing hybridization-mediated OTEs liability associated with ASOs. The most commonly used methodology incorporates Basic Local Alignment Search Tool (BLAST) searches against database of mature, protein-coding genes [32]. However, we have recently shown a high confirmation rate of intronic and exonic OTE associated with 16-mer LNA gapmers [5]. The findings underline the importance of judicious bioinformatic analysis, and the requirement of both sensitive sequence alignment algorithm and appropriate selection of molecular targets against which a search should be performed. The computational guidelines outlined below can also be applied to other oligonucleotide platforms with base pair-driven target recognition and potential for hybridization-mediated OTEs (e.g. splice-modulating ASOs or DNAzymes). A comprehensive software solution ("RNArcher") was recently developed as part of the GlaxoSmithKline/Imperial College London/Public Health England collaboration and applied within GlaxoSmithKline for design and lead optimization of ASO gapmers. Iterative improvements have been made, and the final version of the pipeline will be released to public in the near future.

7.7.1.1 Database Creation

The considerable divergence between siRNA and ASO-silencing mechanisms translates into distinct binding landscapes, which should be reflected in the contents of the corresponding target database. Unintended interactions associated with siRNA are mainly attributed to miRNA-like activity in the cytosol, predominantly targeted against 3′UTR regions of mature mRNAs [72]. This is in contrast to ASOs, which can cleave any transcript within the nucleus or mitochondria, regardless of the target's biotype or stage of RNA processing and can therefore also target introns or noncoding RNAs [15, 26, 73]. ASO-mediated OTE predictions should therefore be performed against a comprehensive set of coding as well as noncoding genes, in both their primary (i.e. unspliced) and mature (i.e. spliced/processed) transcript stages. For a given gene, alternative transcription and splicing can generate a number of distinctive primary and mature sequence variants [74]. While a particular ASO may interact with only a subset of such sequences, it is currently not possible to obtain accurate data on cell and tissue-specific splice variant expression patterns, nor is it possible to assess the implication of such "selective" silencing. Specificity evaluations, performed against "full-length" unspliced transcripts (i.e. overlapping all known exons and introns for a given gene) and all mature variants (covering exon–exon boundaries), followed by experimental evaluation, are currently the most viable strategy. By gathering genomic coordinates of exons, transcription start sites (TSS) and transcription termination sites (TTS), it is possible to differentiate between exonic, intronic, or boundary (i.e. splice site) regions within the primary transcript and assess how many of the primary transcript variants are potentially targeted by the ASO. While requiring an additional processing step, using a single primary transcript per gene, instead of one for each splice variant, markedly reduces the database size and overall computing time. Inclusion of intronic sequences dramatically increases the number of putative hits and is crucial for accurate assessment of selectivity and prediction of theoretical toxicity. With regular iterative improvements being made to the genome builds and annotations, frequent updates to the databases are highly advisable. In addition, integrating genetic polymorphisms into the search space, though currently challenging to implement and computationally intensive to utilize, could provide a platform for estimating patient-to-patient response to ASO therapeutics.

7.7.1.2 Sequence Alignments

Although no threshold has been established in terms of the number of mismatches and/or gaps that a ASO:target duplex can tolerate for a given ASOs' length and chemical modification content, a thorough prediction may be used to rank the leads based on the number of OTEs. We found that for 16-mer LNA–ASOs, 2MM OTEs can show EC_{50}s matching or even exceeding that of the on-target gene [5]. To tackle the combination of short length of the query

(12–24 mer) and low level of sequence identity (e.g. 75%), the sensitivity of off-target detection methods must be greatly increased.

While BLAST is more than adequate for many biological applications, its heuristic design cannot guarantee optimal results. The speed optimization is achieved by initial matches being identified using a short fragment of the query sequence (i.e. seed or word), with local alignment being performed only on the selected list of targets. With speed optimizations based on technologies such as single instruction multiple data (SIMD) parallelization [75], it is now possible to utilize rigorous algorithm such as Smith and Waterman (SW) [76] without a significant increase in computing time. The scoring matrix (i.e. scoring of matches, mismatches, gaps opening, and extension) plays a major role in the ability to detect the alignments of interests and should be carefully tailored to match the application requirements [77]. The commonly used parameters were established to look for evolutionary homology, with significance thresholds put in place to filter results based on the probability of a hit appearing by chance. This is problematic for hits with two or more mismatches and/or gaps, whose estimated score is considerably lowered due to the substantial size of the database paired with the short length of ASO. When possible, statistical calculations and consequent hit filtering should be switched off or modified, so that hits with lower identity are also reported. While most methods are built to report a single, best alignment per sequence, it has to be noted that ASOs can theoretically interact with a given gene at more than one region. On occasions we have encountered examples of very potent OTE genes with more than one predicted target site [5]. A higher number of putative hits per sequence increase the probability of detectable activity against a given target. It is also likely that multiple ASO:transcript interactions within a single sequence may work in an additive or synergistic manner, leading to a high target knockdown [78]. The choice of algorithm for OTEs prediction should therefore also be driven by the ability of the software to report more than one alignment per transcript. Any options, which limit the number of reported hits, such as those based on repeat or low complexity region masking, should be avoided.

7.7.1.3 Cross-species Off-target Homology

Due to significant sequence divergences between the species traditionally used in preclinical assessment, OTEs prediction should be performed separately for each animal model used in experimental safety evaluation. While functional annotation and quality of sequence for lower mammals or primates are not on a par with humans, the OTE analysis may nevertheless add value, especially in cases where a species-specific phenotype is observed. The causative gene(s) that generated the adverse response could potentially be identified by analyzing the difference in OTEs prediction between tested organisms, in particular at the perfect or near-perfect complementary level. Cross-species analysis can also be used to determine whether a safety risk associated with a confirmed human OTE could be studied in any of the available animal models. This is particularly advantageous in cases involving confirmed OTEs, for which potential liability cannot be discharged at the *in vitro* level. Instead of a time-consuming, manual cross-species sequence analysis, the conservation of ASO target regions could be evaluated using the predictions. For each confirmed human OTE gene, the presence, absence, and alignment details of homologous gene(s) could be used to determine the appropriateness of using the particular *in vivo* model in discharging the safety risk.

7.7.1.4 Results Filtering and Annotation

Discharging OTEs based on their alignment pattern, function, genetic redundancy, or expression data is an attractive but practically challenging concept. Baseline expression can support the selection of appropriate cell lines, within which particular OTEs could be studied. *In vivo* distribution of ASO to unintended organs [79], however, limits the ability to deprioritize hits based on target tissue expression. The strategy could potentially be applicable in a few cases, such as more local delivery (e.g. topical, intrathecal, and inhaled) or where ASO tissue levels after systemic administration do not reach meaningful levels (e.g. central nervous system). Even though Gene Ontology (GO) terms and functional annotation help pinpoint particularly problematic genes, a thorough manual safety profile review and experimental verification would still be needed to fully evaluate and confirm the gene knockdown and its potential safety impact. Computational methods based on molecular pathways, RNA structure, or binding energy concepts could support hit prioritization, but currently do not guarantee sufficient accuracy for confident risk discharging. It is possible that certain positions within an ASO, especially those opposite or adjacent to chemical modifications, can have varying levels of mismatch tolerance [15, 80, 81]. While our analysis did reveal a number of patterns, we have also observed potent OTEs with mismatches and gaps across a wide range of positions within the ASO:target duplex [5] (unpublished data). The level of identity, which would translate as a potent interaction is dependent on many factors, including ASO length and its target binding strength. The current most viable *in vitro* approach is to implement a realistic mismatch/gap threshold and experimentally assess the knockdown of the selected genes (using methods such as qPCR or bDNA). Alternatively, genomic methods such as RNA-Seq offer a global view of all expressed genes.

7.7.1.5 RNA Structure and Target Accessibility

Though identifying putative ASO interactions is relatively straightforward, no general method has been established to accurately predict the efficiency of on- or off-target knockdown, equivalent to those created for siRNAs [82]]. A comprehensive hybrid model, developed by Ionis Pharmaceuticals and based on a range of parameters extracted from approximately 10 000 experimentally tested ASOs, failed to provide significant improvement to the ASO design process [2]. Accessibility of intended target sites has to date been one of the most highly correlating features [83–85].

RNA structure prediction algorithms, while not as advanced as those developed for proteins, have improved considerably in recent years [86]. RNA's secondary structure forms with higher thermodynamics stability compared with the tertiary structure, which makes the folding hierarchical and allows the prediction of the former without the knowledge of the latter [87, 88]. Comparative sequence analysis, based on the premise that the structure is more evolutionarily conserved than the sequence, tremendously improves the quality of secondary structure prediction and consequent three-dimensional models [89]. The analysis does however require considerable manual curation and a large number of homologous genes, which are not always available. The Gibbs free energy minimization method, though not as accurate as comparative approaches, enables a fully automated prediction of secondary structures based solely on transcript sequence. Under the assumption that the structure with lowest free energy will be most represented at equilibrium, dynamic programming techniques can be used to implicitly search through all possibilities and build an optimal RNA secondary structure [89, 90]. Partition function can then be used to estimate the probability of a particular base pair to form [90]. However, the computed optimum structures may not always be correct, partly due to limitations in the thermodynamic calculations, presence of many secondary structures with the same free energy, or inability to account for conformational changes driven by interaction with other molecules [91]. Generating alternative, suboptimal structures will provide additional information and markedly increases the average accuracy [92]. Large on-target ASO efficiency screens provide perfect training datasets to establish the software's parameters. Based on available reports [93] and our unpublished data, local folding approaches appear to produce more accurate results relative to global methods.

The algorithms outlined above can be used to predict the accessibility of exonic regions within mature, spliced transcripts. Applying the same approach to primary transcripts is, however, more complex. With observed general correlation in activity between siRNA (targeting mature RNA) and ASO (targeting primary and mature RNA) designed against the same position on the target mRNA [15], it is likely that the RNA-binding molecules and local RNA accessibility affect ASO-mediated exon binding almost immediately after the region is transcribed. Using a prediction algorithm that accounts for cotranscriptional primary RNA folding, it is possible to identify accessible sites within transient secondary structures before they become associated with proteins and RNAs [94]. Nevertheless, there is currently no method that can account for RNA-binding proteins/RNAs, which most likely dictate the observed differences in confirmation between exonic and intronic OTEs [5]. We can however presume that at a certain stage of transcript and/or splicing, intron regions are more accessible to the ASO and hypothesize how different aspects of intronic regions may also determine the efficiency of knockdown. Processing of primary RNA occurs as soon as transcription initiates and the unspliced form exists for a limited period of time [95]. Depending on the actual splice variant, certain introns will be spliced as transcription proceeds, while others will be retained until the sequence is fully transcribed [96]. Fragments that are larger, and require more time to be transcribed and/or spliced out, are potentially a more potent RNase H1 targets. With smaller and more rapidly removed introns, the ASO has a relatively narrow window of opportunity to bind. Gene length and its transcription speed, splicing mechanism, or size and position of ASO binding sites within an intron likely all influence the final knockdown potential.

7.7.1.6 ASO–Target Duplex Thermodynamics

The strength of binding between ASO and the on- or off-target is one of the most fundamental components that determines if, and how, the interaction occurs [5, 97]. Under the nearest neighbor model, the free energy change for forming a base pair is calculated based on the sequence identity of the said base pair and that of its immediate neighbor. Optical melting experiments have been used to establish parameters for various combinations of base pair matches or mismatches, within varying sequence context [98, 99]. While it is relatively straightforward to predict the T_m of DNA–RNA duplexes, the presence of high-affinity chemical modifications greatly impedes such estimation. There are currently few publicly available nearest neighbor energy calculations for modified ASOs [2], let alone ones accounting for the presence of more than one modification within a single sequence (e.g. PS backbone with 2′OMe, MOE, or LNA bases). In addition, the effect of introducing high-affinity bases into the ASO is not additive, with the position pattern playing a large role [22, 100]. Chemical modifications and mismatches may also cause a structural shift that could propagate beyond the neighboring base pairs [81]. Nearest neighbor energy models, a current gold standard in predicting duplex T_m, cannot accurately deal with such cases. The methodology based on molecular dynamics (MD) simulation presented by Shen et al. [101] attempts to address several of these issues. This conceptually elegant method is however not easily accessible to the end user, involves extensive computational run time (~15 h per single duplex using four standard CPUs), and also requires further proof of accuracy for both intended and off-targets. It is however a step in a right direction, given the availability of MD force fields for some of the modifications [102].

Analysis of thermodynamic features has been successfully applied in the case of siRNA, where the binding strength of seed and nonseed regions were found to be correlated with miRNA-like OTEs' potency [70, 103]. Although the correlation between ASO activity and calculated binding free energy was not found to be high [2], at a cost of sensitivity, thermodynamic cutoffs were shown to bring improvement to design efficiency [104]. A more comprehensive model based on cleavage kinetics was recently used to identify target-specific optimal binding affinity, above which a drop in potency is observed [105]. The conceptually simple RNase H1 mechanism is in fact highly complicated to simulate. A complete model must account for several ASO- (binding affinity, stability, DNA core length in gapmers [106], concentration, and distribution), target- (free energy of target unfolding [107], transcript concentration and turnover rate [108], and competition for ASO binding between intended and unintended regions), and RNase H1- (concentration [105], enzyme kinetics of RNase H1 with other terminating mechanisms [2], and binding affinity between the RNase H1 and ASO:target complex) specific factors. Valuable guidelines can be deduced from the structure of both the hybrid-binding domain and catalytic domain of RNase H1, which are now available [59, 60].

7.7.1.7 Computational Framework for OTEs

A comprehensive approach for ASO-mediated off-target identification and verification is summarized in Figure 7.8. Sequence alignment based on a sensitive algorithm and performed against a database of coding and noncoding genes (in both their primary and spliced stage) is currently the only confident method to identify and filter putative OTEs. While current RNA structure prediction methods can add value when used with exonic regions, more work is needed to verify their usefulness when dealing with intronic regions. Cotranscriptional change in local primary transcript structure should also be taken into account. Establishing and incorporating thermodynamics rules and parameters for chemically modified bases will bring substantial improvements to the prediction accuracy. By studying both the ASO-mediated OTEs and the RNase H1 enzyme, it may also be possible to establish specificity rules, similar to those created for siRNA [103]. Following the prediction process, gene annotation can be used to identify appropriate models and experimental approaches to study the activity and potential safety liabilities associated with knockdown of each putative OTE. The methodology outlined above will increase the chances of progressing leads with the highest relative specificity and reduce the probability of OTEs resulting in adverse effects.

7.7.1.8 *In Vitro* Screening for OTEs

Activity for a given ASO is context and sequence dependent, and the computation prediction methods have not yet developed the level of sophistication and accuracy required to state whether a predicted off-target interaction will be observed following delivery of the ASO to cells. Therefore, any predicted OTE must be confirmed experimentally. In principle, the *in vitro* screening for OTEs is simple. Cells in culture are transfected with the ASO and changes in gene expression are measured at different times following transfection. In practice, there are a number of decisions regarding techniques, methodology, and time points that need to be considered. The first consideration is the cell type and method of delivery. The cell type should be amenable to transfection and express the genes of interest. Expression of both the on-target transcripts and any off-target genes allows a therapeutic index between on-target and off-target knockdown to be established. The cell type should also be relevant to tissues in which the target is expressed *in vivo* and to where high levels of oligonucleotide would be expected to accumulate.

There are a number of different ways to deliver the ASO into cells, and these include using transfection reagents, unassisted delivery (also referred to as "gymnosis"), or electroporation. There are advantages and disadvantages with each of the different methods. The route by which ASOs are taken into the tissues and cells *in vivo* has not been fully elucidated, so which *in vitro* method most closely replicates the situation *in vivo* is unclear. Simplistically, unassisted delivery would appear to best mimic the *in vivo* situation of delivery of an ASO in simple saline formulation [109, 110]. Gymnosis experiments are simple to perform, and it has been reported that gymnosis is able to transfect some cells not amenable to transfection using a transfection reagent [110]. Due to the absence of transfection reagent, interpretation of the results should be simpler as there is no change in gene expression mediated by the presence of the transfection reagent.

Transfection reagents come in many different forms and are available from a number of suppliers, e.g. lipid based such as Lipofectamine 2000 and DharmaFECT. A wide range of cell types are amenable to transfection, including many primary cells. Assisted delivery is highly efficient and has a low requirement for ASO and therefore can be performed early in the screening process. Efficient delivery means the experiments can be shorter in duration, which permits greater flexibility in study design. Experimental setup is more complex than with unassisted delivery, and careful optimization may be required, particularly with delicate or difficult to transfect cells. The main concern with assisted delivery using a lipid-based reagent is the transcriptional signal from the transfection reagent itself, which can be mitigated with the appropriate controls. An additional concern is the efficient delivery to the nucleus may overstate the potential liabilities, with some "false-positive" off-targets that would be unlikely to occur *in vivo*. However the correlation between *in vitro* and *in vivo* toxicity shown in Figure 7.7a suggests that assisted delivery can, to some extent, reproduce the situation *in vivo*.

Electroporation, for example, by means of the Amaxa Nucleofector shuttle system (Lonza Ltd.) has been used to deliver ASO into cells [66]. This method requires specialist equipment and consumables and can be used in single tube or 96 well format. It is not gentle to cells but is of utility with some difficult to transfect cells, for example, T cells or neuronal cells [18, 20].

7.7.1.9 Methods for Measuring Gene Expression

There are many different technologies available to measure changes in expression of the target and any off-target genes. These include hybridization-based methods, PCR-based technologies, and RNA-sequencing methods. The methods have different throughput capacities, advantages, and disadvantages and have been described in a number of reviews [111, 112].

Real-time PCR (qPCR), such as Taqman or SYBR green, is widely considered as the gold standard for gene expression measurement due to its high assay specificity and wide linear dynamic range [111, 112]. These are low throughput methods but are relatively low cost, accurate, and flexible. An alternative low capacity method is bDNA, and this is more expensive to run but does not require cDNA synthesis, thereby reducing the potential for interference by any ASO carried through during the RNA purification procedures. Medium throughput systems can use hybridization-based technology, for example, NanoString nCounter. Alternatively, RT-qPCR-based technology may be used, for example, the BioMark HD™ system (Fluidigm Corporation) and OpenArray system (Life Technologies) [113, 114]. These permit numerous genes to be analyzed at the same time, are flexible, and are of reasonable cost. High throughput assay such as microarrays or RNAseq allow the effect of the ASO to be studied on the whole transcriptome [115, 116]. RNAseq has several advantages over microarrays: the detection range of RNaseq is not limited to predetermined probes and can detect expression at the gene, transcript, and exon levels.

The definitive proof demonstrating a phenotypic effect due to off-target knockdown is to measure changes in expression of the off-target protein. This is time consuming and requires suitable reagents to be available, and the stability of protein may mean that changes in protein expression do not always track with changes in mRNA expression.

7.8 Summary

The physiological function and low stringency requirements of RNase H1 mean that specificity issues are an inherent concern with antisense molecules that utilize RNase H1 for activity. Potent off-target activity against sequences within introns has been clearly demonstrated for the latest generation of antisense molecules *in vitro* and *in vivo*, and this activity appears to be responsible for the sequence-dependent toxicities observed with numerous LNA–gapmers *in vivo*. The majority of the studies showing off-target activity and sequence-dependent *in vivo* toxicity have utilized PS–LNA–gapmers; however, there is growing evidence that the off-target interaction is a general phenomenon and is observed with other high-affinity chemical modifications [6, 20]. Off-target activity is a particular issue for therapeutic ASOs but is also of concern if an antisense strategy is employed to determine gene function, as any unintended activity could lead to misinterpretation of the results.

The knockdown activity of ASO for both the intended and unintended targets is the result of a complex interplay of numerous factors. In general, these factors apply to both the intended target and unintended target, but may differ in their significance. For example, higher accessibility of the off-target sequence relative to the on-target sequence may compensate the change in affinity due to the presence of the mismatch or the positional effect of the mismatch on the interaction with RNase H1. Activity for a given ASO is context and sequence dependent, and the computation prediction methods have not yet developed the level of sophistication and accuracy required to state whether a predicted off-target interaction will be observed following delivery of the ASO to cells. Therefore the off-target liability of a therapeutic ASO must be confirmed experimentally.

Compared with more established therapeutic platforms, oligonucleotides provide a unique opportunity to study OTEs in much greater depth and establish parameters for their prediction. A greater understanding of the factors that lead to RNase H1-dependent knockdown of unintended transcripts, including how changes to the spiking pattern can modify the off-target profile, will lead to ASOs of improved design, and minimize off-target activity. Furthermore, with base pair target recognition, it is relatively straightforward to identify potential hybridization-mediated interactions and experimentally verify them in a high throughput manner. This should markedly reduce the risk of encountering unexpected adverse side effects during animal testing and clinical trials. The ability to detect unintended interactions between a biologically active entity and cellular machinery during early stages of drug discovery would ensure that resources are focused on the most selective and promising leads. The application of improved ASO design and the use of comprehensive computational and experimental screening methodologies will permit the identification of molecules with a large therapeutic index, and thus allow the potential of the antisense platform to be realized.

Acknowledgments

We would like to acknowledge members of the Therapeutic Oligonucleotide DPU at GSK, in particular Sofie Eriksson, Paola Malerba, Gemma Turner, Jasmin Menzel, and Aliza Panjwani, for allowing to us to show unpublished data, Satoko Kakiuchi-Kiyota for sharing additional data, and Paul A. Wilson, Joel D. Parry, Nigel J. Gooderham, and Timothy W. Gant for helpful discussion and valuable input on the computational section of the chapter.

References

1 Zamecnik, P.C. and Stephenson, M.L. (1978). Inhibition of Rous sarcoma virus replication and cell transformation by a specific oligodeoxynucleotide. *Proceedings of the National Academy of Sciences of the United States of America* 75: 285–288.

2 Crooke, S.T., Vickers, T., Lima, W., and Wu, H. (2007). *Antisense Drug Technology: Principles, Strategies, and Applications*, 2e, 3–46. Boca Raton, FL: CRC Press.

3 Kole, R., Krainer, A.R., and Altman, S. (2012). RNA therapeutics: beyond RNA interference and antisense oligonucleotides. *Nature Reviews Drug Discovery* 11: 125–140.

4 Zhang, H., Cook, J., Nickel, J. et al. (2000). Reduction of liver Fas expression by an antisense oligonucleotide protects mice from fulminant hepatitis. *Nature Biotechnology* 18: 862–867.

5 Kamola, P.J., Kitson, J.D.A., Turner, G. et al. (2015). *In silico* and *in vitro* evaluation of exonic and intronic off-target effects form a critical element of therapeutic ASO gapmer optimization. *Nucleic Acids Research* 43: 8638–8650.

6 Burel, S.A., Hart, C.E., Cauntay, P. et al. (2015). Hepatotoxicity of high affinity gapmer antisense oligonucleotides is mediated by RNase H1 dependent promiscuous reduction of very long pre-mRNA transcripts. *Nucleic Acids Research*. doi: 10.1093/nar/gkv1210.

7 Stein, C.A. (2000). Is irrelevant cleavage the price of antisense efficacy? *Pharmacology & Therapeutics* 85: 231–236.

8 Summerton, J.E. (2007). Morpholino, siRNA, and S-DNA compared: impact of structure and mechanism of action on off-target effects and sequence specificity. *Current Topics in Medicinal Chemistry* 7: 651–660.

9 Bennett, C.F. and Swayze, E.E. (2010). RNA targeting therapeutics: molecular mechanisms of antisense oligonucleotides as a therapeutic platform. *Annual Review of Pharmacology and Toxicology* 50: 259–293.

10 Cazenave, C., Loreau, N., Thuong, N.T. et al. (1987). Enzymatic amplification of translation inhibition of rabbit b -globin mRNA medicated by anti-messenger oligodeoxynucleotides covalently linked to intercalating agents. *Nucleic Acids Research* 15: 4717–4736.

11 Shuttleworth, J. and Coleman, A. (1988). Antisense oligonucleotide-directed cleavage of mRNA in Xenopus oocytes and eggs. *EMBO Journal* 7: 427–434.

12 Monia, B.P., Lesnik, E.A., Gonzalez, C. et al. (1993). Evaluation of 2′-modified oligonucleotides containing 2′-deoxy gaps as antisense inhibitors of gene expression. *Journal of Biological Chemistry* 268: 14514–14522.

13 Monia, B.P., Johnston, J.F., Geiger, T. et al. (1995). Anti-tumor activity of a phosphorothioate oligodeoxynucleotide targeted against C-raf kinase. *Nature Medicine* 2: 668–675.

14 Freier, S.M. and Watt, A.T. (2007). *Antisense Drug Technology: Principles, Strategies, and Applications*, 2e, 117–141. Boca Raton, FL: CRC Press.
15 Vickers, T.A., Koo, S., Bennett, C.F. et al. (2003). Efficient reduction of target RNAs by small interfering RNA and RNase H-dependent antisense agents. A comparative analysis. *The Journal of Biological Chemistry* 278: 7108–7118.
16 Braasch, D.A., Liu, Y., and Corey, D.A. (2002). Antisense inhibition of gene expression in cells by oligonucleotides incorporating locked nucleic acids: effect of mRNA target sequence and chimera design. *Nucleic Acids Research* 30: 5160–5167.
17 Monia, B.P., Johnston, J.F., Ecker, D.J. et al. (1992). Selective inhibition of mutant Ha-ras mRNA expression by antisense oligonucleotides. *The Journal of Biological Chemistry* 267: 19954–19962.
18 Carroll, J.B., Warby, S.C., Southwell, A.L. et al. (2011). Potent and selective antisense oligonucleotides targeting single-nucleotide polymorphisms in the Huntington disease gene/allele-specific silencing of mutant huntingtin. *Molecular Therapy* 19: 2178–2185.
19 Lee, J.E., Bennett, C.F., and Cooper, T.A. (2012). RNase H-mediated degradation of toxic RNA in myotonic dystrophy type 1. *Proceedings of the National Academy of Sciences of the United States of America* 109: 4221–4226.
20 Ostergaard, M.E., Southwell, A.L., Kordasiewicz, H. et al. (2013). Rational design of antisense oligonucleotides targeting single nucleotide polymorphisms for potent and allele selective suppression of mutant Huntingtin in the CNS. *Nucleic Acids Research* 41: 9634–9650.
21 Skotte, N.H., Southwell, A.L., Østergaard, M.E. et al. (2014). Allele-specific suppression of mutant huntingtin using antisense oligonucleotides: providing a therapeutic option for all Huntington disease patients. *PLoS ONE* 9: e107434.
22 Petersen, M. and Wengel, J. (2003). LNA: a versatile tool for therapeutics and genomics. *Trends in Biotechnology* 21: 74–81.
23 Chan, J.H.P., Lim, S., and Wong, W.S.F. (2006). Antisense oligonucleotides: from design to therapeutic application. *Clinical and Experimental Pharmacology and Physiology* 33: 533–540.
24 Lima, W.F., Wu, H., and Crooke, S.T. (2007). *Antisense Drug Technology: Principles, Strategies, and Applications*, 2e, 47–74. Boca Raton, FL: CRC Press.
25 Wu, H., Lima, W.F., Zhang, H. et al. (2004). Determination of the role of the human RNase H1 in the pharmacology of DNA-like antisense drugs. *Journal of Biological Chemistry* 279: 17181–17189.
26 Suzuki, Y., Holmes, J.B., Cerritelli, S.M. et al. (2010). An upstream open reading frame and the context of the two AUG codons affect the abundance of mitochondrial and nuclear RNase H1. *Molecular and Cellular Biology* 30: 5123–5134.
27 Chon, H., Sparks, J.L., Rychlik, M. et al. (2013). RNase H2 roles in genome integrity revealed by unlinking its activities. *Nucleic Acids Research* 41: 3130–3143.

28 Groh, M. and Gromak, N. (2014). Out of balance: R-loops in human disease. *PLoS ONE* 10: e1004630.

29 Zinker, B.A., Rondinone, C.M., Trevillyan, J.M. et al. (2002). PTP1B antisense oligonucleotide lowers PTP1B protein, normalizes blood glucose, and improves insulin sensitivity in diabetic mice. *Proceedings of the National Academy of Sciences of the United States of America* 99: 11357–11362.

30 Greenberger, L., Horak, I., Filpula, D. et al. (2008). A RNA antagonist of hypoxia-inducible factor-1A, EZN-2968, inhibits tumor cell growth. *Molecular Cancer Therapeutics* 7: 3598–3608.

31 McKay, R.A., Miraglia, L.J., Cummins, L.L. et al. (1999). Characterization of a potent and specific class of antisense oligonucleotide inhibitor of human protein kinase C-a expression. *Journal of Biological Chemistry* 274: 1715–1722.

32 Kakiuchi-Kiyota, S., Koza-Taylor, P.H., Mantena, S.R. et al. (2014). Comparison of hepatic transcription profiles of locked ribonucleic acid antisense oligonucleotides: evidence of distinct pathways contributing to non-target mediated toxicity in mice. *Toxicological Sciences: An Official Journal of the Society of Toxicology* 138: 234–248.

33 Dullea, R., Salatto, C., Sciabola, S. et al. (2014). Study of CRTC2 pharmacology using antisense oligonuceotides. *Nucleic Acid Therapeutics* 24: 127–138.

34 Lindow, M., Vornlocher, H.-P., Riley, D. et al. (2012). Assessing unintended hybridization-induced biological effects of oligonucleotides. *Nature Biotechnology* 30: 920–923.

35 Sharp, P.A. (2009). The centrality of RNA. *Cell* 136: 577–580.

36 Morris, K.V. and Mattick, J.S. (2014). The rise of regulatory RNA. *Nature Reviews Genetics* 15: 423–437.

37 Cunningham, F., Amode, M.R., Barrell, D. et al. (2015). Ensembl 2015. *Nucleic Acids Research* 43 (Database issue): D662–D669.

38 Bartel, D.P. (2009). MicroRNAs: target recognition and regulatory functions. *Cell* 136: 215–233.

39 Wilusz, J.E., Sunwoo, H., and Spector, D.L. (2009). Long noncoding RNAs: functional surprises from the RNA world. *Genes & Development* 23: 1494–1504.

40 Yang, L., Froberg, J.E., and Lee, J.T. (2014). Long noncoding RNAs: fresh perspectives into the RNA world. *Trends in Biochemical Sciences* 39 (1): 35–43.

41 Wahlestedt, C. (2013). Targeting long non-coding RNA to therapeutically upregulate gene expression. *Nature Reviews Drug Discovery* 12: 433–446.

42 Modarresi, F., Faghihi, M.A., Lopez-Toledano, M.A. et al. (2012). Inhibition of natural antisense transcripts *in vivo* results in gene-specific transcriptional upregulation. *Nature Biotechnology* 30: 453–459.

43 Ideue, T., Hino, K., Kitao, S. et al. (2009). Efficient oligonucleotide-mediated degradation of nuclear noncoding RNAs in mammalian cultured cells. *RNA* 15: 1578–1587.

44 Meng, L., Ward, A.J., Chun, S. et al. (2014). Towards a therapy for Angelman syndrome by targeting a long non-coding RNA. *Nature* 518: 409–412.
45 Michalik, K.M., You, X., Manavski, Y. et al. (2014). Long noncoding RNA MALAT1 regulates endothelial cell function and vessel growth. *Circulation Research* 114: 1389–1397.
46 Schaukowitch, K., Joo, J.-Y., Liu, X. et al. (2014). Enhancer RNA facilitates NELF release from immediate early genes. *Molecular Cell* 56: 29–42.
47 Wheeler, T.M., Leger, A.J., Pandey, S.K. et al. (2012). Targeting nuclear RNA for *in vivo* correction of myotonic dystrophy. *Nature* 488: 111–115.
48 Veloso, A., Biewen, B., Paulsen, M.T. et al. (2013). Genome-wide transcriptional effects of the anti-cancer agent camptothecin. *PLoS ONE* 8: e78190.
49 Lima, W.F., Vickers, T.A., Nichols, J. et al. (2014). Defining the factors that contribute to on-target specificity of antisense oligonucleotides. *PLoS ONE* 9: e101752.
50 Blackburn, G.M., Gait, M.J., Loakes, D., and Williams, D.M. ed. (2006). Physical and structural techniques applied to nucleic acids. In: *Nucleic Acids in Chemistry and Biology*, 428. Cambridge: The Royal Society of Chemistry.
51 You, Y., Tataurov, A.V., and Owczarzy, R. (2011). Measuring thermodynamic details of DNA hybridization using fluorescence. *Biopolymers* 95: 472–486.
52 Bonnet, G., Tyagi, S., Libchaber, A., and Kramer, F.R. (1999). Thermodynamic basis of the enhanced specificity of structured DNA probes. *Proceedings of the National Academy of Sciences of the United States of America* 96: 6171–6176.
53 Cardullo, R.A., Agrawal, S., Flores, C. et al. (1988). Detection of nucleic acid hybridization by nonradiative fluorescence resonance energy transfer. *Proceedings of the National Academy of Sciences of the United States of America* 85: 8790–8794.
54 Panjkovich, A. and Melo, F. (2005). Comparison of different melting temperature calculation methods for short DNA sequences. *Bioinformatics* 21: 711–722.
55 Baker, B.F., Lot, S.S., Condon, T.P. et al. (1997). 2'-O-(2-methoxy)ethyl-modified anti-intercellular adhesion molecule 1 (ICAM-1) oligonucleotides selectively increase the ICAM-1 mRNA level and inhibit formation of the iCAM-1 translation initiation complex in human umbilical vein endothelial cells. *Journal of Biological Chemistry* 272: 11994–12000.
56 Wagner, R.W., Matteucci, M.D., Lewis, J.G. et al. (1993). Antisense gene inhibition by oligonucleotides containing C-5 propyne pyrimidines. *Science* 260: 1510–1513.
57 Straarup, E.M., Fisker, N., Hedtjärn, M. et al. (2010). Short locked nucleic acid antisense oligonucleotides potently reduce apolipoprotein B mRNA and serum cholesterol in mice and non-human primates. *Nucleic Acids Research* 38: 7100–7111.
58 Wu, H., Lima, W.F., and Crooke, S.T. (2001). Investigating the structure of human RNase H1 by site-directed mutagenesis. *Journal of Biological Chemistry* 276: 23547–23553.

59 Nowotny, M., Gaidamakov, S.A., Ghirlando, R. et al. (2007). Structure of human RNase H1 complexed with an RNA/DNA hybrid: insight into HIV reverse transcription. *Molecular Cell* 28: 264–276.
60 Nowotny, M., Cerritelli, S.M., Ghirlando, R. et al. (2008). Specific recognition of RNA/DNA hybrid and enhancement of human RNase H1 activity by HBD. *The EMBO Journal* 27: 1172–1181.
61 Lima, W.F., Rose, J.B., Nichols, J.G. et al. (2007). Human RNase H1 discriminates between subtle variations in the structure of the heteroduplex substrate. *Molecular Pharmacology* 71: 83–91.
62 Lima, W.F., Rose, J.B., Nichols, J.G. et al. (2007). The positional influence of the helical geometry of the heteroduplex substrate on human RNase H1 catalysis. *Molecular Pharmacology* 71: 73–82.
63 Henry, S.P., Bolte, H., Auletta, C., and Kornbrust, D.J. (1997). Evaluation of the toxicity of ISIS 2302, a phosphorothioate oligonucleotide, in a four-week study in cynomolgus monkeys. *Toxicology* 120: 145–155.
64 Swayze, E.E., Siwkowski, A.M., Wancewicz, E.V. et al. (2007). Antisense oligonucleotides containing locked nucleic acid improve potency but cause significant hepatotoxicity in animals. *Nucleic Acids Research* 35: 687–700.
65 Seth, P.P., Siwkowski, A., Allerson, C.R. et al. (2009). Short antisense oligonucleotides with novel 2'-4' conformationally restricted nucleoside analogues show improved potency without increased toxicity in animals. *Journal of Medicinal Chemistry* 52: 10–13.
66 Stanton, R., Sciabola, S., Salatto, C. et al. (2012). Chemical modification study of antisense gapmers. *Nucleic Acid Therapeutics* 22: 344–359.
67 Hagedorn, P.H., Yakimov, V., Ottosen, S. et al. (2013). Hepatotoxic potential of therapeutic oligonucleotides can be predicted from their sequence and modification pattern. *Nucleic Acid Therapeutics* 23: 302–310.
68 Burdick, A.D., Sciabola, S., Mantena, S.R. et al. (2014). Sequence motifs associated with hepatotoxicity of locked nucleic acid—modified antisense oligonucleotides. *Nucleic Acids Research* 42: 4882–4891.
69 Laxton, C., Brady, K., Moschos, S. et al. (2011). Selection, optimization, and pharmacokinetic properties of a novel, potent antiviral locked nucleic acid-based antisense oligomer targeting hepatitis C virus internal ribosome entry site. *Antimicrobial Agents and Chemotherapy* 55: 3105–3114.
70 Ui-Tei, K., Naito, Y., Nishi, K. et al. (2008). Thermodynamic stability and Watson–Crick base pairing in the seed duplex are major determinants of the efficiency of the siRNA-based off-target effect. *Nucleic Acids Research* 36: 7100–7109.
71 Naito, Y., Yoshimura, J., Morishita, S., and Ui-Tei, K. (2009). siDirect 2.0: updated software for designing functional siRNA with reduced seed-dependent off-target effect. *BMC Bioinformatics* 10 (1): 392.
72 Jackson, A.L., Burchard, J., Schelter, J. et al. (2006). Widespread siRNA "off-target" transcript silencing mediated by seed region sequence complementarity. *RNA* 12: 1179–1187.

73 Prasanth, K.V., Prasanth, S.G., Xuan, Z. et al. (2005). Regulating gene expression through RNA nuclear retention. *Cell* 123: 249–263.
74 Pal, S., Gupta, R., Kim, H. et al. (2011). Alternative transcription exceeds alternative splicing in generating the transcriptome diversity of cerebellar development. *Genome Research* 21: 1260–1272.
75 Rognes, T. (2011). Faster Smith-Waterman database searches with inter-sequence SIMD parallelisation. *BMC Bioinformatics* 12: 221.
76 Smith, T.F. and Waterman, M.S. (1981). Identification of common molecular subsequences. *Journal of Molecular Biology* 147: 195–197.
77 States, D.J., Gish, W., and Altschul, S.F. (1991). Improved sensitivity of nucleic acid database searches using application-specific scoring matrices. *Methods* 3: 66–70.
78 Vickers, T.A., Freier, S.M., Bui, H.-H. et al. (2014). Targeting of repeated sequences unique to a gene results in significant increases in antisense oligonucleotide potency. *PLoS ONE* 9: e110615.
79 Lendvai, G., Velikyan, I., Bergström, M. et al. (2005). Biodistribution of 68Ga-labelled phosphodiester, phosphorothioate, and 2'-O-methyl phosphodiester oligonucleotides in normal rats. *European Journal of Pharmaceutical Sciences: Official Journal of the European Federation for Pharmaceutical Sciences* 26: 26–38.
80 You, Y., Moreira, B.G., Behlke, M.A., and Owczarzy, R. (2006). Design of LNA probes that improve mismatch discrimination. *Nucleic Acids Research* 34: e60.
81 Owczarzy, R., You, Y., Groth, C.L., and Tataurov, A.V. (2011). Stability and mismatch discrimination of locked nucleic acid-DNA duplexes. *Biochemistry* 50: 9352–9367.
82 Naito, Y. and Ui-Tei, K. (2013). Designing functional siRNA with reduced off-target effects. *Methods in Molecular Biology (Clifton, N.J.)* 942: 57–68.
83 Ding, Y., Chan, C.Y., and Lawrence, C.E. (2004). Sfold web server for statistical folding and rational design of nucleic acids. *Nucleic Acids Research* 32 (suppl 2): W135–W141.
84 Far, R.K.-K. and Sczakiel, G. (2003). The activity of siRNA in mammalian cells is related to structural target accessibility: a comparison with antisense oligonucleotides. *Nucleic Acids Research* 31: 4417–4424.
85 Johnson, E. and Srivastava, R. (2013). Volatility in mRNA secondary structure as a design principle for antisense. *Nucleic Acids Research* 41: e43.
86 Laing, C. and Schlick, T. (2011). Computational approaches to RNA structure prediction, analysis and design. *Current Opinion in Structural Biology* 21: 306–318.
87 Schroeder, R., Barta, A., and Semrad, K. (2004). Strategies for RNA folding and assembly. *Nature Reviews Molecular Cell Biology* 5: 908–919.
88 Greenleaf, W.J., Frieda, K.L., Foster, D.A.N. et al. (2008). Direct observation of hierarchical Folding in single riboswitch aptamers. *Science* 319: 630–633.
89 Mathews, D.H., Moss, W.N., and Turner, D.H. (2010). Folding and finding RNA secondary structure. *Cold Spring Harbor Perspectives in Biology* 2 (12). doi: 10.1101/cshperspect.a003665.

90 Seetin, M.G. and Mathews, D.H. (2012). RNA structure prediction: an overview of methods. *Methods in Molecular Biology (Clifton, N.J.)* 905: 99–122.

91 Wan, Y., Kertesz, M., Spitale, R.C. et al. (2011). Understanding the transcriptome through RNA structure. *Nature Reviews Genetics* 12: 641–655.

92 Mathews, D.H. (2004). Using an RNA secondary structure partition function to determine confidence in base pairs predicted by free energy minimization. *RNA* 10: 1178–1190.

93 Lange, S.J., Maticzka, D., Möhl, M. et al. (2012). Global or local? Predicting secondary structure and accessibility in mRNAs. *Nucleic Acids Research* 40: 5215–5226.

94 Wee, K.B., Pramono, Z.A.D., Wang, J.L. et al. (2008). Dynamics of co-transcriptional pre-mRNA folding influences the induction of dystrophin exon skipping by antisense oligonucleotides. *PLoS ONE* 3: e1844.

95 Audibert, A., Weil, D., and Dautry, F. (2002). *In vivo* kinetics of mRNA splicing and transport in mammalian cells. *Molecular and Cellular Biology* 22: 6706–6718.

96 Brugiolo, M., Herzel, L., and Neugebauer, K.M. (2013). Counting on co-transcriptional splicing. *F1000prime Reports* 5: 9.

97 Sugimoto, N. and Yasumatsu, I. (2001). A new concept for the design of antisense oligonucleotides based on nucleic acid thermostability. *Current Medicinal Chemistry. Anti-Cancer Agents* 1: 95–112.

98 Sugimoto, N., Nakano, S., Katoh, M. et al. (1995). Thermodynamic parameters to predict stability of RNA/DNA hybrid duplexes. *Biochemistry* 34: 11211–11216.

99 Xia, T., SantaLucia, J., Burkard, M.E. et al. (1998). Thermodynamic parameters for an expanded nearest-neighbor model for formation of RNA duplexes with Watson–Crick base pairs. *Biochemistry* 37: 14719–14735.

100 Kurreck, J., Wyszko, E., Gillen, C., and Erdmann, V.A. (2002). Design of antisense oligonucleotides stabilized by locked nucleic acids. *Nucleic Acids Research* 30: 1911–1918.

101 Shen, L., Johnson, T.L., Clugston, S. et al. (2011). Molecular dynamics simulation and binding energy calculation for estimation of oligonucleotide duplex thermostability in RNA-based therapeutics. *Journal of Chemical Information and Modeling* 51: 1957–1965.

102 Condon, D.E., Yildirim, I., Kennedy, S.D. et al. (2014). Optimization of an AMBER force field for the artificial nucleic acid, LNA, and benchmarking with NMR of L(CAAU). *The Journal of Physical Chemistry B* 118: 1216–1228.

103 Kamola, P.J., Nakano, Y., Takahashi, T. et al. (2015). The siRNA non-seed region and its target sequences are auxiliary determinants of off-target effects. *PLoS Computational Biology* 11: e1004656.

104 Matveeva, O.V., Mathews, D.H., Tsodikov, A.D. et al. (2003). Thermodynamic criteria for high hit rate antisense oligonucleotide design. *Nucleic Acids Research* 31: 4989–4994.

105 Pedersen, L., Hagedorn, P.H., Lindholm, M.W., and Lindow, M. (2014). A kinetic model explains why shorter and less affine enzyme-recruiting oligonucleotides can be more potent. *Molecular Therapy--Nucleic Acids* 3: e149.

106 Frieden, M., Christensen, S.M., Mikkelsen, N.D. et al. (2003). Expanding the design horizon of antisense oligonucleotides with alpha-l-LNA. *Nucleic Acids Research* 31: 6365–6372.

107 Mathews, D.H. and Turner, D.H. (2006). Prediction of RNA secondary structure by free energy minimization. *Current Opinion in Structural Biology* 16: 270–278.

108 Larsson, E., Sander, C., and Marks, D. (2010). mRNA turnover rate limits siRNA and microRNA efficacy. *Molecular Systems Biology* 6: 433.

109 Stein, C.A., Hansen, B., Lai, J. et al. (2009). Efficient gene silencing by delivery of locked nucleic acid antisense oligonucleotides unassisted by transfection reagents. *Nucleic Acids Research* 38: e3.

110 Soifer, H.S., Koch, T., Lai, J. et al. (2012). Silencing of gene expression by gymnotic delivery of antisense oligonucleotides. *Methods in Molecular Biology* 815: 333–346.

111 Forreryd, A., Johansson, H., Albrekt, A.-S., and Lindstedt, M. (2014). Evaluation of high throughput gene expression platforms using a genomic biomarker signature for prediction of skin sensitization. *BMC Genomics* 15: 379–396.

112 Arikawa, E., Sun, Y., Wang, J. et al. (2008). Cross-platform comparison of SYBR Green real-time PCR with TaqMan PCR, microarrays and other gene expression measurement technologies evaluated in the MicroArray Quality Control (MAQC) study. *BMC Genomics* 9: 328–340.

113 Canales, R.D., Luo, Y., Willey, J.C. et al. (2006). Evaluation of DNA microarray results with quantitative gene expression platforms. *Nature Biotechnology* 24: 1115–1122.

114 Prokopec, S.D., Watson, J.D., Waggott, D.M. et al. (2013). Systematic evaluation of medium-throughput mRNA abundance platforms. *RNA* 19: 51–62.

115 Guo, Y., Sheng, Q., Li, J. et al. (2013). Large scale comparison of gene expression levels by microarrays and RNAseq using TCGA data. *PLoS ONE* e71462.

116 Wang, Z., Gerstein, M., and Snyder, M. (2009). RNA-Seq: a revolutionary tool for transcriptomics. *Nature Reviews Genetics* 10: 57–63.

8

Class-Related Proinflammatory Effects

Rosanne Seguin

Montreal Neurological Institute, McGill University, Montreal, Quebec, Canada

8.1 Introduction

As drug therapies, antisense oligonucleotides (ASOs) are designed to target specific sequences for effects that are deemed to be both potent and precise. The net negative charge of unmodified ASO results in a very limited ability to penetrate the plasma membrane [1]; therefore uptake is believed to an adsorptive endocytosis process [2]. ASOs have undergone generations of chemical modifications with the goal to enhance nuclease resistance and improve delivery via promotion of adsorptions of the ASO to the cellular surface proteins, leading to higher internalization of the ASO and enhancing efficacy [1]. The chemical modifications also enhance the binding to other proteins, which may lead to unwanted clinical effects.

ASO toxicities generally fall within two broad categories: hybridization-dependent toxicities due to either on-target or off-target pharmacology and hybridization-independent toxicities due to the class effect of ASO [3]. As ASOs are composed of nucleic acids (either unmodified or chemically modified), there may be a perception that these drugs would be less likely to induce an unwanted immune response as compared with peptides or proteins (e.g. monoclonal antibodies). However, considering the capacity of human physiology to respond to invading viruses (DNA and RNA), the idea that these drug candidates can elicit an immune response should not be surprising. In the development of ASO as drug candidates, evidence was accumulated from both animal toxicology studies and early human clinical studies that ASOs exhibit proinflammatory characteristics [3]. These characteristics include infiltration of mononuclear cells not only at the site of administration but also at systemic locations. Changes in spleen and lymph node weight were noted in particular

Oligonucleotide-Based Drugs and Therapeutics: Preclinical and Clinical Considerations for Development, First Edition. Edited by Nicolay Ferrari and Rosanne Seguin.

cases [4]. Preclinical studies in mice observed production of cytokines in response to ASO administration. Further work identified specific oligonucleotide sequences that interact with innate immune response receptors including the Toll-like receptor 9 (TLR9) [5, 6], which function as a first line of defense against microbial pathogens through inducing innate immunity and the release of inflammatory cytokines, which in turn initiate the activation of adaptive immune responses [7]. There are ASOs under development with the aim to leverage the stimulation of these innate receptors (e.g. TLR activation) specifically with the intention of inducing or augmenting immune responses (e.g. vaccine adjuvants) and are the focus of another chapter in this book. For ASO drug candidates designed for gene expression silencing rather than TLR activation, it has been challenging to understand and address these proinflammatory effects. A limited number of ASO products have reached regulatory approval stage, and it is early to generalize conclusions; however proinflammatory effects have been a reason for rejection of a marketing authorization to date (Kyndrisa EPAR 2016) [8]. Proinflammatory responses include production of antibodies to double-stranded DNA [9], activation of the alternate complement pathway [10, 11], stimulation of cytokine production [12–14], infiltration of mononuclear cells into organs [15] including liver and kidney that may play a role in the observed hepato/nephrotoxicity [3, 4], and hematological changes including thrombocytopenia [3, 16].

This chapter will strive to define and characterize these proinflammatory responses, assess the frequency of the responses, describe species-specific responses, and determine future directions when developing ASO as drug candidates. One of the major goals is to decrease the incidence of these immune responses along with any accompanying downstream sequelae as a possible toxicity and not induce an immune response that would limit the efficacy of the ASO.

8.2 Proinflammatory Effects of ASO for Consideration in Drug Development

8.2.1 Activation of the Complement Cascade in Monkeys

Since the 1990s, results from nonclinical toxicology studies demonstrated intravenous administration of phosphorothioate (PS)-modified ASO led to the activation of the alternate complement cascade, with nonhuman primates (NHP) being more sensitive than humans or other species [10, 17–19]. In rhesus monkeys infused with 20 mg/kg of PS-ASO over 10 min, there were moderate to severe bouts of emesis, ataxia, decreased motor activity, changes in blood pressure and heart rate, and neutropenia [10], whereas in animals pretreated with complement inhibitor CAB-2, no signs of toxicity were observed [10] indicating the role of complement activation in the toxicity findings. This activation of the alternate complement pathway was shown to be transient and dependent upon plasma concentrations of PS-ASO [17]. The mechanism of this activation was a result of the interaction of PS-ASO with factor H, which in turn interfered with the inhibitory function factor H exerts on the alternative complement pathway, allowing for the amplification of the proteolytic cascade, leading to production of split product Bb and C3a and C5a [17, 20]. Plasma clearance of the PS-ASO restored the regulatory function of factor H upon the pathway. Infusion rates were adjusted so as to achieve acute plasma levels of ASO below the threshold necessary for complement activation and thus limit toxicity [10]. In long-term dosage studies in cynomolgus or rhesus monkeys, chronic low level activation of the alternate complement pathway eventually leads to depletion of complement, increased sensitivity to infectious disease, and, in cases, damage to the vascular system and kidney [17]. The activation by PS-ASO for the alternate complement pathway appears to be specific to monkeys, with little evidence of activation observed human subjects in clinical trials, although particular attention for monitoring complement activation was enacted [21]. A summary of 20 studies involving a total of 710 NHP dosed intravenously with various PS-ASO chemically modified to contain 2′-methyloxyethyl (2′-MOE) in a gapmer configuration (termed 2′MOE chimeric ASO) showed a dose-dependent increase in the incidence of elevated complement Bb (>2× ULN) [21], whereas results from phase 1 studies with 12 different 2′MOE chimeric ASO in 750 healthy human subjects showed no evidence of complement activation following ASO treatment as measured by increased of split products Bb or C5a [21]. This phenomenon of alternate complement activation with chemically modified ASO in NHP has been well identified, and as such, during screening of ASO, most drug developers perform subchronic (<4 weeks) *in vivo* toxicity studies and *in vitro* complement fixation assays as a step in derisking the candidate selection process [20]. Currently, as there are few approved ASOs, it may be prudent to continue monitoring complement activation in clinical trials until sufficient data have been accumulated to define the actual risk level of this phenomenon in humans.

8.2.2 Cytokine Release

Bacterial DNA containing unmethylated cytosine–phosphorus–guanine (CpG) motifs enable the foreign material to be recognized by TLR9 [5] in immune cells, which in turn leads to the release of cytokines including interleukin (IL)-6, IL-12, and interferon-γ, proliferation of B cells, and activation of T cells and NK cells [13, 22, 23]. Lipopolysaccharide-free synthetic oligonucleotides (ODN), with and without PS modifications, were shown to possess the capacity to activate murine spleen cells to produce IL-6, IL-12, and IFN-γ [13] if the ODN sequences contained CpG motifs. *In vivo*, PS-ASO delivered intraperitoneally to mice led to increased cytokine (IL-6, IL-12, IL-1β, and IL-1Ra) mRNA levels in spleens of the animals and IL-6, IL-12, MIP-1β, and MCP-1 in serum [24]. PS-ASO containing CpG motif induced IL-12 secretion at a dose of 5 m/kg, whereas ASO without CpG failed to induce IL-12 even at 50 mg/kg [24]. ASOs containing CpG motifs are being exploited for immunomodulatory drug indications (e.g. vaccine adjuvants, cancer therapy) by industry [25] and have an entire book chapter dedicated to this potent immunostimulatory character. Induction of a "cytokine storm" whereby immune cells once recruited to a site are activated to release chemokines that induce further cell recruitment and release of proinflammatory cytokines that, in turn, activate immune cells and allow the cycle to continue was raised as a possible undesired effect for those ASOs in development for gene silencing mechanism of activation. When the specific sequences enabling TLR activation were first identified, it was presumed that elimination of the sequences in ASO would reduce the risk of subsequent inflammatory processes.

Younis and colleagues compared the proinflammatory effects produced in liver and peripheral blood mononuclear cells of mice dosed subcutaneously with either CpG ODN (ISIS 12449, containing multiple mouse specific CpG motif AACGTT [14, 23]) 4 mg/kg or non-CpG ODN (ISIS 2302) 50 mg/kg [26]. The different doses were selected with the aim of producing similar elevations of spleen weight as a marker of proinflammation [14, 23]. A single 4 mg/kg dose of CpG ODN increased the weight of spleen 2.3-fold and liver 1.3-fold, whereas for the non-CpG containing ODN, a similar increase of 2.2- and 1.3-fold for spleen and liver, respectively, was observed following dosing for 3 weeks at 50 mg/kg (cumulative dose 550 mg/kg) [26]. Therefore, 140 times more non-CPG ODN were needed to produce equivalent changes in tissue weight. Histological staining showed increased mononuclear cell infiltrate in livers of animals treated with either ODN [26]. In the livers of mice treated with CpG ODN, there was increased gene expression of TNFα, MIP-2β, MIG, IL-1 β, IL-1RA, and IL-1 BCE, whereas in livers of non-CpG ODN-treated mice, only SCYA5 was increased [26]. Therefore, although there were similar increases in liver weight and the histological staining pattern for both ODN, differences in gene expressions suggest differences in the mechanisms of induction of cellular infiltration. Turning to peripheral blood leukocytes isolated from the treated animals, CpG ODN induced a higher fold of gene expression (49-fold increase in chemokine MIG/CXCL9 expression) after a single treatment of CpG ODN 4 mg/kg as compared with the non-CpG ODN (12-fold increase of MIG/CXCL9 after three doses of 50 mg/kg), but it is interesting to note that the non-CpG ODN did induce significant level of chemokine gene expression, albeit not at doses two to fivefold higher than any anticipated clinical dose.

Whether non-CpG ODN interact with innate immune receptors other than TLR9 to activate cells for cytokine and chemokine release, the effect of CpG ODN in TLR9 knockout mice and myeloid differentiation factor 88 (MyD88) knockout animals [27] was assessed. Similar to the study by Younis et al., animals [26] were dosed with non-CpG ODN at a higher dose (50 mg/kg/twice weekly) than with CpG ODN (4 mg/kg/twice weekly) [27]. Even in TLR9 knockout animals, a high dose of non-CpG ODN-activated proinflammatory responses as assessed by increased spleen weight (small but significant increase of 1.2–1.3-fold) and in livers of treated animals elevated expression of inflammatory markers CD86, CD11b, CD11c, and CD68 and increased cytokine levels of IL-1α, IL-18, RANTES, MCP-1, MCP-5, MIP-1α, and MIP-2 but no change in IL-6 [27]. In MyD88 knockout animals, non-CpG ODN increased CD11c and CD68 levels in liver and cytokines/chemokines IL-1α, IL-18, RANTES, MCP-1, and MIP-1α. Regardless of mouse strain tested, the high dose of non-CpG ODN induced cytokines although more of an induction of a chemokine profile than the "classic" Th1 profile of IFN-γ, IL-6, and IL-12 induced by CpG ODN. Increased spleen weight and mononuclear cellular infiltration are still observed in mice treated with non-CpG ODN, and the role chemokine production may have in this response remains to be determined.

As a species, mice appear to be the most sensitive to the immunostimulation of cytokines/chemokines by ASO (either CpG motif or non-CpG motif) as compared with other species such as NHP. Rodents express the TLR9 receptor on more cell types [28] (B cells, dendritic cells, and monocytes) than monkeys and humans; therefore it can be expected that subcutaneous injection of these immunostimulatory ASO led to qualitative different responses in mice compared with humans. In clinical studies of healthy human volunteers with either 30 mg once daily, 70 mg three times weekly or 200 mg one weekly of mipomersen yielded no evidence of systemic inflammatory responses as assessed by monitoring serum levels of cytokines IFN-α, IFN-β, IL-1β, IL-6, IL-13, MCP-1, and MIP1α [29]. Patients in clinical trials have complained of fatigue, GI disorders, chills, and fever – all symptoms similar to release of cytokines [29, 30]. Local cutaneous injection site reactions and flu-like symptoms have been observed in patients and resulted in the discontinuation of therapy by some [30]. Changes in the gauge of needle for injection and frequency and dose of ASO delivered have reduced the number and severity of injection site reactions [21], improving the tolerability of the drug treatment for patients.

In summary, there are steps that can be undertaken to derisk ASO development programs for undesired immunostimulation including exclusion of immunostimulatory sequences in drug candidate ASO and incorporation of *in vitro* and *in vivo* evaluations for identifying sequences that may produce exaggerated proinflammatory effects (cytokine and chemokine release). To date, as it is challenging to correlate definitive levels of induced cytokines/chemokines in animals with clinical manifestations in human trials, it is not routine to measure cytokines/chemokines in IND-enabling toxicology studies. Instead, it has been suggested to bank samples from toxicology studies for future analyses if either pathology findings or clinical symptoms indicate a role of cytokines in the phenomenon studied. In clinical studies, monitoring of adverse events including injection site reactions and flu-like symptoms in patients [21] is incorporated into trial design, and perhaps it may be prudent to also bank samples collected from patients for any future cytokine analyses.

8.2.3 Mononuclear Cellular Infiltrate

Administration of ASO can lead to mononuclear cellular infiltrate at the site of administration whether the route is subcutaneous, intradermal, intramuscular, intraperitoneal, intraocular, inhalation, intrathecal, or intravenous injection [15]. The level of concern for these site reactions depends not only on severity of the infiltrate but also the location of the reaction. For example, a local site reaction following subcutaneous injection may be considered to be tolerable as a safety concern in the clinic [21], whereas when dealing with eyes or lungs, such infiltrate may be dose limiting [1, 3, 16]. The degree of cellular infiltrate is dependent upon sequence of ASO (immunostimulatory or not), the inclusion of chemical modifications (e.g. PS backbone) in the ASO, and the dose. Extensive systemic exposure of ASO can follow parenteral administration (e.g. intradermal, intramuscular, intraperitoneal, intravenous, and subcutaneous), which can result in tissue uptake of ASO and cellular infiltrate in organs removed from the injection site [4, 8, 9, 12, 15, 16].

In rodents following administration of PS-ASO, cellular infiltration in multiple organs has been commonly observed – more pronounced in the major organs of ASO uptake, the kidneys and liver [4, 12]. In mice, repeated administration of high doses of ASO can lead to widespread cellular infiltration including skin, lung, eye, intestines, ovaries, and other tissues [9]. In contrast in monkeys, the cellular infiltration may be absent or limited to the liver [12], especially for ASO that do not contain immunostimulatory sequences.

Immunohistochemistry analyses of kidney and liver show infiltrates accumulating in perivascular regions and often refer to "lymphohistiocytic cell infiltrate" [12]. The affected macrophages have been described as having a "distended, weakly stained and finely vacuolated or granular cytoplasm" [3]. The liver resident macrophages, Kupffer cells, can appear reactive (hypertrophic) or proliferative (hyperplastic). The changes are reversible, but due to the long tissue half-life of the modified PS-ASO, these effects may take weeks to months to resolve [16]. Kupffer cells exhibit granular cytoplasm, which may reflect the uptake of ASO [17, 31] in the endosomal compartment of these cells. Whether uptake of ASO disrupts the phagosome of the macrophages is undetermined. It has been suggested that changes in tissue macrophages result from cellular activation and cytokine production in response to the proinflammatory nature of the ASO [32].

It was presumed that elimination of the immunostimulatory sequences (e.g. CpG motif) in ASO would remove the subsequent inflammatory processes.

While some responses are lessened with non-CpG containing sequences, histologic findings in rodents treated with non-CpG ASO demonstrate mononuclear cellular infiltration of lymphoid hyperplasia and lymphohistiocytic cell infiltrates in multiple organs [12, 15, 17]. Therefore, the specific mechanisms through which the mononuclear cells are recruited into the region or location have yet to be identified. Nor is it understood what is the function of these mononuclear cells once recruited to these locations. Studies *in vitro* with mononuclear cells exposed to nonstimulatory ASO demonstrate that the mononuclear cells change phenotype, exhibiting large vacuoles apparently containing ASO, similar to those observed *in vivo* [33]. *In vitro*, these vacuolated cells do not actively secrete high levels of inflammatory cytokines, and/or the phagocytic capabilities are retained [33] as compared with mononuclear cells treated with vacuole-inducing drugs. Evidence from *in vivo* studies suggests cellular infiltrates are typically not associated with organ fibrosis or parenchymal cell damage [16].

Repeated dosing leads to accumulation of drug in liver and kidneys [3], and the degree of toxicity depends on the concentration of the accumulated ASO. High concentrations can lead to degeneration of the kidney or liver; however there is a suggestion that the toxicity results from lysosomal breakdown or cytokine release by activated Kupffer cells or mononuclear cells [10, 17].

In rodents, toxicities associated with degeneration or necrosis in the liver have been observed with locked nucleic acid (LNA)-modified ASO at doses below those generally accepted to result in accumulation-related hepatic effects [34, 35]. Some mice exposed to LNA-ASO expressed increased levels of plasma aminotransferase (ALT) and aspartate aminotransferase (AST) and exhibited liver necrosis and apoptosis [36]. LNA-ASO toxicity was associated with TGC and TCC sequence motifs in 3-8-3 gapmers [34], and sequences with these motifs tend to exhibit higher binding to mouse liver proteins. A proposed difference for the LNA-ASO is that they bind to intracellular proteins [34, 35, 37], while PS-ASO bind to plasma proteins and extracellular receptor. It has been argued that the LNA-ASO changes are not linked to immunostimulation. The histopathologic lesions of liver toxicity do not show a primary inflammatory response while do show hepatocellular degeneration with single-cell necrosis [34].

In analyzing data from NHP dosed with various 2′MOE chimeric ASO, there was no marked effect of ASO on liver function over time or dose [21]. In 3 out of 375 animals, there was an increase in ALT, and in 2 out of 376 animals an increase in AST, and in 1 out of 476 animals, an increase in total bilirubin [21]. In healthy human volunteers, there was one incidence of ALT >3× ULN that was not associated with the elevated total bilirubin [21], suggesting that the hepatoxocity is even less likely to occur in humans.

Redesigning potentially hepatotoxic ASO sequences to eliminate the suspected sequence motifs may mitigate hepatic effects [35], but as described for other proinflammatory class effects, sometimes base sequences are not only responsible. In some cases, potentially offending sequences had less *in vivo* hepatotoxicity than expected, demonstrating the importance of performing animal toxicology studies as *in vitro* modeling is insufficient.

Renal toxicities with ASO are due to accumulation of ASO within the lysosomes of the proximal tubule, resulting in perturbation of tubular absorptive capacity and tubular proteinuria [17]. However, there are ASOs that demonstrate profound renal toxicities at doses lower than the expected accumulation of ASO indicating another mechanism at work. There is little published data to date, yet oral discussions at workshops concur that as yet no specific sequences have been identified and that ASOs with different chemical modifications yield this effect.

Glomerulonephritis is most often associated with PS-ASO in animal studies over 3 months. Glomerulonephritis in monkeys is a preclinical proinflammatory lesion of concern [3]. Monkeys have had lesions with enlarged glomeruli, increased cellularity of the tufts, increased mesangium, and occasional infiltration of inflammatory cells. Although electron microscopy of the subendothelial glomerular basement membrane showed immune-mediated dense deposits in monkeys treated with 2′MOE-PS-ASO, immunofluorescent staining for anti-C3a showed complement fragments along vascular tufts, suggesting that in monkeys, the interaction of ASO with factor H leading to the activation of the alternative complement pathway is the major contributing factor to the initial injury to the epithelium [3], whereas the mechanism in rodents is less clear. Chronic subcutaneous administration of drisapersen, a 2′OME-PS-ASO, resulted in tubular accumulation, secondary tubular degeneration, and glomerulopathy in both mice and monkeys [38]. Glomerular changes in monkey were less pronounced, and the lesions stained positively for complement C3 fragments, which could be linked to the activation of the alternative complement pathway by ASO in monkeys. Yet, serum MCP-1 levels were increased in monkeys at week 4 of treatment, and by week 26 most monkeys dosed 6 mg/kg/injection or higher had marked increases in serum inflammatory markers including CRP and MCP-1 [38], suggesting proinflammatory responses that may or may not be linked to the activation of complement. Glomerular changes described in mice 56 days after weekly injections included increased glomerular cellularity and inflammatory cells or nuclear debris mechanism of changes considered related to immune pathogenesis and local inflammatory activity in the kidney [38]. In mice, treatment with 300 mg/kg of drisapersen increased expression of CD68 by monocytes and correlated with increased CD68 mRNA [38]. The analysis of the renal cortex in the mice showed at least a twofold change in message normalized to 18S and beta-actin of IL-10, CCL3, CCL2, IL-6, IL-1α, CXCL10, IL-17, TNF-α, IL-1β, CD4, ccl5, IL-12, and CTLA4, among others, in the histologically affected treatment animals, with a similar trend but less magnitude in the treated histologically unaffected animals [38]. Systemically, in mice treated for 27 weeks, there was a dose-related increase in MCP-1, while IL-6 was unaffected [38]. Taken together, this evidence suggests that in mice, an immune-based mechanism may be playing a role in glomerulopathy. The initial site of injury in the mouse glomerulus is the glomerular endothelial cell, which is a potential target for circulating cytokine-mediated injury [3].

Treatment of cynomolgus monkeys with 10–40 mg/kg/week with different 2′OME chimeric ASO also showed histopathologic changes in proximal tubular epithelial cells (basophilic granules in the cytoplasm up to at higher doses degenerative changes) and showed ASO in kidney but failed to show a change in renal function as assessed by blood urea nitrogen (BUN), serum creatinine, creatinine clearance, or urine specific gravity [39].

As the kidney-toxicity-inducing ASO candidates are often screened out before human trials, cases of human nephrotoxicity are rare. In the first human trial of a LNA-ASO targeting proprotein convertase subtilisin/kexin type 9 (PCSK9) for the indication of lipoprotein cholesterol reduction, there was one case of acute tubular injury [40]. Acute kidney injury was noted 5 days after the patient received her third weekly subcutaneous dose of drug – serum creatinine was increased, which coincided with the presence of white blood cells, granular casts, and hematuria on urine microscopy [40]. A biopsy of the kidney showed evidence of ASO accumulation. The patient fully recovered and the kidney function returned to normal [40]. The reason for this ASO effect was deemed to not be target mediated as loss-of-function mutations of PCSK9 are not associated with kidney function [41].

In another study with an anticancer 2′MOE-PS-ASO, a patient with metastatic melanoma had acute tubular necrosis after 74 weekly doses of 10 mg/kg [42]. These renal changes were reversible and complete recovery of renal function was noted 6 months after drug discontinuation [42].

In the compilation analyses of NHP treated with various 2′MOE, no dose-dependent changes in renal parameters even at high doses [21] were observed. In the same paper, in healthy human volunteers, there initially appeared to be a dose-dependent trend on the incidence of increased creatinine in humans, but upon further investigation, these increases occurred in a single observation and were not confirmed by a repeat test. No dose-dependent changes in urinary protein, mean levels of creatinine or BUN, or mean glomerular filtration rate were observed in response to the treatment with the 2′MOE chimeric ASO; therefore the authors concluded no drug-associated effects on renal function [21].

In both preclinical and clinical studies with ASO, proteinuria is commonly noted and attributed to tubular changes rather than glomerular disease. It is important to distinguish tubular toxicity that is more medically manageable from the more deleterious glomerular injury when identifying proteinuria in a patient [3]. Large increases or the presence of high molecular weight proteins may aid in identifying glomerular origin and risk of glomerulonephritis, whereas small or trace amounts of protein would be consistent with tubular effect and less concern clinically.

In summary, the mononuclear cellular infiltration by PS-ASO in rodents is generally considered to be greater and of a more extensive nature than in other species and as such is not regarded as predictive of human sensitivity. In monkeys, most inflammation results from the activation of the alternative complement pathway [10, 16], and the mononuclear cell infiltrate is more muted as a proinflammatory response. While for some chemical modifications (e.g. LNA) there appear to be noted sequences that induce proinflammatory responses, it is less clear for others which sequences are at fault, highlighting the importance of screening in animals when selecting ASO candidates. To date, data from human trials have not shown the similar incidence of responses; nevertheless it remains key to monitor patients for kidney and liver changes that may result either from ASO accumulation in those organs or the cellular infiltrate in response to the ASO treatment.

8.2.4 Hematological Changes

The most common hematological changes observed in animals following repeated dosing with PS-ASO include anemia and thrombocytopenia. Thrombocytopenia or reduced platelet counts have been observed in studies with rodents and monkeys. Although the mechanism for cytopenia is undefined and may not be caused by proinflammatory responses, this is a class effect of ASO and should be addressed. The reductions were transient and reversible over courses of weeks or months depending on the half-life of the ASO. For approximately 10% of 2′MOE-PS-ASO, platelet reduction has been observed in monkeys at high doses (>20 mg/kg/week) [17]. A compilation dataset evaluated 40 2′MOE chimeric ASOs and identified a few ASOs that produced sustained effects on platelet counts, suggesting that individual sequences may be problematic [21] rather than target specificity of the ASO. Further confounding the phenomenon is that individual monkeys can have reduction in platelets in a sporadic fashion, neither dose nor sequence related. In one example, a single monkey had profound reduction in counts over treatment course with no obvious dose-dependent reductions.

In human trial with anticancer drug to patients with tumors, a transient thrombocytopenia was observed only during the first course of therapy, after which platelet count normalized and there were no episodes of bleeding [43]. A phase 1 study of ASO targeting PKC for cancer observed dose-limiting toxicities of thrombocytopenia at doses of 3 mg/kg/day [44]. Similarly, a phase 1 study in BCLS for thrombocytopenia was dose limiting [45]. In a review of the Ionis integrated safety database that includes over 2600 human subjects dosed with 16 different 2′MOE chimeric ASO via different routes including subcutaneous and intravenous infusion and for which 1000 subjects had been treated with ASO for more than 3 months with some up to 4.6 years, no clinical meaningful effect (below 50 K/μl) on platelets has been observed [46]. In some cases, platelets returned to normal levels after drug withdrawal but would fall again when drug was returned.

The mechanism of the thrombocytopenia is not defined. It has been proposed that that mechanism could be different between species and for mild versus serious events. One theory is that inappropriate binding of platelet factor 4, which is released after platelet activation and can bind to negatively charged molecules, may be occurring. A similar phenomenon happens with heparin and is responsible for heparin-induced thrombocytopenia in humans [47].

8.2.5 Immunogenicity

As there is evidence in the evolution of the human physiology capacity to respond in an immunological manner to invading viruses (DNA and RNA), the concept that these drug candidates can produce immune responses should not be surprising. From the beginning of ASO drug development, there rose the question of immunogenicity of ASO or the ability of ASO to induce immune responses, specifically memory responses. As discussed extensively in this chapter, there are specific sequences that enable ASO to interact with TLR or other innate immune receptors; however, the downstream effects of these interactions are considered to fall under the umbrella of innate immune responses as opposed to adaptive memory immune responses. The adaptive immune response involves recognition by either T-cell or B-cell receptors. To date, administration of ASO alone (i.e. not conjugated) through a variety of routes has not generated ASO sequence specific antibodies or generated memory T cells that recognize ASO. If antibodies to ASO were generated, it would be expected to observe a reduction in activity with ASO over time. No change in pharmacokinetic or pharmacological activity has been reported with subchronic or chronic administration of ASO in animals or humans that would be indicative of an inhibitory antibody response. Oral discussions among drug developers at meetings indicate much effort on behalf of companies to search for these responses yet not finding any as yet.

Another possible aspect of immunogenicity would be the induction of immune responses to nucleotides in general. Autoimmunity refers to the inappropriate recognition of self-antigens and the mounting of an immune response targeting this antigen. Although chemically synthesized, ASO are nucleotides, and there have been concerns of a generation of antibodies targeting nucleotides (ssDNA or dsDNA). Monitoring has been performed both in animal studies and in clinical studies for the detection of antibodies targeting either ssDNA or dsDNA. Some animal studies have shown an increase in antibodies targeting dsDNA [9], while others did not [48]. No scale has been established for the interpretation of these results relative to biological effects, nor does the presence of anti-dsDNA antibodies alone necessarily correlate with autoimmune disease. Overall, the risk of immunogenicity of ASO appears to be low; however monitoring in clinical trials or in clinic with approved medications should continue to determine whether over time, a decrease in pharmacokinetics or pharmacodynamics may be attributed to the induction of memory immune responses to ASO.

8.3 Conclusions

The concept that ASO drug candidates can produce proinflammatory responses has been observed since early animal studies [4, 12]. Modifications of chemistry (PS backbone, 2′MOE, etc.) have been enacted to improve ASO efficacy either through resistance to proteases or enhanced binding to target sequences. The identification of the sequences that interact with various innate immune receptors leading to the induction of innate immune responses was deemed to explain the mechanism of action for the observed proinflammatory responses. While removal of these immunostimulatory sequences reduced many of the proinflammatory responses observed, not all were eradicated, nor as shown from the work in the TLR9 knockout animals, were all mechanisms fully defined. Ongoing assessment in the screening process, including avoidance of any known immunostimulatory sequences from the ASO candidate selection and inclusion of different *in vitro* assessments (e.g. induction of cytokines and/or chemokines from splenocytes/leukocytes) as well as *in vivo* assessments performed in those species known to be most sensitive to the inflammatory response under investigation, contributes to the successful removal of "bad actors" from the ASO candidates. Key considerations for preclinical screening and testing of ASO are the nature of the immune responses that differ between species and which species would be the best surrogates for drug development. Monkeys appear to be the optimal model correlating with humans in regard to the pharmacokinetics of ASO uptake and removal to observing changes in liver, kidney, and hematology with the exception of complement activation for which monkeys appear to be the most sensitive species and have little effect observed in humans. It is expected as more ASO therapies reach regulatory approval and once sufficient numbers of patients are treated with this drug class for prolonged periods of time that it will be possible to determine the importance of the animal toxicological findings for human clinical outcomes and in particular the importance of the proinflammatory responses observed in the animal studies. It is evident from the clinical studies that humans do react to ASO with a degree of inflammatory response as measured by the incidence of local site reactions and reports of flu-like symptoms. The described changes to the screening and selection procedures described earlier will function to limit the number and severity of these adverse events in patients. The overall goal is for improved potency and tolerability of ASO therapies in humans.

References

1 Chan, J.H., Lim, S., and Wong, W.S. (2006). Antisense oligonucleotides: from design to therapeutic application. *Clin. Exp. Pharmacol. Physiol.* 33 (5–6): 533–540.

2 Lysik, M.A. and Wu-Pong, S. (2003). Innovations in oligonucleotide drug delivery. *J. Pharm. Sci.* 92 (8): 1559–1573.

3 Frazier, K.S. (2015). Antisense oligonucleotide therapies: the promise and the challenges from a toxicologic pathologist's perspective. *Toxicol. Pathol.* 43 (1): 78–89.

4 Henry, S., Stecker, K., Brooks, D. et al. (2000). Chemically modified oligonucleotides exhibit decreased immune stimulation in mice. *J. Pharmacol. Exp. Ther.* 292 (2): 468–479.

5 Hemmi, H., Takeuchi, O., Kawai, T. et al. (2000). A Toll-like receptor recognizes bacterial DNA. *Nature* 408 (6813): 740–745.

6 Ewald, S.E., Lee, B.L., Lau, L. et al. (2008). The ectodomain of Toll-like receptor 9 is cleaved to generate a functional receptor. *Nature* 456 (7222): 658–662.

7 Iwasaki, A. and Medzhitov, R. (2004). Toll-like receptor control of the adaptive immune responses. *Nat. Immunol.* 5 (10): 987–995.

8 Mustonen, E.K., Palomaki, T., and Pasanen, M. (2017). Oligonucleotide-based pharmaceuticals: non-clinical and clinical safety signals and non-clinical testing strategies. *Regul. Toxicol. Pharmacol.* 90: 328–341.

9 Levin, A.A., Henry, S.P., Monteith, D., and Templin, M.V. (2001). Toxicity of antisense oligonucleotides. In: Antisense Drug Technology (ed. S.T. Crooke), 201–267. New York: Marcell Dekker.

10 Henry, S.P., Beattie, G., Yeh, G. et al. (2002). Complement activation is responsible for acute toxicities in rhesus monkeys treated with a phosphorothioate oligodeoxynucleotide. *Int. Immunopharmacol.* 2 (12): 1657–1666.

11 Galbraith, W.M., Hobson, W.C., Giclas, P.C. et al. (1994). Complement activation and hemodynamic changes following intravenous administration of phosphorothioate oligonucleotides in the monkey. *Antisense Res. Dev.* 4 (3): 201–206.

12 Henry, S.P., Templin, M.V., Gillett, N. et al. (1999). Correlation of toxicity and pharmacokinetic properties of a phosphorothioate oligonucleotide designed to inhibit ICAM-1. *Toxicol. Pathol.* 27 (1): 95–100.

13 Klinman, D.M., Yi, A.K., Beaucage, S.L. et al. (1996). CpG motifs present in bacteria DNA rapidly induce lymphocytes to secrete interleukin 6, interleukin 12, and interferon gamma. *Proc. Natl. Acad. Sci. U. S. A.* 93 (7): 2879–2883.

14 Krieg, A.M. (2002). CpG motifs in bacterial DNA and their immune effects. *Annu. Rev. Immunol.* 20: 709–760.

15 Henry, S.P., Taylor, J., Midgley, L. et al. (1997). Evaluation of the toxicity of ISIS 2302, a phosphorothioate oligonucleotide, in a 4-week study in CD-1 mice. *Antisense Nucleic Acid Drug Dev.* 7 (5): 473–481.

16 Levin, A.A. and Henry, S.P. (2008). Toxicology of oligonucleotide therapeutics and understanding the relevance of toxicities. In: *Preclinical Safety Evaluation of Biopharmaceuticals. A Science-Based Approach to Facilitating Clinical Trials* (ed. J.A. Cavagnaro), 538–574. Hoboken, NJ: Wiley.
17 Henry, S.P., Kim, T.-W., Kramer-Strickland, K. et al. (2008). Toxicologic properties of 2′-methyloxyethyl chimeric antisense inhibitors in animals and man. In: *Antisense Drug Technology: Principles, Strategies and Applications* (ed. S.T. Crooke), 327–363. Carlsbad, CA: CRC Press.
18 Shen, L., Frazer-Abel, A., Reynolds, P.R. et al. (2014). Mechanistic understanding for the greater sensitivity of monkeys to antisense oligonucleotide-mediated complement activation compared with humans. *J. Pharmacol. Exp. Ther.* 351 (3): 709–717.
19 Henry, S.P., Giclas, P.C., Leeds, J. et al. (1997). Activation of the alternative pathway of complement by a phosphorothioate oligonucleotide: potential mechanism of action. *J. Pharmacol. Exp. Ther.* 281 (2): 810–816.
20 Henry, S.P., Seguin, R., Cavagnaro, J. et al. (2016). Considerations for the characterization and interpretation of results related to alternative complement activation in monkeys associated with oligonucleotide-based therapeutics. *Nucleic Acid Ther.* 26 (4): 210–215.
21 Crooke, S.T., Baker, B.F., Kwoh, T.J. et al. (2016). Integrated safety assessment of 2′-O-methoxyethyl chimeric antisense oligonucleotides in nonhuman primates and healthy human volunteers. *Mol. Ther.* 24 (10): 1771–1782.
22 Yamamoto, S., Yamamoto, T., Shimada, S. et al. (1992). DNA from bacteria, but not from vertebrates, induces interferons, activates natural killer cells and inhibits tumor growth. *Microbiol. Immunol.* 36 (9): 983–997.
23 Krieg, A.M., Yi, A.K., Matson, S. et al. (1995). CpG motifs in bacterial DNA trigger direct B-cell activation. *Nature* 374 (6522): 546–549.
24 Zhao, Q., Temsamani, J., Zhou, R.Z., and Agrawal, S. (1997). Pattern and kinetics of cytokine production following administration of phosphorothioate oligonucleotides in mice. *Antisense Nucleic Acid Drug Dev.* 7 (5): 495–502.
25 Krieg, A.M., Yi, A.K., Schorr, J., and Davis, H.L. (1998). The role of CpG dinucleotides in DNA vaccines. *Trends Microbiol.* 6 (1): 23–27.
26 Younis, H.S., Vickers, T., Levin, A.A., and Henry, S.P. (2006). CpG and Non-CpG oligodeoxynucleotides induce differential proinflammatory gene expression profiles in liver and peripheral blood leukocytes in mice. *J. Immunotoxicol.* 3 (2): 57–68.
27 Senn, J.J., Burel, S., and Henry, S.P. (2005). Non-CpG-containing antisense 2′-methoxyethyl oligonucleotides activate a proinflammatory response independent of Toll-like receptor 9 or myeloid differentiation factor 88. *J. Pharmacol. Exp. Ther.* 314 (3): 972–979.
28 Puig, M., Tosh, K.W., Schramm, L.M. et al. (2012). TLR9 and TLR7 agonists mediate distinct type I IFN responses in humans and nonhuman primates *in vitro* and *in vivo*. *J. Leukoc. Biol.* 91 (1): 147–158.

29 Flaim, J.D., Grundy, J.S., Baker, B.F. et al. (2014). Changes in mipomersen dosing regimen provide similar exposure with improved tolerability in randomized placebo-controlled study of healthy volunteers. *J. Am. Heart Assoc.* 3 (2): e000560.
30 Hovingh, K., Besseling, J., and Kastelein, J. (2013). Efficacy and safety of mipomersen sodium (Kynamro). *Expert Opin. Drug Saf.* 12 (4): 569–579.
31 Monteith, D.K., Horner, M.J., Gillett, N.A. et al. (1999). Evaluation of the renal effects of an antisense phosphorothioate oligodeoxynucleotide in monkeys. *Toxicol. Pathol.* 27 (3): 307–317.
32 Engelhardt, J.A., Fant, P., Guionaud, S. et al. (2015). Scientific and regulatory policy committee points-to-consider paper*: drug-induced vascular injury associated with nonsmall molecule therapeutics in preclinical development: part 2. Antisense oligonucleotides. *Toxicol. Pathol.* 43 (7): 935–944.
33 Brasey, A., Igue, R., Djemame, L. et al. (2011). The effect of *in vitro* exposure to antisense oligonucleotides on macrophage morphology and function. *J Nucleic Acids Invest.* 2 (1): 77–84.
34 Burdick, A.D., Sciabola, S., Mantena, S.R. et al. (2014). Sequence motifs associated with hepatotoxicity of locked nucleic acid – modified antisense oligonucleotides. *Nucleic Acids Res.* 42 (8): 4882–4891.
35 Hagedorn, P.H., Yakimov, V., Ottosen, S. et al. (2013). Hepatotoxic potential of therapeutic oligonucleotides can be predicted from their sequence and modification pattern. *Nucleic Acid Ther.* 23 (5): 302–310.
36 Swayze, E.E., Siwkowski, A.M., Wancewicz, E.V. et al. (2007). Antisense oligonucleotides containing locked nucleic acid improve potency but cause significant hepatotoxicity in animals. *Nucleic Acids Res.* 35 (2): 687–700.
37 Kakiuchi-Kiyota, S., Koza-Taylor, P.H., Mantena, S.R. et al. (2014). Comparison of hepatic transcription profiles of locked ribonucleic acid antisense oligonucleotides: evidence of distinct pathways contributing to non-target mediated toxicity in mice. *Toxicol. Sci.* 138 (1): 234–248.
38 Frazier, K.S., Sobry, C., Derr, V. et al. (2014). Species-specific inflammatory responses as a primary component for the development of glomerular lesions in mice and monkeys following chronic administration of a second-generation antisense oligonucleotide. *Toxicol. Pathol.* 42 (5): 923–935.
39 Henry, S.P., Johnson, M., Zanardi, T.A. et al. (2012). Renal uptake and tolerability of a 2′-O-methoxyethyl modified antisense oligonucleotide (ISIS 113715) in monkey. *Toxicology* 301 (1–3): 13–20.
40 van Poelgeest, E.P., Swart, R.M., Betjes, M.G. et al. (2013). Acute kidney injury during therapy with an antisense oligonucleotide directed against PCSK9. *Am. J. Kidney Dis.* 62 (4): 796–800.
41 Zhao, Z., Tuakli-Wosornu, Y., Lagace, T.A. et al. (2006). Molecular characterization of loss-of-function mutations in PCSK9 and identification of a compound heterozygote. *Am. J. Hum. Genet.* 79 (3): 514–523.
42 Herrington, W.G., Talbot, D.C., Lahn, M.M. et al. (2011). Association of long-term administration of the survivin mRNA-targeted antisense

oligonucleotide LY2181308 with reversible kidney injury in a patient with metastatic melanoma. *Am. J. Kidney Dis.* 57 (2): 300–303.
43 Cuddihy, A.R., Li, S., Tam, N.W. et al. (1999). Double-stranded-RNA-activated protein kinase PKR enhances transcriptional activation by tumor suppressor p53. *Mol. Cell. Biol.* 19 (4): 2475–2484.
44 Yuen, A.R., Halsey, J., Fisher, G.A. et al. (1999). Phase I study of an antisense oligonucleotide to protein kinase C-alpha (ISIS 3521/CGP 64128A) in patients with cancer. *Clin. Cancer Res.* 5 (11): 3357–3363.
45 Waters, J.S., Webb, A., Cunningham, D. et al. (2000). Phase I clinical and pharmacokinetic study of bcl-2 antisense oligonucleotide therapy in patients with non-Hodgkin's lymphoma. *J. Clin. Oncol.* 18 (9): 1812–1823.
46 Crooke, S.T., Baker, B.F., Witztum, J.L. et al. (2017). The effects of 2′-O-methoxyethyl containing antisense oligonucleotides on platelets in human clinical trials. *Nucleic Acid Ther.* 27 (3): 121–129.
47 Jaax, M.E., Krauel, K., Marschall, T. et al. (2013). Complex formation with nucleic acids and aptamers alters the antigenic properties of platelet factor 4. *Blood* 122 (2): 272–281.
48 Henry, S.P., Levin, A.A., White, K., and Mennear, J.H. (2006). Assessment of the effects of ISIS 2302, an anti-sense inhibitor of human ICAM-1, on cellular and humoral immunity in mice. *J. Immunotoxicol.* 3 (4): 199–211.

9

Exaggerated Pharmacology

Alain Guimond[1] *and Doug Kornbrust*[2]

[1] *InSymbiosis, Montreal, Quebec, Canada*
[2] *Preclinsight, Reno, NV, USA*

9.1 Introduction

Exaggerated pharmacology (EP) can be defined as potential adverse effects caused by an "exaggerated" degree of the intended pharmacologic activity of a drug. It is typically documented in a drug development program during the preclinical safety evaluations (IND-enabling GLP toxicology studies) required for first-in-human clinical trials. EP is not to be mistaken for adverse effects resulting from the mechanism of action against unintended targets (i.e. off-target effects; see Chapter 7). For oligonucleotide (ON) therapeutics, the evaluation of EP poses an extra degree of complexity since, for most of them, they are directed against a species-specific target sequence. Thus, this target sequence may not necessarily be homologous across species, unless the developer included this parameter in the selection process of the candidate ON. However the time, effort, and resources required to optimize candidate ONs are simply not available to most companies. On the other hand, when there is cross-species homology, the affinity and/or the extent of target gene inhibition is not easily measurable.

Most ON therapeutics fall under one of the four major subclasses according to their mechanisms of action: (i) antisense oligonucleotides (ASOs), (ii) small interfering RNAs (siRNAs) (RNA interference [RNAi] mechanism), (iii) aptamers, and (iv) immunostimulatory (IS) ONs. ASOs and siRNAs are characterized by their ability to affect gene expression via interaction with specific sequences of the target mRNA, resulting in the inhibition of a target gene product. The aptamer subclass is designed to bind to a protein target via a specific tridimensional conformation, and for this reason, it is more prone to

Oligonucleotide-Based Drugs and Therapeutics: Preclinical and Clinical Considerations for Development, First Edition. Edited by Nicolay Ferrari and Rosanne Seguin.

EP, since the likelihood of cross-species reactivity is greater especially for nonhuman primates. The IS subclass is designed or selected for its ability to stimulate the immune system of the host; it conveys its effects via cellular-stimulating pathways, rather than specific hybridization to a target sequence. Other types of ONs (e.g. ribozymes, microRNA, hairpin, decoys, etc.) will likely fall under the subclasses described above with respect to how EP assessments would be performed. When planning the preclinical assessment of ONs, it is important to avoid misconstruing toxicological findings that are unrelated to EP; therefore several factors that are largely dependent on the subclass of ONs that is being developed must be considered. In this chapter, we will describe the major challenges encountered and discuss the current thinking and strategies for evaluating EP for ON therapeutics.

9.2 Regulatory Expectations

When a new class of molecule with a new mechanism of action emerges, it is always challenging to devise a preclinical development plan to evaluate its safety. The relatively small body of published information available at an early stage or the known effects of the molecular class, combined (in the case of ONs) with the complexity of the required assessment, may explain the difficulty for developers to identify a pathway for proper safety assessment. This would in part also explain the difficulty for regulators to lay out clear guidance. Hence, for the first ONs directed against human target(s), the regulatory expectations were largely unclear [1, 2]. The early ASO programs for which EP was assessed (often using rodent-active analogues) did not reveal dramatic or consistent toxicity that could clearly be attributed to EP. Hence, the initial level of regulatory attention directed toward EP assessment for ONs appeared to be less than that applied to biotherapeutics, for which most adverse effects related to EP are more commonly observed.

Over time, an increasing degree of scrutiny was directed to the assessment of EP for ONs, and questions were typically posed during early-stage regulatory interactions (e.g. pre-IND meetings) regarding cross-species activity and the sponsor's plans for investigation of EP (i.e. for those ONs targeting human gene products). Particularly, for sponsors advancing ASOs, the degree of uncertainty about regulatory expectations regarding the scope of EP assessment has been high, some of which will be summarized in the next paragraph.

The nature of the selection process for identifying ASOs with optimal activity in a human system often yields a lead molecule with little or no cross-species activity. One reason for this human specificity is that optimally active ASOs often target untranslated regions of the mRNA or other regions that are not well conserved across species (personal communication). Furthermore, the effectiveness of the hybridization-based destruction of the antisense target mRNA can be substantially undermined by minor nonhomology (i.e. "mismatches") in the mRNA target sequence between the animal species and humans (personal communication). Depending on the length of the ON, even one mismatch can reduce pharmacologic potency, and two or more mismatches can reduce hybridization activity by 10–100-fold, resulting in the total absence of antisense activity and pharmacologic potency in the animal species. For many of the early antisense programs, it was not uncommon for the target sequence in nonhuman primates (i.e. cynomolgus or rhesus monkeys) to differ from that in humans by one or two nucleotides. Because monkeys are commonly used as the nonrodent species for ON safety evaluation, this circumstance presented resource issues in terms of documenting pharmacologic activity of the ON in monkeys or identifying an analogue that would be active in monkeys. In some cases, the nucleotide sequence of the human target gene has not been elucidated in the animal species selected for toxicity studies, which raised the daunting prospect of having to perform gene sequencing in an animal species for some molecular targets or documenting pharmacologic potency of the animal analogues. In addition to these challenges, the validity of using animal-active analogues to document EP could not always be considered the best approach to the problem, and more often, the strategy needed to be tailored on a case-by-case basis (see Section 9.4.1).

9.3 Scope of EP Assessment

Proper development of ON therapeutics requires careful planning of safety evaluation. Fortunately, several ON developers have generated a large amount of information over the last two decades. However, the preponderance of this information was derived from programs focused on ASO development, and this biased experience is reflected in the content of the discussion below.

9.3.1 Species Selection

A key question to be considered when assessing EP is how many and which animal species should be used. Characterization of cross-species pharmacologic activity is important for species commonly used in toxicology assessments and particularly for those species that have been used historically for safety assessment of ONs, such as nonhuman primates [3, 4]. However, the above-described species specificity of ONs often precludes ubiquitous activity in the commonly used animal species. In some cases, activity can be documented or predicted (based on sequence homology) for one species, e.g. monkey, but may be lacking in other common laboratory species, and there may be no cross-species activity. In many cases, the information is simply not available and would need to be generated using time-consuming and expensive resources not always available for small companies. Under such circumstances, the usefulness and validity of employing animal-active analogues would be considered (see below). However, prior to considering the use of analogues, the likelihood of encountering adverse effects stemming from EP should be considered, since there are very few examples where ON EP was observed in animals that translated into significant toxicity. Furthermore, such expression of EP has not been commonly identified as a key safety issue that impacted the selection of the starting clinical dose (personal communication). The absence of serious toxicity attributable to EP with ONs may be related, at least in part, to the often incomplete or modest impact on gene expression from ASOs. In addition, the toxicity profile of many types of ONs is often dominated by nonspecific "class effects," particularly for ONs that contain chemically modified backbones that impart a strong anionic character, most notably, those with a phosphorothioate (PS) modification [5–7]. For such molecules, the class effects typically manifest at lower dose levels than those required to produce complete inhibition of gene product expression, thereby obscuring or precluding any adverse effects stemming from EP.

More intensive investigation of EP might be justified based on the nature of the target and from well-founded concern over the consequences of knockdown of the specific target or from the complete absence of information of the consequences of such knockdown. The pharmacologic potency of the ON should also be considered, since in many cases the level of target inhibition measured in an *in vitro* assay may not be a good predictor of the *in vivo* potency. The newer generations of ASOs as well as the newer subclasses targeting gene product expression (such as siRNAs and microRNAs) appear to be generally more pharmacologically potent, owing to greater *in vivo* stability and/or to their mechanism of action. For these newer types of ONs, greater attention to assessment of EP may be warranted. Furthermore, the route of administration (and the likelihood of extensive systemic exposure), the expression profile of the specific target, the dosing frequency and duration, and finally the clinical indication (risk–benefit considerations) should be carefully evaluated, and the level of effort directed toward assessing the safety implications of EP with ONs should be based on the available body of information stated above on a case-by-case basis. Finally, the level of concern about possible EP-related toxicity may be elevated when the ON is targeting a ubiquitous key regulatory protein, whereas there may be less concern when the ON is directed against a protein that is expressed only in diseased tissue or abnormal cells (e.g. cancer cells). Similarly, there may be more concern about the consequences of EP when the ON is delivered systemically at relatively high doses and is widely distributed into tissues, as opposed to an ON that is administered topically to a discrete area of skin or into a "contained organ" such as the eye and/or is not absorbed systemically.

Depending on the available information for each ON, it may be appropriate to consider the options for addressing EP in two species (i.e. both the rodent and nonrodent species). However, while there may be cases where ONs act by novel mechanisms and/or where the toxicity profile of the subclass is not well characterized, for those ON subclasses mentioned above, investigation of EP in one species should suffice, unless further investigation is warranted by compelling theoretical concerns or by the results of general toxicity studies that revealed a novel toxicity apparently stemming from EP. Apart from those reservations, assessment in one species is consistent with ICH S6 (R1) where similar limitations in cross-species activity may exist, although dose-limiting toxicity of biopharmaceuticals is most often an extension of pharmacologic activity, rarely has this been the case with ONs.

9.3.2 Determination of Pharmacologic Relevance

Another important point to consider is the level of information that is required to "validate" a species for the investigation of EP, which raises questions about the theoretical potency of the ON based on sequence homology. It is scientifically accepted that when there is 100% sequence homology between the human and animal target mRNA sequences, the human ON will very likely be active in the animal species. Under such circumstances (i.e. with 100% animal–human homology), documentation of the activity of the ON in the animal species is probably not needed. However, when the degree of nonhomology is moderate (e.g. one or two mismatches), the animal species may still be a valid model for the assessment of EP (with the human ON), but this uncertainty warrants investigation to experimentally document pharmacologic activity, either by demonstrating decreased target gene expression or some other measure of intended activity reflecting decreased gene expression (e.g. efficacy in a relevant animal model). Pharmacologic activity could be confirmed by a relatively simple *in vitro* assay, such as demonstration of reduced target mRNA or protein expression following incubation of the ON in a relevant *in vitro* system that expresses the target gene (e.g. a monkey peripheral blood leukocyte preparation). The potency of the human ON in the whole animal or *in vitro* system need not precisely match that of the ON in a human system to justify using that species for assessment of EP. This view is supported by the fact that the ON will usually be tested at high clinical-multiple doses in toxicity studies, such that target gene product inhibition will likely be achieved [5]. However, for those programs in which a high clinical-multiple dose level cannot be evaluated in the toxicity studies (e.g. ocular studies), the pharmacologic potency of the human ON in the animal species should be considered in making a judgment about whether EP can be adequately assessed in that species.

When no activity can be documented for a human ON in any of the animal species that are commonly used for toxicity studies, the use of animal-active analogues (surrogates) for assessment of EP should be considered in one species, as discussed below.

9.4 EP Evaluation Strategies

EP evaluation strategies differ depending on the degree of cross-species reactivity of the candidate ON and on the availability of surrogate or analogue ONs. Many cases are intermediate between the extremes of ubiquitous cross-species activity for the human ON and no cross-species activity, and several of these circumstances are discussed below.

One scenario commonly encountered is when the human ON has 100% sequence homology (i.e. complementarity) with the target mRNA region for one animal species (e.g. monkey) and/or documentation of pharmacologic activity in this nonrodent species, and when the sponsor additionally has utilized a rodent analogue for pharmacology investigations. In this circumstance, documentation of activity in a single species (e.g. monkey) is probably sufficient. Inclusion of the rodent analogue in the GLP toxicity study is not needed, as EP of the clinical candidate would be adequately assessed in the nonrodent study. This view is consistent with ICH S6 (R1). However, for several reasons, including the lack of clear regulatory guidance and concern on the part of sponsors about not providing adequate EP evaluation in their preclinical packages, several developers have set a precedent for investigating the toxicity of a rodent analogue. However, this more extensive investigation typically has not revealed additional information regarding the risks of EP (i.e. no new manifestations of toxicity stemming from EP were identified with the rodent analogue [personal communication]). In some instances, unclear results were obtained that were difficult to interpret for clinical decision-making (e.g. no adverse findings with the clinical candidate in a relevant nonhuman primate and toxicity with the rodent analogue in a rodent study).

When the human ON is inactive in the nonrodent species, but active in rodents, the potential adequacy of a one species assessment of EP would be generally applicable, such that EP investigation in the nonrodent species should not be required. However, in such cases where a nonrodent species analogue is available (e.g. because it had been developed for use in pharmacology studies), the inclusion of that analogue in the nonrodent toxicity study may be warranted.

In the case of an ON that is not pharmacologically active in the nonrodent species (e.g. nonhuman primate) or in rodents, but a rodent analogue is available, the utilization of the rodent analogue in the GLP toxicity study to address EP (i.e. tested in parallel with the human sequence) may be warranted, but the developers should give careful consideration to the concerns discussed above and outlined below regarding the use of analogues.

9.4.1 Concerns About the Use of Animal-active Analogues

Animal-active analogues are generally developed as a pharmacology tool and may ultimately be used for safety assessment in the same species in a GLP toxicity study [5]. These analogues often have a substantially or completely different nucleotide sequence than the human ON and, hence, they are distinct molecular entities. One concern is that the intensity of class effects can vary substantially among ONs of different sequences, which can result in different manifestations of toxicity between the human ON and analogue that may be construed as evidence of EP. However, those differences may actually only be reflection of some nonpharmacology-based sequence-related difference. Developers with extensive experience in the safety assessment of ASOs have learned over the years that certain ONs exhibit dramatically greater toxicity than other sequences for reasons unrelated to EP, and this anomalous toxicity may occur in approximately 10–20% of the ONs tested (personal communication). The structure–activity relationship for this type of toxicity is not well understood, and, hence, it would behoove the sponsors to select active analogues that do not exhibit such properties before including them in a side-by-side comparison with the human ON in toxicity studies. However, because of the sense of urgency for most ON development programs, many sponsors elect to forgo preliminary toxicity screening of animal-active analogues.

In addition, the fact that the analogue possesses a different sequence than the human ON poses a risk that the analogue could, by chance, elicit some type of mechanism-based (e.g. antisense) effect, i.e. an "off-target" inhibition of another gene product that translates into toxicity. If unexpected toxicity occurred with an analogue, discerning whether the effect reflected EP or an off-target effect may be challenging and time consuming. There are other possible differences in properties between the human ON and an analogue that are unrelated to EP such as differences in pharmacokinetics, biodistribution, and tissue stability that could yield differential toxicity and possibly be construed as evidence of EP. Although the ADME profile is typically similar among ONs that are structurally related, there is some influence of nucleotide sequence on ON disposition that could affect toxicologic potency.

In the event that the human molecule was not species cross-reactive and EP needed to be evaluated in at least one species, it may generally be more appropriate to conduct focused safety evaluations with analogues in a disease model in which the target may be overexpressed, with specific investigations of endpoints that would be expected to be affected by the EP, rather than subject the analogue to a general toxicity evaluation in parallel with the human ON. Studies in disease models might be challenging to accomplish, if available, and often cannot be performed under GLP-compliant conditions. However, the absence of a GLP setting should not be a deterrent from conducting such studies, as long as the execution and record-keeping practices for such studies are sound. Nevertheless, even with more focused safety studies, the potential to encounter off-target toxicities with analogues exists and could yield results that are difficult to interpret.

9.4.2 Animal-active Analogues in Reproductive and/or Carcinogenicity Studies

For those human ONs that have activity in nonhuman primates, but not rodents, a rodent analogue may be considered when there is concern or uncertainty about whether inhibited expression of the targeted gene product would be expected to affect reproductive performance. Therefore, while the use of animal-active analogues for general toxicity investigations should be approached with caution, it is recognized that such analogues may be the best option for addressing requirements for reproductive toxicity testing. In addition, such analogues may have a role in the evaluation of carcinogenicity, considering that these studies are conducted only in rodents; however, such an undertaking should be carefully considered as per the criteria discussed above (i.e. regarding the level of concern about the consequences of target inactivation and other factors).

9.4.3 Other Considerations for Use of Animal Analogues

When evaluating the safety of an animal-active analogue, there are a number of tactical considerations to think about carefully. At the outset, it is important to understand the additional resources needed to manufacture and characterize a second test item. The most common question is whether to include one or more groups to be treated with the analogue in the GLP toxicity study, in parallel with the human sequence. The sponsors may elect to test only one dose level of the analogue, typically matching the high or middle dose level of the human sequence. However, the evaluation of a single-dose level presents a risk that, should the group exhibit a unique or more pronounced toxicity that is regarded as a manifestation of EP, a regulatory issue may arise from not having characterized a no-observed-adverse-effect level (NOAEL) for that effect. Hence, careful consideration needs to be given to the number of groups required for proper evaluation of an analogue, or the selection of the analogue would have included a screening step to exclude any unpredictable manifestation of EP. Prescreening, using a range of doses to aid this decision process is an option, but it may be more time and cost efficient to simply include multiple groups in a GLP toxicity study.

9.4.4 The Use of Inactive Analogues as Control Articles

Another strategy has been the use of inactive analogues to permit the distinction of EP-related effects from nonpharmacologic effects. This approach is becoming increasingly prevalent for siRNA programs, particularly for those programs in which the siRNA is delivered via a specialized formulation, and formulation-based toxicity is expected (see Section 9.4.5). This circumstance presents challenges in distinguishing between the effects of the formulation excipients, the nonspecific ON class effects, and effects stemming from EP. In such cases, the inclusion of an inactive analogue can be very informative to "subtract out" the effect that could be observed with the active candidate. However, the concerns expressed about the use of active analogues are relevant to the use of inactive analogues, i.e. anomalous toxicity related to off-target mechanism-based activity or other sequence-dependent effects could manifest and could confound interpretation of the study data. In those cases where the human ON is devoid of activity in the animal species and an analogue is to be used for testing, the human ON could serve as an inactive control ON and may be the best choice for such a control because it avoids the addition of another inactive sequence to the study design. However, it is important to ensure that the human ON is truly inactive in the animal species.

9.4.5 The Role of Formulations

Many ON development programs are moving forward with specialized delivery systems that could impact the likelihood of encountering significant toxicity related to EP. Formulations that afford targeted delivery to specific cell types could dramatically increase the potential for adverse manifestations of EP in those cell types. However, if such targeted delivery is directed against cancer cells or cells containing nonhost pathogens, and if the pharmacologic objective is cellular destruction, adverse consequences of EP in those cell types may be moot. For several types of specialized formulations, it appears that the toxicity stemming from excipients or other properties of the formulation apart from the ON content may be much more pronounced than any EP effect of the ON, particularly if the ON (active ingredient) possesses little or no xenobiotic chemical modification. However, it is difficult to know a priori whether the formulation-based toxicity or the ON-related toxicity will be predominant. Hence, the increasing use of delivery formulations presents new considerations about the strategies and scope of EP assessment, and the sponsors should examine the circumstances of their program(s) and make appropriate proposals for regulatory interactions.

9.4.6 Aptamer Oligonucleotides

Aptamers are a unique subclass of ONs that are designed to interact with a protein, as opposed to mRNA. The tertiary structure of ONs is sufficiently complex to provide a rich array of conformations for interacting with proteins. Aptamer design uses a complex progressive selection process to yield molecules with very high affinity for the target protein, such that binding of the aptamer ON to the protein inhibits the function of the protein. Hence, the antihuman aptamers that are selected possess a unique structure that is highly specific for the target protein. In this respect, aptamers behave like monoclonal antibodies, in that their cross-species activity may be limited if conservation of the protein sequence and structure across species is poor. Therefore, most aptamers exhibit activity in primate species but typically not in rodents or other nonprimate species such as rabbit and dog.

The leading developers in this space have often relied on pharmacology data from human or nonhuman primate models, and they have occasionally developed a rodent analogue aptamer to enable *in vivo* pharmacology studies when nonhuman primate models are not available or feasible. However, because aptamer discovery is a sequential selection process rather than a molecular design process, the analogue will always be substantially different in length, composition, and structure than the antihuman aptamer, and the only common properties may be the homologous target binding and basic nucleic acid structure. The analogue is a valuable tool for performing pharmacology studies of the target biology, but it may not completely mimic the clinical candidate and may also not be suitable for toxicology evaluations for reasons previously described.

Many of the aptamers are PEGylated to confer desirable pharmacokinetic properties. PEGylation can profoundly affect the activity of an aptamer. However, since the rodent analogue that is employed for target biology investigations is typically not PEGylated, there may be significant differences in the property of the analogue that preclude its valid use in toxicity studies. With PEGylated aptamers, the primary chemistry-related toxicity that is observed is the vacuolation of various cells, reflecting the uptake of the PEGylated molecule. Hence, the toxicity profile of a non-PEGylated rodent analogue aptamer cannot be compared directly with the PEGylated human aptamer.

For all of these reasons, the use of analogue aptamers in toxicity studies has historically been avoided, and the developers have relied mainly on the testing done with the human aptamer in nonhuman primate toxicity studies (i.e. one species evaluation of EP). As discussed for the other subclasses, judgments about the appropriate scope of EP assessment for aptamers should be made on a case-by-case basis.

9.4.7 Immunostimulatory Oligonucleotides

IS ONs are much less species specific than other subclasses and can exhibit EP across species. The IS ONs interact with Toll-like receptors (TLRs) on immunocompetent cells and trigger intracellular events that translate into various cytokine-mediated responses. Common anatomic pathology findings stemming from the intended IS activity include injection site reactions, lymphoid hyperplasia, cellular infiltration in various organs, and related responses. These effects invariably dominate the toxicity profile of IS ONs and are typically dose limiting. Thus, EP is quite evident from pharmacology and toxicity studies of IS ONs, and the main challenge is to determine which animal species is the most appropriate model for human responses.

In this regard, recent information has been gathered about the differences between rodent and primate species with respect to the cellular distribution of TLRs, the downstream cytokine responses and other sequelae, and the structure–activity relationships for TLR activation [8]. However, there is a growing body of evidence indicating that the severity and diversity of the anatomic pathology elicited by IS ONs in rodents may not be representative of human responses. Hence, nonhuman primates appear to be a better model for human responses to IS ONs, and it is questionable whether the rodent species provides any additional value in the context of general toxicity assessment and first-in-human dose selection [9].

9.4.8 MicroRNA

Several new subclasses of ONs are emerging that have unique properties and will present new issues for assessment of EP. One of the most promising new subclasses is microRNA (miR). Various strategies to deliver anti-miRs or miR mimetics are being employed. The mechanism for the modulation of gene regulation by miRs shares some features with the RNAi pathway, but with some important distinctions. The introduction of a miR mimetic or anti-miR typically affects a constellation of largely interrelated gene products. The direct effect of an anti-miR is the derepression of expression (increase), but downstream genes may change in a positive or negative direction (quite commonly in both directions for different gene products). The aim of miR-based therapies is to produce some degree of overall modulation of a biological response by mimicking or inhibiting a disease-related miR. Another important distinction between miR-based therapy and siRNAs, and antisense ONs is that miRs function by fine-tuning levels of expression of interrelated genes, as opposed to the more dramatic inhibition for ONs that affect gene expression. Thus far, the conservation of miRs across species, combined with their mechanism of action, translates into robust cross-species activity, such that most animal pharmacology studies with miRs exhibit the desired changes in biomarkers in species commonly used for safety assessment of ONs (i.e. monkeys and rodents). Hence, animal-active analogues are generally not needed to assess EP for this subclass.

The extent to which the pattern of miR-induced modulation of gene product expression will be analogous across species is uncertain, although, theoretically, there should be considerable commonality. Because of the broader spectrum of mRNA targets for an individual miR, it may be difficult to distinguish whether any specific change in gene expression reflects intended targeting versus off-targeting versus a compensatory response of the cell. Microarrays and other means of characterizing changes in gene product expression can be used to obtain snapshots of the modulation profile, as such profiling is central to understanding the miR's pharmacologic response. From such information, one might also glean insights into whether certain ancillary effects may be anticipated based on the observed pattern of alterations in gene expression. However, it is questionable whether this type of information should serve as an impetus for special, dedicated toxicity investigations apart from standard regulatory safety studies.

In summary, the current approach to identifying potential safety issues associated with the intended pharmacologic action of miRs needs to be considered, particularly because of the novelty of such molecules and the uncertainty about the array of gene targets affected and any downstream consequences of such effects. However, because the miRs currently under development have shown cross-species activity, and because they typically produce only a moderate degree of modulation in gene product expression, there should not be a heightened concern about the likelihood of encountering EP with these molecules, and there is no reason to expect that conventional toxicity studies will fail to reveal effects in the EP category that would be relevant to human safety.

9.5 Conclusions

EP assessment for ONs requires careful planning and close attention to several factors to yield relevant safety data and minimize the risk of producing unrelated effects that can be perceived as EP. Regulatory guidance in this area has been lacking, and most developers are generally relying on prior experience and case-by-case dialogue with regulatory agencies to guide their strategies. Thus, this chapter provides some examples of past experience that should help ON therapeutic developers design their safety assessment programs in a manner that will ensure adequate evaluation of the potential for the ON candidate to produce adverse effects stemming from EP. Unlike small molecules that are generally active across species, ONs are often species specific by design and require specific selection processes to identify drug candidates that are active in more than one species. The selection of the species for EP evaluation is also a complex process requiring information on cross-species target sequence homology and/or pharmacologic activity. When selected ON candidates do not have robust cross-species activity, strategies to evaluate EP are based on considerations about the nature of the target and consequences of exaggerated targeting, as well as several other factors such as the availability of animal-active analogues and the relevance of deploying such analogues during safety assessment. The use of animal-active analogues is often accompanied by uncertainties regarding the value of experiments performed with completely distinct molecular entities and the possibility of revealing some toxicities that would be perceived as EP but could actually be unrelated to the inactivation of the target. Hence, the use of species-specific analogues to assess EP should be approached with caution, with some attention to screening of the analogue against anomalous toxicity unrelated to EP. With proper attention to species selection and/or relevant use of analogues, assessment of EP-related toxicity for ON therapeutics can be successfully accomplished. For newer subclasses of ON therapeutics, or for those that are delivered in complex formulations, experience in assessing EP is limited, and the proper course of action has not yet been well defined. However, it is likely that the considerations for approaching this challenge with novel subclasses and delivery technology will be similar to the strategies proposed in this chapter.

References

1 Black, L., Farrelly, J., Cavagnaro, J. et al. (1994). Regulatory considerations for oligonucleotide drugs: updated recommendations for pharmacology and toxicology studies. *Antisense Res. Dev. 4*: 299–301.

2 Ahn, C. and Degeorge, J. (1997). Preclinical development of antisense oligonucleotide therapeutics for cancer: regulatory aspects. In: *Clinical Trials of Genetic Therapy with Antisense DNA and DNA Vectors* (ed. E. Wickstrom), 39–52. New York: Marcel Dekker.

3 Levin, A., Monteith, D., Leeds, J. et al. (1998). Toxicity of oligonucleotide therapeutic agents. In: *Handbook Exp. Pharmacol* (ed. G.V.R. Born et al.), 169–215. Berlin: Springer-Verlag.

4 Monteith, D. and Levin, A. (1999). Synthetic oligonucleotides: the development of antisense therapeutics. *Toxicol. Pathol. 27*: 8–13.

5 Levin, A., Henry, S., and Monteith, D. (2001). Toxicity of antisense oligonucleotides. In: *Antisense Drug Technology, Principles, Strategies and Applications* (ed. S.T. Crooke), 201–267. Basel/New York: Marcel Dekker.

6 Geary, R., Yu, R., and Levin, A. (2001). Pharmacokinetics of phosphorothioate antisense oligonucleotides. *Curr. Opin. Investig. Drugs 2*: 562–573.

7 Levin, A. (1999). A review of the issues in the pharmacokinetics and toxicology of phosphorothioate antisense oligonucleotides. *Biochim. Biophys. Acta 1489*: 69–84.

8 Campbell, J.D., Cho, Y., Foster, M.L. et al. (2009). CpG-containing immunostimulatory DNA sequences elicit TNF-α-dependent toxicity in rodents but not in humans. *J. Clin. Invest. 119* (9): 2564–2576.

9 Kornbrust, D., Cavagnaro, J., Levin, A. et al. Oligo safety working group exaggerated pharmacology subcommittee consensus document. *Nucleic Acid Ther. 23*: 21–28.

10

Genotoxicity Tests for Novel Oligonucleotide-Based Therapeutics

Cindy L. Berman[1], Scott A. Barros[2], Sheila M. Galloway[3], Peter Kasper[4], Frederick B. Oleson[5], Catherine C. Priestley[6], Kevin S. Sweder[7], Michael J. Schlosser[8], and Zhanna Sobol[9]

[1] *Berman Consulting, Wayland, MA, USA*
[2] *Sage Therapeutics, Cambridge, MA, USA*
[3] *MRL, Merck & Co., Inc., West Point, PA, USA*
[4] *Federal Institute for Drugs and Medical Devices (BfArM), Bonn, Germany*
[5] *Independent Consultant, Concord, MA, USA*
[6] *Innovative Medicines & Early Development, AstraZeneca, Cambridge, UK*
[7] *Forensic and National Security Sciences Institute, Syracuse University, Syracuse, NY, USA*
[8] *MSR Pharma Services, Inc., Lincolnshire, IL, USA*
[9] *Pfizer Inc., Groton, CT, USA*

10.1 Introduction

This document is prepared by the Genetox Subcommittee of the OSWG, a group of industry, academic, and regulatory scientists involved in the development and regulation of therapeutic oligonucleotides (ONs) [1]. The purpose is to outline the experience with genotoxicity testing of ONs and to provide scientific recommendations regarding the appropriateness of genotoxicity testing and on the type of testing, if indicated. The approach to genotoxicity testing of ONs should be tailored to the scientific cause for concern and the historical experience with this class of therapeutics.

Several different chemical classes of ONs with varying modes of action are in development. In general, the chemical structures of these ONs are not dissimilar to native DNA or RNA (Table 10.1). Antisense, CpG, and anti-miR ONs are typically single-stranded (ss) molecules and are composed primarily of native DNA or RNA nucleotides with phosphorothioate (PS) rather than phosphodiester (PD) linkages, and often some fraction (or all) of the nucleotides are modified with either 2′-*O*-methyl (2′-*O*-Me) or 2′-*O*-methoxyethyl (2′-MOE) substituents. Some subclasses of antisense ONs use chemistries such as bridged

Oligonucleotide-Based Drugs and Therapeutics: Preclinical and Clinical Considerations for Development, First Edition. Edited by Nicolay Ferrari and Rosanne Seguin.

Table 10.1 Classes of ONs and their structures and mechanisms.

Class	Structure	Typical length	General mechanism of action	Specific mechanism of action
Antisense	ssDNA	13–25	Hybridization to complementary mRNA leads to target mRNA degradation and/or reduced translation	RNAse H1-mediated mRNA cleavage
Ribozyme/ deoxyribozyme	ssRNA or ssDNA	>20		Catalytically active; cause RNA cleavage
siRNA	Duplex RNA or ssRNA	19–21		RISC-mediated mRNA cleavage
miRNA	ssRNA	20–25		RISC-mediated mRNA cleavage
Decoy	dsDNA	>20	High affinity interaction to target protein	Gene expression inhibition via binding to the DNA binding site of transcription factor
Aptamer	ss or ds RNA, DNA	30–40		Target protein inhibition
Immunomodulatory (CpGs)	ssDNA	>20		Toll-like receptor (TLR) agonism

DNA, deoxyribonucleic acid; ds, double stranded; RNA, ribonucleic acid; RISC, RNA-induced silencing complex; mRNA, messenger ribonucleic acid; siRNA, small interfering RNA ss, single stranded.

nucleic acid (BNA) or phosphorodiamidate morpholino oligomers (PMOs) that are distinct from native DNA. Small interfering RNA (siRNA) ONs are double-stranded (ds) and predominantly have a natural PD backbone with 2′-*O*-Me or 2′-fluoro sugar modifications. ON aptamers are often a mixture of native DNA and RNA with limited presence of the PS linkages or 2′-ribomodifications listed above. In addition, ONs may be conjugated to molecules such as polyethylene glycol (PEG) to increase stability in circulation or to ligands to facilitate cellular uptake via receptor-mediated endocytosis. Thus, the chemical modifications across different classes of ONs could be quite broad.

The nonclinical safety evaluation of ON therapeutics has generally followed the regulatory guidelines for small molecules, rather than the International Conference on Harmonisation (ICH) S6 guidance for biotechnology-derived products [2] because they are chemically synthesized. As a result, numerous ON therapeutics have been tested in the standard battery of *in vitro* and *in vivo* genotoxicity assays for clastogenicity (chromosome breakage) and mutagenesis (gene mutations). ONs have been uniformly negative for these genotoxicity assays across the chemical classes tested. For ONs that have been submitted to the FDA for IND review as of March 2012, 47 compounds had been tested in at least one genetic toxicity study, and all were found to be negative for genotoxicity [3]. The compounds were primarily from the two most common chemical classes of ONs (ssPS oligodeoxyribonucleotides +/− 2′-alkyl modification and dsRNA ONs), which suggests that ONs of these chemical classes are not genotoxic. Nonetheless, new modifications to nucleic acids could result in genotoxicity if monomers (altered nucleosides) are released. In addition, ONs may have unique properties that could result in genotoxicity that is not detectable by the standard battery. These issues were discussed in an EMEA reflection paper [4], which stated that consideration should be given to the potential of ONs to cause mutagenesis through sequence-specific mechanisms (i.e. hybridization-dependent triplex formation) that are not addressed by the standard battery.

The goal of this paper is to summarize the technical aspects and historical experience of testing ONs in genotoxicity assays and to provide a science-based recommendation for genotoxicity testing of ONs. After more than 20 years of experience, a broad spectrum of ON compounds has been through the standard genotoxicity battery for chromosomal damage and gene mutations. Aspects unique to this class of compound that bear on the technical and scientific validity of testing in the standard battery of assays are reviewed. Recommendations are provided based on historical experience from the most evaluated chemical classes of ONs, with consideration of additional chemical classes, conjugated and/or formulated ONs, where experience is maturing. The genotoxicity testing recommendations outlined herein are intended to assess the potential for ONs to induce stable and transmissible changes in DNA (e.g. gene mutations and chromosomal damage). These recommendations are the consensus opinion of the Subcommittee and do not necessarily reflect the current expectations of regulatory authorities.

10.1.1 History of Regulatory Guidance on Genotoxicity Testing

The battery of genotoxicity tests recommended in the ICH S2A (1996) [5] and ICH S2B (1997) [6] guidance documents was based largely on experience with small molecule chemicals and their metabolites that react with DNA via direct interactions, such as covalent binding and intercalation, and on the correlation of the test results with carcinogenicity as demonstrated by rodent bioassays. These guidance documents do not specifically refer to ON therapeutics. The recommended tests included:

1) An *in vitro* assay for "point" mutations involving one or a few base pairs; the Ames microbial mutagenicity test was selected because of its historically high positive predictability for rodent carcinogens (e.g. Ref. [7]).
2) An *in vitro* assay in mammalian cells for large-scale chromosomal damage, because some mechanisms that operate in mammalian cells are not found

in bacterial cells. Although mutagens generally induce both small- and large-scale damage, some are more effective at the induction of point mutations, whereas others are more effective at the induction of chromosome breakage and rearrangements. The ICH S2A/B battery [5, 6] included either a test for chromosomal aberrations scored at metaphase or a mouse lymphoma (L5178Y) cell thymidine kinase (*tk*) gene mutation assay.

3) An *in vivo* rodent assay for chromosomal damage, typically the bone marrow micronucleus assay or chromosome aberration assay. The *in vivo* assay was included because certain genotoxic carcinogens were known to be detectable only in *in vivo* genotoxicity assays.

The guidance was subsequently updated (ICH S2(R1), 2011) [8] and now has two options for testing:

Option 1: An updated version of the ICH S2 battery that includes a third alternative for the *in vitro* mammalian cell assay, i.e. the *in vitro* micronucleus assay.

Option 2: A battery that omits the *in vitro* mammalian cell assay but requires *in vivo* assessment in two different tissues. Typically, the *in vivo* assays are the rodent micronucleus test as in (c), and a DNA strand break assay, usually in the liver as it is the site expected to have optimal exposure and metabolism.

The ICH S6(R1) guideline [2] for biotechnology-derived (generally cell derived) pharmaceuticals states that the range and type of genotoxicity studies conducted routinely for pharmaceuticals are not applicable to biotechnology-derived pharmaceuticals and are not needed. Although the guidance states that the principles may also be applicable to ON drugs, most regulatory agencies treat ONs as synthetically manufactured drugs and require the full battery of genotoxicity tests. Therefore, companies developing therapeutic ONs have generally followed the ICH S2A [5] and S2B [6] guidelines.

10.1.2 Relevance of the Standard Genotoxicity Test Battery to ONs

The current ICH guideline, ICH S2(R1) [8], clarifies that its primary focus is small molecule pharmaceuticals. The standard tests were selected based on many years of experience with hundreds of small molecules, and the batteries were designed to be sensitive for the detection of carcinogens, in particular rodent carcinogens. Thus, these tests aim to characterize direct interactions with DNA (covalently bound adducts, intercalation, cross-linking, oxidative damage) or indirect effects on DNA synthesis, cell cycle, or the mitotic process and chromosome segregation. For example, chemically reactive molecules, such as alkylating agents, bind covalently to DNA throughout the genome and cause mutations that can be detected in the standard "reporter" genes, such as the *his* operon in the *Salmonella* strains used in the Ames test or the *tk* or hypoxanthine–guanine phosphoribosyltransferase (*hprt*) genes in mammalian cell models. The combination of the Ames test and the *in vivo* assay for chromosomal damage (chromosome aberrations or micronuclei) has been shown to be effective in detecting a high proportion of the known human carcinogens [9–11]. The battery also detects known germ cell mutagens [12–14] and, thus, is considered effective in screening for both genotoxic carcinogens and potential human mutagens.

In contrast to small molecules, ONs are not expected to interact chemically (e.g. by covalent binding) with DNA or other cellular targets such as mitotic microtubules. However, the metabolism of ONs can result in the release of modified monomer nucleosides that could enter the pool of nucleoside precursors used for DNA synthesis. The liberation of monomers after *in vivo* administration of ONs is well known to occur. After single administration of Macugen, 2′-fluorouridine was found in the plasma and urine [15]. In addition, sponsors routinely analyze for and often detect the N-1 metabolite and other metabolites in plasma (see Refs. [16–19]), which presupposes that the terminal base is released. It is likely that other chain-shortening metabolites, such as N-2, N-3, etc., also result in the release of monomers. *In vitro* degradation of ONs to the N-1 metabolite has also been demonstrated in plasma and cell culture systems [18, 20].

Theoretically, monomeric metabolites could cause pool imbalance and/or could be phosphorylated with subsequent incorporation into nuclear DNA. An imbalance in the nucleotide pools might result in chromosome breakage or mutation [21, 22], effects that could be detectable in any of the three *in vitro* mammalian cell assays in the current ICH S2(R1) battery. However, pool imbalance is unlikely to occur with ONs because the rate of degradation of ONs into mononucleotides is slow and most therapeutic ONs contain a reasonably balanced mix of nucleotides. Should any modified nucleotides be incorporated, they could potentially cause strand breakage due, for example, to chain termination during replication or could lead to mispairing or errors of replication or repair and, thus, to mutation. Incorporation into nascent DNA (or RNA) has been demonstrated following subchronic administration of 2′FC and 2′FU [23, 24]. In addition, interaction with human DNA polymerases and potential for incorporation of 2′-modified nucleosides commonly used in therapeutic ONs have also been evaluated in *in vitro* DNA primer extension assays [25]. For certain antiviral nucleoside analogues, toxicity and/or genotoxicity (chromosomal damage and mutation) are also correlated with incorporation into DNA [26, 27]. However, it is important to note that, in contrast to antiviral and anticancer nucleoside analogues, mononucleotide components of ONs are neither designed nor intended to interact with DNA or RNA processing. The effect of modifications of nucleotide analogues on the ability to inhibit

DNA polymerases or to serve as substrates for DNA synthesis is reviewed in Wright and Brown [28]. Some modified nucleotides, including PS monophosphates contained in some ONs, are known to be poor substrates for cellular kinases, which decreases the likelihood of their incorporation into DNA [29]. Nonetheless, this hypothetical mechanism of genotoxicity must be given consideration. The effects of incorporation would be detected in the standard battery of genotoxicity assays [8].

Potential genotoxicity has to be considered not only for DNA ONs but also for RNA-based molecules. Hypothetically, modified ribonucleotides could be converted into dNTPs by ribonucleotide reductase and get incorporated into DNA. Also, in lower eukaryotes, some direct incorporation of ribonucleotide triphosphates (rNTPs) into DNA has been demonstrated (reviewed in Caldecott [30]). However, the likelihood of such incorporation is low due to the abundance of normal rNTPs, and if it did occur, effects would be similar to those of modified deoxyribonucleotides and would be detectable in the standard battery of genotoxicity tests.

Theoretically, ONs might also interact with DNA in a highly sequence-dependent fashion to form triple helices through electrostatic interaction (i.e. Hoogsteen base pairing). The potential mutagenic mechanism of triplex formation of ONs with homologous DNA sequences is discussed in Section 10.4. In contrast to the mechanisms discussed above, the standard assays of genotoxicity would not be informative for assessing mutation caused by triple helix formation, but the probability of such a mechanism occurring with therapeutic ONs is very low.

Although not the focus of this chapter, a separate consideration is genotoxicity related to intended pharmacology of the ONs. For example, inhibition of DNA replication or repair may result in DNA damage, whereas alteration of cell cycle checkpoints or inhibition of certain kinases, including mitotic kinases, may result in potential induction of aneuploidy. These effects may be detectable in the standard battery if the pharmacologic target is present in the test systems.

10.2 Experience with ONs in the Standard Battery

The Subcommittee recognized that substantial data exist on ONs as assessed in the standard genotoxicity battery. Well over 200 ON therapeutics have been in development and over half have entered clinical trials. The first ONs in development consisted of naturally occurring structures, which are considered to be nongenotoxic. Subsequent ONs contained chemical modifications to improve stability and/or pharmacological activity. The Subcommittee has created a database of the candidate ONs, their chemical modifications, and the results of genotoxicity testing based on available information from the public domain or provided by sponsors. A synopsis of this database is provided in Table 10.2.

Table 10.2 Summary of oligonucleotide genotoxicity testing database.

Modification/ moiety	Estimated no. of ONs in development	Total no. of ONs with Genetox results	No. of ONs with results				
			In vitro gene mutation (Ames)	*In vitro* mammalian cell (chromosomal aberrations)	*In vitro* mammalian cell (gene mutation)	*In vivo* rodent micronucleus	Other (specify)
Backbone							
PS	≥61	24	24	12	10	17	1 (Comet)
PMO	≥16	12	12	12		12	
Sugar							
2'MOE	≥24	13	13	8	9	11	1 (Comet)
2'OMe	≥24	14	14	10	3	8	1 (Comet), 1 (SHE CTA)
2'F	13	6	6	2	4	4	1 (Comet), 1 (SHE CTA)
LNA	≥16	9	4		5 (2 Different assays)	2	
cEt	4	4	4	3	1	1	1 (Comet)

2′MOE, 2′-*O*-methoxyethyl; 2′OMe, 2′-*O*-methyl; 2′F, 2′-fluoro; cEt, constrained ethyl; LNA, locked nucleic acid; ONs, oligonucleotides; PMO, phosphorodiamidate morpholino oligomer; PS, phosphorothioate; SHE CTA, Syrian hamster embryo cell transformation assay.

10.2.1 ON Chemical Classes Tested for Genotoxicity

Backbone modifications: The most common modification, the hallmark of first-generation ONs, is the PS linkage in the backbone. PS linkages have been included in antisense and immunomodulatory CpG ONs, ribozymes, siRNA, and aptamers. Over 60 compounds containing PS linkages have been in development, with no reports of a compound that gave positive results in the standard genotoxicity battery (*in vitro* and/or *in vivo* tests). In particular, negative results were obtained with ISIS 2302 [18], ISIS 301012 (also known as mipomersen or Kynamro) [31, 32], ISIS 396443 (also known as nusinersen or Spinraza) [33], and PRO051 (also known as drisapersen) [34]. At least 12 different PMOs proprietary to Sarepta Therapeutics (formerly AVI Bio) have been tested in the standard battery, and negative results were obtained with all, including AVI-4658 (also known as eteplirsen or Exondys 51) [35].

Sugar modifications: A variety of sugar modifications has been employed across the subclasses of ONs. The most common sugar modification is 2′-MOE ribose, the hallmark of second-generation antisense ONs. Another common sugar modification is 2′-*O*-Me ribose, which has been used in not only antisense ONs but also aptamers, siRNA, and ribozymes. At least 24 compounds containing 2′-MOE and 24 compounds containing 2′-*O*-Me have been tested in the standard genotoxicity battery (*in vitro* and/or *in vivo*), and all have been nongenotoxic, including PRO501, PRO044, PRO045, and PRO053 [34, 36, 37] and at least six *N*-acetylgalactosamine (GalNAc)-conjugated siRNAs containing 2′-*O*-Me sugar modifications [38]. In addition, Macugen (pegaptanib sodium), a PEGylated aptamer containing 2′-*O*-Me sugar modifications, was negative in all genotoxicity assays, and the 2′-*O*-Me adenosine and 2′-*O*-Me guanosine mononucleotides were negative for bacterial mutagenesis and chromosomal aberrations [38].

2′-Fluoro (2′F) sugar modifications are commonly found in aptamers (including Macugen) and are also used in siRNA or antisense ONs, and at least 13 ONs containing this modification have entered development. Macugen, which contains 2′F-cytidine (2′-FC) and 2′F-uridine (2′-FU), was negative in all genotoxicity assays. Upon separate testing of the 2′F mononucleotides, both 2′-FC and 2′-FU were negative for chromosomal aberrations in human peripheral blood lymphocytes (HPBLs), for cell transformation in the Syrian hamster embryo cell assay and for mutagenicity in the *Salmonella* strains; however, they "yielded a marginal but reproducible positive response" for mutagenicity in *Escherichia coli* WP2*uvrA* [39]. In its review, the EMEA indicated that 2′-FC and 2′-FU "produced a small increase in revertant frequency with no relationship to dose" and concluded that neither was genotoxic [40]. The Subcommittee has reviewed the publically available data and agrees with this interpretation. Although double-strand breaks were reported after transfection of HeLa cells with 2′F-, PS-modified ONs [41], it is unclear whether the genotoxicity was caused by the observed cytotoxicity. Studies by Janas et al. [42] suggested that *in vitro* cytotoxicity and double-strand breaks in HeLa cells associated with ONs were influenced by the delivery method (transfection) and determined primarily by single-strandedness and PS content, and not by 2′F modification alone. As noted above, numerous 2′F-modified GalNAc-conjugated siRNAs were negative in the ICH battery of genotoxicity assays [38]. In addition, all other ONs containing 2′F-modified nucleotides have been negative for genotoxicity.

BNA chemistry includes locked nucleic acid (LNA) and constrained ethyl (cEt) modifications of the sugar. These sugar modifications increase the affinity of antisense ONs to target mRNA. At least 20 ONs containing such BNA modifications have been in development, and all were negative in the assays conducted to date (primarily *in vitro* assays) [43].

Conjugates: The conjugation of an ON to another molecule has been utilized to protect the ON from exonucleases, or to improve pharmacokinetic properties and/or delivery to a particular cell type by linking to ligands for specific cell surface receptors. For instance, aptamers are often conjugated to PEG; such PEGs are typically 20–40 kDa, similar to those used in PEGylated biologics. PEGs are not directly genotoxic [44]. ONs may be conjugated to peptides or naturally occurring ligands, neither of which is likely to be genotoxic. However, the linkers used to conjugate molecules (e.g. PEG, peptide, or ligand) to the ON may be unique and should be evaluated for genotoxicity as part of the drug product.

Formulations: Certain ON chemical classes such as siRNA are frequently formulated with delivery systems such as lipid nanoparticles to enhance their cellular uptake. Although formulated siRNAs and other ONs have been tested for genotoxicity, these formulations are usually proprietary and little information is available publicly regarding their genotoxic potential.

10.2.2 Conclusions Based on the Database

The ON genotoxicity database compiled by the Subcommittee indicates that ONs, within the chemical classes listed (Table 10.2), are consistently negative for genotoxicity when tested in the standard battery as outlined in ICH S2A and B [5, 6]. Such negative results could be indicative of the absence of intrinsic genotoxic potential and/or the lack of intracellular uptake and, thus, nuclear exposure. With documentation of uptake, the test results are considered to be adequate for determining the genotoxic potential.

Evidence of cellular uptake exists for several classes of ONs. The uptake of ssPS ONs both into bacteria (*Salmonella* TA 98) and mammalian cells, under conditions similar to the *in vitro* genotoxicity test systems, has been confirmed for a representative antisense ON at high concentrations tested in those assays [18]. Although specific documentation of exposure in the standard *in vitro* genotoxicity assays is very limited (especially in the case of the Ames assay) at high concentrations of the ON tested (i.e. up to 5000 μg/mL), cellular uptake of ssPS ONs will occur even in the absence of a transfection reagent. In mammalian cells, uptake of PS ONs that may contain additional modifications including 2′-MOE, 2′-OMe, LNA, and/or cEt have been documented in a number of independent experiments in a variety of cell lines (HCT-15, A549, T98G, HepG2, HUVEC, and CaCo-2 cells) at concentrations less than or equal to those used in mammalian cell genetic toxicity testing [45–47]. In addition, uptake of LNA ONs has been demonstrated in L5178Y$tk^{+/-}$ cells used for the mouse lymphoma assay [43]. Thus, uptake is also likely in cells such as CHO cells used in genotoxicity testing. Although published reports of uptake of ONs into HPBLs are lacking, sequence-specific pharmacological activity in these cells has been demonstrated with certain ss antisense or CpG ONs, as well as formulated ds siRNAs, which suggests that such molecules can be taken up by these cell types [48]. Cellular uptake of GalNAc-conjugated siRNA was confirmed in bacterial strains, CHO cells, HPBLs, and rat bone marrow [38]. Uptake has also been documented *in vitro* and *in vivo* for ONs that are formulated in cationic liposomes, both by imaging of labeled molecules and by demonstrated pharmacological activity [49–52], but genotoxicity test data for these are not yet available.

The negative genotoxicity data together with evidence of uptake of certain types of ONs into the genotoxicity test systems or similar cells support the conclusion that ONs have low genotoxic potential. Because intracellular metabolism of the ON is likely to occur (see Section 10.3.4.2), the demonstration of ON uptake into cells implies that metabolites or breakdown products such as the chemically modified mononucleotides are being evaluated for genotoxic potential. The Subcommittee recognizes that potential adverse effects related to modified monomers could differ across chemistries. Therefore, the recommendations for genotoxicity testing were considered separately for chemistries for which sufficient genotoxicity and uptake data exist versus those for which sufficient information is unavailable to the Subcommittee.

10.3 OSWG Recommendation for Genotoxicity Testing of ONs

10.3.1 Recommended Test Battery

The Subcommittee recommends a test battery for assessing the genotoxicity of novel ONs that includes the following assays (see Table 10.3):

1) An *in vitro* mammalian cell assay for chromosomal damage (chromosome aberrations or micronuclei).
2) An *in vitro* cell assay for gene mutation (with test system selection as discussed below).
3) An *in vivo* assay for chromosomal damage (e.g. rodent bone marrow micronucleus test).

Table 10.3 Genotoxicity testing recommendations for ONs.

		Oligonucleotide	
		Well characterized	**Novel**
Delivery system/conjugate/linker	**Well characterized**	No testing required	Test[a] drug product with evidence of uptake of ON[b]
	Novel	Test[a] drug product	Test[a] drug product with evidence of uptake of ON

Definitions:

[a] **Test battery**: As defined in Section 10.3.1, modified from ICH S2(R1), Option 1. Cells selected for *in vitro* and *in vivo* evaluation of ONs should be appropriate for the uptake of ON.

Drug product: ON formulated with its delivery system or conjugate/linker.

Novel ON, delivery formulation/conjugate/linker: Any chemistry that does not occur naturally in humans and that has not been tested previously for genotoxicity in accordance with these recommendations.

Well characterized: Any chemistry that occurs naturally in humans or has been negative for genotoxicity when tested according to these recommendations.

Evidence of uptake: Options for providing evidence of uptake of ON in Ames test bacteria and appropriate mammalian cells are described in Section 10.3.2. Modifications may be required such as testing a novel ON separately from its formulation components to enable uptake; e.g. a lipid nanoparticle formulation may enable uptake of the ON in mammalian cells, but prevent uptake in bacteria.

[b] In this case, where the delivery system/conjugate/linker is known to be nongenotoxic, the formulation components would be included to enable ON uptake in mammalian cells.

The recommended battery is the same for all ONs, regardless of their pharmacologic mode of action. However, as discussed in Section 10.3.3, testing is recommended only for ONs with novel chemistry where genotoxicity data are lacking.

The *in vitro* mammalian cell assay for chromosomal damage is considered to be the most informative test for assessing the genotoxic potential of ONs. Chromosomal damage is the primary concern for ONs because of the possible liberation of chemically modified mononucleotides during metabolism and the possible incorporation of such mononucleotides into genomic DNA. Hypothetically, incorporation could cause genotoxicity by a similar mechanism like antiviral and anticancer nucleoside analogues, which have been shown to cause chromosome damage (see Section 10.1.2). However, it is important to note that, in contrast to nucleoside analogues, mononucleotide metabolites of ONs are neither designed nor intended to interact with DNA or RNA processing. Nonetheless, the *in vitro* mammalian cell assay for chromosomal damage is recommended based on its sensitivity to clastogenicity of nucleoside analogues and the documented cellular uptake and/or activity of ONs in mammalian cells. The specific assays recommended include a test for chromosome aberrations or micronuclei (e.g. in HPBLs or hamster cells), as described in Option 1 of the ICH S2(R1) [8].

The Subcommittee recommends that, even though gene mutations are an unlikely mechanism of genotoxicity for ONs, testing for gene mutations should be conducted. The preference would be to use a mammalian cell assay, such as the mouse lymphoma L5178Y$tk^{+/-}$ assay, CHO/*hgprt* assay, or human lymphoblastoid TK6 cell *tk* assay. Although the bacterial mutagenicity assay is generally considered to be the most predictive genotoxicity test for carcinogenicity of chemicals in rodents, the Subcommittee considers it to have less utility due to poor uptake of ONs. The bacterial strains used in the standard mutagenicity assay have a "deep rough" cell wall mutation that enhances uptake of small molecule mutagens [53]. However, in *Salmonella* and *E. coli* strains that lack endogenous competence for uptake of large molecules of naked DNA, substantial uptake usually requires techniques such as $CaCl_2$ permeabilization, electroporation, or phage delivery [54–56]. In addition, the processing of monomers (via kinases, phosphoribosyl transferases, etc.) is different in bacteria than in mammalian cells [e.g. Ref. 57], which further supports use of mammalian systems for assessing the genotoxic potential of ONs. Furthermore, for nucleoside analogues, the Ames test (i.e. in bacteria) does not appear to have added value above that of a mammalian cell test. Those that were positive in the Ames test (which is generally limited to anticancer agents as opposed to antiviral agents) were also positive in *in vitro* mammalian cell assay and/or the *in vivo* micronucleus test [26, 27]. The Subcommittee is not aware of any nucleoside analogues that were mutagenic only in bacterial assays. For ONs that contain additional components (e.g. delivery systems, conjugates, and/or linkers), the Ames test is considered informative, just as it is for testing small molecule chemicals. In cases where bacterial mutagenicity assays are performed to assess the ON, uptake of the ON into bacteria should be demonstrated (see Section 10.3.2).

An acceptable alternative may be to assess the chromosomal damage and gene mutation end points in a single *in vitro* test system, such as the mouse lymphoma L5178Y$tk^{+/-}$ assay or the human TK6 cell *tk* assay appropriately modified to detect large-scale events (by longer colony growth periods). Common mammalian cell mutation assays use target genes involved in nucleic acid precursor metabolism, the *tk* and *hprt* genes, and selection of mutants with base/nucleoside analogues. Therefore, in tests of monomers, the potential exists for false positive results through unpredictable interactions. As an example, FIAU [fialuridine, or 1-(2-deoxy-2-fluoro-β-D-arabinofuranosyl)-5-iodo-2,4(1*H*,3*H*)-pyrimidinedione] produced apparent increases in *tk* mutation in the L5178Y mouse lymphoma cell mutation assay that were associated with selective growth inhibition of the wild-type ($tk^{+/-}$) cell compared with $tk^{-/-}$ cell [58] and, thus, did not indicate induction of mutation. In addition, the reported increase in TFT-resistant colonies in TK6 cells treated with a monomer of a PS ON, dAMPαS [16], was later shown not to reflect mutation as the apparently resistant colonies did not survive when rechallenged with TFT [59]. The studies showed that dAMPαS induces cell cycle arrest, so that during the prolonged period of selection in TFT, as the cytotoxic effectiveness of the selective agent diminishes, nonmutant cells grow into colonies, giving a false positive result [59]. Thus, at least two examples exist where marked effects on the growth rate, or selective toxicity to wild-type cells, led to false positive results in *tk* mutation assays. The Subcommittee recommends that when apparent increases in mutants are seen, mutational events should be confirmed in rechallenge experiments with selection agents, and susceptibility to toxicity should be compared in wild-type and mutant cells.

The *in vivo* genotoxicity test should assess chromosome damage in tissues with dividing cells in which exposure has been demonstrated and cell uptake is expected (see Section 10.3.2). In general, for ONs that are absorbed systemically, this requirement is satisfied by measuring micronuclei, or chromosomal aberrations in metaphase cells, in rodent blood or bone marrow, which is consistent with ICH S2(R1) [8]. This approach is based on the hypothetical mechanism by which ONs could cause genotoxicity, i.e. incorporation of modified mononucleotides into DNA during DNA synthesis. Because *in vivo* assays involve tissue distribution and metabolism, they provide critical information in the weight-of-evidence assessment of the ONs genotoxic potential. DNA damage could also be assessed *in vivo* via DNA strand break assays, such as the comet assay, which are included in Option 2 of ICH S2(R1). In principle, the *in vivo* comet assay should be effective for the detection of DNA strand breaks caused by modified mononucleotide metabolites of ONs so long as the assay was conducted in a tissue with replicating cells and has documented exposure *in vivo*. The Subcommittee acknowledges the paucity of published data with ONs or nucleoside analogues in the comet assay; nonetheless, in certain cases, such *in vivo* testing might provide useful information, especially for ONs intended for distribution to specific tissues and known to have limited potential for uptake into bone marrow cells.

Other modifications to the battery might well be justifiable based on the known properties of the molecule in question, in particular its cellular uptake, breakdown products, *in vivo* distribution, and half-life.

The genotoxicity testing recommendations outlined above are intended to assess the potential for ONs to induce stable and transmissible changes in DNA (e.g. gene mutations and chromosomal damage). They are not considered appropriate for evaluating epigenetic changes, that is, either heritable changes in gene expression that occur without a change in DNA sequence or stable, long-term alterations in transcriptional potential that are not necessarily heritable. Such changes were beyond the scope of the Subcommittee's discussions.

10.3.2 Requirement for Evidence for Uptake

The Subcommittee recommends that uptake of the ON be demonstrated in bacteria used for the mutation assay and in mammalian cell types selected for the *in vitro* genotoxicity tests or in comparable mammalian cells. The recommendation for demonstration of cellular uptake to support valid genotoxicity tests of ONs is a key difference from testing "small molecule" pharmaceuticals that, for the large part, are expected to enter cells based on known properties including small size, solubility, and data such as cytotoxicity or efficacy.

The methodology used to demonstrate cellular uptake of an ON can be either qualitative or quantitative. Qualitative methods include a concentration-dependent biological response that requires intracellular localization, e.g. pharmacological activity, such as effects on a gene product or cytokine induction related to endosomal toll-like receptor (TLR) activation. Imaging techniques such as immunohistochemistry are considered sufficient as long as intracellular localization is confirmed. Quantitative methods include capillary gel electrophoresis [18] and the use of radiolabeling [45] and fluorescence labeling [60]. However, because labeling of ONs can hypothetically affect uptake and/or stability and the label may become associated with molecules other than the ONs after ON metabolism, the rationale for selection of the methodology should be reported. For mammalian cells, high test concentrations and/or delivery systems (such as lipid nanoparticles) may accomplish uptake (see Section 10.3.3.2). If uptake is known to occur in mammalian cells by a mechanism that is independent of cell type, this information may be used to support the presumption of uptake in the mammalian cells commonly used for the genotoxicity assays. If uptake occurs through a cell-specific mechanism, such as for ONs designed with delivery systems or conjugates that target a specific receptor, then demonstration of uptake should be conducted in the cell systems selected for genotoxicity testing. Other test strategies can be considered, including *in vivo* testing in an appropriate cell type where feasible (see discussion of *in vivo* testing in Section 10.3.1). Demonstration of uptake can be incorporated into the genotoxicity study or conducted as a supportive study.

Even if no uptake is detected (e.g. for a product that is not intended to enter cells *in vivo*), an effort should be made to assess genotoxicity because some uptake cannot be ruled out in selected tissues *in vivo*. Thus, *in vitro* genotoxicity testing should still be conducted, in an attempt to achieve uptake at high concentrations evaluated in these tests (see also Section 10.3.4.2).

Evidence of uptake into bacterial and mammalian cells exists for ssPS ONs, for PS ONs that also contain certain sugar chemical modifications, and for GalNAc-conjugated siRNAs (see Section 10.2.2). Therefore, further demonstration of uptake into these test systems for such ONs is not necessary, but is warranted for all other ONs. However, once a new chemical modification of a nucleic acid (or a new [untested] linker or component of a delivery system) has been tested thoroughly according to these recommendations with evidence of uptake of the ON, the results would be considered adequate to justify not repeating such testing of the same modification/molecule in future candidate ONs.

10.3.3 Need for Testing of ONs

10.3.3.1 Nonconjugated ONs in Simple Aqueous Formulations

The Subcommittee considers that genotoxicity testing is not warranted for ON therapeutics that use simple aqueous formulations and contain only the following chemistries:

1) Any unmodified or modified base that occurs naturally in humans. Examples of naturally occurring modified nucleosides include 5-methylcytosine, 5-hydroxymethylcytosine, N^6-methyladenosine, and 7-methylguanosine. To date, over 100 naturally occurring modified nucleosides have been identified. The most common of which is pseudouridine. The Subcommittee considers that continued testing of ONs that contain only unmodified (natural) bases and/or modifications described above has little or no scientific merit and should not be required.
2) Chemical modifications that have been tested sufficiently in the standard genotoxicity battery with demonstration of uptake of the ON into the test systems. For example, Henry et al. demonstrated that ISIS 2302, an ssDNA ON with PS modifications, was nongenotoxic in the standard battery with evidence of uptake in all cell types evaluated [18]. The EMEA has indicated

that, based on data from a 21 mer PS ON, "further studies to assess the mutagenic potential of phosphorothioate nucleotides from other phosphorothioate oligonucleotides are deemed not necessary"[4]; the Subcommittee agrees with this position and extends it to all PS ONs regardless of pharmacologic mechanism of action (e.g. ss antisense ONs, dsRNA molecules used for siRNA applications, ON aptamers, decoy DNA binding site, etc.). In addition, the Subcommittee believes that ONs containing 2′-MOE and 2′-*O*-Me sugar modifications have been tested sufficiently. This recommendation is based on the consistently negative genotoxicity results for the extensive set of 2′-alkyl-modified PS ssDNA or RNA ON drug candidates that have been tested in the traditional ICH S2A/B battery *in vitro* and *in vivo* (see Section 10.2.1), which suggests that they do not interact with DNA in a fashion that causes chromosome damage or mutations. The negative results for numerous 2′-*O*-Me-modified siRNAs are supported by evidence of uptake into the genotoxicity test systems [38].

The Subcommittee recommends that ONs containing any chemical modifications other than those listed in Point 2 above be tested as indicated (see Section 10.3.1). Although uptake of such ONs may occur in the mammalian cell genotoxicity test systems (for instance, because of the high concentration of ON tested and/or based on known biological effects of the molecules in other cell types), uptake has not been examined systematically in parallel with the genotoxicity data. Furthermore, when uptake has been demonstrated only in specific cell types (e.g. receptor-mediated uptake), adequate uptake cannot be assumed in the genotoxicity test systems. As genotoxicity testing experience with these modifications increases and data become available, an assessment of their overall genotoxic potential will enable recommendations regarding further testing.

An exception to the recommendation that certain chemistries need no further testing is ONs that pharmacologically target DNA replication processes or any other mechanisms affecting genomic integrity (see Section 10.1.2). ONs with such intended functions would require testing for genotoxicity regardless of their chemistry. Such tests would also serve to confirm the pharmacologic activity of the ONs.

10.3.3.2 ONs in Complex Formulations or Conjugates

The Subcommittee recommends testing of the drug product in order to provide the most clinically relevant assessment of genotoxic potential. That is, ONs that are formulated in a delivery system or conjugated to molecules to form a new chemical entity (e.g. ON, conjugate, and linker molecule) should be tested as the drug product, not just the ON portion. Testing of the ON portion alone or the other component(s) alone may provide additional information, but is not considered sufficient. In cases where the formulation delivery systems or conjugates used are well studied and their use in humans has precedence, additional testing may not be warranted. However, testing is appropriate if the drug product contains novel excipients (e.g. liposome components), a novel conjugate/linker, or an ON containing novel chemistry (Table 10.3).

In addition, the cellular uptake of the conjugated ONs or ON formulated in a delivery system must be considered. For example, an ON encapsulated in lipid nanoparticles that is effectively delivered into mammalian cells may not be available to bacterial cells. Similarly, if ON conjugation is designed for selective uptake into specific cell types (e.g. via receptor-mediated endocytosis), but not into cells used in the genotoxicity test battery, even at high concentrations, alternative test designs should be considered. With appropriate selection of the test cells, assessment in two *in vivo* tests, i.e. Option 2 under ICH S2(R1), may be the best approach for formulated and conjugated ONs. On the other hand, modification via conjugation or formulation does not necessarily prohibit uptake into the genotoxicity test systems, as demonstrated by Janas et al. [38] for GalNAc-conjugated siRNAs.

10.3.4 Recommended Test Conditions

10.3.4.1 Top Concentration for *In Vitro* Tests

With regard to the selection of the top concentration for *in vitro* genotoxicity testing of ONs, the Subcommittee recommends that the ICH S2(R1) guideline should be followed. For the Ames test, the standard top concentration is 5000 μg of ON/plate. For *in vitro* mammalian cell assays, a maximum ON concentration of 1 mM or 500 μg/mL, whichever is lower, should be used unless limited by cytotoxicity or solubility in the culture medium. The justification for the concentrations recommended in the ICH S2(R1) guideline includes the fact that they are higher than typical exposures to small molecule pharmaceuticals *in vivo*. In certain cases, for small molecules and for ONs, use of higher concentrations might be justified.

For formulated or conjugated ONs, the Subcommittee recommends that the top concentration be calculated based on the ON portion of the drug product only because complex formulations and conjugates can constitute a significant portion of the drug product by weight; this approach is more conservative than that proposed by the OSWG Formulated Oligonucleotide Subcommittee [61] and is in line with the approach taken for other classes of therapeutics. Regardless, the test article should be the complete drug product if at all possible. If non-ON component(s) alone is also tested, the top concentration tested should be in accordance with the ICH S2(R1) guideline.

10.3.4.2 Use of S-9 in *In Vitro* Tests

The Subcommittee recommends that *in vitro* genotoxicity testing of ONs should be conducted in the presence and absence of an exogenous metabolic activation system, consistent with ICH S2(R1). The standard metabolic activation system used in *in vitro* genotoxicity tests is post-mitochondrial fraction (S-9) from the liver of rats treated with P450 inducers such as araclor or phenobarbital with β-naphthoflavone. This S-9 fraction contains various cytochrome P450 activities and some cytosolic enzymes such as esterases and is included to assess the genotoxic potential of test article metabolites. ONs are not substrates for cytochrome P450s. In general, ONs are metabolized by nucleases that are present in all cells and most biological fluids, including blood. Endonucleases cleave ONs internally to produce shorter chain lengths, whereas exonucleases metabolize from the ends of ONs to liberate mononucleotides. The metabolic pathway is the same for DNA- and RNA-based ONs. The primary difference across the structural classes of ONs is the rate of nuclease-mediated degradation and, thus, the amounts of liberated nucleotides or nucleotide fragments.

Although P450-mediated metabolism is not applicable to ONs, S-9 has utility in the testing of ON drug products.

1) P450-mediated metabolism of components of drug product.
   The use of S-9 is recommended in testing ON drug products containing novel non-ON components, such as linker molecules in conjugated ONs, or components of a delivery system, which might be liable to P450-mediated metabolic activation.
2) Metabolism of ONs.
   The use of S-9 may also be justified when testing ONs themselves because S-9 appears to contain some nuclease activity. In the Subcommittee's experience, metabolism of ONs can occur in the presence of S-9, as shorter fragments/monomers are detectable by liquid chromatography–mass spectrometry following <24-h incubations (Alnylam Pharmaceuticals, Inc., unpublished data). In principle, the shorter fragments/monomer metabolites of ONs may be taken up by the cells more easily than the parent ONs. Such degradation may be particularly useful when uptake of ONs into the genotoxicity test systems is otherwise minimal, as positive genotoxicity test results for nucleoside analogues attest to the fact that mononucleosides/tides can be taken up into bacteria and mammalian cells. Alternative preparations to supply nuclease activity, as described by Gilar et al. [20], could be considered; however, it should be noted that extracellular degradation is not typically the preferred option for providing evidence of uptake (see Section 10.3.2). Testing in the absence of S-9 is still recommended as, in mammalian cell systems, it has the advantage of providing a prolonged treatment period, which may enhance the likelihood of uptake.

10.3.4.3 *In Vivo* Tests

For *in vivo* genotoxicity testing, the recommendations in the ICH S2(R1) guideline are considered appropriate for ONs. These *in vivo* genotoxicity studies can be stand-alone studies or integrated into existing single- or repeat-dose toxicity tests if appropriate. The timing of sampling after treatment should take into consideration the persistence of the test material in the test animals. For example, current guidance for small molecule pharmaceuticals recommends sampling for bone marrow micronucleus analysis at 24 h after the final dose in a repeat-dose study. However, certain ONs (e.g. those with long tissue half-lives) may be administered infrequently, with euthanasia several days after the final administration. This design may be acceptable for genotoxicity analysis, provided that exposure to the tissue to be analyzed is maintained during the critical period prior to sampling, e.g. 1 to 2 days for micronucleus assays or last few hours for the comet DNA strand break assay. The doses tested should typically fulfill the recommendations listed in the ICH S2(R1) guideline.

10.4 Triplex Formation

The capacity of the standard genotoxicity battery to detect all potential genotoxic mechanisms of ONs was discussed in a 2005 EMEA reflection paper [4]. The EMEA highlighted the concern that the standard battery would not detect sequence-specific and hybridization-dependent ON interactions with ds genomic DNA (known as triplex or triple helix formation). Thus, they concluded that the potential for triplex formation should be considered in the genetic toxicity risk assessment of ONs intended for clinical development. Triple helices can be formed when an ON binds in a sequence-specific manner with dsDNA, and such interactions have been shown to produce site-specific mutations in an engineered reporter DNA construct [62]. Currently, no validated test exists to evaluate the potential for triplex formation across the entire genome. This section will briefly review the current published literature on sequences shown to form triple helical DNA structures and describe the potential genetic toxicity associated with such interactions. In general, the potential for triplex formation by ONs that are not intended to bind to DNA (i.e. antisense, siRNA, aptamers, and immunostimulatory CpG ONs, among others) is very low because such interactions are unfavorable under physiological conditions, and the homology and general sequence requirements (≥10 purine or pyrimidine-rich nucleotide stretches) are typically absent from therapeutic ONs.

10.4.1 Biochemical Requirements for Triplex Formation

An ssON binding to a dsDNA target was initially reported based on biochemical analyses [63, 64]. This discovery, in part, led to an effort to identify the potential for triple helix-forming oligonucleotides (TFOs) to modulate gene transcription as possible therapeutics. To this end, the requirements for triple helix formation have been investigated [65, 66]. Briefly, third strand binding to dsDNA can occur when polypyrimidine or polypurine ONs form Hoogsteen or reverse-Hoogsteen hydrogen bonds to the purine strand along the major groove of DNA [65, 67]. As a whole, TFO–DNA triple helices are not stable interactions due to negative charge repulsion of the third-strand phosphates with the phosphates of the DNA duplex. In addition, access to the major groove of DNA by a third strand is not an energetically favorable reaction, in part due to the packaging of DNA into chromatin [68]. TFOs with a neutral backbone (e.g. peptide nucleic acid) have improved binding affinity to dsDNA [66]. Incremental improvements in binding affinity can also be made by replacing C with 5-methyl-C or neutralizing the PD backbone by addition of 2′O-Me or 2′-aminoethyl on the ribose [68, 69].

The properties of a TFO that can form a triple helix with DNA have been identified based on the biochemical characterization of TFO–DNA triple helix formation in cell-free systems. Studies suggest that at least a 12–14 polypurine or polypyrimidine stretch in a TFO is required for triplex formation [65]. A minimum of a 10-base stretch of polypurines is required in the target DNA [69]. A single nucleotide mismatch (i.e. switch from a purine to pyrimidine or vice versa) decreases the binding affinity of a TFO to its target DNA by over 100-fold, but such TFOs are still capable of forming a triple helical structure [70–74]. The TFO–DNA triplex is favored to form at 25 °C and, in the case of polypyrimidine TFOs, at $pH \leq 6$. Thus, TFO interactions are disfavored under physiological conditions where binding affinity for triplex formation can be reduced up to 1000-fold [66]. The charge on the TFO is also a contributing factor for triple helix formation [75].

In experiments aimed at developing methods for site-specific mutagenesis, specifically targeted recombinant constructs that contain the biochemical criteria for TFO formation have resulted in mutagenicity following triplex interaction between TFO and dsDNA [62, 76]. These data suggest a potential concern with ON therapeutics and, thus, it is important to consider the recombinant construct used to elicit mutagenicity. An optimized target DNA sequence was generated by inserting a polypurine motif into the *supFG1* reporter gene within the simian virus 40 (SV40) vector and transfected into monkey COS cells. Induction of point mutations and small deletions in the *supFG1* gene were observed when COS cells were treated with a 30-mer purine motif TFO and to a lesser extent with a corresponding 20-mer TFO [62]. Similar mutations were observed *in vitro* with a sequence-specific TFO, and with a TFO with cationic phosphoramidate linkages, in the *supFG1* transgene in a chromosomal locus in mouse cells [75, 76] and *in vivo* following subcutaneous injection in mice that contained the *supFG1* transgene inserted into genomic DNA [76]. These TFOs are postulated to induce mutagenesis at the *supFG1* gene through errors during replication, nucleotide excision repair (NER), and transcription-coupled repair [77, 78]. Because the *supFG1* locus is a mutation hotspot [79], increases in the frequency of spontaneous mutations relative to genomic DNA may lead to enhanced mutation susceptibility by TFO-mediated triplex formation. Mutagenicity resulting from the interaction of an unmodified TFO (i.e. not conjugated to a DNA-reactive agent) with unmodified genomic DNA has not been reported in the published literature.

A recent study using a TFO targeting genomic DNA failed to induce mutations *in vitro*. Reshat et al. [80] showed that a TFO (27 nucleotides in length) targeting the genomic coding region of the *hprt* locus in human lymphoblastoid TK6 cells failed to induce mutations at the targeted *hprt* locus as well as the off-target *tk* locus at ON concentrations up to 10 μM. In this test system, an increase in mutation frequency at the *hprt* locus was observed with the TFO when cells were cotreated with the genotoxicant methyl methanesulfonate (MMS) to induce DNA repair processes. However, the increased mutation frequency under these conditions was not clearly associated with triplex mechanism, and it is not clear how MMS-induced DNA damage influences TFO–DNA interactions.

The majority of studies with TFOs have been with ssDNA and RNA ONs; however, the potential for ds ONs or aptamers to participate in triplex formation cannot be excluded.

10.4.2 Assessment of New ONs for Triplex Formation

The Subcommittee believes that the formation of triplexes with genomic DNA is highly unlikely based on the available mechanistic information and because ON therapeutics are not designed to bind to DNA. Such interactions are unfavorable under physiological conditions. The optimal sequence requirements (≥12 polypurine or polypyrimidine nucleotide stretches) are typically absent from therapeutic ONs, and triplex formation to unmodified genomic DNA has not been demonstrated. Therefore, experimental testing of ONs for triplex formation is not recommended, especially as no validated test exists. However, the Subcommittee recommends that the potential of a therapeutic ON to form a triplex be assessed by sequence analysis. The sequence should be examined for the presence of a polypurine or polypyrimidine stretch of at least 12 nucleotides. If the sequence does not contain such a stretch, the formation of a triplex is highly unlikely and, therefore, genotoxicity via this mechanism is not a concern. In those cases where a polypurine or polypyrimidine stretch exists, the ON sequence should be assessed further for the likelihood of triplex formation with regard to ON length, modifications to the sugar and backbone of the molecule, and availability of a target sequence in the human genome. If the weight of evidence suggests that triplex formation is plausible, biological relevance can be assessed on a case-by-case basis, and the approach should be discussed with regulatory authorities.

10.5 Impurities

Impurities may be considered either product related or process related. Product-related impurities are those that are structurally related to the ON (i.e. used in its synthesis or occurring as breakdown products). Process-related impurities are those that are not structurally related to the ON but are potential chemical contaminants or by-products of the manufacturing process and are considered in a similar way to process-related impurities of "small molecule" drugs.

10.5.1 ON-Related Impurities

ON-related impurities are molecules that are structurally very similar to the parent compound or its metabolites. Metabolites are typically shorter fragments of the parent molecule or mononucleotides, but are not chemically modified as part of metabolism. Likewise, most synthetic impurities contain structural elements (e.g. shorter lengths or presence of PD linkages) that are found in the parent ON or in naturally occurring genetic material and, thus, would not have an increased risk of genotoxicity. Occasionally, process-related impurities with structural elements not contained in the parent molecule are seen at low levels, but these typically do not possess the classical electrophilic structural alerts for genotoxic compounds [81]. *In silico* analysis using rule-based structural analysis software for prediction of mutagenicity has been conducted for a few of these impurity structures. One example of commonly observed impurities is an apurinic ON. In these, the base is missing on a single purine residue, which results in the potential for an aldehyde functionality on the ribose. These apurinic sites exist in native DNA and contain the same structural fragment as 2-deoxyribose-5-phosphate, which is a normal component of nucleotide biosynthesis. Therefore, these impurities are not thought to increase genotoxic risk. Similarly, impurities containing a trichloroethylene modification to a single internucleotide PS linkage were subjected to *in silico* analysis and were negative for mutagenic structural alert.

10.5.2 Potentially Mutagenic Impurities

The potential for the production of DNA-reactive and/or mutagenic process impurities during the manufacturing process should be assessed as outlined in the ICH M7 (R1) [82]. In accordance with the ICH S6(R1) guideline on biotechnology-derived products, the Subcommittee considers that the use of standard genotoxicity studies on the final product is not appropriate for assessing the genotoxic potential of process contaminants. Analytical methods used to characterize the drug product should be capable of delineating any residual reagents or impurities that are likely in the drug product. Any materials used in preparing an ON (or its delivery system) that are known DNA-reactive mutagens or have "structural alerts" for DNA-reactive mutagenicity (e.g. by *in silico* assessment) should be examined closely. Such chemicals will generally be eliminated from the ON product during purification. However, if persistence into the final product is likely, they would have to be controlled based on the above guideline for mutagenic impurities, taking into account the likely duration of treatment and the therapeutic indication.

10.6 Conclusions

The Subcommittee considers that ONs have low potential for genotoxicity based on data available to date. However, testing of new ONs should be conducted due to the potential for modified monomers derived from the breakdown of the ON to become incorporated into DNA and cause chain termination, miscoding, and/or errors of replication or repair. Further, the overall experience and data on potential for ON-related genotoxicity across all classes of ONs is not extensive and, therefore, genotoxicity testing of the ON therapeutics is warranted.

The recommended battery is similar to Option 1 of ICH S2(R1) guideline except that the *in vitro* assay for gene mutations should be conducted in mammalian cells. Mammalian cell assays are generally considered more relevant than bacterial assays because of the known responsiveness of mammalian cells to nucleoside analogues and because uptake is more likely. However, the bacterial mutagenicity ("Ames") test is preferred for testing of all ONs that contain non-ON components, such as delivery formulations or conjugate/linkers; this recommendation is based on the greater predictability of the bacterial system. A second difference from ICH S2(R1) guideline is the documentation of cellular uptake. Evidence for uptake in the cell types used for genotoxicity testing is considered an essential aspect of testing ONs; due to their large size and chemistry, uptake of ONs cannot be assumed. Options for demonstrating the uptake of the parent ONs into the mammalian and/or bacterial cell system include qualitative and/or quantitative methodologies. The recommended test conditions for ONs are similar to those described in ICH S2(R1). The use of S-9 is considered appropriate for ONs primarily because of its nuclease activity. In addition, its cytochrome P450 activity is critical for ONs that contain non-ON components.

The Subcommittee considers that genotoxicity testing is not necessary for ONs that contain only naturally occurring chemistries and utilize simple aqueous formulations. This view is similar to that of Thybaud et al. [83] regarding the necessity for genotoxicity testing of peptides containing exclusively natural amino acids. In addition, testing is no longer warranted for ONs that contain only selected chemistries (e.g. PS linkages, 2′-MOE, and 2′-*O*-Me sugar modifications) because the uptake of such ONs into test cells has been demonstrated and the results of the *in vitro* and *in vivo* standard battery of genotoxicity assays have been consistently negative. Genotoxicity testing is recommended for all ONs that contain nonnatural chemical modifications that have not been well tested and for those that employ components such as conjugates/linkers or delivery systems that are not naturally occurring. Tests should be performed on the complete drug product to provide the most clinically relevant assessment of genotoxic potential. Once a new chemical modification of an ON (or a new formulation component or linker) has been tested thoroughly according to these recommendations with evidence supporting uptake of the ON into the test cells, the Subcommittee concludes that multiple repeat testing of the same components is not warranted.

The potential for sequence-dependent interaction, such as triple helix formation between ONs and genomic DNA, is considered to be highly unlikely. Stringent sequence and biochemical conditions required for triplex interaction are not compatible with ON therapeutics. However, even though the potential for triplex interactions by development candidates is low, it should be assessed by sequence analysis without experimental testing.

These recommendations are the consensus opinion of the Subcommittee and do not necessarily reflect the current expectations of regulatory authorities.

Acknowledgments

The authors wish to thank S. Henry, H. Younis, S.-L. Lee, Y. Tessier, and C. den Besten for their contributions to the discussion; Isis, Santaris, Alnylam, Prosensa, and Sarepta for their contributions to the database; and C. Horvath and J. Cavagnaro for their thoughtful reviews.

References

1 Schubert, D., Levin, A.A., Kornbrust, D. et al. (2012). The Oligonucleotide Safety Working Group (OSWG). *Nucleic Acid Ther.* 22: 211–212.
2 ICH S6(R1) (1997. Addendum added 2011). Preclinical safety evaluation of biotechnology-derived pharmaceuticals.
3 Lee, S.-L., Brown, P., Wang, J., and Dorsam, R.T. (2013). Nonclinical safety assessments and clinical pharmacokinetics for oligonucleotides therapeutics: a regulatory perspective. In: *Advanced Delivery and Therapeutic Applications of RNAi* (ed. K. Cheng and R.I. Mahato), 63–81. Chichester: Wiley.
4 EMEA (2005). CHMP SWP Reflection paper on the assessment of the genotoxic potential of antisense oligodeoxynucleotides. Doc. Ref.EMEA/

CHMP/SWP/199726/2004. http://www.ema.europa.eu/docs/en_GB/document_library/Scientific_guideline/2009/09/WC500003149.pdf (accessed 24 January 2018).

5 ICH S2A (1996). Specific aspects of regulatory genotoxicity tests for pharmaceuticals. http://www.fda.gov/downloads/Drugs/GuidanceCompliance RegulatoryInformation/Guidances/ucm074925.pdf (accessed 24 January 2018).

6 ICH S2B (1997). Genotoxicity: a standard battery for genotoxicity testing of pharmaceuticals. http://www.fda.gov/downloads/Drugs/GuidanceCompliance RegulatoryInformation/Guidances/ucm074929.pdf (accessed 24 January 2018).

7 Tennant, R.W., Margolin, B.H., Shelby, M.D. et al. (1987). Prediction of chemical carcinogenicity in rodents from in vitro genetic toxicity assays. *Science 236*: 933–941.

8 ICH S2(R1) (2011). Guidance on genotoxicity testing and data interpretation for pharmaceuticals intended for human use. http://www.ich.org/fileadmin/Public_Web_Site/ICH_Products/Guidelines/Safety/S2_R1/Step4/S2R1_Step4.pdf (accessed 24 January 2018).

9 Shelby, M.D. and Zeiger, E. (1990). Activity of human carcinogens in the salmonella and rodent bone-marrow cytogenetic tests. *Mutat. Res. 234*: 257–261.

10 Tinwell, H. and Ashby, J. (1991). Activity of the human carcinogen MeCCNU in the mouse bone marrow micronucleus assay. *Environ. Mol. Mutagen. 17*: 152–154.

11 Nersessians, A.K. (1992). Activity of human carcinogens in the Salmonella and rodent bone marrow cytogenetic tests. *Mutat. Res. 281*: 1239–1243.

12 Shelby, M.D. (1996). Selecting chemicals and assays for assessing mammalian germ cell mutagenicity. *Mutat. Res. 352*: 159–167.

13 Waters, M.D., Stack, H.F., Jackson, M.A. et al. (1994). The performance of short-term tests in identifying potential germ cell mutagens: a qualitative and quantitative analysis. *Mutat. Res. 341*: 109–131.

14 Holden, H.E. (1982). Comparison of somatic and germ cell models for cytogenetic screening. *J. Appl. Toxicol. 2*: 196–200.

15 Macugen [package insert] (2011). San Dimas: Gilead Science Inc.

16 Glover, J.M., Leeds, J.M., Mant, T.G.K. et al. (1997). Phase I safety and pharmacokinetic profile of an intercellular adhesion molecule-1 antisense oligonucleotide (ISIS 2302). *J. Pharmacol. Exp. Ther. 282*: 1173–1180.

17 Geary, R.S., Leeds, J.M., Fitchett, J. et al. (1997). Pharmacokinetics and metabolism in mice of a phosphorothioate oligonucleotide antisense inhibitor of C-raf-1 kinase expression. *Drug Metab. Dispos. 25*: 1272–1281.

18 Henry, S.P., Monteith, D.K., Matson, J.E. et al. (2002). Assessment of the genotoxic potential of ISIS 2302: a phosphorothioate oligodeoxynucleotide. *Mutagenesis 17*: 201–209.

19 Wei, X., Dai, G., Liu, Z. et al. (2006). Metabolism of GTI-2040, a phosphorothioate oligonucleotide antisense, using ion-pair reversed phase

high performance liquid chromatography (HPLC) coupled with electrospray ion-trap mass spectroscopy. *AAPS J. 8*: E743–E755.

20 Gilar, A., Belenky, A., Smisek, D.L. et al. (1997). Kinetics of phosphorothioate oligonucleotide metabolism in biological fluids. *Nucleic Acids Res. 25*: 2615–2620.

21 Kunz, B.A., Kohalmi, S.E., Kunkel, T.A. et al. (1994). International Commission for Protection Against Environmental Mutagens and Carcinogens report. Deoxyribonucleoside triphosphate levels: a critical factor in the maintenance of genetic stability. *Mutat. Res. 318*: 1–64.

22 Mathews, C.K. (2006). DNA precursor metabolism and genomic stability. *FASEB J. 20*: 1300–1314.

23 Richardson, F.C., Tennant, B.C., Meyer, D.J. et al. (1999). An evaluation of the toxicities of 2′-fluorouridine and 2′-fluorocytidine-HCl in F344 rats and woodchucks (Marmota monax). *Toxicol. Pathol. 27*: 607–617.

24 Richardson, F.C., Zhang, C., Lehrman, S.R. et al. (2002). Quantification of 2′-fluoro-2′-deoxyuridine and 2′-fluoro-2′-deoxycytidine in DNA and RNA isolated from rats and woodchucks using LC/MS/MS. *Chem. Res. Toxicol. 15*: 922–926.

25 Richardson, F.C., Kuchta, R.D., Mazurkiewicz, A., and Richardson, K.A. (2000). Polymerization of 2′-fluoro and 2′O-methyl-dNTPs by human DNA polymerase α, polymerase γ, and primase. *Biochem. Pharmacol. 59*: 1045–1052.

26 Wutzler, P. and Thust, R. (2001). Genetic risks of antiviral nucleoside analogues – a survey. *Antivir. Res. 49*: 55–74.

27 Negishi, K., Bessho, T., and Hayatsu, H. (1994). Nucleoside and nucleobase analog mutagens. *Mutat. Res. 318*: 227–238.

28 Wright, G.E. and Brown, N.C. (1990). Deoxyribonucleotide analogs as inhibitors and substrates of DNA polymerases. *Pharmacol. Ther. 47*: 447–497.

29 Reshat, R., Priestley, C.C., and Gooderham, N.G. (2012). Mutagenesis by an antisense oligonucleotide and its degradation product. *Toxicol. Sci. 130*: 319–327.

30 Caldecott, K.W. (2014). Ribose- an internal threat to DNA. *Science 343*: 260–261.

31 FDA (2013). Center for Drug Evaluation and Research, Application Number 203568Origi1s000, Pharmacology Review(s). http://www.accessdata.fda.gov/drugsatfda_docs/nda/2013/203568Orig1s000PharmR.pdf (accessed 24 January 2018).

32 Kynamro [package insert] (2013). Cambridge, MA: Genzyme Corporation.

33 Spinraza [package insert] (2016). Cambridge, MA: Biogen Inc.

34 EMA (2016). Assessment report for Kyndrisa (drisapersen). http://www.ema.europa.eu/docs/en_GB/document_library/Application_withdrawal_assessment_report/2016/09/WC500212619.pdf (accessed 24 January 2018).

35 Sazani, P., Weller, D.L., and Shrewsbury, S.B. (2010). Safety pharmacology and genotoxicity evaluation of AVI-4658. *Int. J. Toxicol. 29*: 143–156.

36 Ponstein-Samarro Doorten, A.Y., De Kimpe, S.J. (2011). Absence of genotoxicity by antisense oligonucleotide PRO044. American College of Toxicology 32nd Annual Meeting Poster #P411.
37 Ponstein-Samarro Doorten A.Y., Den Besten C., De Kimpe S. (2013). Absence of genotoxicity in a series of 2′O-methyl phosphorothioate oligonucleotides. Society of Toxicology 52nd Annual Meeting Poster #556.
38 Janas, M.M., Jiang, Y., Duncan, R.G. et al. (2016). Exposure to siRNA-GalNAc conjugates in systems of the standard test battery for genotoxicity. *Nucleic Acids Ther. 26*: 363–371.
39 Food and Drug Administration (2004). Center for Drug Evaluation and Research, Application Number 21-756, Pharmacology Review(s). http://www.accessdata.fda.gov/drugsatfda_docs/nda/2004/21-756_Macugen_pharmr.pdf (accessed 24 January 2018).
40 EMEA (2006). Scientific Discussion for Macugen (pegaptanib). http://www.ema.europa.eu/docs/en_GB/document_library/EPAR_-_Scientific_Discussion/human/000620/WC500026218.pdf (accessed 24 January 2018).
41 Shen, W., Liang, X.H., Sun, H., and Crooke, S. (2015). 2′-Fluoro-modified phosphorothioate oligonucleotide can cause rapid degradation of P54nrb and PSF. *Nucleic Acids Res. 43*: 4569–4578.
42 Janas, M.M., Jiang, Y., Schlegel, M.K. et al. (2017). Impact of oligonucleotide structure, chemistry, and delivery method on in vitro cytotoxicity. *Nucleic Acids Ther. 27* (1): 11–22.
43 Guérard, M., Andreas, Z., Erich, K. et al. (2017). Locked nucleic acid (LNA): based single-stranded oligonucleotides are not genotoxic. *Environ. Mol. Mutagen. 58*: 112–121.
44 Fruijtier-Pölloth, C. (2005). Safety assessment on polyethylene glycols (PEGs) and their derivatives as used in cosmetic products. *Toxicology 214*: 1–38.
45 Nakai, D., Seita, T., Terasaki, T. et al. (1996). Cellular uptake mechanism for oligonucleotides: involvement of endocytosis in the uptake of phosphodiester oligonucleotides by a human colorectal adenocarcinoma cell line, HCT-15. *J. Pharmacol. Exp. Ther. 278*: 1362–1372.
46 Bennett, C.F., Chiang, M.Y., Chan, H. et al. (1992). Cationic lipids enhance cellular uptake and activity of phosphorothioate antisense oligonucleotides. *Mol. Pharmacol. 41*: 1023–1033.
47 Koller, E., Vincent, T.M., Chappell, A. et al. (2011). Mechanisms of single-stranded phosphorothioate modified antisense oligonucleotide accumulation in hepatocytes. *Nucleic Acids Res. 39*: 4795–4807.
48 Younis, H.S., Machemer, T., Norris, D.A. et al. (2013). Evaluation of antisense oligonucleotides in human PBMCs and association of release of IL-6 in vitro to constitutional symptoms. *The Toxicologist 132* (1): AB# 1816.
49 Pei, Y., Hancock, P.J., Zhang, H. et al. (2010). Quantitative evaluation of siRNA delivery in vivo. *RNA 16*: 2553–2563.

50 Basha, G., Novobrantseva, T.I., Rosin, N. et al. (2011). Influence of cationic lipid composition on gene silencing properties of lipid nanoparticle formulations of siRNA in antigen-presenting cells. *Mol. Ther.* *19* (12): 2186–2200.

51 Yang, X.Z., Dou, S., Sun, T.M. et al. (2011). Systemic delivery of siRNA with cationic lipid assisted PEG-PLA nanoparticles for cancer therapy. *J. Control. Release* *156*: 203–211.

52 Wang, X., Yu, B., Wu, Y. et al. (2011). Efficient down-regulation of CDK4 by novel lipid nanoparticle-mediated siRNA delivery. *Anticancer Res.* *31*: 1619–1626.

53 Ames, B.N., Lee, F.D., and Durston, W.E. (1973). An improved bacterial test system for detection and classification of mutagens and carcinogens. *Proc Nat Acad Sci U.S.A.* *70*: 782–786.

54 Chen, I. and Dabnau, D. (2004). DNA uptake during bacterial transformation. *Nat. Rev. Microbiol.* *2*: 241–249.

55 Overballe-Petersen, S., Harms, K., Orlando, L.A.A. et al. (2013). Bacterial natural transformation by highly fragmented and damaged DNA. *Proc. Natl. Acad. Sci. U. S. A.* *110*: 18960–18965.

56 Sinha, S. and Redfield, R.J. (2012). Natural DNA uptake by Escherichia coli. *PLoS ONE* *7* (4): 1–6.

57 Sandrini, M.P., Clausen, A.R., On, S.L. et al. (2007). Nucleoside analogues are activated by bacterial deoxyribonucleoside kinases in a species-specific manner. *J. Antimicrob. Chemother.* *60*: 510–520.

58 Kirby, P.E., Back, A.M., and Oleson, F.B. (1985). L5178Y $tk^{+/-}$ mouse lymphoma assay of FIAU, a nucleoside analogue: a false positive response. *Environ. Mutagen.* *7* (suppl 3): 9–10.

59 Saleh, A.F., Priestley, C., Gooderham, N.J., and Fellows, M. (2015). Re-evaluation of the mutagenic response to phosphorothioate nucleotides in human lymphoblastoid TK6 cells. *Toxicol. Sci.* *145*: 169–176.

60 Praus, P., Kočišová, E., Mojzeš, P. et al. (2012). Study of cellular uptake of modified oligonucleotides by using time-resolved microspectrofluorimetry and florescence imaging. *Spectrosc Int J.* *27* (4): 415–419.

61 Marlowe, J.L., Akopian, V., Karmali, P. et al. (2017). Recommendations of the Oligonucleotide Safety Working Group's Formulated Oligonucleotide Subcommittee for the Safety Assessment of Formulated Oligonucleotide-Based Therapeutics. *Nucleic Acid Ther.* *27* (4): 183–196.

62 Wang, G., Seidman, M.M., and Glazer, P.M. (1996). Mutagenesis in mammalian cells induced by triple helix formation and transcription-coupled repair. *Science* *271*: 802–805.

63 Felsenfeld, G. and Rich, A. (1957). Studies on the formation of two- and three-stranded polyribonucleotides. *Biochim. Biophys. Acta.* *26*: 457–468.

64 Felsenfeld, G., Davies, D.R., and Rich, A. (1957). Formation of a three-stranded polynucleotide molecule. *J. Am. Chem. Soc.* *79*: 2023–2024.

65 Kuan, J.Y. and Glazer, P.M. (2004). Targeted gene modification using triplex-forming oligonucleotides. *Methods Mol. Biol. 262*: 173–194.
66 Hansen, M.E., Bentin, T., and Nielsen, P.E. (2009). High-affinity triplex targeting of double stranded DNA using chemically modified peptide nucleic acid oligomers. *Nucleic Acids Res. 37*: 4498–4507.
67 Beal, P.A. and Dervan, P.B. (1991). Second structural motif for recognition of DNA by oligonucleotide-directed triple-helix formation. *Science 251*: 1360–1363.
68 Seidman, M.M., Puri, N., Majumda, A. et al. (2005). The development of bioactive triple helix-forming oligonucleotides. *Ann. N. Y. Acad. Sci. 1058*: 119–127.
69 Jain, A., Wang, G., and Vasquez, K.M. (2008). DNA triple helices: biological consequences and therapeutic potential. *Biochimie 90*: 1117–1130.
70 Aich, P., Ritchie, S., Bonham, K., and Lee, J.S. (1998). Thermodynamic and kinetic studies of the formation of triple helices between purine-rich deoxyribo-oligonucleotides and the promoter region of the human c-src proto-oncogene. *Nucleic Acids Res. 26*: 4173–4177.
71 Cheng, A.J. and Van Dyke, M.W. (1994). Oligodeoxyribonucleotide length and sequence effects on intermolecular purine-purine-pyrimidine triple-helix formation. *Nucleic Acids Res. 22*: 4742–4747.
72 Cheng, Y.K. and Pettit, B.M. (1992). Stabilities of double- and triple-strand helical nucleic acids. *Prog. Biophys. Mol. Biol. 58*: 225–257.
73 Gowers, D.M. and Fox, K.R. (1997). DNA triple helix formation at oligopurine sites containing multiple contiguous pyrimidines. *Nucleic Acids Res. 25*: 3787–3794.
74 Orson, F.M., Klysik, J., Bergstrom, D.E. et al. (1999). Triple helix formation: binding avidity of acridine-conjugated AG motif third strands containing natural, modified and surrogate bases opposed to pyrimidine interruptions in a polypurine target. *Nucleic Acids Res. 27*: 810–816.
75 Vasquez, K.M., Dagle, J.M., Weeks, D.L., and Glazer, P.M. (2001). Chromosome targeting at short polypurine sites by cationic triplex-forming oligonucleotides. *J. Biol. Chem. 276*: 38536–38541.
76 Vasquez, K.M., Narayanan, L., and Glazer, P.M. (2000). Specific mutations induced by triplex-forming oligonucleotides in mice. *Science 290*: 530–533.
77 Chin, J.Y., Schleifman, E.B., and Glazer, P.M. (2007). Repair and recombination induced by triple helix DNA. *Front. Biosci. 12*: 4288–4297.
78 Vasquez, K.M., Wang, G., Havre, P.A., and Glazer, P.M. (1999). Chromosomal mutations induced by triplex-forming oligonucleotides in mammalian cells. *Nucleic Acids Res. 27*: 1176–1181.
79 Hegan, D.S., Narayanan, L., Jirik, F.R. et al. (2006). Differing patterns of genetic instability in mice deficient in the mismatch repair genes Pms2, Mlh1, Msh2, Msh3, and Msh6. *Carcinogenesis 27*: 2402–2408.

80 Reshat, R., Priestley, C.C., and Gooderham, N.G. (2012). A triple-helix forming oligonucleotide targeting genomic DNA fails to induce mutation. *Mutagenesis 27*: 713–719.
81 Ashby, J. and Paton, D. (1993). The influence of chemical structure on the extent and sites of carcinogenesis of 522 rodent carcinogens and 55 different human carcinogen exposures. *Mutat. Res. 286*: 3–74.
82 ICH (International Conference on Harmonisation of Technical Requirements for Registration of Pharmaceuticals for Human Use) M7(R1) (2017). Assessment and Control of DNA Reactive (Mutagenic) Impurities in Pharmaceuticals to Limit Potential Carcinogenic Risk, step 4. http://www.ich.org/fileadmin/Public_Web_Site/ICH_Products/Guidelines/Multidisciplinary/M7/M7_Step_4.pdf. http://www.ich.org/fileadmin/Public_Web_Site/ICH_Products/Guidelines/Multidisciplinary/M7/M7_R1_Addendum_Step_4_31Mar2017.pdf (accessed 3 April 2018)
83 Thybaud, V., Kasper, P., Sobol, Z. et al. (2016). Genotoxicity assessment of peptide/protein-related biotherapeutics: points to consider before testing. *Mutagenesis 31*: 375–384.

11

Reproductive and Developmental Toxicity Testing Strategies for Oligonucleotide-Based Therapeutics

Tacey E.K. White[1] and Joy Cavagnaro[2]

[1] *Aclairo Pharmaceutical Development Group, Inc., Vienna, VA, USA*
[2] *Access BIO, Boyce, VA, USA*

11.1 Introduction

Preclinical developmental and reproductive toxicity (DART) studies are required for the development of most therapeutic products. The overall approach for these evaluations is similar across all product classes, but the designs can be customized to take into account specific product attributes for each type of therapy. There are well-characterized differences in product attributes between chemically synthesized small molecules (aka, new chemical entities (NCEs), drugs, pharmaceuticals) and biologically derived large molecules (aka, new biological entities (NBEs), biologics, biopharmaceuticals) as shown in Table 11.1, which have led to customization of DART study designs. For example, once-weekly dosing regimens have been used for NBEs with long half-lives (e.g. monoclonal antibodies). Additionally, compared with NCEs, NBEs are more likely to have species-specific pharmacologic activity, and their toxicity is more often related to exaggerated pharmacology [1], leading to conduct of DART studies for NBEs only in pharmacologically active species. The difference between pharmaceuticals and biopharmaceuticals can be viewed as a product continuum based upon size and complexity of molecular structure, particularly with the advent of newer technologies (e.g. mAb fragments, protein mimetics, and modifications to extend the half-life of drugs), which blur the lines between these product categories.

Oligonucleotide (ON) therapies have historically been regulated as NCEs because they are chemically synthesized [2]. In addition, they may contain modified chemical moieties not found in nature, and their toxicity has typically been associated with these chemical moieties [3]. However, ON-based

Oligonucleotide-Based Drugs and Therapeutics: Preclinical and Clinical Considerations for Development, First Edition. Edited by Nicolay Ferrari and Rosanne Seguin.

Table 11.1 Comparison of product attributes across product classes.

Product attribute	New chemical entities (NCEs)	Oligonucleotide (ON) therapies	New biological entities (NBEs)
Molecular weight	Low (<1 kDa)	Mid (>6 kDa)	High (>30 kDa)
Manufacture	Chemical synthesis	Chemical synthesis	Biologically derived
Structure	Single entity	Single entity	Heterogeneous
Tissue distribution	• Intra- and extracellular • Wide distribution	• Intra- and extracellular • Selected distribution	• Largely extracellular • Limited distribution
PK/ADME	• Species-specific metabolites • Short half-life	• Catabolized to nucleotides • Long acting	• Catabolized to amino acids • Long half-life
Species specificity	Less likely	More likely	Often
Off-target toxicity	Often	Sometimes	Rarely

Adapted from Schubert et al. [4].

therapeutics have attributes of both NCEs and NBEs [1, 3] (Table 11.1), and these product-specific attributes should be taken into account when planning DART studies with ONs. For example, like NBEs, ON products tend to show species specificity, have structural similarities to endogenous molecules (i.e. DNA and/or RNA), and have longer half-lives and/or pharmacodynamic (PD) effects than NCEs. On the other hand, like NCEs, ONs are chemically synthesized, and their chemical structure is specifically defined, in contrast to the complex and heterogeneous chemical structure of NBEs. Therefore, a case-by-case approach is recommended for the DART assessment of ONs [5], taking into account regulatory guidelines for both NCEs and NBEs [1, 6, 7].

This chapter will provide a brief overview of DART study designs and then outline points to consider for a case-by-case DART assessment of ONs, based on current international regulatory guidelines for the assessment of both pharmaceuticals and biopharmaceuticals and the unique product attributes of ONs. Topics will include the selection of the appropriate animal model(s), dose selection, dosing regimen, and exposure assessments, and the importance of understanding the pharmacologic activity of the ON as it relates to potential for reproductive or developmental effects. Different subclasses of ONs have distinct product attributes, and subclass-specific considerations will also be discussed. The points to consider for ONs in this chapter are the output of the Reproductive Subcommittee of the Oligonucleotide Safety Working Group

(OSWG), the findings of which were previously published [5]. The recommendations in this chapter are offered as points to consider and should not be construed as regulatory guidance.

11.2 General Design of Reproductive and Developmental Toxicity Studies

DART studies are conducted to identify hazards associated with the use of a therapeutic product during pregnancy and lactation and in women and men of childbearing potential. The results of these studies are used to support enrollment of women of childbearing potential (WOCBP) into clinical trials, support marketing applications, and provide guidance to clinicians when prescribing to and counseling WOCBP and pregnant women (pre- and postmarketing). DART studies are required for all classes of therapeutics, including ONs, and are of particular importance because studies to understand developmental and reproductive risks are generally not conducted in human populations premarketing. Therefore, results from nonclinical DART studies are presented in the final product label and often serve as the only available information to assess risks in human populations.

The International Conference on Harmonisation (ICH) guideline S5(R2) provides recommendations for the evaluation of reproductive and developmental toxicity of pharmaceuticals [8]. This guideline recommends that all aspects of the reproductive cycle be evaluated from gamete development through sexual maturation of the offspring and recommends a general strategy for this assessment using three separate study designs (Figure 11.1). However, as emphasized in the guideline, the use of these specific designs is not mandatory, and considerable flexibility is given to sponsors for alternative testing strategies, provided that all aspects of the reproductive cycle are evaluated. The three studies described in ICH S5(R2) include the fertility and early embryonic development (FEED), the embryo–fetal development (EFD), and the pre-/postnatal development (PPND) study, also historically referred to as segment I, II, and III studies, respectively. These studies are described briefly below. For more information, readers are referred to ICH S5(R2) and recent review articles [8–11].

The FEED study (Figure 11.1) is typically conducted in rodents and is designed to evaluate reproductive function and early development of the embryo through implantation [9]. The drug is administered to adult males and females prior to conception and during the mating period and to females through implantation of the embryo. The study evaluates effects of the therapeutic on pregnancy rate, estrous cyclicity, hormonal control of mating, ovulation, fertilization, implantation of fertilized eggs into the uterus, and development of the embryo through the time of implantation. There is an

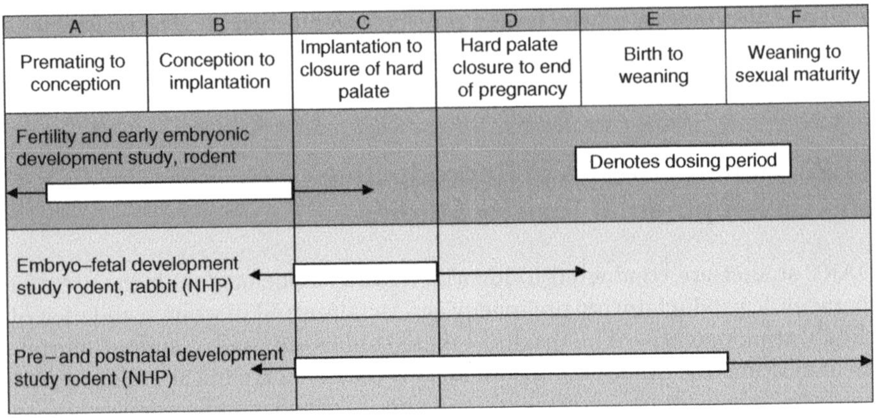

Figure 11.1 Overview of reproduction toxicity evaluations described in ICH guideline S5(R2): Detection of Toxicity to Reproduction for Medicinal Products and Toxicity to Male Fertility. A–F denote the stages in the complete reproductive life cycle from conception in one generation through conception in the following generation, as described in the guideline. The study types shown are standard study types described in the guideline as "the most probable option" for ensuring that all phases of the life cycle are evaluated and include studies for fertility and early embryonic development, embryo–fetal development, and pre- and postnatal development, including maternal function. Species listed are the typical species for each study type. NHP is shown in parentheses as one of several acceptable alternative species.

option to conduct spermatogenic assessments and histologic evaluations of reproductive organs, but these evaluations are not a routine in the FEED. Therefore, this study can be considered the functional complement to the histologic evaluation of the male and female reproductive tract typically conducted in repeat-dose general toxicity studies.

The EFD study (Figure 11.1) is traditionally conducted in one rodent species and one nonrodent species, typically the rabbit [10]. Therapeutics are administered to pregnant females from the day of implantation through the major period of organogenesis (equivalent to the first trimester in primates), and cesarean sections are conducted near the end of gestation. The main endpoints are survival, growth, and morphology of the fetuses as assessed by examination of their external, visceral, and skeletal features. Additionally, the study is intended to assess the maternal maintenance of pregnancy.

The PPND study (Figure 11.1) is generally conducted in rodents and is designed to evaluate the postnatal functional consequences of *in utero* and lactational exposure [11]. This study also evaluates the ability of the mother to carry a pregnancy to term, give birth, nurse, and properly rear offspring to weaning. The drug is administered to pregnant females from the time of implantation of the embryo, through parturition, and throughout the lactational period. A subset of pups from each litter is selected to continue into adulthood without further drug exposure and is tested for effects on learning and memory, reflex development, motor activity, and reproductive performance.

As alluded to above and outlined in ICH S6(R1) [12], alternative DART study designs are commonly used to accommodate the specific product attributes of biopharmaceuticals highlighted in Table 11.1 [13–16]. For example, because the toxicity of biopharmaceuticals is most often related to exaggerated pharmacology, emphasis has been placed on the use of a pharmacologically active species for DART testing of NBEs. This has led to the acceptance of a single species for the EFD study, if two pharmacologically active species are not available, and the use of the nonhuman primate (NHP) in place of more standard species [17]. Additional refinements of NHP DART studies have also included combining the EFD and PPND studies into a single-enhanced pre- and postnatal development (ePPND) study [16]. Traditional fertility assessments are not practical in NHPs; therefore, an enhanced reproductive organ assessment in place of the FEED study has also been deemed an acceptable alternative if the therapeutic is only active in the NHP. These examples for NBEs illustrate ways in which standard DART studies can be altered to take into account product-specific attributes for all types of therapeutics and serve as a backdrop for discussion of the points to consider for DART testing of ONs.

Although the use of the NHP model is acceptable under particular circumstances, the use of the standard species (rodent and rabbit) is still considered to afford the most complete evaluation of DART effects for several reasons. These species have large litters rather than singleton births. This, coupled with larger group sizes (N=20–25/group) for studies in rats and rabbits, increases the power to detect rare malformations. Very well-established study designs and techniques exist for these species, including large historical control databases. Finally, a much more comprehensive fertility assessment is possible in the rodent compared with the NHP.

The timings of DART studies relative to clinical milestones for NCEs and NBEs are set forth in ICH M3(R2) [18], S9(R1) [19], and S6(R1) [12], and these general timings are also applicable to ON therapies. EFD are generally conducted to support enrollment of WOCBP into early clinical trials, FEED studies are conducted to support Phase 3 trials, and PPN studies are expected to be completed to support the marketing application. Recent changes in ICH M3(R2) allow some flexibility in timing of DART studies, depending on the size and duration of clinical trials, but some information on developmental and reproductive risk is generally expected prior to starting clinical trials in WOCBP [18].

11.3 Product Attributes of Oligonucleotide Drugs

As described in the "Exaggerated Pharmacology" chapter of this book (Chapter 9), ONs fall into four main subclasses: (i) antisense oligonucleotides (ASOs); (ii) immunostimulatory (IS) ONs, predominantly those containing

CpG sequences; (iii) silencing RNAs (siRNAs) and microRNA antagonists/mimetics; and (iv) aptamers. The product attributes of these ON subclasses have similarities to both NCEs and NBEs (Table 11.1), and those attributes relevant for planning of DART studies will be introduced in this section and expanded upon in subsequent sections.

As with NCEs, DART effects of ONs can theoretically be driven by chemical structure, as well as both on-target and off-target pharmacology. While most NCEs are pharmacologically active in at least one animal species, this may not be the case for ONs. ASOs and siRNAs require direct hybridization with nucleic acid targets, in contrast to aptamers and IS ONs, which involve interactions with protein targets. As such, ASOs and siRNAs may be more likely to have species-specific pharmacologic activities. This species specificity could complicate the selection of the appropriate animal model for certain ON therapies, particularly if the goal is to evaluate effects of both the chemical structure and the pharmacology.

Another attribute of many ONs is that they have prolonged PD effects and, therefore, do not require daily administration in the clinic. Examples would be ONs that cause a prolonged reduction of message or protein levels, or immunomodulatory ONs which initiate a cascade of events taking place over several days. For these types of ONs, nonclinical toxicity studies are generally designed to mimic the clinical dosing regimen in order to avoid side effects related to overdosing (e.g. exaggerated pharmacology or nonspecific toxicity). However, developmental toxicity studies in rodents and rabbits with NCEs are generally conducted using daily dosing regimens in order to ensure that all periods of embryo–fetal development are adequately assessed. Therefore, as described below, care is needed in selecting a dosing regimen that ensures that the chemical structure is adequately evaluated across all developmental windows, while minimizing the risk of overdosing.

The chemical structure of ONs can affect their toxicity and pharmacokinetics (PK). Structural subclasses based on single-stranded DNA and double-stranded RNA (dsRNA) have varying degrees of PS, 2′-*O*-alkyl, 2′-fluoro, or other modifications [20]. Modifications can be made to the ribose sugar, such as with synthetic morpholinos, or to the interunit linkages by substitution with PS or phosphorodiamidate moieties. These differences in chemical structure can have significant impact on the PK, metabolic stability, nonspecific interactions, and the possible need for sophisticated formulations, in addition to producing unique patterns of toxicity [3]. ONs with minimal chemical modifications, such as certain siRNAs, are often administered in complex formulations. Components in these formulations can cause toxicity and can affect the PK. Therefore, administration of ONs in their final formulation is advisable. Given the unique attributes of ON therapies, a single study design is not necessarily appropriate for all ON subclasses. Instead, the product attributes of each individual subclass of ONs will dictate the scientifically valid approach.

11.4 The Role of Intended Pharmacology in Reproductive and Developmental Effects

On-target and off-target pharmacology can be an important predictor of DART effects for all classes of pharmaceuticals. Genes can have very different functions during development than in adults, and modulation of a number of different pharmaceutical targets (e.g. hormone receptors, endothelin receptors, cation channels) has been associated with developmental effects [21]. Even before the initiation of animal studies, information can be obtained from the literature, which will help predict potential DART effects, and can help to determine how best to approach DART testing. For most ONs, this assessment would generally focus on the impact of decreased expression or function of the targeted protein. For microRNAs, the focus might be on the impact of increased expression of a family of RNAs. It is helpful if the function of the specific target is known, and there is literature on reproductive functions in related knockout (KO) or transgenic mouse models, as well as in human populations deficient in, or overexpressing, the target. If available information suggests a potential concern, or if information is unavailable, it will be important to ensure that the effects of pharmacology modulation are thoroughly studied in DART studies through the use of appropriate animal models.

The expression profile for the specific target during development is also useful in assessing the level of concern for a particular target. If the target is not expressed during development, there is less concern that target modulation would directly affect development. The opportunity for a direct effect on the target in the embryo or fetus may also be limited if exposure of the fetus to ONs is low. For example, placental transfer of certain PS-modified ONs has been shown to be very limited [22–24]. However, for other PS ONs, quantifiable ON concentrations have been detected [25], and modulation of target expression in the fetus has been demonstrated [26, 27].

Finally, it should be recognized that pharmacology-related effects on embryo–fetal development can occur even when a therapeutic agent does not cross the placenta and reach the embryo/fetus directly. Therapeutic-driven changes to maternal physiology or nutritional status not only can affect implantation of the embryo and maintenance of pregnancy but also have the potential to cause fetal abnormalities even if the therapeutic doesn't interact directly with the fetus [28]. Examples include agents that cause hypoxia, angiogenesis inhibitors, compounds whose pharmacologic effect leads to hyper- or hypoglycemia, and a host of other responses. Additionally, placental toxicity or changes to placental function can adversely affect development. The use of a pharmacologically relevant model is especially important when pharmacologic modulation has the potential to affect reproduction or development.

11.5 Selection of Animal Species

An ideal animal model(s) should be representative of humans with regard to PK, metabolism, sensitivity to toxicity, and pharmacologic effects. Selection of the animal species should first consider whether the clinical candidate has pharmacologic activity in rodents and/or rabbits. For clinical candidates that are pharmacologically active in rodents and/or rabbits, or have targets that are not endogenous to any animal species (e.g. antimicrobial ONs), standard rodent and rabbit studies can be conducted, as is generally done for small molecules. In the latter case, the purpose of DART studies would be to evaluate the effects of chemical structure on reproduction and development. Consistent with recommendations for assessment of exaggerated pharmacology for general toxicity studies [29] and with DART testing of NBEs (ICH S6(R1)) [12], the use of one pharmacologically relevant species for reproductive toxicity testing should be sufficient to evaluate potential pharmacology-related effects on development, if only one species is available. However, in contrast to NBEs, in those cases where the clinical candidate is active in only one of the standard species used for EFD studies (rat or rabbit), the EFD study in the second species is still warranted to thoroughly assess effects related to ON chemistry. If the ON is pharmacologically inactive in both the rodent and rabbit, studies in these standard species are still considered of value for testing effects of chemical structure, particularly for ONs whose chemistry has not been tested previously in reproductive toxicity studies. However, when the clinical ON candidate lacks activity in either rats or rabbits, developers will need to consider other options to assess reproductive and developmental effects stemming from pharmacology, such as animal-active analogues, as discussed in the next section.

In general, NHPs have not been used in the DART assessment of ONs, and it is the opinion of the Reproductive OSWG Subcommittee [5] that NHPs should be used only for unique cases where there is a clear cause for concern, the clinical candidate is only active in NHPs, and animal-active analogues are not available. This is different than current regulatory guidance for NBEs as stated in ICH S6(R1), where studies in NHPs using the clinical drug candidate are preferable over use of an animal-active analogue [12]. The difference for ONs is that, as described above, there is value in using standard DART species to assess the potential effects of chemical structure of ONs, so addition of an animal-active analogue to the same study is the most practical means of also evaluating effects of target pharmacology. Additionally, gauging the comparability of an animal-active analogue to the clinical candidate is more straightforward for ONs with defined chemical structures than for NBEs with heterogeneous structures.

11.5.1 Design and Use of Animal-active Analogues

Use of an animal-active analogue is an option for addressing reproductive effects related to the intended pharmacology if the clinical candidate has an endogenous target but lacks homology and pharmacologic activity in either rodent or rabbit. This approach allows evaluation of the impact of the expected pharmacology on pregnancy and embryo–fetal development [30] and the effects related to chemical structure in one robust study. Typically, this approach follows the same expectations and strategy as for general toxicity studies, where pharmacological assessment is conducted in at least one species [29]. Most often, this assessment would be conducted in a rodent species because a mouse or rat analogue would likely have been used in the pharmacology and general toxicity studies. An advantage of the use of the rodent analogue is that all aspects of the reproductive cycle, including fertility and postnatal development, can be evaluated. In the selection of the analogue, the structure of the clinical candidate and the analogue should be as close as possible in both the chemical composition (length and modifications) and chemical class to facilitate interpretation of the data [29]. The use of an analogue is an acceptable strategy for ONs because it allows evaluation of effects related to both the chemical structure and the pharmacologic activity of the ON in larger numbers of animals, allowing a better characterization of risk than a corresponding smaller study in NHPs. Although animal-active analogues have the same chemical backbone as the clinical candidate, they may sometimes produce a different pattern of toxicity unrelated to target pharmacology [29]. Thus, when using analogues, the toxicity and TK profile of the surrogate should be characterized and shown to be similar to the clinical candidate.

One strategy for conducting reproductive toxicity assessments with rodent analogues is to conduct rodent studies using multiple doses of the clinical candidate to establish effect and no-effect dose levels related to the chemical structure and also include a satellite group treated with a single dose of the analogue (for examples, see Ref. [5]). The EFD study in the second species would only evaluate the clinical candidate. The dose level selected for the analogue in rodents should be at a level that causes the expected PD effect or exaggerated PD effect. If the pharmacologically active ON has no effects on reproductive or developmental endpoints, then the evaluation is complete for the rodent study. If an effect on reproduction or development is seen with evaluation of only a single-dose level of the analogue in the rodent, more work would likely be needed to evaluate dose–response relationships and establish a no-effect level. Therefore, if there is a high level of concern for a given target with regard to reproductive toxicity, it might be more useful to characterize the dose–response for the analogue up front to ensure that PD effects are well characterized. In that case, a targeted assessment of the clinical candidate at just the highest dose level(s) might be possible. This could be an especially good strategy for candidates from one of the more well-characterized structural classes for which the reproductive toxicity potential of the chemical structure has already been thoroughly evaluated.

11.6 Justification of Dosing Regimen

For DART studies, the dosing regimen is generally selected to produce adequate PK and PD exposure throughout the period of organogenesis and some degree of maternal toxicity at the high dose (ICH S5(R2)) [8]. The standard DART species (rodents and rabbits) have very short gestational periods (3–4 weeks), and daily dosing has typically been employed in these species to ensure that all periods of embryo–fetal development are adequately assessed. For NCEs, a daily dosing regimen is often consistent with their PK and PD half-lives and with the clinical dosing frequency. However, selecting a dose regimen for ONs in DART studies can be challenging. For many ONs, plasma half-lives are short, but tissue and PD half-lives tend to be quite long, and clinical dosing may be infrequent (e.g. weekly). Dosing regimens in nonclinical general toxicity studies generally mimic the clinical regimen. Infrequent dosing in DART studies might not produce adequate exposure to the fetus and the mother, but daily dosing could be problematic with ONs. Dose-limiting toxicities are often related to accumulation of ONs within tissues, and daily dosing could cause unacceptable levels of maternal toxicity [3, 31]. In addition, if the PD effect is prolonged, more frequent dosing could lead to exaggerated pharmacology or even disrupt the intended pharmacology in a way that would not be representative of the intended effect in humans. On the other hand, if the test species is not pharmacologically responsive to the test article (often true for the rabbit), the goal of the study would primarily be to evaluate toxicity of the chemical structure, and questions could be raised about the adequacy of exposure if daily dosing is not employed. In practice, the dosing regimen for DART studies of ONs should be selected to achieve a balance between adequate exposure during gestation, evaluation of prolonged PD effects, and management of maternal toxicity (related to excessive tissue accumulation of the ON).

For ONs whose PD requires a once-daily clinical dosing regimen, daily dosing should be used for the DART studies as well. Daily dosing should also be considered for ONs whose chemical structure is not well understood and ONs that show relatively low toxicity with repeated administration, regardless of the clinical dosing regimen. For ONs that are of a structural class that is metabolically stable and has a long tissue half-life and extended PD effects (e.g. PS ONs), where infrequent (e.g. weekly) dosing is used in the general toxicity and clinical studies, a dose range-finding study is recommended to evaluate whether daily dosing can be achieved in DART studies without eliciting excessive maternal toxicity. In cases where no toxicity occurs upon daily dosing, the highest dose could be a multiple of the highest anticipated human dose as is the case for biologics. In cases where daily dosing would cause unacceptable toxicity, one possibility is to administer the same weekly dose as in the general toxicity studies but as smaller doses given on a more frequent basis (daily or several times per week). This would ensure repeated exposure in the plasma compartment as frequently as possible throughout organogenesis while limiting tissue accumulation and adverse effects on PD. Additional work, such as measurement of target organ tissue concentrations, might be needed to ensure that the dosing regimen provides a similar weekly maternal tissue exposure as the single weekly dose (for examples, see Ref. [5]). One drawback to a fractionated dosing scheme is that it could result in a lower maternal plasma maximum concentration (Cmax) relative to the same dose given on a weekly basis; hence, understanding the relationship between known or predicted toxicities and plasma Cmax versus tissue exposure is important.

For ONs whose PD and/or toxicity is related to plasma concentration, a dosing regimen involving dose fractionation may not be advisable. In that case, alternative study designs could be considered. For example, the use of multiple cohorts with staggered dosing at dose levels used in the repeat-dose toxicity studies may be more appropriate, that is, each cohort receives the full dose at the clinical dosing interval to achieve the Cmax, and different cohorts are dosed on different gestation days or ranges of days to ensure dosing at each point during organogenesis. However, this study design is complicated and resource intensive, so this approach should be taken only if clearly warranted. Regardless of the type of ON, the objective is to achieve adequate exposure to both the PD effects and the chemical structure at each stage of organogenesis, and the approach should be justified.

11.7 Exposure Assessment

Assessment of exposure in the reproductive toxicity studies may be warranted to enable a correlation with general toxicity studies and to document the expected exposure in the study, but the optimal measure of exposure may differ for specific structural classes. In the case of stable single-stranded PS ONs, measurement of (maternal or paternal) target organ concentration (the liver and/or kidney) might be the most relevant means of assessing this exposure. For these ONs, the PK and tissue distribution properties are quite consistent from one sequence to another within a given chemical structural subclass. A detailed assessment of area under the plasma concentration versus time curve (AUC) may not be necessary if data already exist from other studies in the same species. The measurement of target organ concentration is also a useful way to compare the total exposure in the DART study with the general toxicity studies if the dose regimens are different. For other classes of ONs (such as PEGylated aptamers) or those using sophisticated delivery formulations (such as liposome-formulated siRNAs), the exposure measure should be tailored to the specific PK properties.

Depending on the nature of the ON product, the concentration of ON in placenta or fetal tissues could also be assessed, if this information was determined to be important for interpretation of study results. Placental transfer has been determined for several of the early single-stranded antisense PS ONs because the transplacental transfer properties were not known. These studies revealed poor placental transfer with generally low concentrations measured in the placenta and less than detectable exposure to the fetus [22, 23]. Therefore, pharmacological effects in fetal tissues were predicted to be low for similar structural and chemical subclasses. Because of the consistent sequence-independent PK properties, such measurements may not be needed for all classes or may be determined initially for a specific chemical class, and the results applied to other members of the class. It should be noted that ICH S5(R2) does not make any specific reference to the need for evaluating fetal exposure to therapeutic products, and fetal exposure is not routinely evaluated for NCEs. In this context, exposure assessment might be most useful as a confirmatory exercise when the intended pharmacology of the ON predicts that effects on embryo–fetal development would occur, but no effects are observed in DART studies.

11.8 Subclass-specific Considerations

Different subclasses of ONs have their own specific attributes, which are also important to consider when designing DART studies. This section reviews considerations specific to particular subclasses of ONs and provides examples from the published literature, where available. Additional examples of approaches taken with subclasses of ON therapeutics can be found in the OSWG Reproductive Subcommittee report [5].

The points to consider and subclass-specific recommendations given below seek to balance evaluation of potential DART effects related to chemical structure with those related to target pharmacology. In this context, it is important to leverage all available information about the ON, including general toxicity, PK and PD profiles, and information about predecessor ON molecules. There are specific patterns of toxicity related to specific ON chemical backbone structures, and it is likely that this also holds true for DART effects; therefore, experience with other related molecules should also be considered. For example, if a sponsor is developing several ONs with the same chemical backbone structure, it is possible that information gained from testing one of the ONs in a full DART package using three dose levels in two species might be able to partially support the development of other chemically related ONs. In such circumstances, DART studies with subsequent closely related ONs could be designed to focus on effects specific to the target pharmacology.

11.8.1 Single-stranded DNA Antisense Oligonucleotides

ASOs are designed to enter cells, bind mRNA, and decrease the level of protein expression through degradation of the mRNA. Although some mRNA targets have sequence homology between human and standard laboratory animal species, most do not. Therefore, alternative designs, such as the use of animal-active analogues, will likely be needed to adequately assess PD effects on development and reproduction. For some ASOs, particularly second- and third-generation products with enhanced stability, PD effects can be long lasting, and clinical dosing regimens are typically not daily, requiring careful planning of dosing regimens in DART studies.

An ASO inhibitor of human ICAM-1 has been evaluated for reproductive and developmental toxicity in mice and rabbits [22, 23]. This ASO did not have sequence homology to rodent or rabbit, and a mouse-active analogue was employed to evaluate PD effects on development. Three dose levels of the analogue were compared with the same dose levels of the clinical therapeutic in a combined FEED/EFD study design in order to distinguish potential effects related to chemical structure from effects related to pharmacology. No reproductive or developmental effects were produced that could be attributed to the pharmacology of the ASO, and fetal exposure of the ASO was very limited, justifying the use of the clinical candidate alone in a subsequent mouse PPND study. The clinical ASO was also evaluated in a rabbit EFD study, as no rabbit-active analogue was available. The clinical dosing regimen was three times weekly; in the DART studies, daily doses were administered in order to maximize exposure. Maternal toxicity occurred in both studies consistent with class effects for PS ASOs but was not severe enough to preclude daily dosing. No developmental toxicity was seen other than a reduction in fetal body weights in rabbits in the presence of maternal toxicity. These results were consistent with the very low fetal exposure measured in both species [22, 23] and in the lack of concern for DART effects related to target modulation based on available KO mouse information [22]. In other studies by this group using longer-acting ASOs (once-weekly dosing in the clinic and general toxicity studies), dosing in DART studies was modified to an every-other-day dosing schedule, using fractionated doses that in total equaled the weekly doses in general toxicity studies [5]. In most cases, studies were designed with a full dose–response of the human ASO and only a single dose of the mouse-active analogue at a pharmacologically active level. As mentioned above, the choice of whether to evaluate a full dose–response of the clinical candidate, the animal-active analogue, or both, will depend on the amount of information already generated for the chemical backbone of the clinical candidate as well as the concerns for DART effects related to target pharmacology.

Another known class effect of ASO is their proinflammatory properties, leading to classic toxicities including vasculitis and glomerulonephritis [3, 32].

Proinflammatory activities have the potential for affecting the maintenance of pregnancy (see below). However, there is little evidence so far, either in the published literature or for a recently approved ASO therapy (Mipomersen), to suggest that reproductive effects related to immune modulation by ASOs have occurred in animal DART studies.

11.8.2 CpG and Immunostimulatory (IS) Oligonucleotides

CpG ONs (also known as CpG oligodeoxynucleotides [CpG ODNs]) are designed to stimulate the immune system through activation of toll-like receptor (TLR)-9 and subsequent induction of cytokines and chemokines [33]. They are generally active in humans and multiple animal species; therefore, animal-active analogues are generally not required for testing pharmacologic activity.

The mechanism of action of CpG ONs raises concerns for potential effects on reproduction and development. The activation of TLR9 in B cells, plasmacytoid dendritic cells (pDC), and other immune cells elicits a rapid and massive production of interferon (IFN), which ultimately leads to production of a broad range of cytokines by T cells, B cells, and monocytes (reviewed in Ref. [34]). High levels of IFN are known to interfere with the maintenance of pregnancy, as exemplified by the abortifacient activity reported for marketed IFN biopharmaceuticals in several species including primates (see prescribing information for Betaseron [IFN β1b], Actimmune [IFNγ], Intron A [IFNα], and Avonex [IFN β1a]). Additionally, developmental effects have been noted for Aldara, which is a proinflammatory adjuvant that stimulates production of IFNs through activation of TLR7. A small number of studies have also linked IFN treatment with adverse pregnancy outcomes in clinical populations [35, 36]. This information suggests that the pharmacology of therapeutic CpG ONs could have adverse effects on maintenance of pregnancy.

Although the number of therapeutic CpG ONs evaluated for DART effects is limited, an academic research study in mice suggests that CpG ONs can affect reproduction and development through TLR9 activation. Administration of a CpG ON to mice on gestational day (GD) 10–14 resulted in fetal resorption and/or preterm birth [37, 38], which were dependent on the presence of functional TLR9 in the mice [38]. Cranial and distal limb malformations were also produced in these mice when the CpG ON was administered GD 6 or GD 14 [37, 39], which were dose dependent and not produced by a control ON [37, 39, 40].

Nonclinical species are likely to respond to CpG ON-mediated immune stimulation, but there is some evidence that species differences could exist in this response [41]. As a result of a broader cellular distribution of TLR9, activation in rodents produces a broader spectrum of cytokines than in humans and NHP, which may result in overestimating the risk associated with CpG ONs [42]. In contrast, rabbits may be less responsive to CpG ON-dependent TLR9 activation [43]. Because rabbits can exhibit quantitatively or quantitatively different responses to IS ONs, as compared with rodents, performing an advance screen on rabbits may be prudent to ensure a relevant pharmacologic response to the CpG ONs prior to selecting this species for use in DART studies.

The mode of action of CpG ONs depends on a highly coordinated series of activities, including initial activation of cells, massive production of IFN, and a number of downstream events. CpG ONs are typically not delivered daily, in order to give the immune system time to reset and to achieve maximum responses. This profile of activity could be challenging for designing an appropriate dosing regimen for DART studies. As discussed previously, an infrequent dosing schedule may not provide adequate exposure throughout the period of organogenesis. On the other hand, a fractionated dose could lower the Cmax, thereby dampening the initial PD response and subsequent downstream responses, which could underestimate the risk of adverse effects. It is, therefore, important that sponsors use all available PD and PK information for their CpG ON and other related molecules to plan DART studies that balance evaluation of chemical structure with that of target pharmacology. It would also be helpful to leverage information on other ONs within this class when assessing DART risks. As mentioned in the "Dosing Regimen" section, it is possible that study designs such as staggered dosing cohorts could be used. However, these designs are complex and should only be used if clearly warranted.

11.8.3 microRNA Mimetics/Antagonists and siRNAs

microRNA mimetic/antagonist and siRNA ONs have been less well studied, and information on their DART evaluations is limited [5]. However, the same principles outlined for other ON therapies would also apply to these therapeutics (i.e. evaluation of chemical structure and pharmacologic action; maximization of exposure during development). Both types of ONs operate through engagement of the cytoplasmic RISC complex responsible for processing of mRNA [32, 44, 45]. However, their intended pharmacologies differ, and they comprise a relatively broad array of chemical structural modifications and complex formulations, which will need to be considered when designing DART evaluations for these molecules.

microRNA-based ON therapeutics are being designed as both antagonists and mimetics. For microRNA antagonists, the mechanism of action is to hybridize with endogenous microRNAs and inhibit their function. These compounds are often structurally similar to traditional antisense ONs. Specifically, they are generally 12–20 nucleotides in length, fully modified on the 2′-position and have PS linkages [45]. Therefore, the PK, tissue distribution, and toxicity profile are generally similar to the classes of PS ONs that have been well studied, and similar considerations for testing of chemical structure would apply.

Unlike PS ASOs, the sequence homology for microRNAs ON antagonists is often well conserved across species [46], allowing target pharmacology effects to be evaluated using the clinical candidate in at least one of the standard species used for embryo–fetal developmental toxicity studies.

MicroRNA mimetics differ structurally from antagonists in that they are double-stranded molecules, which do not have the same chemical modifications as microRNA antagonists or PS ASOs; thus, different toxicities related to chemical structure would be expected. microRNA mimetics are not intended to reduce gene expression but, rather, to modulate a broad array of targets linked to the specific microRNA. Since natural microRNAs do not show perfect sequence homology with the mRNA sequences they inhibit [47], it would be reasonable to assume that the microRNA mimetics would exhibit cross-species activity. However, this would need to be considered on a case-by-case basis.

siRNAs are also double stranded and exert their effects via reduction in gene expression [48]. Some of the siRNAs in development are delivered as the free ON (unformulated), in which case they may be chemically modified. These applications are typically limited to local delivery (e.g. intraocular or intrapulmonary) and would have limited distribution to reproductive tissues. However, for many of the current siRNA development programs, particularly those involving systemic delivery, the siRNA is delivered via a complex formulation, and the ON chemistry is generally natural (unmodified) [48]. Toxicity of these formulated siRNAs has been observed and is believed to be largely attributable to the formulation excipients (i.e. cationic lipids in the formulation). Although some examples of pharmacology-based toxicity exist, the profile is dominated mainly by the excipient effects. In addition, the lead siRNA candidates chosen for these programs often exhibit activity in either rat or rabbit (or both), and, hence, reproductive effects related to pharmacology can potentially be assessed in one or both species. Therefore, the main challenge to proper conduct of reproductive toxicity studies with such products is to determine the appropriate dosing frequency, as is the case for other subclasses of ONs that are not administered daily. Although experience to date is minimal, the highest dose levels that can be tested in reproductive toxicity studies with formulated siRNAs are predicted to be limited by the excipient-related toxicity, and large clinical multiple-dose levels may not be achievable. Despite this limitation, DART testing should be done with the final formulated product rather than the free siRNA, as the blood PK and tissue distribution will be dramatically different from those of the formulated siRNA (e.g. the plasma half-life may be several minutes for unformulated siRNAs versus hours for the formulated product).

As with ASO, siRNAs and microRNA mimetics/antagonists may have prolonged PD effects and may not be administered daily in the clinic or in nonclinical studies. Therefore, although species sequence homology would allow the use of standard species in DART testing, some thought will need to be given to the dosing regimen to balance adequate cover during gestation with the risks of overdosing.

An additional consideration with siRNAs and microRNA mimetics/antagonists is the potential for immunostimulation and the consequences of this on reproduction. microRNA ONs with PS-modified backbones would have a similar theoretical risk for immunostimulation as described previously for the ASO. The double-stranded siRNA and microRNA mimetics have the potential to stimulate the innate immune system through unintentional activation of TLR3, 7, and 8. Lin et al. demonstrated that dsRNA can increase the resorption rate in mice due to TLR3 stimulation [49]. Similar results were obtained with a synthetic dsRNA, poly(I:C) [50, 51]. Treatment of pregnant mice with TLR3, 7, or 8 agonists can cause pregnancy-dependent hypertension and maternal changes similar to preeclampsia [52]. Potential IS properties of siRNA and microRNA therapies should be evaluated prior to the planning and conduct of DART in order to incorporate this information into the design of DART studies, if warranted.

11.8.4 Aptamer Oligonucleotides

Aptamers are a unique subclass of ONs with respect to their mechanism of action and structure (reviewed in Refs. [31, 48]). Aptamers are generally single-stranded RNA molecules that may have structural alterations, such as PEGylation, in order to enhance binding and increase stability. Their mode of action involves direct binding to proteins or other macromolecules, inhibiting or otherwise altering their activity. Thus, although these molecules are composed of nucleotide sequences, they do not function through hybridization with genetic materials inside the cell. Some aptamers are being developed as single-dose agents (e.g. those that target clotting proteins) or as intermittent local therapies (e.g. intraocular administration on a monthly basis), and such programs may not warrant reproductive toxicity studies. Those aptamers intended for repeated systemic administration in WOCBP should be evaluated for reproductive effects.

As with other ONs, if the aptamer exhibits robust cross-species pharmacologic activity, conducting the studies in one or more standard species should suffice for evaluating potential PD-related DART effects. However, if cross-species activity is not demonstrated with the clinical candidate, evaluating pharmacology-related effects of aptamers may be more challenging than for other ON therapeutics. Chemical modifications are frequently used to alter binding characteristics. Therefore, if an animal-active analogue exists, it may have both a different chemical structure and a different nucleotide sequence to the clinical candidate. If this is the case, then use of an animal-active analogue may not be as scientifically sound as use of an analogue for other types of ON therapeutics. Under these circumstances, consideration could also be given to conducting DART studies in NHPs.

One therapeutic aptamer has been approved by the FDA. Macugen (pegaptanib) is a PEGylated modified ON that binds to VEGF and is approved for the treatment of wet age-related macular degeneration (AMD) by intravitreal (IVT) injection. Initially, a limited reproductive toxicology program was conducted due to the advanced age of the patient population (median age is over 70 years old) and the limited systemic exposure after IVT injection (for more details, see Ref. [5]). Due to a potential change in indication and DART concerns over the importance of VEGF in embryo–fetal development, regulatory authorities subsequently requested an EFD. Macugen showed pharmacologic activity in mice, rats, rabbits, dogs, and NHPs, and these species were used in acute, subchronic, and chronic studies toxicity studies using subcutaneous, IVT, and/or intravenous (IV) dosing routes. These studies demonstrated that Macugen was well tolerated, following both systemic and IVT administration, and a maximum-tolerated dose was not established. Based on this information, Macugen was able to be tested directly in an IV mouse EFD study using a daily dosing regimen. Although Macugen was shown to be able to cross the placenta, it did not produce teratogenicity or effects on embryo–fetal survival at up to 7000 times the human dose after IVT administration (Macugen pharmaceutical label) [53].

11.9 Conclusions

ONs have attributes that are similar to NCEs and NBEs, and both sets of guidelines can be useful when planning DART studies. The approach to DART testing should take into account both the potential effects of chemical structure and the intended pharmacology of the ON. Additionally, testing should be designed to adequately assess exposure throughout organogenesis, even for ONs that are typically not administered daily. Standard reproductive toxicity species (rodent/rabbit) have been historically used to evaluate DART effects of ONs, and this is appropriate given that toxicity of ON therapeutics tends to be related to chemical structure rather than pharmacology. However, inclusion of study designs that allow testing of effects related to the pharmacology target are also very important in evaluating DART effects of the ON. Animal-active analogues can be useful to assess potential reproductive effects related to pharmacology in addition to effects related to the chemical backbone. The choice of the relevant animal model, the dosing regimen, and whether to use the clinical candidate or an animal analogue will all need to be considered carefully based on the specific product attributes of the ON. Information from general toxicity studies and previous experience with similar ONs can help to inform these decisions. For various reasons outlined above, NHP studies should be used only if needed to answer specific questions or when animal-active analogues are not practical. Dosing regimens should be tailored to ensure adequate exposure throughout the period of organogenesis without compromising the ability to assess PD- and PK-related effects.

The lack of historical precedent in the assessment of these novel products allows for creativity and scientific expertise to play a major role in the design and interpretation of studies. Although ICH guidelines provide a common international basis for the basic reproductive toxicity protocol development process, they also allow for the necessary flexibility required in order to evaluate ONs in a scientifically valid manner. Ultimately, the successful development of ONs will depend on the use of an appropriate experimental model in each case to enable the predictive value of the preclinical safety evaluation, thus providing optimal safety information for the patients.

Acknowledgments

The authors would like to thank the other members of the Reproductive Subcommittee of the Oligonucleotide Safety Working Group: Cynthia Berman, Doug Kornbrust, Sarah Campion, and Scott Henry. Their collective input, experience, and knowledge of DART evaluations of ON therapies formed the basis of the OSWG reproductive toxicity white paper [5] and this chapter.

References

1 Cavagnaro, J.A. (2008). The principles of ICH S6 and the case-by-case approach. In: *Preclinical Safety Evaluation of Biopharmaceuticals: A Science-based Approach to Facilitating Clinical Trials*, 45–65. Hoboken: Wiley.
2 Black, L.E., Farrelly, J.G., Cavagnaro, J.A. et al. (1994). Regulatory considerations for oligonucleotide drugs: updated recommendations for pharmacology and toxicology studies. *Antisense Research and Development* 4: 299–301.
3 Frazier, K.S. (2015). Antisense oligonucleotide therapies: the promise and the challenges from a toxicologic pathologist's perspective. *Toxicologic Pathology* 43: 78–89.
4 Schubert, D., Levin, A., Kornbrust, D. et al. (2012). The Oligonucelotide Safety Working Group (OSWG). *Nucleic Acid Therapeutics* 22: 211–212.
5 Cavagnaro, J., Berman, C., Kornbrust, D. et al. (2014). Considerations for assessment of reproductive and developmental toxicity of oligonucleotide-based therapeutics. *Nucleic Acid Therapeutics.* 24: 313–325.
6 Cavagnaro, J.A. (2002). Preclinical safety evaluation of biotechnology-derived pharmaceuticals. *Nature Reviews Drug Discovery* 1: 469–475.
7 Cavagnaro, J. (2010). Considerations in the Preclinical Development of Biopharmaceuticals. In: Comprehensive Toxicology, Toxicity Testing & Evaluation, 2e, vol. 3 (ed. C.S. McQueen), 29–52. Elsevier.

8 ICH (2005). ICH Topic S5(R2): Detection of toxicity to reproduction for medicinal products & toxicity to male fertility. http://www.ich.org/fileadmin/Public_Web_Site/ICH_Products/Guidelines/Safety/S5/Step4/S5_R2__Guideline.pdf (accessed 6 February 2018).

9 Lerman, S.A., Hew, K.W., Stewart, J. et al. (2009). The nonclinical fertility study design for pharmaceuticals. *Birth Defects Research B Developmental Reproductive Toxicology* 86: 429–436.

10 Wise, L.D., Buschmann, J., Feuston, M.H. et al. (2009). Embryo-fetal developmental toxicity study design for pharmaceuticals. *Birth Defects Research B Developmental Reproductive Toxicology* 86: 418–428.

11 Bailey, G.P., Wise, L.D., Buschmann, J. et al. (2009). Pre- and postnatal developmental toxicity study design for pharmaceuticals. *Birth Defects Research B Developmental Reproductive Toxicology* 86: 437–445.

12 ICH (2011). ICH Topic S6(R1): Addendum to S6 Preclinical Safety Evaluation of Biotechnology-derived Pharmaceuticals. http://www.ich.org/fileadmin/Public_Web_Site/ICH_Products/Guidelines/Safety/S6_R1/Step4/S6_R1_Guideline.pdf (accessed 6 February 2018).

13 Chellman, G.J., Bussiere, J.L., Makori, N. et al. (2009). Developmental and reproductive toxicology studies in nonhuman primates. *Birth Defects Research B Developmental Reproductive Toxicology* 86: 446–462.

14 Martin, P., Breslin, W., Rocca, M. et al. (2009). Considerations in assessing the reproductive and developmental toxicity potential of biopharmaceuticals. *Birth Defects Research B Developmental Reproductive Toxicology* 33: 176–203.

15 Martin, P. and Weinbauer, G. (2010). Developmental toxicity testing of biopharmaceuticals in nonhuman primates: previous experiences and future directions. *International Journal of Toxicology* 29: 552–568.

16 Weinbauer, G.F., Fuchs, A., Niehaus, M., and Luetjens, C.M. (2011). The enhanced pre- and postnatal study for nonhuman primates: update and perspectives. *Birth Defects Research C Embryo Today* 93: 324–333.

17 Henck, J.W., Hilbish, K.G., Serabian, M.A. et al. (1996). Reproductive toxicity testing of therapeutic biotechnology agents. *Teratology* 53: 185–195.

18 ICH International Conference on Harmonisation of Technical Requirements for Registration of Pharmaceuticals for Human Use) (2009) ICH Topic M3(R2): Guidance on nonclinical safety studies for the conduct of human clinical trials and marketing authorization for pharmaceuticals. http://www.ich.org/fileadmin/Public_Web_Site/ICH_Products/Guidelines/Multidisciplinary/M3_R2/Step4/M3_R2__Guideline.pdf (accessed 6 February 2018).

19 ICH (2011). ICH Topic S9: Nonclinical Evaluation for Anticancer Pharmaceuticals. http://www.ich.org/fileadmin/Public_Web_Site/ICH_Products/Guidelines/Safety/S9/Step4/S9_Step4_Guideline.pdf (accessed 6 February 2018).

20 Levin, A.A. and Henry, S.P. (2008). Toxicology of oligonucleotide therapeutics and understanding the relevance of the toxicities. In: Preclinical Safety Evaluation of Biopharmaceuticals (ed. J.A. Cavagnaro), 537–574. Hoboken: Wiley.

21 Faustman, E.M., Gohlke, J.M., Ponce, R.A. et al. (2006). Experimental approaches to evaluate mechanisms of developmental toxicity. In: *Developmental and Reproductive Toxicology, A Practical Approach*, 2e (ed. R.D. Hood). Boca Raton: CRC Press.

22 Henry, S.P., Denny, K.H., Templin, M.V. et al. (2004a). Effects of human and murine antisense oligonucleotide inhibitors of ICAM-1 on reproductive performance, fetal development, and post-natal development in mice. *Birth Defects Research B Developmental Reproductive Toxicology* 71: 359–367.

23 Henry, S.P., Denny, K.H., Templin, M.V. et al. (2004b). Effects of an antisense oligonucleotide inhibitor of human ICAM-1 on fetal development in rabbits. *Birth Defects Research B Developmental Reproductive Toxicology* 71: 368–373.

24 Gaudette, M.F., Hampikian, G., Metelev, V. et al. (1993). Effect on embryos of injection of phosphorothioate-modified oligonucleotides into pregnant mice. *Antisense Research and Development* 3: 391–397.

25 Agrawal, S., Temsamani, J., and Tang, J.Y. (1991). Pharmacokinetics, biodistribution, and stability of oligodeoxynucleotide phosphorothioates in mice. *Proceedings of the National Academy of Sciences* 88: 7595–7599.

26 Soucy, N.V., Riley, J.P., Templin, M.V. et al. (2006). Maternal and fetal distribution of a phosphorothioate oligonucleotide in rats after intravenous infusion. *Birth Defects Research B Developmental Reproductive Toxicology* 77: 22–28.

27 Driver, S.E., Robinson, G.S., Flanagan, J. et al. (1999). Oligonucleotide-based inhibition of embryonic gene expression. *Nature Biotechnology* 17: 1184–1187.

28 Daston, G.P. (1994). Relationships Between Maternal and Developmental Toxicity, In: Developmental Toxicology, 2nd ed., C.A. Kimmel and J Buelke-Sam, eds., Raven Press, Ltd., New York, 189–212.

29 Kornbrust, D., Cavagnaro, J., Levin, A. et al. (2013). Oligo safety working group exaggerated pharmacology subcommittee consensus document. *Nucleic Acid Therapeutics* 23: 21–28.

30 Treacy, G. (2000). Using and analogous monoclonal antibody to evaluate the reproductive and chronic toxicity potential for a humanized anti-TNF alpha monoclonal antibody. *Human Experimental Toxicology* 19: 226–228.

31 Sundaram, P., Kurniawan, H., Byrne, M.E., and Wower, J. (2013). Therapeutic RNA aptamers in clinical trials. *European Journal of Pharmaceutical Sciences* 48: 259–271.

32 Frazier, K.S., Sobry, C., Derr, V. et al. (2014). Species-specific inflammatory responses as a primary component for the development of glomerular lesions in mice and monkeys following chronic administration of a second-generation antisense oligonucleotide. *Toxicologic Pathology* 42: 923–935.

33 Bauer, S., Kirschning, C.J., Häcker, H. et al. (2001). Human TLR9 confers responsiveness to bacterial DNA via species-specific CpG motif recognition. *Proceedings of the National Academy of Sciences U S A* 98: 9237–9242.

34 Lui, Y.J. (2005). IPC: professional type 1 interferon-producing cells and plasmacytoid dendritic cell precursors. *Annual Review of Immunology* 23: 275–306.

35 Kremenchutzky, M., Morrow, S., and Rush, C. (2007). The safety and efficacy of IFN-beta products for the treatment of multiple sclerosis. *Expert Opinions in Drug Safety* 6: 79–288.

36 Sandberg-Wollheim, M., Frank, D., Goodwin, T.M. et al. (2005). Pregnancy outcomes during treatment with interferon beta-1a in patients with multiple sclerosis. *Neurology* 65: 802–806.

37 Prater, M.R., Johnson, V.J., Germolec, D.R. et al. (2006). Maternal treatment with a high dose of CpG ODN during gestation alters fetal craniofacial and distal limb development in C57BL/6 mice. *Vaccine* 24: 263–271.

38 Scharfe-Nugent, A., Corr, S.C., Carpenter, S.B. et al. (2012). TLR9 provokes inflammation in response to fetal DNA: mechanism for fetal loss in preterm birth and preeclampsia. *Journal of Immunology* 188: 5706–5712.

39 Thaxton, J.E., Romero, R., and Sharma, S. (2009). TLR-9 activation coupled to IL-10 deficiency induced adverse pregnancy outcomes. *Journal of Immunology* 183: 1144–1154.

40 Ito, S.-I., Ishii, K.J., Shirota, H., and Klinman, D.M. (2004). CpG oligodeoxynucleotides improve the survival of pregnant and fetal mice following Listeria monocytogenes infection. *Infection and Immunity* 72: 3543–3548.

41 Campbell, J.D., Cho, Y., Foster, M.L. et al. (2009). CpG-containing immunostimulatory DNA sequences elicit TNF-α–dependent toxicity in rodents but not in humans. *Journal of Clinical Investigation* 119: 2564–2576.

42 Liu, J., Xu, C., Liu, Y.L. et al. (2012). Activation of rabbit TLR9 by different CpG-ODN optimized for mouse and human TLR9. *Comparative Immunology, Microbiology & Infectious Diseases* 35: 443–451.

43 Chuang, T.-H., Lai, C.-Y., Tseng, P.-H. et al. (2014). Development of CpG-oligodeoxynucleotides for effective activation of rabbit tlr9 mediated immune responses. *PLoS ONE* 9 (9): e108808.

44 Carthew, R.W. and Sontheimer, E.J. (2009). Origins and Mechanisms of miRNAs and siRNAs. *Cell* 136: 642–655.

45 Stenvang, J., Petri, A., Lindow, M. et al. (2012). Inhibition of microRNA function by antimiR oligonucleotides. *Silence* 3: 1–17.

46 van Rooij, E., Purcell, A.L., and Levin, A.A. (2012). Developing microRNA therapeutics. *Circulation Research* 110: 96–507.

47 Deleavey, G.F. and Damha, M.J. (2012). Designing chemically modified oligonucleotides for targeted gene silencing. *Chemistry & Biology* 19: 937–954.

48 Burnett, J.C. and Rossi, J.J. (2012). RNA-based therapeutics: current progress and future prospects. *Chemistry & Biology* 19: 60–71.

49 Lin, Y., Liang, Z., Chen, Y., and Zeng, Y. (2006). TLR3-involved modulation of pregnancy tolerance in double-stranded RNA-stimulated NOD/SCID mice. *Journal of Immunology* 176: 4147–4154.

50 Zhang, J., Wei, H., Wu, D., and Tian, Z. (2007). Toll-like receptor 3 agonist induces impairment of uterine vascular remodeling and fetal loses in CBA × DBA/2 mice. *Journal of Reproductive Immunology* 74: 61–67.

51 Koga, K. and Mor, G. (2010). Toll-like receptors at the maternal-fetal interface in normal pregnancy and pregnancy disorders. *American Journal of Reproductive Immunology* 63: 587–600.
52 Chatterjee, P., Weaver, L.E., Doersch, K.M. et al. (2012). Placental toll-like receptor 3 and toll-like receptor 7/8 activation contributes to preeclampsia in humans and mice. *PloS ONE* 7 (e41884): 1–9.
53 Macugen U.S. Pharmaceutical Label, NDA 21-756/S-018, Revised 07/2011, https://www.accessdata.fda.gov/drugsatfda_docs/label/2011/021756s018lbl.pdf (accessed 6 February 2018).

12

Specific Considerations for Preclinical Development of Inhaled Oligonucleotides

Nicolay Ferrari

Centre de recherche du CHUM – Tour Viger, Montreal, Quebec, Canada

12.1 Background

Airways and lung diseases are ideal for the use of therapeutic oligonucleotides. Administration to the site of action via inhalation permits direct exposure to oligonucleotides to the diseased tissue and target cells. In principle, this approach also permits the use of lower doses, which can aid in limiting systemic delivery and, consequently, reduce the likelihood of undesirable side effects.

Preclinical safety studies needed to support administration of inhaled oligonucleotides in clinical trials are no different than for other drugs and other routes of administration. The development plans normally follow the recommendations outlined in the relevant ICH guidelines. The inhalation route of administration should be employed in repeat-dose toxicology studies to mimic the intended clinical route of administration and to ensure the characterization of any potential adverse effects locally (i.e. respiratory tract) and to define the systemic effects of the oligonucleotide. However, it is generally acceptable to conduct selected studies such as carcinogenicity, reproductive toxicology, or certain safety pharmacology assessments using alternate route of administration, typically the intravenous route of administration. Oligonucleotides are rather large water-soluble molecules, ~7000 Da for single-stranded antisense oligonucleotides (ASO), and evidence suggest that the tissue of major accumulation following inhaled delivery of oligonucleotides is the lung, with limited amounts of the drug measured in plasma and other organs [1–3]. Therefore, the utilization of alternative route of administration may be prescribed when needing to maximize systemic exposure.

As mentioned above, changes in preclinical species following inhalation of oligonucleotides are predominantly respiratory tract findings that can include irritancy of the larynx and nasal cavity, particularly in rodents. The limited

Oligonucleotide-Based Drugs and Therapeutics: Preclinical and Clinical Considerations for Development, First Edition. Edited by Nicolay Ferrari and Rosanne Seguin.

systemic exposure of inhaled oligonucleotides presents some advantages. However, it places greater emphasis on the accumulation of the drug in the lungs and its elimination by macrophages. Consequently, an increase in macrophage numbers is often noted. In the absence of any other findings, this increase is generally considered to be a nonadverse, representing the physiological response to an inhaled particulate [4]. Other findings in the lung that can be observed following inhalation of oligonucleotides can include an inflammatory response and/or epithelial hyperplasia. This is believed to be resulting from irritancy or particulate (oligonucleotide) overload. An inflammatory response is a safety concern mainly because it is considered not monitorable in humans [5] (reviewed in other chapters). For such findings, however, safety margins can be obtained. These are important factors that should be taken into account when designing preclinical studies or programs for inhaled oligonucleotides.

Toxicity studies using the inhalation route present unique challenges compared with those conducted via other routes of administration. Regardless of the drug class, oligonucleotides or small molecules, dosing by the inhalation route of administration is particularly complex. For example, the efficacy of a drug may be affected by its deposition locale in the respiratory tract, its delivered dose, and the disease it may be trying to treat. Therefore, inhalation toxicity studies require considerable preliminary work to generate suitable formulations (for example, nebulized solutions/suspensions, powdered blends, or metered dose inhaler) and reproducible test atmospheres. Because of the use of specific delivery systems, significant time (at least 2 weeks for nonhuman primates) is required to allow the animals to be accustomed to the animal restraint procedure and dosing apparatus (e.g. a mask and aerosol delivery tube). Furthermore, during the dosing phase of a study, chamber concentrations and particle size need to be controlled regularly. These extra procedures ultimately make the studies labor intensive and consequently more expensive. Many other factors affecting potential toxicity of inhaled oligonucleotides are common to other routes of administration and include biological (strain, age, and health of animals) and environmental (temperature, housing, humidity) factors.

This chapter aims to provide some guidance when planning the preclinical assessment of inhaled oligonucleotide therapeutics. The chapter will focus on repeat-dose toxicity and safety pharmacology studies. The potential for genotoxic effects of oligonucleotides is being discussed in a separate chapter.

12.2 Oligonucleotide Delivery Systems

12.2.1 Inhalation Exposure Systems

There are different dosing systems available for the preclinical assessment of inhaled oligonucleotides. To date, most oligonucleotides developed for the treatment of respiratory diseases have been delivered to the lungs successfully as aerosols generated from nebulized solutions, e.g. saline for inhalation, USP (0.9%). Inhalation systems are typically monitored by determination of aerosol concentrations, particle size distribution, oxygen concentration, relative humidity, and temperature on representative atmosphere samples. Close monitoring is necessary as all these factors may affect particle size and shape. The therapeutic effect of aerosolized oligonucleotides, as for any other inhaled therapy, is dependent upon the dose deposited and its distribution within the lung. Aerosol particle size is one of the most important variables in defining the dose deposited and the distribution of drug aerosol in the lung. It can be generally considered that particles with a mass median aerodynamic diameter (MMAD) higher than 10 μm are deposited in the oropharynx, those measuring between 5 and 10 μm in the central airways, and those from 0.5 to 5 μm in the small airways and alveoli. Therefore, for topical respiratory treatment it is best to use particles with an MMAD between 0.5 and 5 μm although it is desirable to target 1–3 μm with a geometric standard deviation (GSD) (an index of dispersity) of 2–3 μm [6, 7].

Another challenge during the preclinical development of inhaled oligonucleotides, unlike most other routes of administration, is that the exact dose administered by inhalation (mg kg^{-1}) is not known but is estimated. To estimate the delivered dose, the respiratory minute volume (RMV) (l min^{-1}), airstream concentration of the test material (mg l^{-1}), duration of daily exposure (minutes), body weight (kg), and inhaled fraction (%) are taken into account using the following formula:

$$\text{Dose} = \frac{\text{RMV} \times \text{concentration} \times \text{duration} \times \text{inhaled fraction}}{\text{body weight}}$$

Of these parameters, airstream concentration, body weight, and exposure time will be known. Some laboratories have the capability to determine RMV during a study; however, more often it is estimated based on a function of body weight [8, 9]. For simplicity, the inhaled fraction (that is, particles smaller than 7 μm) is assumed to be 100%.

12.2.2 Intratracheal Aerosol Instillation

For various reasons, the inhalation route of administration cannot always be used, and the direct instillation of a test material into the lungs via the trachea has been employed in many studies as an alternative exposure procedure [10]. There are distinct differences in the distribution, clearance, and retention of materials when administered by intratracheal instillation compared with inhalation. The former, therefore, may not be an optimal method that should be used routinely to characterize the toxicity of inhaled oligonucleotides in preclinical studies. This method is nonphysiological as it involves invasive delivery to anesthetized animals. In addition, the distribution of the test compound throughout the lungs may not be uniform. Further, the dose and/or dose rate are usually substantially greater than would be observed after inhalation, and the upper respiratory tract is bypassed [10]. When these limitations are well understood, intratracheal instillation can be a useful and cost-effective (may be used to deposit a precise amount of oligonucleotides directly into the lungs) procedure for addressing specific questions regarding the respiratory toxicity of chemicals. An important advantage of this method of delivery is the requirement of significantly smaller amounts of oligonucleotides. Therefore, while it may not completely replicate the same pulmonary distribution after inhalation, delivery via intratracheal aerosol instillation using a microsprayer could be useful in providing information on the relative toxicity between oligonucleotides and over a range of doses [11, 12]. For example, intratracheal aerosol instillation can be used to evaluate toxicity potential of various oligonucleotide sequences or various chemistries (Table 12.1).

12.3 Repeat-dose Toxicity

12.3.1 General Principles

The general principles of the design for inhaled oligonucleotides toxicology studies are the same as for any other route of administration and should include an evaluation of both systemic and local (i.e. respiratory tract) effects of the drug. One potential difference is that acute toxicity studies for inhaled drugs are generally not performed. The duration of repeat-dose toxicity studies depends on the duration of the proposed therapeutic use in humans and should be consistent with the regulatory guidance ICH M3(R2) and ICH S4B for chronic toxicity testing.

Regulatory requirements for the assessment of extremely high doses can be obtained from dose-escalating or dose-ranging studies, which are performed to aid in the selection of the high doses for subsequent subacute and chronic repeat-dose studies. Similar to other drug therapies, toxicology studies for inhaled oligonucleotides should be performed in one rodent and a nonrodent species and should include both sexes. For inhaled oligonucleotides, as for other routes of administration with this drug class, nonhuman primates are often selected as the nonrodent species. Due to some inherent properties of oligonucleotides (for example, the proinflammatory effects of the phosphorothioate class, reviewed elsewhere in this book), the selection of the rodent species, either rat or mouse, should be carefully considered.

Exposure should mimic the intended clinical frequency of administration and the toxicological/pharmacokinetic/pharmacodynamic profile of the oligonucleotide compound, normally once-daily dosing would be adequate. In some cases more frequent administration in animals than what is anticipated in the

Table 12.1 Summary of changes in bronchoalveolar lavage and lung tissue following intratracheal aerosol instillation of antisense oligonucleotides.

	PBS	TOP1572	TOP1572-PO7-DAP		TOP1731	TOP1731-PO7-DAP	TOP1769-PO7-DAP	TOP1777-PO7-DAP
Dose ($mg\,kg^{-1}$)	—	2.5	2.5	10	2.5	2.5	2.5	2.5
BAL								
Total cells	2.7×10^5 (6.4×10^4)	1.1×10^6 (2.7×10^5)	3.4×10^5 (7.9×10^4)	4.0×10^5 (8.6×10^4)	4.0×10^6 (1.1×10^6)	3.0×10^5 (5.5×10^4)	3.1×10^5 (9.7×10^4)	2.7×10^5 (1.7×10^5)
Neutrophils	4.3×10^3 (2.7×10^3)	8.4×10^4 (6.9×10^4)	1.7×10^4 (5.3×10^3)	5.8×10^4 (2.1×10^4)	8.9×10^5 (6.2×10^5)	3.9×10^3 (2.6×10^3)	2.2×10^4 (1.1×10^4)	3.2×10^3 (2.6×10^3)
Macrophages	2.6×10^5 (5.8×10^4)	9.1×10^5 (3.1×10^5)	3.1×10^5 (8.1×10^4)	3.2×10^5 (5.6×10^4)	2.4×10^6 (4.7×10^5)	3.0×10^5 (7.0×10^4)	3.0×10^5 (8.7×10^4)	1.3×10^5 (2.4×10^4)
Lymphocytes	8.1×10^3 (1.1×10^4)	1.9×10^5 (5.4×10^4)	1.1×10^4 (5.0×10^3)	2.1×10^4 (1.0×10^4)	6.8×10^5 (2.5×10^5)	9.4×10^3 (3.8×10^3)	2.0×10^4 (1.5×10^4)	3.2×10^4 (1.4×10^4)
% large macrophages	6.0 (1.8)	61.0 (4.3)	18.0 (4.4)	46.8 (9.1)	55.0 (4.1)	37.5 (5.2)	18.2 (5.4)	15.7 (5.9)
Lung								
Infiltrate, mixed cell, perivascular								
Minimal	—	1/4	—	—	1/4	—	—	—
Mild	—	1/4	—	—	1/4	—	—	—
Moderate	—	2/4	—	—	2/4	—	—	—
Inflammation, alveolar (chronic active)								
Minimal	—	-	—	2/4	—	1/4	—	—

(Continued)

Table 12.1 (Continued)

	PBS	TOP1572	TOP1572-PO7-DAP		TOP1731	TOP1731-PO7-DAP	TOP1769-PO7-DAP	TOP1777-PO7-DAP
Dose (mg kg^{-1})	—	**2.5**	**2.5**	**10**	**2.5**	**2.5**	**2.5**	**2.5**
Mild	—	1/4	—	—	1/4	—	—	—
Moderate	—	1/4	—	—	3/4	—	—	—
Severe	—	2/4	—	—	—	—	—	—
Infiltrate, macrophages alveolar								
Minimal	—	—	4/4	—	—	3/4	1/4	2/4
Mild	—	—	—	4/4	4/4	—	3/4	—

BAL, Bronchoalveolar lavage differential cell counts 24 h after the last dosing, values represent mean (±SEM); % large macrophages (±SEM) as defined by the macrophages having a cytoplasm to nucleus ratio ≥3. Lung, values represent number of animals with findings/total animals; —, means no findings. PBS, phosphate buffered saline (0.9%). Antisense oligonucleotide sequences: TOP1572, 5′-ggttgctcagitctgcaca-3′; TOP1572-P07-DAP, 5′-$g_sg_st_ot_og_sc_st_oc_ox_sg_si_ot_oc_st_sg_oc_ox_sc_sx$-3′; TOP1731, 5′- tcatgagtggcagctgcaatt-3′; TOP1731-PO7-DAP, 5′- $t_sc_sx_ot_og_sx_sg_ot_og_sg_sc_sx_og_oc_st_sg_sc_ox_ox_st_st$-3′; TOP1769-PO7-DAP, 5′- $x_sc_sx_og_ot_sg_sc_ox_ot_sg_sg_sc_oc_ot_sg_sx_sg_ot_ox_sx_sc$-3′; TOP1777-P07-DAP, 5′- $t_sx_sx_og_ot_sx_sx_oc_ox_sg_st_sg_oc_ox_st_sg_sg_oc_oc_st_sg$-3′; lowercase letters = DNA; x = 2-amino-2′-deoxyadenosine; i = inosine; lower script S = phosphorothioate linkages; lower script O = phosphodiester linkages.

clinic may be suitable. In general, treatment should include appropriate control group(s), a low dose at a level sufficient to produce a pharmacodynamic effect, an intermediate dose, and a high dose, selected to enable characterization of target organ toxicity (lung) as well as other nonspecific toxicity. For oligonucleotide therapeutics, the evaluation of exaggerated pharmacology can represent an extra degree of complexity since, for most of them, they are directed against a species-specific target sequence, human. Thus, this target sequence may not necessarily be homologous across species and therefore not produce the desired therapeutic effect. The specific issue of exaggerated pharmacology is reviewed in a separate chapter.

As for any drug, careful examinations of animals for clinical signs should be performed periodically during the dosing period including food consumption, body weight, clinical pathology (blood chemistry, hematology, urinalysis), ophthalmoscopy, electrocardiography (in nonrodents), and toxicokinetics. At the end of the treatment period, a full necropsy should be conducted, and organ weights (for example, close monitoring of lung weights may be a useful indicator for identifying acute lung toxic effect) and the presence of gross lesions recorded. A full microscopic examination should be performed, with a careful attention to the examination of the respiratory tract. This means all lung lobes (including main bronchi), nasal cavity/nasopharynx (with examination of the squamous, nonciliated and ciliated epithelium, olfactory epithelium, and draining lymphatic tissue), and trachea along bronchial lymph nodes.

12.3.2 Recovery Phase

The addition of recovery group animals is important to understand whether the effects observed at the end of the dosing phase persist, partially recover, or, ideally, fully recover. The duration of the recovery period should be long enough to meet the program needs. In some instances, partial reversibility may be acceptable, while in other cases, full reversibility may be required. Consideration of existing pharmacology, toxicology, and pharmacokinetic data should play a key role in determining the duration of recovery. Also, the duration of recovery may not be the same in rodents and nonrodent species. For oligonucleotide drug candidates with relatively short half-lives (1 day or less), recovery periods of 2–4 weeks in initial toxicity studies may suffice. For molecules with longer half-lives, such as second-generation oligonucleotides or those with chemical modifications, or for studies with longer dosing phases, prolonged recovery periods may be warranted (recovery periods of 5–7 half-lives could be used for oligonucleotides with half-lives longer than 1 week).

Another reason for inclusion of recovery (often referred as dose-free period) groups has been to determine whether the test article has the potential to produce delayed toxicity after dosing has ended. The term dose-free period is used because it should be recognized that part of this time will have continued exposure to the oligonucleotide and will not represent a true recovery period. In this context, for oligonucleotides with long half-lives, the possibility exists for toxic effects to appear after cessation of the dosing period, related to the prolonged exposure. The possibility for continued oligonucleotide exposure, or of its metabolites, in at least part of the recovery phase may need to be factored into determining the duration of recovery and the interpretation of the data.

For example, results from 2-week inhalation studies in mice and monkey of a phosphorothioate 2′-deoxy-2′-fluoro-beta-D-arabinonucleic acid (2′F-ANA) gapmer ASOs highlighted the importance of including a recovery phase. Dose-dependent histomorphologic changes in the liver (increased hepatocellular mitotic figures with or without hepatocellular cytomegaly/karyomegaly) and kidney (tubular degeneration/necrosis, dilatation, and glomerulonephropathy) were observed in the mice. These changes were observed only in animals of the high dose group (2.5 mg $(kg\,day)^{-1}$; at the end of a 4-week recovery phase, such findings were not present at the end of the dosing phase, suggesting a delayed onset. Similar delayed findings were not observed in monkeys (2-week inhalation study with a 4-week recovery phase) (Table 12.2).

Studies using the subcutaneous route of administration to begin investigate the mechanism of the observed delayed effects indicated that the timing of appearance of the findings in the kidney and liver varied between species and likely induced by the metabolism of the 2F′-ANA containing ASOs. Mice were treated for 14 days by daily subcutaneous administration of a phosphodiester all 2F′-ANA ASOs (selection of a phosphodiester backbone was to maximize the degradation of the ASO) and sacrificed at various time points of the dose-free recovery phase. Results showed that delayed liver and kidney findings observed in the inhalation studies could be reproduced when the 2′F-ANA-containing ASO was administered by the subcutaneous route. Kidney tubular degeneration/necrosis, dilatation, and glomerulonephropathy could not be detected in mice after 2 weeks of recovery but appeared 4 weeks after the end of the treatment period (Table 12.3). These findings increased in severity after a longer dose-free recovery phase (8 weeks). Similarly, hepatocellular cytomegaly/karyomegaly could not be observed 2 weeks after the end of the treatment period but were present at 4 weeks. In contrast to the kidney changes, the liver findings were absent at 8 weeks, suggesting recovery (Table 12.3).

When rats were treated daily (subcutaneous administration) for 14 days with a phosphorothioate 2′F-ANA gapmer ASOs, comparable changes were observed in the liver and kidney during the dose-free recovery period. However, the changes appeared after an even longer recovery period compared with mice, 6 weeks for the hepatocellular cytomegaly/karyomegaly and kidney glomerulonephropathy (Table 12.4). Furthermore, these changes seem to be rodent specific as they couldn't be observed in rabbit, even after a 16-week recovery phase (data not shown). Altogether, the results from these studies underscore the importance of carefully considering duration of the dose-free phase in the toxicology studies.

Table 12.2 Summary of delayed histopathologic changes in the liver and kidneys in recovery mice and monkeys following daily administration via inhalation exposure for 14 days.

	Recovery mice (day 42) ($mg\,kg^{-1}$)		Recovery monkey (day 42) ($mg\,kg^{-1}$)	
	0	2.5	0	2.5
Liver				
Cytomegaly/karyomegaly, hepatocellular				
Minimal	—	6/12	—	—
Mild	—	2/12	—	—
Increased mitotic figures, hepatocellular				
Minimal	—	4/12	—	—
Mild	—	6/12	—	—
Kidney				
Dilatation, tubular				
Minimal	—	7/12	—	—
Mild	—	3/12	—	—
Glomerulonephropathy				
Minimal	—	6/12	—	—
Mild	—	1/12	—	—
Moderate	—	2/12	—	—
Severe	—	3/12	—	—
Mitotic figures: tubular				
Minimal	—	6/12	—	—
Infiltrate, mixed cell:				
Minimal	—	2/12	—	—

Values represent number of animals in which change was observed per number of animals examined; —, means no findings. Other changes were observed in main phase animals but only findings observed in recovery animals are reported here. ASO sequences: 5′-**GG**ttgctcagitct**GCACA**-3′ and 5′-**TC**atgagtggcagctgc**AATT**-3′; lowercase letters = DNA; bold capital letters = 2′F-ANA; i = inosine; all phosphorothioate linkages.

12.4 Toxicokinetics

Information on systemic exposure of animals during repeat-dose toxicity studies is essential for the interpretation of the results. The principles involved for determining the toxicokinetics of a drug during inhalation of oligonucleotides are the same as for the other routes of administration and should follow ICH

Table 12.3 Summary of delayed histopathologic changes in the liver and kidneys in recovery mice treated with a phosphodiester 2F′-ANA antisense oligonucleotide (s.c.).

	Recovery mice								
	Day 15 ($mg\,kg^{-1}$)			Day 42 ($mg\,kg^{-1}$)			Day 57 ($mg\,kg^{-1}$)		
	0	1	5	0	1	5	0	1	5
Liver									
Cytomegaly/karyomegaly									
Minimal	—	—	—	—	2/6	2/6	—	—	2/4
Mild	—	—	—	—	1/6	1/6	—	—	—
Increased mitotic figures									
Minimal	—	—	—	—	2/6	1/6	—	—	—
Mild	—	—	—	—	1/6	3/6	—	—	—
Moderate	—	—	—	—	—	—	—	—	—
Kidney									
Glomerulonephropathy									
Minimal	—	—	—	—	3/6	4/6	—	1/4	—
Mild	—	—	—	—	1/6	—	—	2/4	—
Moderate	—	—	—	—	1/6	1/6	—	—	—
Severe	—	—	—	—	—	—	—	—	3/4
Degeneration/necrosis tubular									
Minimal	—	—	—	—	1/6	2/6	—	1/4	—
Mild	—	—	—	—	3/6	2/6	—	1/4	—
Moderate	—	—	—	—	1/6	1/6	—	1/4	1/4
Severe	—	—	—	—	—	—	—	—	2/4

Values represent number of animals in which change was observed per number of animals examined; —, means no findings. Other changes were observed in main phase animals. Only delayed findings are reported. ASO sequence: 5′-$T_oC_oC_oC_oG_oC_oC_oT_oG_oT_oG_oA_oC_oA_oT_oG_oC_oA_oT_oT$-3′; bold capital letters = 2′F-ANA; lower script O = phosphodiester linkages.

S3A guidelines. These objectives may be achieved by the derivation of one or more pharmacokinetic parameters from measurements (plasma or whole blood or serum concentrations for the parent compound and/or metabolite[s]) made at appropriate time points during the course of the individual studies. The toxicokinetics profile usually consists of the most commonly used parameters in assessing exposure such as AUC and C_{max}.

Table 12.4 Summary of delayed histopathologic changes in the liver and kidneys in recovery rats treated with a phosphorothioate 2F′-ANA gapmer antisense oligonucleotide (s.c.).

	Day 28 ($mg\,kg^{-1}$)		Day 44 ($mg\,kg^{-1}$)	
	0	10	0	10
Liver				
Cytomegaly/karyomegaly				
Minimal	—	—	—	4/8
Mild	—	—	—	4/8
Increased mitotic figures				
Minimal	—	—	—	2/8
Mild	—	—	—	1/8
Moderate	—	—	—	—
Kidney				
Glomerulonephropathy				
Minimal	—	—	—	4/8
Mild	—	—	—	4/8
Moderate	—	—	—	—
Severe	—	—	—	—
Degeneration/necrosis tubular	—	—	—	—
Minimal	2/6	6/12	2/4	5/8
Mild	—	—	—	3/8
Moderate	—	—	—	—
Severe	—	—	—	—

Values represent number of animals in which change was observed per number of animals examined; —, means no findings. Other changes were observed in main phase animals. Only delayed findings are reported. ASO sequence: 5′-**TC**ccgcctgtgacat**GCATT**-3′; lower case letters = DNA; bold capital letters = 2′F-ANA; all phosphorothioate linkages.

The main challenge is that systemic exposure following pulmonary delivery of oligonucleotides using the inhalation route of administration is expected to be relatively low as compared with other drug classes (e.g. small molecules) for a given dose. Systemic exposure following inhalation of oligonucleotides can be as low as 2–10% of the delivered dose [1–3, 13]. Quantification of plasma levels as well as tissue levels will therefore require the use of bioanalytical methods with suitable sensitivity, such as hybridization-based ELISA, to measure oligonucleotide concentrations in the low $ng\,ml^{-1}$ range [1, 13].

Systemic absorption of inhaled oligonucleotides is generally rapid but as mentioned above limited. Careful selection of sampling time is therefore critical as plasma concentrations can be often below the limit of quantitation and C_{max} and t_{max} can be as early as 10–15 min postdose. In addition, results indicate little or no systemic accumulation with repeated exposures [1, 13].

Similarly, tissue disposition, except lung and upper airways (trachea), is also limited particularly following a single exposure. In contrast to plasma concentrations, oligonucleotide tissue levels can increase following repeat exposure, and minimal amounts of oligonucleotides can be measured, principally in the kidney and liver following pulmonary delivery.

The deliberate generation of an aerosol and the wide range of doses, combined with the very low limits of quantification (e.g. 0.1 ng ml^{-1}) required to detect the oligonucleotide in biological fluids at very low doses, means that cross contamination between dose groups (including controls) can be a potential issue. However, a variety of procedures to limit transfer of oligonucleotides between groups via equipment and clothing, along with the use of individually ventilated cages, can be employed to avoid excessive contamination and potential invalidation of a study.

12.5 Safety Pharmacology

Safety pharmacology studies of oligonucleotides are regulated under several guidances, including ICH M3(R2), ICH S7A, and ICH S7B. The primary objective of safety pharmacology studies is the identification of possible acute, undesirable functional effects on the major physiological systems. In general, the core battery of studies includes the assessment of effects on cardiovascular, central nervous, and respiratory systems. Other dedicated safety pharmacology studies, beyond the core battery, may provide limited information for inhaled oligonucleotides. Since accumulation in the kidney and liver, evaluation of functional changes in these organs, as well as gastrointestinal, may be best evaluated during repeat-dose toxicity studies. Although safety pharmacology studies have been traditionally conducted as separate single-dose stand-alone studies, more recent guidance encourages incorporation of the functional endpoints into general toxicity studies in order to reduce animal use. Separating the safety assessment may overcome the potential disadvantage of traditional approach in that a compromise in terms of the timing of the cardiovascular, central nervous system, and respiratory assessments, as they cannot be measured concomitantly without affecting the integrity of the endpoints, has not to be made. For example, optimal time points for performing a neurological evaluation could compromise cardiovascular endpoints due the stress induced by the examination. On the other hand, performing a neurological evaluation at an arbitrary time point postdose in order to capture cardiovascular endpoints may question the value of the neurological data. Incorporating the safety assessments into the repeat-dose toxicity studies has also the potential advantage to monitor effects following both single and repeated doses.

The general approaches for safety pharmacology of oligonucleotides should be no different from those used for small molecule drugs, and strategies used including those to species relevance, select dose levels, and the route of administration have been reviewed recently [14]. As for toxicology studies, safety pharmacology studies for inhaled oligonucleotides can be typically done in rodent and/or nonhuman primates. The selection of evaluation time points and washout periods in between doses is particularly important and should reflect the pharmacodynamics and pharmacokinetics (including tissue persistence) of the oligonucleotide as with any drug candidate. The ideal window of time for capturing safety pharmacology endpoints should encompass the period during which plasma concentration (e.g. for cardiovascular and central nervous system safety pharmacology), target tissue concentration (e.g. for respiratory safety pharmacology), and/or pharmacological activity are maximum.

12.5.1 Respiratory System

In general, the level of concern for pulmonary effects related to parenterally administered noninhaled oligonucleotides is considered to be very low. In contrast, the level of concern over pulmonary effects for inhaled oligonucleotides is expectably greater, in part due to potential irritant and inflammatory effects observed with this route of administration [4, 15]. With repeated exposure to inhaled oligonucleotides, inflammation associated with particulate accumulation and degradation and/or pharmacological effects may also occur [1–3]. Thus, the respiratory safety pharmacology strategy for inhaled oligonucleotides may be different from that for parenterally administered oligonucleotides.

There are a variety of methodologies and measurement endpoints available for detecting and characterizing drug-induced respiratory dysfunction in animal models. Current practice in respiratory safety pharmacology follows the guidance and, in general, will focus on measures of pulmonary ventilation. For example, respiratory rate, tidal volume, minute volume, and/or a measure of arterial blood gases determined on conscious animals are parameters that will provide a measure of ventilation. In addition, other ventilatory parameters (inspiratory and expiratory times and flows and apneic time), which can provide mechanistic insight, may also be considered.

One strategy that has been used successfully for inhaled oligonucleotides is incorporation of respiratory parameters into a comprehensive repeat-dose toxicity study (using a pharmacologically responsive species for added benefit) instead of a prototypical stand-alone safety pharmacology study that incorporates cardiovascular, central nervous system, and respiratory assessment.

Respiratory assessment in the context of a repeat-dose study may represent the best and most valuable approach to assess pharmacology-related effects of inhaled oligonucleotides because of the concern that exists regarding cumulative nonspecific effects in the respiratory tract. As for other routes of administration, measurements should be made within the standard 24-h postdose monitoring period on three occasions, prior to initiation of treatment, after the first dose, and close to the end of the study. The end of treatment time points might be particularly important as lung toxicity and potential effects on pulmonary functions are expected to be cumulative and occur as the result of oligonucleotide accumulation in the lung.

12.5.2 Cardiovascular and Central Nervous Systems

Due to the limited systemic exposure following pulmonary delivery of oligonucleotides via inhalation, the assessment of cardiovascular safety and effects on the central nervous system may be a challenge as oligonucleotide delivery using this route limits achieving sufficiently high exposures for adequate safety assessment. This means that in contrast to respiratory safety, incorporation of the cardiovascular and neurological assessment into the repeat-dose toxicity studies may not allow the generation of the required information about potential effect of the oligonucleotide candidate on these systems.

To achieve higher systemic exposure, duration of inhalation exposure and/or concentration of dosing formulations need to be increased. Both approaches, however, have limitations. Animal welfare issues regarding the safe duration for which animals can be restrained physically during inhalation exposure may limit achieving sufficiently high exposures for safety assessment. In addition, while solubility of oligonucleotides in dosing formulations is generally not an issue, other parameters such as viscosity may affect aerosol properties. Viscosity of high concentration solutions (i.e. $>100\,\mathrm{mg\,ml^{-1}}$) may also represent a challenge for maintaining some key exposure system parameters such as flow rate and target aerosol concentrations.

When duration of exposure is an issue, the inhalation dose may be supplemented with parenteral dosing to achieve adequately high systemic exposures. Furthermore, it may be arguable that a parenteral route (typically intravenous) will provide a better assessment of the adequacy of cardiovascular or nervous system function than would occur with inhalation exposure. For example, the study plan for the cardiovascular and central nervous system studies could include two or three dose levels (low, mid, and high dose) as well as control group animals that receive a single inhalation administration and an additional group of animals that receive a single high dose intravenous administration. For phosphorothioate oligonucleotides, parenteral dose levels in nonhuman primates may need to be, however, adjusted due to potential effect on complement activation.

Cardiovascular studies typically include *in vitro* and *in vivo* studies. *In vitro* studies to assess the risk of cardiovascular effects typically focus on characterizing the potential to cause electrocardiographic (ECG) disturbances. As recommended by ICH S7B, the hERG assay evaluates test article effects on the repolarizing potassium current that is most commonly implicated in drug-induced QT prolongation. In general, oligonucleotides do not appear to interact directly with the hERG channel, and, therefore, arguments could be made about the need for this type of assay for oligonucleotides [16]. Furthermore, concern about the likelihood of ECG abnormalities in relation to the pharmacologic activity of the oligonucleotide is relatively low.

The main objective of *in vivo* cardiovascular safety pharmacology studies has traditionally been the assessment of acute effects after single-dose administration. For phosphorothioate oligonucleotides and other highly polyanionic molecules, the most commonly observed changes in cardiovascular parameters are hemodynamic effects stemming from complement activation [17, 18]. Because of the potential for phosphorothioate oligonucleotides to activate the alternative pathway of complement, nonhuman primates, with surgically implanted radiotelemetry transmitters, have commonly been used for conducting *in vivo* cardiovascular safety pharmacology studies with most subclasses of therapeutic oligonucleotides. Parameters that can continuously be measured, typically for a period up to 24 h postdose, can include systolic, diastolic, and mean arterial blood pressure, heart rate, respiratory rate, body temperature, and ECG parameters (PR, QT, QTc, RR, QRS intervals).

Concern regarding central nervous system safety of oligonucleotides-based therapeutics has been relatively low because oligonucleotides delivered systemically, particularly those with charged backbones, are cleared rapidly from the blood and typically do not cross the blood–brain barrier to a quantitatively significant extent nor been shown to produce central nervous system pathology or serious adverse events [16, 19]. The prototypic approach to central nervous system safety pharmacology evaluations is a stand-alone neurofunctional assessment in rodents, consisting of performing a functional observational battery (FOB) [20] or Irwin test [21] at one or more time points following single-dose administration of the drug candidate. Neurological examination can also be performed in nonhuman primates. For oligonucleotides, nonhuman primates may represent the most pharmacologically relevant species. In addition to the recommended neurological assessment, a quantitative measure of locomotor activity may be performed to determine potential acute effects on central and peripheral nervous system function. As for the cardiovascular assessment, measurements for the neurological evaluation can be performed for a period up to 24 h postdose.

12.6 Additional Testing

Additional testing may be carried out as part of the repeat-dose toxicity studies to address specific concerns and/or obtain additional information about the safety of oligonucleotide therapeutics.

12.6.1 Complement Activation

The activation of the alternative complement pathway is relatively common for systemically administered phosphorothioate oligonucleotides but has been more an issue in monkey toxicology studies, which have been shown to be more sensitive than man, and has been shown to occur primarily after rapid infusion of the oligonucleotide drug [17, 22–25]. In monkey, complement activation has been characterized with a well-defined threshold of phosphorothioate oligonucleotide plasma concentration required for activation. Evidence of complement activation is characterized by the reduction of serum hemolytic potential (CH_{50} analysis) and concomitant increases in the liberation of complement split products Bb, C3a, and C5a. Bb is a well-accepted marker for the alternative pathway as it is relatively stable and readily measured. The split product C5a, however, is the most highly biologically active split product that mediates almost all of the possible adverse sequelae in monkeys. On the other hand, C4a is a marker of the classical pathway, and the absence of C4a increase is consistent with the activation of complement through the alternative pathway and not the classical complement pathway.

The likelihood of triggering complement activation in monkeys with nonsystemic routes of administration is expectedly very low. This is a blood level-related effect that is seen with phosphorothioate oligonucleotides when plasma concentrations exceed ~30–40 $\mu g\,ml^{-1}$ [17, 18, 22, 24], and typically the plasma levels are far below that range with inhalation dosing [1, 13].

The alternative pathway activation is an acute and transient phenomenon, occurring within 30 min after intravenous dosing [17] and within hours after subcutaneous dosing [26], and the split products formed are eliminated within minutes (e.g. C5a) to hours (e.g. Bb). C5a is also extremely labile *in vivo* and will only accumulate if there is a substantial burst of complement activation. The sample collection time is therefore critical. The steps taken to show no complement activation following inhalation dosing need to be accompanied by toxicokinetics sampling at the same time points in order to document the maximal exposure (plasma levels) under these conditions that do not produce complement activation.

Complement testing for Bb, C4a, and C5a could be performed in monkeys as part of the repeat-dose toxicology studies. Testing could be done after single-dose and at the end of the dosing period on blood samples collected prior to dosing and at different time points postdosing. For inhaled oligonucleotides, a stepwise approach for the testing of complement activation may be recommended, beginning with the analysis of control and high dose samples for Bb and C4a. Lack of increases of Bb and C4a split products would suggest that there is no effect on complement activation, and therefore the analysis for C5a may not be required.

12.6.2 Proinflammatory Effects

Proinflammatory effects in animals have been well described for oligonucleotides therapeutics (mainly those with a charged backbone and persistence in tissues) (reviewed in other chapters), so evaluation of unintended effects on the immune system may be warranted for this type of molecules. A common observation following administration of ASO is the infiltration of mononuclear cells in various tissues, most often the kidneys and liver. Similar findings occur in the lung of rodents following inhalation administration of ASOs. In the lung, two observations are generally made: the accumulation of alveolar macrophages and interstitial infiltrate in the lung. Regarding the first observation, it has been suggested that the accumulation of alveolar macrophages is a general response to any inhaled material and not specific for ASO. In contrast, the interstitial infiltrate has been suggested to be a specific response to oligonucleotides delivery. At high dose levels of inhaled oligonucleotides, in particular those with modified backbones, the infiltration extends beyond the lungs and is observed in the pharynx, larynx, and bronchial lymph nodes [1, 3].

Of concern with infiltrating macrophages is their capacity to release proinflammatory cytokines (IL-1, TNF-a, IL-6) and chemokines (CXCL8, CCL2, CCL3) following activation. It is unclear whether the migration of macrophages into tissue is sufficient for cell activation or whether the macrophage responses following uptake of inhaled material are proinflammatory in nature. While cytokine levels can be readily assessed in the blood circulation following ASOs inhalation, it is more challenging to accurately assess the local levels of cytokines released in the lung. In animal studies, lavage of the bronchial regions can be performed, and cytokines can be measured in this collected material by various methods, including ELISA and Luminex assays. Furthermore, cytokine mRNA levels produced by the cells collected in the bronchoalveolar lavage can be quantified by RT-PCR. Protein levels can be detected in cells utilizing either ELISPOT or intracellular cytokine staining assessed by flow cytometry. Interpretation of the findings is challenging as there are no defined levels of proinflammatory cytokines that can be attributed as either safe or dangerous. It is therefore potentially risky to commence these studies as the interpretation may not be forthcoming. Furthermore, in clinical studies, assessing macrophage activation in human lungs is difficult [5]. Sputum has been utilized in some studies; however, not all patients can readily produce sputum for a single sample, yet alone to produce serial samples over time. As yet unknown is the optimal time to assess macrophage infiltration and macrophage activation. In the clinical studies to date, there has not been observations of induction of adaptive or memory immune responses following inhalation of ASOs [27–33].

12.7 Conclusion

In summary, pharmacokinetic and pharmacodynamic properties of inhaled oligonucleotides in animals support the lung as the likely target organ in humans. At present, preclinical and clinical data do not suggest that inhaling oligonucleotides presents any more, or less, risk of toxicity than has been demonstrated with other novel classes of drugs. Thus, the preclinical testing for inhaled oligonucleotides should follow current best practice as applied to the assessment of other inhaled drug classes. A plan to adequately test their preclinical and then clinical safety, in discussion with the respective regulatory authorities, needs to be developed. Finally, as various inhaled oligonucleotide drug candidates advance through development, unexpected toxicity may be encountered, which will further guide future evaluations.

References

1 Guimond, A., Viau, E., Aubé, P. et al. (2008). Advantageous toxicity profile of inhaled antisense oligonucleotides following chronic dosing in non-human primates. *Pulm. Pharmacol. Ther.* 21 (6): 845–854.

2 Ali, S., Leonard, S.A., Kukoly, C.A. et al. (2001). Absorption, distribution, metabolism, and excretion of a respirable antisense oligonucleotide for asthma. *Am. J. Respir. Crit. Care Med.* 163 (4): 989–993.

3 Templin, M.V., Levin, A.A., Graham, M.J. et al. (2000). Pharmacokinetic and toxicity profile of a phosphorothioate oligonucleotide following inhalation delivery to lung in mice. *Antisense Nucleic Acid Drug Dev.* 10 (5): 359–368.

4 Forbes, B., O'Lone, R., Allen, P.P. et al. (2014). Challenges for inhaled drug discovery and development: induced alveolar macrophage responses. *Adv. Drug Deliv. Rev.* 71: 15–33.

5 Alton, E.W., Boushey, H.A., Garn, H. et al. (2012). Clinical expert panel on monitoring potential lung toxicity of inhaled oligonucleotides: consensus points and recommendations. *Nucleic Acid Ther.* 22 (4): 246–254.

6 Labiris, N.R. and Dolovich, M.B. (2003). Pulmonary drug delivery. Part I: physiological factors affecting therapeutic effectiveness of aerosolized medications. *Br. J. Clin. Pharmacol.* 56 (6): 588–599.

7 Jackson, L.D., Polygenis, D., McIvor, R.A., and Worthington, I. (1999). Comparative efficacy and safety of inhaled corticosteroids in asthma. *Can. J. Clin. Pharmacol.* 6 (1): 26–37.

8 Bide, R.W., Armour, S.J., and Yee, E. (2000). Allometric respiration/body mass data for animals to be used for estimates of inhalation toxicity to young adult humans. *J. Appl. Toxicol.* 20 (4): 273–290.

9 Alexander, D.J., Collins, C.J., Coombs, D.W. et al. (2008). Association of Inhalation Toxicologists (AIT) working party recommendation for standard delivered dose calculation and expression in non-clinical aerosol inhalation toxicology studies with pharmaceuticals. *Inhal. Toxicol.* 20 (13): 1179–1189.

10 Driscoll, K.E., Costa, D.L., Hatch, G. et al. (2000). Intratracheal instillation as an exposure technique for the evaluation of respiratory tract toxicity: uses and limitations. *Toxicol. Sci.* 55 (1): 24–35.

11 Wong, J., Korcheva, V., Jacoby, D.B., and Magun, B. (2007). Intrapulmonary delivery of ricin at high dosage triggers a systemic inflammatory response and glomerular damage. *Am. J. Pathol.* 170 (5): 1497–1510.

12 Pauluhn, J. (2000). Repeated pulmonary function measurements in rats after intratracheal instillation of a polyamine textile pigment component. *Inhal. Toxicol.* 12 (7): 591–604.

13 Fey, R.A., Templin, M.V., McDonald, J.D. et al. (2014). Local and systemic tolerability of a 2′*O*-methoxyethyl antisense oligonucleotide targeting interleukin-4 receptor-α delivery by inhalation in mouse and monkey. *Inhal. Toxicol.* 26 (8): 452–463.

14 Berman, C.L., Cannon, K., Cui, Y. et al. (2014). Recommendations for safety pharmacology evaluations of oligonucleotide-based therapeutics. *Nucleic Acid Ther.* 24 (4): 291–301.

15 Owen, K. (2013). Regulatory toxicology considerations for the development of inhaled pharmaceuticals. *Drug Chem. Toxicol.* 36 (1): 109–118.

16 Kim, T.W., Kim, K.S., Seo, J.W. et al. (2014). Antisense oligonucleotides on neurobehavior, respiratory, and cardiovascular function, and hERG channel current studies. *J. Pharmacol. Toxicol. Methods* 69 (1): 49–60.

17 Galbraith, W.M., Hobson, W.C., Giclas, P.C. et al. (1994). Complement activation and hemodynamic changes following intravenous administration of phosphorothioate oligonucleotides in the monkey. *Antisense Res. Dev.* 4 (5): 201–206.

18 Henry, S.P., Giclas, P.C., Leeds, J.M. et al. (1997). Activation of the alternative pathway of complement by a phosphorothioate oligonucleotide: potential mechanism of action. *J. Pharmacol. Exp. Ther.* 281 (2): 810–816.

19 Boado, R.J., Tsukamoto, H., and Pardridge, W.M. (1998). Drug delivery of antisense molecules to the brain for treatment of Alzheimer's disease and cerebral AIDS. *J. Pharm. Sci.* 87 (11): 1308–1315.

20 Moser, V.C. and MacPhail, R.C. (1992). International validation of a neurobehavioral screening battery: the IPCS/WHO collaborative study. *Toxicol. Lett.* 64–65: 217–223.

21 Irwin, S. (1968). Comprehensive observational assessment: Ia. A systematic, quantitative procedure for assessing the behavioral and physiologic state of the mouse. *Psychopharmacologia* 13 (3): 222–257.

22 Henry, S.P., Beattie, G., Yeh, G. et al. (2002). Complement activation is responsible for acute toxicities in rhesus monkeys treated with a phosphorothioate oligodeoxynucleotide. *Int. Immunopharmacol.* 2 (12): 1657–1666.

23 Henry, S.P., Bolte, H., Auletta, C., and Kornbrust, D.J. (1997). Evaluation of the toxicity of ISIS 2302, a phosphorothioate oligonucleotide, in a four-week study in cynomolgus monkeys. *Toxicology* 120 (2): 145–155.

24 Henry, S.P., Jagels, M.A., Hugli, T.E. et al. (2014). Mechanism of alternative complement pathway dysregulation by a phosphorothioate oligonucleotide in monkey and human serum. *Nucleic Acid Ther.* 24 (5): 326–335.

25 Shen, L., Frazer-Abel, A., Reynolds, P.R. et al. (2014). Mechanistic understanding for the greater sensitivity of monkeys to antisense oligonucleotide-mediated complement activation compared with humans. *J. Pharmacol. Exp. Ther.* 351 (3): 709–717.

26 Geary, R.S., Leeds, J.M., Henry, S.P. et al. (1997). Antisense oligonucleotide inhibitors for the treatment of cancer: 1. Pharmacokinetic properties of phosphorothioate oligodeoxynucleotides. *Anticancer Drug Des.* 12 (5): 383–393.

27 Ball, H.A., Sandrasagra, A., Tang, L. et al. (2003). Clinical potential of respirable antisense oligonucleotides (RASONs) in asthma. *Am. J. Pharmacogenomics* 3 (2): 97–106.

28 DeVincenzo, J., Cehelsky, J.E., Alvarez, R. et al. (2008). Evaluation of the safety, tolerability and pharmacokinetics of ALN-RSV01, a novel RNAi antiviral therapeutic directed against respiratory syncytial virus (RSV). *Antivir. Res.* 77 (3): 225–231.

29 DeVincenzo, J., Lambkin-Williams, R., Wilkinson, T. et al. (2010). A randomized, double-blind, placebo-controlled study of an RNAi-based therapy directed against respiratory syncytial virus. *Proc. Natl. Acad. Sci. U. S. A.* 107 (19): 8800–8805.

30 Gauvreau, G.M., Boulet, L.P., Cockcroft, D.W. et al. (2008). Antisense therapy against CCR3 and the common beta chain attenuates allergen-induced eosinophilic responses. *Am. J. Respir. Crit. Care Med.* 177 (9): 952–958.

31 Gauvreau, G.M., Hessel, E.M., Boulet, L.P. et al. (2006). Immunostimulatory sequences regulate interferon-inducible genes but not allergic airway responses. *Am. J. Respir. Crit. Care Med.* 174 (1): 15–20.

32 Gottlieb, J., Zamora, M.R., Hodges, T. et al. (2016). ALN-RSV01 for prevention of bronchiolitis obliterans syndrome after respiratory syncytial virus infection in lung transplant recipients. *J. Heart Lung Transplant* 35 (2): 213–221.

33 Gauvreau, G.M., Pageau, R., Séguin, R. et al. (2011). Dose–response effects of TPI ASM8 in asthmatics after allergen. *Allergy* 66 (9): 1242–1248.

13

Lessons Learned in Oncology Programs

Cindy Jacobs, Monica Krieger, Patricia S. Stewart, Karen D. Wisont, and Scott Cormack

OncoGenex Pharmaceuticals, Bothell, WA, USA

13.1 Introduction

For more than three decades, synthetic antisense oligonucleotide (ASO) structures have been considered as potential drug candidates due to their specific ability to inhibit production of target proteins. Cancer cells produce an abundance of mRNA to generate necessary proteins that promote proliferation and migration and enhance their likelihood of survival [1]. Using ASO technology, a specific complementary or "antisense" strand of single-stranded DNA designed to bind to mRNA strands can lead to degradation of the mRNA before translation can occur, thereby reducing levels of the target protein [2].

In animal models, ASO strategies designed to target genes implicated in oncogenesis and disease progression have demonstrated an ability to specifically inhibit mRNA expression, resulting in clear antitumor activity [2–8]. However, early clinical investigations with ASO therapies were hampered due to rapid ASO degradation *in vivo*, requiring continuous or daily infusions in order to achieve therapeutic levels. Modifications in the chemical structure of ASOs have led to slower degradation, better pharmacokinetic profiles, improved binding affinity, and increased potency [9]. Clinical trials evaluating several ASO compounds utilizing these modified chemistries are currently underway in oncology settings.

Development of treatment resistance is a common feature of solid tumor malignancies and is the underlying basis for most deaths in cancer patients. Treatment resistance results from various stepwise changes in DNA structure and gene expression, arising in part from the selective pressures of oncology treatments. In most solid tumors this evolutionary process cannot be

Oligonucleotide-Based Drugs and Therapeutics: Preclinical and Clinical Considerations for Development, First Edition. Edited by Nicolay Ferrari and Rosanne Seguin.

attributed to singular genetic events but instead involves many cumulative changes in gene structure and expression that facilitate cancer cell growth, migration, and survival. In advanced prostate cancer, for example, treatment resistance is manifested by progression to castrate-resistant prostate cancer (CRPC) via mechanisms attributed to reactivation of the androgen receptor axis [10], alternative mitogenic growth factor pathways [11–13], stress-induced prosurvival genes [14–16], and protective chaperone networks [17, 18].

Molecular chaperones help tumor cells cope with stress-induced protein misfolding, aggregation, and denaturation and play prominent roles in cellular signaling and transcriptional regulatory networks. Chaperones act as genetic buffers, stabilizing the phenotype of various cells at times of environmental stress and enhancing the ability of the cells to withstand formerly toxic treatments [19, 20]. Of special relevance to treatment-resistant cancers are those chaperones that are upregulated by anticancer therapies and function to inhibit treatment-induced tumor cell death, such as clusterin (CLU), heat shock protein 27 (Hsp27), and heat shock protein 90 (Hsp90) [17, 18, 21–24]. These molecular chaperones are not traditional therapeutic targets because either they exist in high concentrations in blood but operate intracellularly to effect tumor biology or they are exclusively intracellular, making them "nondruggable." However, an antisense inhibitor targeting the mRNA of a specific exon for a chaperone protein would have potential as a therapeutic product.

13.2 Clinical Development of First-generation ASOs

ASO structures have been evaluated as drug candidates for several decades, beginning with "first-generation" compounds. Numerous compounds were evaluated in phase 1 and phase 2 studies, but none were successful in phase 3 development; this was due in part to technical challenges dictated by the chemical structure of these compounds (e.g. nuclease degradation, CpG motif activity) and to the lack of an adequate risk/benefit profile. To successfully inhibit a stress-induced target protein, the ASO should be administered in a manner to achieve consistent intratumor cell levels and subsequent mRNA binding. For first-generation products susceptible to rapid degradation, this meant daily intravenous infusions, which were inconvenient and posed additional safety risks. Despite the requirement for continuous administration, two first-generation ASOs reached advanced levels of clinical development in oncology: aprinocarsen and oblimersen.

13.2.1 Aprinocarsen

Clinical development of the Isis Pharmaceuticals, Inc. (Carlsbad, CA) product aprinocarsen (Isis 3521/LY900003) demonstrated the limitations of first-generation ASO chemistry. The 20-base ASO aprinocarsen (trade name AFFINITAK™) was designed to block protein kinase C-alpha (PKC-α); PKC-α was identified as a therapeutic target in oncology because it has a pivotal role in tumor cell signal transduction, is involved in drug resistance, and is abnormally expressed in human tumors.

Clinical development of aprinocarsen began in 1996 with a phase 1, single-arm, dose-escalation safety study in 21 patients; treatment was delivered over a period of 21 days by continuous infusion, followed by a 7-day rest period. A maximum tolerated dose (MTD) of 2 mg/kg/day was identified based on dose-limiting toxicities (DLTs) of thrombocytopenia and fatigue [25]. Evidence of tumor response with a duration of up to 11 months was observed in three of four patients with ovarian cancer, and this evidence of antitumor activity provided the rationale for subsequent phase 2 studies in ovarian cancer as well as other malignancies [25].

In a phase 1/2 single-arm clinical study evaluating 53 patients with nonsmall cell lung cancer (NSCLC), aprinocarsen was administered as a continuous infusion at 2 mg/kg/day over 14 days of each 21-day cycle in combination with carboplatin and paclitaxel. Partial or complete responses (CR) were observed in 42% of 48 evaluable patients, and median overall survival was 19 months (actuarial 1-year survival of 75%). Neutropenia and thrombocytopenia were reported as the major toxicities [26].

Two additional single-arm phase 1/2 studies investigated the safety and efficacy of the combination of aprinocarsen with chemotherapy in patients with advanced NSCLC as well as other advanced cancers. The first study evaluated a regimen of aprinocarsen (2 mg/kg/day) plus gemcitabine and cisplatin. In the phase 2 portion of the study, a total of 55 patients with NSCLC were treated at two different gemcitabine doses. Fourteen of 39 evaluable patients had a combined response rate of 36%. Intent to treat median survival time was 8.9 months. Forty-eight percent of patients experienced catheter-related events, which included venous thrombosis, pulmonary embolism, bacteremia/sepsis, localized infection, and erythema/pain [27]. The second study evaluated a regimen of aprinocarsen (2 mg/kg/day) with gemcitabine and carboplatin ($N = 36$); median overall survival was 8.3 months. Thrombocytopenia (78%) and neutropenia (50%) were the major grades 3–4 toxicities. Enrollment in this study was terminated due to failure of a concurrent phase 3 study in NSCLC, as discussed below [28].

In mid-2001, Eli Lilly and Company (Indianapolis, IN) licensed aprinocarsen for development as a treatment for lung cancer. Two randomized phase 3 trials with aprinocarsen were conducted by Eli Lilly and Company, one with aprinocarsen (2 mg/kg/day), gemcitabine, and cisplatin ($N = 670$) and one with aprinocarsen, carboplatin, and paclitaxel ($N = 616$). The company reported in 2003 that the combination of aprinocarsen with carboplatin and paclitaxel did not demonstrate significant differences in survival or in any other efficacy measures [29, 30]. Based on these results, the study of aprinocarsen in combination with gemcitabine and cisplatin was terminated early; analysis of data from this study also indicated no significant difference in survival with aprinocarsen treatment, and grade 3 and 4 toxicities, particularly thrombocytopenia, were significantly increased among patients receiving aprinocarsen [31]. The fact that other clinical studies with non-ASO inhibitors of PKC-α also failed, however, suggests that this target might not have influenced tumor biology as originally postulated

13.2.2 Oblimersen

Oblimersen sodium (trade name Genasense®) was an ASO compound evaluated by Genta Incorporated (Berkeley Heights, NJ). Oblimersen was designed to block production of the Bcl-2 protein, an antiapoptotic protein that inhibits the release of cytochrome *C*. Oblimersen was evaluated in combination with dacarbazine (DTIC) in a phase 1/2 dose-escalation study in patients with advanced malignant melanoma expressing Bcl-2; drug resistance in melanoma has been partially attributed to overexpression of Bcl-2. The combination regimen appeared well tolerated in the dose-escalation study evaluating doses up to 6.5 mg/kg/day, with no observed DLTs. Six of 14 patients had an antitumor response, and the estimated median survival of all patients exceeded 12 months [5]. Based on the strength of clinical response and the observation of 40% median decreases in Bcl-2 protein in tumor tissue, the company began a phase 3 study in melanoma, and in 2002, Genta signed a development and commercialization agreement with Aventis (part of the Sanofi–Aventis Groupe, Paris, France).

The phase 3 study in metastatic melanoma evaluated oblimersen (7 mg/kg/day) in patients who had not received prior chemotherapy ($N=771$). The primary endpoint of the trial was overall survival. After a minimum of 24 months follow-up, median survival in the oblimersen–DTIC arm was not significantly longer than in the DTIC-only arm, although median progression-free survival (PFS) was significantly different between the arms [32]. The oblimersen–DTIC arm was associated with increased toxicity and discontinuations due to adverse events (AEs); 67% of patients on the oblimersen–DTIC arm experienced at least one grade 3 or 4 event versus 43% on the DTIC-only arm and 6.2% versus 2.2% discontinued treatment due to a treatment-related AE. Neutropenia and thrombocytopenia were the most important AEs (grades 3–4 neutropenia with infection, 4.3% vs. 2.8%, respectively), and there was an increase in bleeding events in the oblimersen–DTIC arm (13.7% vs. 9.2%). Genta presented results of this phase 3 trial in melanoma to the US Food and Drug Administration (FDA) Oncologic Drugs Advisory Committee (ODAC) in 2004. In the absence of achieving the primary endpoint of increased survival, the committee voted that there was not sufficient evidence of clinical benefit to outweigh the risk of oblimersen toxicity. Subsequently, Aventis discontinued its development agreement with Genta.

Genta also evaluated oblimersen in patients with chronic lymphocytic leukemia (CLL). Results of a phase 3 study in CLL ($N = 241$) showed a 10% positive difference in the rate of CR plus nodular partial response (nPR) with oblimersen treatment (3 mg/kg/day); however there was no significant difference in overall response rate, time to progression, survival, or symptomatic benefits [33]. Nausea was the most common nonhematologic AE and occurred more frequently in the oblimersen group; laboratory AEs of thrombocytopenia, febrile neutropenia, and hypotension were also more frequent with oblimersen treatment. In 2006, the FDA advisory committee recommended against approval of oblimersen in CLL due to lack of "substantial" evidence of effectiveness, and Genta failed to receive FDA approval for oblimersen. Although the company returned to the FDA in 2008 with data on 5 years of follow-up from this study [34], the agency declined to approve oblimersen for treatment of CLL.

In 2009, Genta completed accrual to a new phase 3 trial of oblimersen in patients with melanoma ($N = 314$) using lactate dehydrogenase (LDH) as a biomarker and stratification factor. The trial did not show a statistically significant benefit in PFS with oblimersen (7 mg/kg/day) versus placebo [35]. In addition to studies in melanoma and CLL, Genta also conducted phase 3 studies in multiple myeloma and acute myeloid leukemia; neither trial met its primary endpoint.

13.2.3 Challenges Associated with First-generation ASOs

Clinical development of first-generation ASOs frequently met with failure in cancer trials for several possible reasons, including limitations of the first-generation ASO chemistry, misunderstanding of target relevance, lack of consistent clinical benefit, and inadequate dosing. Collectively, these issues resulted in products that did not demonstrate sufficient efficacy to justify approval for use.

Inherent in the initial first-generation ASO chemistry was the requirement for administration via continuous or daily infusion due to rapid degradation by nucleases. The continuous presence of the ASO as well as degraded fragments in plasma led to accumulation in the kidney, liver, spleen, and bone marrow, resulting in significant toxic effects such as thrombocytopenia, which was notable in most studies. In contrast, the 2′-*O*-(2-methoxyethyl) (2′-MOE) modifications at the ends of the second-generation compounds were designed to decrease nuclease degradation (before hybridizing to the target mRNA), which resulted in significantly improved ASO tissue half-life *in vivo* [36]. Therefore, second-generation ASO compounds could be administered less frequently as weekly infusions or as subcutaneous injections, with lower toxicity.

Misunderstanding of target relevance also contributed to failure of early ASO compounds in the clinic; this was a result of the drug target not being as fully elucidated as early data suggested. For example, Bcl-2 was the target of the first-generation compound oblimersen, but it became appreciated later in its development that other proteins can provide similar biologic functionality; consequently, reduction of Bcl-2 led to upregulation of other pathways, which rendered the Bcl-2 pathway obsolete to the cancer cells' response. Similarly, drugs targeted to PKC-α also failed because that target was not as relevant to cancer cell survival pathways as originally thought.

Lack of consistent clinical benefit may have resulted from study design and patient variables that can impact measures of both safety and effectiveness, leading to statistically inconclusive results. For example, most first-generation ASOs were administered using indwelling catheters, which increased susceptibility to thrombotic events and infections in the ports; these events confounded safety observations and increased the risk profile. Infections at the ports may have also contributed to trial infidelity by causing interrupted dosing and/or early study treatment discontinuation. Also detrimental may have been a lack of stratification criteria based on relevant survival prognostic factors. Accounting for such factors in the trial designs may have revealed survival benefits in subgroups. Clinical programs might also have benefitted from larger phase 2 studies in various patient populations in order to identify any inclusion criteria that would have optimized phase 3 study designs.

In addition, early ASO compounds may have been evaluated at doses that were inadequate to demonstrate efficacy. There were no human pharmacodynamics data to accurately determine whether the ASO was reaching the intracellular mRNA target and effectively inhibiting production of the targeted protein. Single dose-escalation studies in small numbers of patients may have also led to trial doses that were not optimized for clinical benefit.

Together these challenges led to ASO product candidates that lacked an approvable benefit–risk ratio. At the doses tested (2–7 mg/kg/day), the ASO products may not have achieved effective levels before resulting in toxicity, and imbalances or differences among trial populations may have kept any potential benefits from being observed.

13.3 Clinical Development of Second-generation ASOs

Over the past 15 years, considerable effort has been made to improve the *in vivo* potency of ASOs by modifying the phosphodiester linkage and the heterocyclic structure of the sugar. Advances in modified nucleic acid chemistry [9, 37, 38] yielded "first-generation" ASO products with a nonbinding oxygen substituted for a sulfur, producing the phosphorothioate backbone.

Further modifications resulted in "second-generation" ASO products that have a 2′-MOE modification at the 2′ position of the carbohydrate moiety on both ends of the oligonucleotide. These structural changes result in improved target binding affinity, better tissue pharmacokinetics, and enhanced resistance to nuclease degradation; the latter led to increased ASO tissue half-life and, consequently, potency (as observed in cell culture and animal models) [9].

The improved affinity of the second-generation ASOs is attributable to both design and composition. In particular, second-generation compounds are composed of both RNA-like and DNA-like nucleotides, while first-generation compounds were entirely DNA-like. Because RNA hybridizes more tightly to RNA than to DNA, the second-generation compounds have a greater affinity for mRNA targets and therefore greater potency. The second-generation compounds have decreased binding affinity to exonucleases *in vivo* [36], allowing for longer duration of action and less frequent dosing via weekly 2-h infusion or subcutaneous injection instead of continuous or daily IV infusions [38].

Clinical development of second-generation ASOs in cancer trials to date has shown evidence of targeted protein suppression within cancer cells, an improved safety profile, and more potential clinical benefit compared with first-generation ASOs. ASO products currently in clinical development for oncology indications are listed in Table 13.1.

13.3.1 Custirsen

Custirsen (OGX-011; TV-1011) targets the chaperone protein clusterin and was the first second-generation ASO to be evaluated in the field of oncology. By mid-2013, custirsen had been evaluated in over 900 patients in clinical studies sponsored by OncoGenex Pharmaceuticals (Bothell, WA) and Teva Pharmaceutical Industries, Ltd (Netanya, Israel) under license with Isis Pharmaceuticals. Phase 2 and phase 3 clinical studies evaluating custirsen are summarized in Table 13.2 (i.e. for studies with >40 patients per study). Clinical development of custirsen began with a phase 1, dose-escalation study evaluating weekly doses of 40–640 mg of custirsen in combination with neoadjuvant hormone therapy in subjects with localized prostate cancer prior to radical prostatectomy ($N=25$) [39]. The main purpose of this study was to assure that custirsen, when given via a 2-h infusion once weekly, had an acceptable safety profile, resulted in measurable custirsen levels within tumor cells, and led to suppressed translation of the clusterin protein. This study was designed specifically to address critical, unanswered questions about ASO technology that resulted from the disappointing outcomes of first-generation ASOs, namely, could ASOs be delivered systemically, would they accumulate in nonclearance organs, would they be taken up by tumor cells, and most importantly, would they be effective enough to downregulate the target protein sufficiently to disrupt the intended tumor biology? Pharmacokinetics and pharmacodynamics

Table 13.1 ASO products currently in clinical development for oncology indications.

Product	Target	Company	Indications	Stage
Custirsen (OGX-011)	Clusterin	OncoGenex Pharmaceuticals	Prostate cancer	Phase 3
			Lung cancer	Phase 3
Apatorsen (OGX-427)	Heat shock protein 27 (Hsp27)	OncoGenex Pharmaceuticals	Bladder cancer[a]	Phase 2
			Prostate cancer	Phase 2
			Lung cancer	Phase 2
			Pancreatic cancer	Phase 2
LY2275796 (ISIS-E1F4E$_{Rx}$)	Eukaryotic initiation factor-4e (eIF-4E)	Eli Lilly and Company (licensed from Isis Pharmaceuticals)	NSCLC	Phase 2
			Prostate cancer	
AZD9150 (ISIS-STAT3$_{Rx}$)	Signal transducer and activator of transcription 3 (STAT3)	AstraZeneca (licensed from Isis Pharmaceuticals)	Hematological cancers	Phase 1/2
EZN-2968	HIF-1α	Santaris Pharma A/S (partnered with Enzon)	Solid tumors, lymphoma	Phase 1
EZN-3042 (SPC3042)	Survivin	Santaris Pharma A/S (partnered with Enzon)	Solid tumors, lymphoma, pediatric acute lymphoblastic leukemia	Phase 1
EZN-4176	Androgen receptor	Santaris Pharma A/S (partnered with Enzon)	Prostate cancer	Phase 1

[a] Denotes lead indication.

Table 13.2 Clinical development of custirsen (OGX-011): Phase 2 and phase 3 studies with >40 patients.

Study	Phase	Indication	Design	Objectives	Status	Reference
011-03	2	mCRPC, 1st line	Open-label, randomized study of OGX-011 in combination with docetaxel/prednisone versus docetaxel/prednisone alone ($N = 82$)	*Primary*: Determine efficacy as measured by PSA response *Secondary*: Determine objective response in duration in subjects with measurable disease at baseline; determine tolerability and toxicity of weekly OGX-011 plus q 3 weekly docetaxel; measure effect on serum clusterin levels; describe TTP and OS	Completed	[40]
011-05	1/2	NSCLC, 1st line	Open-label study of OGX-011 in combination with a gemcitabine/platinum-based regimen ($N = 81$)	*Primary*: Estimate ORR; establish recommended dose; determine safety and tolerability *Secondary*: Estimate PFS, OS; determine PK profile in combination with chemotherapy; measure effect on clusterin in serum and within tumors; assess relationship between clusterin staining in tissue, TTP, and response rates	Completed	[47]
011-07	2	mCRPC, 2nd line	Open-label, randomized study of OGX-011 in combination with docetaxel/prednisone or in combination with mitoxantrone/prednisone ($N = 69$)	*Primary*: Evaluate safety in combination with chemotherapy *Secondary*: Assess feasibility of treatment regimens based on PSA response and time to pain progression; evaluate relationship between serum clusterin levels and changes in PSA, time to pain progression	Completed	[48]

(Continued)

Table 13-2 (Continued)

Study	Phase	Indication	Design	Objectives	Status	Reference
011-11 (SYNERGY)	3	mCRPC, 1st line	Open-label, randomized study of OGX-011 in combination with docetaxel/prednisone versus docetaxel/prednisone alone ($N = 1000$, planned)	*Primary*: Evaluate OS *Secondary*: assess proportion of patients alive without event at Day 140 milestone assessment	Enrollment completed	
011-12 (AFFINITY)	3	mCRPC, 2nd-line	Open-label, randomized study of OGX-011 in combination with cabazitaxel/prednisone versus cabazitaxel/prednisone alone ($N = 630$, planned)	*Primary*: Evaluate OS *Secondary*: Assess proportion of patients alive without event at milestone Day 140 assessment	Ongoing	
TV1011-LC-303 (ENSPIRIT)	3	NSCLC, 2nd line	Open-label, randomized study of OGX-011 in combination with docetaxel versus docetaxel alone ($N = 1100$, planned)	*Primary*: Evaluate OS *Secondary*: Evaluate PFS, ORR, duration of objective response, disease control rate, and duration of disease control; assess safety	Ongoing	

DLT, dose-limiting toxicity; mCRPC, metastatic castrate-resistant prostate cancer; NSCLC, nonsmall cell lung cancer; ORR, objective response rate; OS, overall survival; PFS, progression-free survival; PK, pharmacokinetic; PSA, prostate-specific antigen; TTP, time to progression.

were evaluated in this study at each dose cohort, with custirsen administered as three loading dose, 2-h infusions within 9 days followed by four weekly 2-h infusions. Prostatectomy was performed within 7–14 days after the last infusion of custirsen, which allowed for pharmacodynamic assessment of custirsen levels as well as inhibition of clusterin production within prostate cancer cells.

During this phase 1 study, no subjects experienced a DLT or withdrew from study therapy due to an AE. No serious AEs (SAEs) occurred during the protocol therapy or prior to prostatectomy. The most frequently reported AEs in all dose cohorts combined were fever, fatigue, and chills, predominantly following the loading doses and were self-limiting; the majority of these AEs were grade 1. A dose-relationship pattern was seen, with an increased incidence of the more common AEs in the higher dose cohorts, especially for chills, fatigue, and fever. Only 4 of 25 subjects experienced grade 3 treatment-emergent AEs, and no grade 4 events were reported. There were no grade 3 or 4 laboratory AEs reported. These early results suggested a substantial improvement in safety parameters versus first-generation ASOs.

Pharmacodynamic results from the phase 1 study showed proportional increases in custirsen tissue concentrations between study cohorts with dose escalation to the 640-mg dosage ($P < 0.001$) and corresponding dose-dependent decreases in clusterin expression within prostate cancer cells. At the highest dose of custirsen (640 mg), clusterin mRNA had decreased at the time of prostatectomy by more than 90% compared with untreated historical controls. Evaluation by immunohistochemistry (IHC) also verified that custirsen decreased clusterin protein levels in prostate cancer tissue. In summary, results from this study suggested that treatment with second-generation ASO technology had an acceptable safety profile and could effectively reduce its target within tumor cells when delivered systemically at the 640-mg dosage via a loading dose interval followed by a once weekly dosing schedule. A phase 2 development program was then initiated.

The early phase 2 development program for custirsen encompassed trials in prostate cancer, breast cancer, and NSCLC. The lead indication of metastatic castrate-resistant prostate cancer (mCRPC) was chosen based on an open-label, randomized phase 2 study ($N = 82$) [40]. In this study, men with mCRPC were randomized to receive standard first-line docetaxel with or without 640-mg custirsen. All patients also received oral prednisone (5 mg twice daily) throughout the study.

Safety results in the randomized phase 2 study were acceptable, with the majority of AEs being grade 1 or 2. Ten subjects in each arm (25% in custirsen arm, 24% in control arm) experienced at least one SAE. Grade 3 or 4 neutropenia and leukopenia were the most frequent laboratory toxicities in both arms. The incidence of grade 3 or 4 lymphopenia was higher in the custirsen arm (45%) than in the control arm (22%), primarily due to grade 3 values; there appeared to be no relationship between grade 3 lymphopenia and occurrence of infections. Thrombocytopenic events were comparable between the arms, with 28% of subjects in the custirsen arm having grade 1 or 2 events compared with 20% of subjects in the control arm; one subject in the custirsen arm had a grade 3 event. Serum chemistry testing was mostly comparable between the arms with the exception of grade 1 or 2 elevations of creatinine in 30% of subjects on the custirsen arm and 10% on the control arm. Thus, minimal increased toxicity occurred by adding custirsen to standard docetaxel/prednisone treatment in men with mCRPC [40].

In this phase 2 study, median survival was 23.8 months (95% CI, 16.2–infinity) for the custirsen plus docetaxel/prednisone arm and 16.9 months (95% CI, 12.8–25.8) for the docetaxel/prednisone control arm, with an unadjusted hazard ratio (HR) of 0.61 (95% CI, 0.36–1.02; $p = 0.06$), as shown in Figure 13.1.

A multivariate analysis of survival defined prior to the analysis included major prognostic variables (performance status, hemoglobin, LDH, pain, visceral metastases) and minor prognostic variables (absolute prostate-specific antigen [PSA], alkaline phosphatase). Predictive for longer survival on multivariate analysis were good performance status ($p < 0.0001$), no presence of visceral metastasis ($p = 0.01$), and custirsen treatment assignment (HR = 0.49 [95% CI, 0.28–0.85], $p = 0.01$). A phase 3 development program was initiated based on the 6.9-month survival benefit and acceptable safety profile, with the design of the confirmatory phase 3 study primarily mirroring that of the

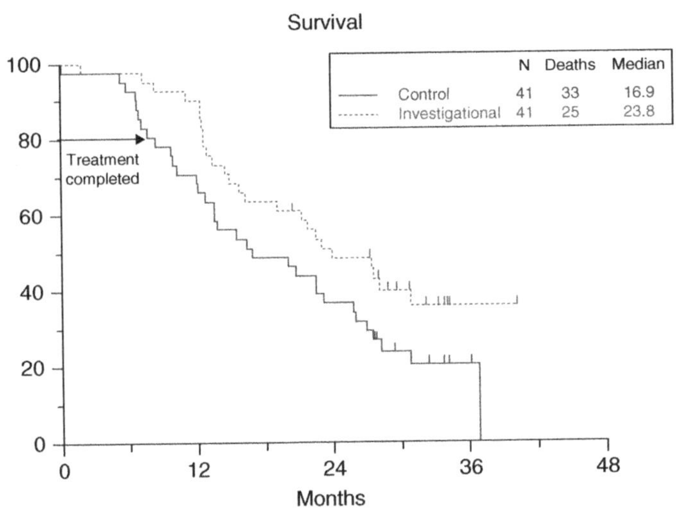

Figure 13.1 Randomized phase 2 study in mCRPC: Survival results for custirsen/docetaxel/prednisone (investigational, dotted line) versus docetaxel/prednisone (control, solid line). Adapted from Chi et al. [40].

phase 2 study. The phase 3 study has completed enrollment of 1022 subjects, and results should be available in 2014.

Other phase 2 studies evaluating custirsen with chemotherapy in mCRPC and NSCLC led to two additional phase 3 studies, both administering custirsen in combination with standard second-line chemotherapy. Thus, a total of three randomized phase 3 trials are underway to evaluate the risk/clinical benefit of adding 640-mg custirsen to standard chemotherapy in over 2700 subjects. Results from these phase 3 trials could provide further validation of second-generation antisense technology as a therapeutic tool for reducing cancer treatment resistance.

13.3.2 Lessons Learned from Custirsen Clinical Development

OncoGenex gained valuable insights during the phase 1 and 2 development of custirsen that helped shape the clinical development program of its second product, apatorsen (OGX-427), another second-generation ASO. During development of custirsen, OncoGenex found it beneficial to have:

- Evaluated pharmacodynamics of the ASO in a phase 1 study to verify that knockdown of the target protein occurred, as evidenced by either inhibition of mRNA expression or suppression of protein production following treatment.
- Evaluated the ASO in combination with chemotherapy as well as androgen ablation during multiple phase 1 studies to better define the safety profile of custirsen, especially for neutropenia and thrombocytopenia.
- Conducted a randomized, controlled, phase 2 study that allowed for comparative benefit/risk assessments between the arms, including survival outcome, for later trial size calculations and phase 3 study design.

Therefore, the custirsen phase 3 development program was built on a solid foundation, based on a pharmacodynamic study to define the dose and schedule that regulated the target protein, multiple treatment combinations, and a randomized phase 2 study that assessed preliminary results on survival outcome. Regardless, lessons were learned from the development of custirsen that have been incorporated into the development plans for apatorsen, including:

- Evaluating higher ASO doses to better understand the benefit/risk potential of apatorsen as well as to assess for any safety limitations based on either ASO class- or target-specific toxicity.
- Evaluating the safety profile and potential clinical benefit for apatorsen treatment alone and in combination with chemotherapy within the first phase 1 study.
- Evaluating potential survival benefit early in randomized phase 2 trials. As in many phase 2 oncology programs, most of the phase 2 studies of custirsen

were single-arm, uncontrolled studies, which made it difficult to estimate a potential survival benefit for a phase 3 trial design. All of the phase 2 trials with apatorsen are randomized, controlled trial designs, with survival as the primary or secondary outcome in the majority of the studies. This strategy should help to increase the likelihood of a successful phase 3 program.

- Conducting multiple randomized, controlled phase 2 studies in various indications with different chemotherapy regimens for the purpose of assessing an optimal phase 3 development program for apatorsen, even beyond the lead indication.

Thus, a broad phase 1 and 2 clinical development program for apatorsen has been shaped by these insights, which will hopefully result in a successful phase 3 development program and expedite the market-approval strategy for apatorsen.

13.3.3 Apatorsen

Apatorsen is a 2′-MOE-modified ASO designed to bind to Hsp27 mRNA and decrease production of the Hsp27 protein, which functions as a molecular chaperone. Hsp27 is situated as a "hub" and influences the intrinsic, extrinsic, and other pathways that regulate the response of a cell to stress. As a result, targeted inhibition of Hsp27 could affect multiple pathways implicated in cancer progression and resistance, as opposed to targeting a single pathway. A strategy using an antisense compound to inhibit Hsp27 and target multiple pathways might be expected to have a greater benefit given the heterogeneity and redundancy that exists in cancer cells. In addition to these pathways, Hsp27 plays a role in androgen receptor transcriptional activity [41] and in epithelial– mesenchymal transition (EMT), which is an important determinant in enhancing tumor cell metastasis [42]. Many tumors have been shown to express Hsp27, including prostate, lung, bladder, pancreatic, ovarian, breast, head and neck, endometrial, gastric, esophageal, and hepatic cancers plus multiple myeloma, leukemia, and osteosarcoma [43].

The first phase 1 study with apatorsen, OGX-427-01, enrolled 64 patients with prostate, breast, ovarian, lung, and bladder cancer. This phase 1 study evaluated escalating doses of apatorsen up to 1000 mg, first as a monotherapy (200–1000 mg) and then in combination with docetaxel chemotherapy (800- and 1000-mg doses). Results from the study demonstrated that doses of apatorsen up to 1000-mg IV were tolerable when administered alone or with docetaxel chemotherapy. Although increased toxicity was observed at the 800- and 1000-mg doses, an MTD was not reached. The majority of AEs were grade 1 or 2 (88%). The major nonlaboratory toxicity was infusion reactions (rigors, pruritus, flushing, pyrexia, hypertension, and erythema), which were seen in ~80% of patients; infusion reactions appeared to occur at an increased frequency and severity at the highest doses (800 and 1000 mg) and mainly during the loading doses and cycle 1. The most common grade 3 and 4 laboratory AEs included lymphopenia and transient prolongation of PTT; these events showed a trend to increase at the two highest doses administered (800 and 1000 mg). There were no clinical sequelae related to these laboratory findings. Preliminary evidence of activity was indicated by reductions in tumor markers in patients with prostate and ovarian cancer (PSA and CA-125, respectively) and by declines of ≥50% in total circulating tumor cells (CTCs) and Hsp27^{+} CTCs in ≥50% of patients. Potential biological activity of apatorsen as monotherapy was observed at doses ≥400 mg.

The second phase 1 study BL-01 was an open-label, investigator-sponsored trial designed for safety and pharmacodynamic assessments while dose-escalating apatorsen intravesically in patients with superficial bladder tumor or carcinoma *in situ* prior to a transurethral resection; patients with muscle-invasive disease prior to radical cystectomy were also included. Apatorsen was administered intravesically on four nonconsecutive days prior to surgery. Cohorts of three patients were treated at each dose level, and Hsp27 expression was evaluated in tumor tissues. Preliminary results include:

- Trend toward decreased Hsp27 levels and increased tumor cell death rates as apatorsen dosing increased.
- CR based on having no pathologic evidence of disease in postsurgical tissue among 5/15 (33%) of patients treated with apatorsen.
- No significant drug-related AEs and no DLTs; grade 1 gross hematuria occurred in one patient within 24 h of apatorsen administration, which spontaneously resolved.

Results from this phase 1 study in superficial bladder cancer supported the decision to proceed with development of apatorsen in metastatic bladder cancer.

An OncoGenex-sponsored phase 2 study (Borealis-1™) was designed to evaluate the safety and efficacy of apatorsen at two dose levels (600 and 1000 mg) versus placebo when combined with a standard first-line regimen of gemcitabine and cisplatin in patients with metastatic or locally unresectable transitional cell carcinoma of the urinary tract. This indication was chosen not only based on the preliminary finding that apatorsen could inhibit growth of bladder cancer cells but also because it is an indication with an unmet medical need. There are few ongoing clinical trials in advanced bladder cancer, and no therapeutic products have been approved for use in advanced bladder cancer in the past two decades. Therefore, the Borealis-1 trial was designed and is being conducted as a pivotal phase 2 study, with key features in place for the study to be a supportive phase 2 study in a new drug application. The primary endpoint is overall survival; the sample size of 180 patients has ~90% power to show a critical HR of 0.70 (H_A: HR = 0.5). A significant survival difference or a survival trend close to an observed HR of 0.70 would be clinically relevant for the treatment of bladder cancer. In addition, the trial is blinded and placebo controlled, allowing for less criticism regarding possible biased results due to either the sites or the sponsor. Two stratification factors are used at randomization: performance status and presence or absence of visceral disease. Two dose levels (600 and 1000 mg) are being evaluated to allow a clinical benefit/safety risk analysis of which dose level would be best to evaluate in a phase 3 trial. The phase 3 trial would then be the same trial design as the phase 2 study, with two arms: placebo control and the dose of apatorsen with the best clinical benefit and lowest safety risk.

Beyond the Borealis-1 study in bladder cancer, the clinical development program for apatorsen was expanded during 2012–2013 by initiating six other phase 2 studies: one additional study supporting treatment in second-line bladder cancer and five studies supporting treatment paradigms in nonsquamous and squamous NSCLC, pancreatic cancer, and prostate cancer. Clinical studies evaluating apatorsen to date are listed in Table 13.1. These additional studies, conducted as investigator-sponsored studies with key opinion leaders and oncology group networks, will help to define and prioritize which indications warrant further phase 3 development. The tumor types, brief descriptions, and rationale behind these studies are as follows.

13.3.4 Bladder Cancer

The Borealis-2™ study is a randomized, open-label phase 2 clinical trial to evaluate whether using 600-mg apatorsen in combination with docetaxel can prolong survival time compared with docetaxel alone in patients with advanced bladder cancer that has relapsed after, or is refractory to, a platinum-containing regimen. Approximately 200 subjects (100 in each arm) will be enrolled using two stratification criteria at randomization: time from prior systemic chemotherapy and Bellmunt criteria [44]. The study is powered to detect a survival benefit in the docetaxel + apatorsen arm corresponding to an HR of 0.667 (docetaxel + apatorsen/ docetaxel). This study has been designed to further support an indication for apatorsen in combination with chemotherapy in bladder cancer.

13.3.5 Lung Cancer

Two investigator-sponsored studies will evaluate apatorsen as a first-line therapy in subjects with squamous and nonsquamous NSCLC. The phase 2 randomized, placebo-controlled Spruce™ study will evaluate the effect of 600-mg apatorsen in combination with carboplatin and pemetrexed in subjects with untreated, advanced, nonsquamous NSCLC ($N = 155$). Two stratification factors are used at randomization: histology (adenocarcinoma vs. large cell carcinoma) and smoking status. The primary endpoint for the study is PFS; additional analyses will evaluate overall survival and other measures of safety and efficacy.

A second study in subjects with advanced, squamous NSCLC (Cedar™ study) will evaluate 600-mg apatorsen in combination with gemcitabine and carboplatin (*N* = 140). This investigator-sponsored study is randomized and open label. Two stratification factors are used at randomization: Stage (IIIB vs. IV vs. recurrent disease) and Eastern Cooperative Oncology Group (ECOG) performance status (0 or 1 vs. 2). The primary endpoint is PFS, and additional analyses will evaluate overall survival and other measures of safety and efficacy.

13.3.6 Pancreatic Cancer

One investigator-sponsored study will evaluate 600-mg OGX-427 as a first-line therapy in subjects with metastatic pancreatic cancer. The Rainier™ study is a phase 2, randomized, placebo-controlled study to evaluate apatorsen in combination with Abraxane® (nab-paclitaxel) and gemcitabine (*N* = 130). The primary endpoint is overall survival; additional analyses will evaluate PFS and other measures of safety and efficacy.

13.3.7 Prostate Cancer

Two studies will evaluate apatorsen in men with prostate cancer, in combination with prednisone and in combination with Zytiga® (abiraterone acetate; Janssen Biotech, Inc., Horsham, PA). A phase 2, open-label, randomized, crossover study in men with CRPC (Study PR-01) has completed enrollment. This study was designed to evaluate the antitumor effects of 1000-mg apatorsen with prednisone compared with prednisone alone in men with CRPC who have not previously received chemotherapy for metastatic disease. The primary endpoint is the proportion of patients without disease progression at 12 weeks after start of study treatment. A total of 74 patients have been randomized, and preliminary safety data have been presented for 65 randomized subjects (treatment arm, $n = 32$; control arm, $n = 33$) [45]. The most frequently reported nonlaboratory AEs among subjects treated with apatorsen were chills, diarrhea, fatigue, and nausea; the majority of AEs (93%) were grade 1 or 2. The most common laboratory AEs among subjects treated with apatorsen were lymphopenia, hyperglycemia, and elevated creatinine. Because of the high frequency of infusion reactions in the phase 1 OGX-427-01 study, especially at the higher 800- and 1000-mg apatorsen doses, the loading dose for all subjects in study PR-01 was decreased to 600 mg, and subjects received prophylactic medications. Infusion reactions were substantially reduced with these changes.

A second investigator-sponsored phase 2, open-label, randomized study in prostate cancer (Pacific™ study) was designed to evaluate the effect of adding

800-mg apatorsen to ongoing standard therapy with abiraterone and prednisone in men with metastatic CRPC who have evidence of PSA progression but no clinical evidence of symptomatic or radiographic progression ($N = 80$). Because most patients without clinical progression continue treatment with abiraterone despite biochemical PSA progression, this study presents an opportunity to assess whether apatorsen can inhibit or reverse abiraterone treatment resistance. The primary objective for this study is PFS at the milestone Day 60 assessment. In order to minimize between-arm imbalance, randomization is stratified by having prior chemotherapy and by prior PSA response to abiraterone.

13.4 Regulatory Considerations

Neither the US FDA nor the European Medicines Agency (EMA) has issued guidance specifically pertaining to ASOs, so regulatory considerations for clinical development of ASOs can be derived only from general regulations governing drug development and prior rulings on ASO compounds. A working group of development scientists has published recommended specifications for quality aspects of oligonucleotides that are being evaluated as active pharmaceutical ingredients [46], but no recommendations have been issued regarding clinical development programs.

To date, only two ASO-based drug products have been approved by the FDA, neither of which is indicated for the treatment of cancer. The first product, fomivirsen (Vitravene®), was developed by Isis Pharmaceuticals and CIBA Vision (Duluth, GA) and was approved by the FDA in <6 months in 1998. CIBA Vision filed and obtained approval of a marketing authorization application in Europe. Vitravene was used as an antiviral drug in the treatment of cytomegalovirus retinitis in immunocompromised patients, including those with acquired immunodeficiency syndrome (AIDS). Vitravene has not been sold since 2005 because human immunodeficiency virus (HIV) retinopathy became almost nonexistent with the introduction of newer anti-HIV drugs.

The second oligonucleotide approved by the FDA is mipomersen sodium (KYNAMRO™; Genzyme Corporation, Cambridge, MA), which was approved in early 2013 for patients with homozygous familial hypercholesterolemia (HoFH). Mipomersen is a second-generation ASO inhibitor targeted to apolipoprotein B-100 (apoB-100). It was developed by Isis Pharmaceuticals through phase 2 and subsequently licensed to the Genzyme Corporation, who conducted the phase 3 studies and obtained FDA approval.

Mipomersen was reviewed by an FDA advisory committee in late 2012, at which time data were presented on 586 subjects receiving mipomersen at the indicated dose of 200 mg weekly by subcutaneous injection. The committee was asked to evaluate concerns regarding drug-induced liver injury, risk for cardiovascular morbidity/mortality, and increased incidence of neoplasms in animal and human studies. Despite divided opinions among the advisory committee, the FDA approved mipomersen for use but required additional safety precautions. The KYNAMRO label carries a "black box warning" regarding the risk of hepatotoxicity based on the potential for elevations in transaminases and increased hepatic fat. Further, FDA required the manufacturer to establish a risk evaluation and mitigation strategy (REMS) program that includes elements to ensure safe use, dispensing by certified pharmacies, and restriction to HoFH patients only.

In contrast to the US FDA, the EMA Committee for Medicinal Products for Human Use (CHMP) twice issued a negative opinion on mipomersen and declined to approve the product. Main concerns of the CHMP included high rates of drug discontinuation due to side effects, possible hepatic toxicity with long-term use, and higher rates of cardiovascular events among treated patients.

Any clinical benefit/safety risk assessment is made within the context of the indication, affected patient population, robustness of the outcome, and the outcome itself. Significant improvement in survival in a patient population with advanced cancer has been the gold standard for clinical benefit and is difficult to achieve. However, improved survival outcome would allow for a reasonable tolerance of safety risks.

13.5 Future Opportunities for ASOs as Therapeutic Agents for Cancer Treatment

Innovation in ASO-based therapies will be driven by further modifications in chemical structure, which offer continuing opportunities for development of clinical compounds. These modifications in ASO chemistry are expected to result in pharmacodynamic benefits that improve dosing, enhance safety, and improve efficacy. Advances in chemistry are also expected to lead to improved delivery options, and efforts to adapt the molecular structures to allow more patient-friendly delivery options, such as subcutaneous self-injection or oral dosing, will remain important.

With advances in molecular biology, new pathways and molecular targets will be identified for which RNA-based medications are uniquely suited. These advances will also further the use of ASOs for therapeutic benefit in oncology and other disease states.

References

1 Chi, K.N. and Gleave, M.E. (2004). Antisense approaches in prostate cancer. *Expert. Opin. Biol. Ther.* 4: 927–936.

2 Gleave, M., Nelson, C., and Chi, K. (2003). Antisense targets to enhance hormone and cytotoxic therapies in advanced prostate cancer. *Curr. Drug Targets* 4: 209–221.
3 Heere-Ress, E., Thallinger, C., Lucas, T. et al. (2002). Bcl-X(L) is a chemoresistance factor in human melanoma cells that can be inhibited by antisense therapy. *Int. J. Cancer* 99: 29–34.
4 Jansen, B., Schlagbauer-Wadl, H., Brown, B.D. et al. (1998). bcl-2 antisense therapy chemosensitizes human melanoma in SCID mice. *Nat. Med.* 4: 232–234.
5 Jansen, B., Wacheck, V., Heere-Ress, E. et al. (2000). Chemosensitisation of malignant melanoma by BCL2 antisense therapy. *Lancet* 356: 1728–1733.
6 Monia, B.P., Johnston, J.F., Geiger, T. et al. (1996). Antitumor activity of a phosphorothioate antisense oligodeoxynucleotide targeted against C-raf kinase. *Nat. Med.* 2: 668–675.
7 Wacheck, V., Heere-Ress, E., Halaschek-Wiener, J. et al. (2001). Bcl-2 antisense oligonucleotides chemosensitize human gastric cancer in a SCID mouse xenotransplantation model. *J. Mol. Med.* 79: 587–593.
8 Ziegler, A., Luedke, G.H., Fabbro, D. et al. (1997). Induction of apoptosis in small-cell lung cancer cells by an antisense oligodeoxynucleotide targeting the Bcl-2 coding sequence. *J. Natl. Cancer Inst.* 89: 1027–1036.
9 Zellweger, T., Miyake, H., Cooper, S. et al. (2001). Antitumor activity of antisense clusterin oligonucleotides is improved in vitro and in vivo by incorporation of 2′-O-(2-methoxy)ethyl chemistry. *J. Pharmacol. Exp. Ther.* 298: 934–940.
10 Knudsen, K.E. and Scher, H.I. (2009). Starving the addiction: new opportunities for durable suppression of AR signaling in prostate cancer. *Clin. Cancer Res.* 15: 4792–4798.
11 Miyake, H., Nelson, C., Rennie, P.S., and Gleave, M.E. (2000). Overexpression of insulin-like growth factor binding protein-5 helps accelerate progression to androgen-independence in the human prostate LNCaP tumor model through activation of phosphatidylinositol 3′-kinase pathway. *Endocrinology* 141: 2257–2265.
12 Culig, Z. (2004). Androgen receptor cross-talk with cell signalling pathways. *Growth Factors* 22: 179–184.
13 Craft, N., Shostak, Y., Carey, M., and Sawyers, C.L. (1999). A mechanism for hormone-independent prostate cancer through modulation of androgen receptor signaling by the HER-2/neu tyrosine kinase. *Nat. Med.* 5: 280–285.
14 Gleave, M., Tolcher, A., Miyake, H. et al. (1999). Progression to androgen independence is delayed by adjuvant treatment with antisense Bcl-2 oligodeoxynucleotides after castration in the LNCaP prostate tumor model. *Clin. Cancer Res.* 5: 2891–2898.
15 Miyake, H., Tolcher, A., and Gleave, M.E. (1999). Antisense Bcl-2 oligodeoxynucleotides inhibit progression to androgen-independence after castration in the Shionogi tumor model. *Cancer Res.* 59: 4030–4034.

16 Miayake, H., Tolcher, A., and Gleave, M.E. (2000). Chemosensitization and delayed androgen-independent recurrence of prostate cancer with the use of antisense Bcl-2 oligodeoxynucleotides. *J. Natl. Cancer Inst.* 92: 34–41.
17 Rocchi, P., So, A., Kojima, S. et al. (2004). Heat shock protein 27 increases after androgen ablation and plays a cytoprotective role in hormone-refractory prostate cancer. *Cancer Res.* 64: 6595–6602.
18 Miyake, H., Pollak, M., and Gleave, M.E. (2000). Castration-induced up-regulation of insulin-like growth factor binding protein-5 potentiates insulin-like growth factor-I activity and accelerates progression to androgen independence in prostate cancer models. *Cancer Res.* 60: 3058–3064.
19 Li, N., Zoubeidi, A., Beraldi, E., and Gleave, M.E. (2013). GRP78 regulates clusterin stability, retrotranslocation and mitochondrial localization under ER stress in prostate cancer. *Oncogene* 32: 1933–1942.
20 Whitesell, L. and Lindquist, S.L. (2005). HSP90 and the chaperoning of cancer. *Nat. Rev. Cancer* 5: 761–772.
21 Miyake, H., Nelson, C., Rennie, P.S., and Gleave, M.E. (2000). Acquisition of chemoresistant phenotype by overexpression of the antiapoptotic gene testosterone-repressed prostate message-2 in prostate cancer xenograft models. *Cancer Res.* 60: 2547–2554.
22 Zoubeidi, A. and Gleave, M. (2012). Small heat shock proteins in cancer therapy and prognosis. *Int. J. Biochem. Cell Biol.* 44: 1646–1656.
23 Lamoureux, F., Thomas, C., Yin, M.J. et al. (2011). Clusterin inhibition using OGX-011 synergistically enhances Hsp90 inhibitor activity by suppressing the heat shock response in castrate resistant prostate cancer. *Cancer Res.* 71: 5838–5849.
24 Bagatell, R. and Whitesell, L. (2004). Altered Hsp90 function in cancer: a unique therapeutic opportunity. *Mol. Cancer Ther.* 3: 1021–1030.
25 Yuen, A.R., Halsey, J., Fisher, G.A. et al. (1999). Phase I study of an antisense oligonucleotide to protein kinase C-alpha (ISIS 3521/CGP 64128A) in patients with cancer. *Clin. Cancer Res.* 5: 3357–3363.
26 Yuen, A., Halsey, J., Fisher, G.A. et al. (2001). Phase I/II trial of ISIS 3521, an antisense inhibitor of PKC-alpha, with carboplatin and paclitaxel in non-small cell lung cancer [abstract]. *Proc ASCO* 20: 1234.
27 Villalona-Calero, M.A., Ritch, P., Figueroa, J.A. et al. (2004). A phase I/II study of LY900003, an antisense inhibitor of protein kinase C-alpha, in combination with cisplatin and gemcitabine in patients with advanced non-small cell lung cancer. *Clin. Cancer Res.* 10: 6086–6093.
28 Ritch, P., Rudin, C.M., Bitran, J.D. et al. (2006). Phase II study of PKC-alpha antisense oligonucleotide aprinocarsen in combination with gemcitabine and carboplatin in patients with advanced non-small cell lung cancer. *Lung Cancer* 52: 173–180.
29 Lynch, T.J. Jr., Raju, R., Lind, M. et al. (2003). Randomized phase III trial of chemotherapy and antisense oligonucleotide LY900003 (ISIS 3521) in patients with advanced NSCLC: initial report. *Proc. Am. Soc. Clin. Oncol.* 22: 2504.

30 Lynch, T. J. (2003). Progress in the treatment of metastatic NSCLC, Medscape Oncology. http://www.medscape.org/viewarticle/457540 (accessed 18 January 2018).

31 Paz-Ares, L., Douillard, J.Y., Koralewski, P. et al. (2006). Phase III study of gemcitabine and cisplatin with or without aprinocarsen, a protein kinase C-alpha antisense oligonucleotide, in patients with advanced-stage non-small-cell lung cancer. *J. Clin. Oncol.* 24: 1428–1434.

32 Bedikian, A.Y., Millward, M., Pehamberger, H. et al. (2006). Bcl-2 antisense (oblimersen sodium) plus dacarbazine in patients with advanced melanoma: the Oblimersen Melanoma Study Group. *J. Clin. Oncol.* 24: 4738–4745.

33 O'Brien, S., Moore, J.O., Boyd, T.E. et al. (2007). Randomized phase III trial of fludarabine plus cyclophosphamide with or without oblimersen sodium (Bcl-2 antisense) in patients with relapsed or refractory chronic lymphocytic leukemia. *J. Clin. Oncol.* 25: 1114–1120.

34 O'Brien, S., Moore, J.O., Boyd, T.E. et al. (2009). 5-year survival in patients with relapsed or refractory chronic lymphocytic leukemia in a randomized, phase III trial of fludarabine plus cyclophosphamide with or without oblimersen. *J. Clin. Oncol.* 27: 5208–5212.

35 Bedikian, A. Y., Lebbe, C., Robert, C., Agarwala, S. S., Cousin, T., Wu, J., Garbe, C., and Grob, J. J. (2011) Survival in a phase III, randomized, double-blind study of dacarbazine with or without oblimersen (Bcl-2 antisense) in patients with advanced melanoma and low-normal serum lactate dehydrogenase (LDH; AGENDA), *ASCO Meeting Abstracts* 29, 8531.

36 Gleave, M., Miyake, H., Zangemeister-Wittke, U., and Jansen, B. (2002). Antisense therapy: current status in prostate cancer and other malignancies. *Cancer Metastasis Rev.* 21: 79–92.

37 Henry, S., Stecker, K., Brooks, D. et al. (2000). Chemically modified oligonucleotides exhibit decreased immune stimulation in mice. *J. Pharmacol. Exp. Ther.* 292: 468–479.

38 Henry, S.P., Geary, R.S., Yu, R., and Levin, A.A. (2001). Drug properties of second-generation antisense oligonucleotides: how do they measure up to their predecessors? *Curr. Opin. Investig. Drugs* 2: 1444–1449.

39 Chi, K.N., Eisenhauer, E., Fazli, L. et al. (2005). A phase I pharmacokinetic and pharmacodynamic study of OGX-011, a 2′-methoxyethyl antisense oligonucleotide to clusterin, in patients with localized prostate cancer. *J. Natl. Cancer Inst.* 97: 1287–1296.

40 Chi, K.N., Hotte, S.J., Yu, E.Y. et al. (2010). Randomized phase II study of docetaxel and prednisone with or without OGX-011 in patients with metastatic castration-resistant prostate cancer. *J. Clin. Oncol.* 28: 4247–4254.

41 Zoubeidi, A., Zardan, A., Beraldi, E. et al. (2007). Cooperative interactions between androgen receptor (AR) and heat-shock protein 27 facilitate AR transcriptional activity. *Cancer Res.* 67: 10455–10465.

42 Shiota, M., Bishop, J.L., Nip, K.M. et al. (2013). Hsp27 regulates epithelial mesenchymal transition, metastasis, and circulating tumor cells in prostate cancer. *Cancer Res.* 73: 3109–3119.

43 Thor, A., Benz, C., Moore, D. 2nd et al. (1991). Stress response protein (srp-27) determination in primary human breast carcinomas: clinical, histologic, and prognostic correlations. *J. Natl. Cancer Inst.* 83: 170–178.

44 Bellmunt, J., Choueiri, T.K., Fougeray, R. et al. (2010). Prognostic factors in patients with advanced transitional cell carcinoma of the urothelial tract experiencing treatment failure with platinum-containing regimens. *J. Clin. Oncol.* 28: 1850–1855.

45 Chi, K. N., Hotte, S. J., Ellard, S., Gingerich, J. R., Joshua, A. M., Yu, E. Y., and Gleave, M. E. (2012) A randomized phase II study of OGX-427 plus prednisone (P) versus P alone in patients (pts) with metastatic castration resistant prostate cancer (CRPC), *ASCO Meeting Abstracts* 30, 4514.

46 Capaldi, D., Ackley, K., Brooks, D. et al. (2012). Quality Aspects of Oligonucleotide Drug Development: Specifications for Active Pharmaceutical Ingredients. *Drug Inf. J.* 46: 611–626.

47 Laskin, J.J., Nicholas, G., Lee, C. et al. (2012). Phase I/II trial of custirsen (OGX-011), an inhibitor of clusterin, in combination with a gemcitabine and platinum regimen in patients with previously untreated advanced non-small cell lung cancer. *J. Thorac. Oncol.* 7: 579–586.

48 Saad, F., Hotte, S., North, S. et al. (2011). Randomized Phase II Trial of Custirsen (OGX-011) in Combination with Docetaxel or Mitoxantrone as Second-line Therapy in Patients with Metastatic Castrate-Resistant Prostate Cancer Progressing after First-line Docetaxel: CUOG Trial P-06c. *Clin. Cancer Res.* 17: 5765–5773.

14

Inhaled Antisense for Treatment of Respiratory Disease

Gail M. Gauvreau[1], Beth E. Davis[2], and John Paul Oliveria[1]

[1] *Department of Medicine, McMaster University, Hamilton, Ontario, Canada*
[2] *Department of Medicine, University of Saskatchewan, Saskatoon, Saskatchewan, Canada*

14.1 Introduction

Asthma and chronic obstructive pulmonary disease (COPD) are increasing in prevalence and pose a significant burden to health care [1, 2]. A recent report indicates that in 2010, the economic burden of asthma and COPD in Canada was approximately 6 billion dollars with a projected increase to about 13.6 billion by 2030 [3]. In the United States, direct costs alone for asthma and COPD in 2009 were reported to be 64.2 billion [4]. Improvements in therapeutic options are therefore necessary, and respirable antisense oligonucleotides (RASONs) represent a promising new class of drug for the treatment of respiratory disorders.

14.2 Atopic Asthma

Asthma is a complex heterogeneous chronic respiratory disorder in which a growing number of clinical phenotypes have been described [5]. In all cases, the airway is prone to episodes of reversible obstruction with underlying inflammation and tissue remodeling. Atopic (often referred to as allergic) asthma, for example, describes a phenotype in which the individual has a propensity for the development of immunoglobulin E (IgE)-mediated sensitivities. Exposure to an allergenic trigger results in the characteristic symptoms of asthma: shortness of breath, chest tightness, and wheezing. Writings describing atopic asthma, in this case asthma triggered by grass allergy, date back to 1819 wherein a practitioner, John Bostock, described his own "periodical affection of the eyes and chest" after exposure to grass pollen [6]. Now, nearly one and a

Oligonucleotide-Based Drugs and Therapeutics: Preclinical and Clinical Considerations for Development, First Edition. Edited by Nicolay Ferrari and Rosanne Seguin.

half decades into the twenty-first century, we have therapies to treat and minimize the occurrence of asthma symptoms and exacerbations, but the precise mechanism(s) underlying atopic asthma remain obscure.

In carefully controlled research settings, we can induce and quantify the airway response to inhaled allergen using the allergen inhalation challenge model [7]. This model has proven to be an invaluable tool for identifying our current understanding of the mechanisms underlying allergic asthma and for assessing therapeutic efficacy, both in proof of concept and comparative pharmacological investigations. Using this model, we have learned that allergen inhalation increases airway responsiveness to methacholine [8] and leads to leukocyte recruitment [9, 10]. The response is IgE mediated and predominantly a T-helper type 2 cell (Th2) inflammatory phenotype of which the inflammatory cytokines IL-4, IL-5, and IL-13 have a major role.

14.2.1 Pharmacotherapy of Asthma

Asthma treatment guidelines generally present an algorithm of treatment options based on the level of disease control [11]. Most available treatments fall into two broad categories: controller medications and rescue medications. Rescue medications provide quick relief of bronchoconstriction and include the short-acting β_2-receptor agonists salbutamol and terbutaline. Excessive, regular use of conventional β_2 agonists resulting in tolerance to its bronchoprotective effects has also been shown to increase the airway response to allergen [12, 13] and may contribute to airway inflammation [14]. Controller medications target airway inflammation and include the gold standard, inhaled glucocorticosteroids (e.g. budesonide and fluticasone). Chronic and/or high-dose inhaled corticosteroid (ICS) can increase the risk of undesirable side effects including thrush, cataracts, bruising, decreased bone density, and suppression of the hypothalamic–pituitary–adrenal axis, particularly in the young and old [15]. In cases where asthma control remains suboptimal with stable dose ICS and infrequent beta-agonist use, the addition of leukotriene receptor antagonists and/or long-acting β-agonists (LABA) is recommended. In extreme cases, omalizumab, an anti-IgE monoclonal antibody, or systemic steroid may be utilized. Importantly, a small percentage of asthma sufferers cannot achieve an acceptable level of control using currently available treatments. It is this group that accounts for a large percentage of the overall socioeconomic burden of asthma and necessitates new therapeutic options.

Recent advances in treatment strategies have been limited to extending the duration of action of existing therapies or combining existing therapies (e.g. ICS/LABA) for delivery from a single inhaler. The seemingly slow progress is likely twofold. First, the current "gold standard" treatments (i.e. β_2 agonists and ICS) are extremely effective in the majority of patients. Second, treatment advances are driven by our level of understanding of disease mechanisms, which in turn are limited by the technologies available to study them. Over the last decade, extensive clinical research has been undertaken to develop monoclonal antibodies targeting Th2-mediated airway inflammation, and a number of molecules blocking IL-4, IL-5, and IL-13 signaling have led the way.

14.2.2 Anti-IL-5 Monoclonal Antibodies

Interleukin-5 is a key cytokine involved in the maturation, migration, activation, and survival of eosinophils. Various clinical pathologies are known to have elevated levels of eosinophils, including the recognized phenotype of "eosinophilic asthma." This suggests the potential for clinical benefit through IL-5 antagonism. Interestingly however, one of the first human studies of the effects of an anti-IL-5 monoclonal antibody (SB-240563/mepolizumab) on airway responses to inhaled allergen failed to document inhibition of the EAR, LAR, or increased airway responsiveness to histamine in mild allergic asthmatics but did dose dependently reduce circulating eosinophils [16]. A similar outcome occurred in a population of severe persistent asthmatics receiving high-dose inhaled or oral steroids wherein a different, more potent anti-IL-5 molecule (SCH55700) failed to provide meaningful clinical benefit despite significantly reducing circulating eosinophils [17]. Even when eosinophil numbers in the bronchial mucosa, bone marrow, and blood were significantly decreased following mepolizumab (750 mg intravenously once a month for 3 months), there was no improvement in clinical outcomes of airway hyperresponsiveness to histamine, FEV_1, or peak flow in mild atopic asthmatics [18]. Blocking IL-5 was subsequently shown to have differential outcomes depending on eosinophil developmental stage and tissue location. Two weeks after the last of three monthly intravenous doses of 750 mg mepolizumab, there were significant decreases in both mature eosinophils and eosinophils in the later stages of development in the bone marrow of mild asthmatics. This study also showed fewer early progenitor cells in bronchial mucosal tissue after anti-IL-5 treatment, but there was no effect on early progenitors in the blood or bone marrow at 2 weeks' post-dose [19]. The authors concluded that IL-5 was probably not acting independently in eosinophil differentiation and this might explain the lack of clinical efficacy in previous studies. Looking beyond the IL-5/eosinophil relationship, Buttner et al. [20] assessed whether blocking IL-5 would alter T-cell numbers or the Th1/Th2 cytokine expression pattern (i.e. levels of IL-2, 4, 5, 10, 13, and IFN-gamma). They showed mepolizumab had no effect on the levels of T cells, T-cell cytokines, or other circulating innate immune cells, but confirmed blocking IL-5 significantly decreases peripheral blood eosinophils and serum ECP.

Flood Page et al. [21] hypothesized that the decrease in lung tissue eosinophils might coincide with a decrease in eosinophil-derived growth factors and extracellular matrix protein deposition (e.g. tenascin, lumican, and procollagen

III) and this could, at least in theory, decrease airway tissue remodeling. They used lung biopsy and bronchoalveolar lavage samples collected from a previous clinical trial in which three doses of mepolizumab had been administered and showed significant decreases in tissue eosinophils expressing TGF-beta mRNA and TGF-beta protein following mepolizumab treatment. These data suggest that blocking IL-5 may lead to a reduction in tissue remodeling directly impacting the persistent component of airway hyperresponsiveness, decline in lung function, and increase in exacerbations. A subsequent investigation by this group looked at the effect of mepolizumab on clinical outcomes in a large cohort of 362 mostly atopic individuals with moderate persistent asthma currently using ICS [22]. A total of three intravenous doses of 250 or 750 mg were administered once a month, and endpoints of blood eosinophils, lung function, beta-agonist use, symptom score, exacerbation rate, and quality of life were assessed. No statistically significant effects on clinical outcomes were identified, but a signal for efficacy on the rate of exacerbations was evident in the group receiving the higher, 750 mg dose (50% decrease compared with placebo). Not surprisingly, both doses significantly decreased eosinophil counts in peripheral blood and in sputum of a small subset of participants.

In 2009, two independent reports in the *New England Journal of Medicine* tested the hypotheses that mepolizumab may be beneficial in refractory eosinophilic asthma [23] and prednisone-dependent asthma with sputum eosinophilia [24]. Both studies used doses of 750 mg. The Halder investigation administered 12 doses (monthly intervals), and the Nair study administered five doses (monthly intervals). Significantly fewer exacerbations occurred in the active treatment groups in both investigations. In addition, the Haldar study documented significant improvement in quality of life, and the Nair study showed a prednisone sparing effect in the mepolizumab-treated group. As expected, the blood and sputum eosinophils were significantly decreased by active treatment in both studies. The findings of these studies were confirmed in a subsequent global multicenter trial that enrolled 621 individuals with recurrent severe asthma exacerbations and eosinophilic airway inflammation, 75% of whom had concomitant allergy and were, if necessary, being currently treated with oral steroid [25]. Participants were randomized to one of four treatments: 75, 250, 750 mg mepolizumab, or placebo. Thirteen infusions were administered at monthly intervals. The primary endpoint was the rate of clinically significant asthma exacerbations. The rate in the placebo group (2.4) was significantly higher than all active treatment groups (1.24 for the 75 mg group, 1.46 in the 250 mg group, and 1.15 in the 750 mg group). Positive clinical outcomes (i.e. decreases in dose of oral corticosteroid and exacerbations as well as improvement in symptom control and quality of life) have also been achieved following subcutaneous administration (100 mg monthly) [26]. Mepolizumab received regulatory approval as an add-on therapy in severe asthma with an eosinophilic phenotype in 2015.

Reslizumab, another anti-IL-5 monoclonal antibody has shown similar clinical benefit and has also been approved (2016) as an add-on therapy in severe asthma with an eosinophilic phenotype [27–29].

A third anti-IL-5 monoclonal antibody, benralizumab, is under development. Benralizumab differs from mepolizumab and reslizumab in that it binds the receptor for IL-5 and IL-5Rα. The IL-5 receptor is located predominantly not only on eosinophils but also on basophils. An added benefit to binding the receptor is believed to be enhanced antibody-dependent cell cytotoxicity. A recent multicenter investigation documented decreased eosinophil counts in the airway mucosa, sputum, bone marrow, and blood, suggesting possible clinical benefit in eosinophilic asthma [30]. This benefit was later reported in an adult population with uncontrolled eosinophilic asthma (blood eosinophils greater than 300 cells per microliter) wherein the exacerbation rate was significantly decreased following 1 year of monthly subcutaneous dosing [31]. An additional benefit may be the use of a single intravenous dose of benralizumab when administered prior to discharge following presentation to an emergency room setting for treatment of an acute exacerbation [32].

While the benefit of anti-IL-5 treatment as add-on therapy in severe uncontrolled eosinophilic asthma is clear, the future or extent of its usefulness may be limited due to rebound eosinophilia [33] and limited cost-effectiveness [34, 35].

14.2.3 Anti-IL-4/13 Monoclonal Antibodies

Interleukin-4 and interleukin-13 are Th2 cytokines that have key roles in orchestrating the development of the airway response to inhaled allergen by mediating the underlying mechanisms of IgE production, leukocyte recruitment, tissue remodeling, mucus hypersecretion, and airway hyperresponsiveness. As such, various anti-IL-4/anti-IL-13 molecules have been developed.

One of the first molecules was altrakincept, a soluble recombinant human IL-4 receptor, shown to be beneficial in treating moderate asthma requiring ICS following a single 1500 μg nebulized dose. Versus placebo and a 500 μg dose, improvements were observed in lung function, beta$_2$-agonist use, and symptom scores following ICS withdrawal [36]. A subsequent study also in moderate persistent asthmatics requiring ICS confirmed clinical benefit after steroid withdrawal, using a 3.0 mg dose, nebulized once a week for 12 weeks [37]. More specifically, the FEV_1 in the 3.0 mg group remained similar to study entry, whereas a significant decrease in FEV_1 occurred in the placebo-treated group. No benefit was realized in the groups dosed with 0.75 or 1.5 mg.

The functional redundancy between IL-4 and IL-13 and the unopposed activity of IL-13 may play a role in the overall efficacy of blocking IL-4 alone. This was addressed using pitrakinra, a recombinant human interleukin-4 variant, which competitively inhibits the binding of both IL-4 and IL-13 with

IL-4Rα receptor subunit. Two separate allergen challenge studies were performed in individuals with mild atopic asthma. In one study, 25 mg of pitrakinra was administered subcutaneously once a day for 4 weeks, and in the other study 60 mg of pitrakinra was inhaled twice a day for 4 weeks. There was no significant change in the LAR in the group receiving subcutaneous pitrakinra; however, the decrease in LAR following inhaled pitrakinra was highly significant [38]. Although not directly tested, these data support the rationale that blocking both IL-4 and IL-13 signaling may be superior to blocking IL-4 alone. However, in a cohort of moderate to severe asthmatics using ICS anti-IL-4Rα monoclonal antibody failed to show an improvement in asthma control across three dosing regimens [39]. Conversely, the efficacy of 300 mg (once a week for 12 weeks) of dupilumab, also a monoclonal antibody to IL-4Rα, was studied in moderate to severe asthmatics with increased blood and sputum eosinophilia despite treatment with ICS and LABA [40]. The primary endpoint was the occurrence of asthma exacerbations, and a number of secondary endpoints related to asthma control were also assessed. Blocking IL-4 and IL-13 signaling with dupilumab showed a marked reduction (87%) in clinically relevant asthma exacerbations that was maintained after ICS and LABA therapy were withdrawn. Dupilumab treatment also increased percent predicted and absolute FEV_1, improved asthma control and symptom scores, and decreased FeNO, TARC, eotaxin 3, and IgE levels. A subsequent study in a population with uncontrolled persistent asthma showed similar efficacy following a 200 mg dose of dupilumab administered every 2 weeks for 24 weeks. Exacerbation rates significantly decreased, and lung function improved, on average, by nearly 250 ml [41]. Efficacy was greater in those with elevated blood eosinophils (≥300 cells/μl).

The effect of blocking IL-13 and leaving IL-4 signaling unopposed was tested in the allergen challenge model through the AllerGen NCE Inc. Clinical Investigator Collaborative in two independent studies. One study blocked the binding of IL-13 to IL-4Rα, and the other blocked the binding of IL-13 to IL-13R1 and IL-13R2 [42]. IL-13 signal inhibition via the IL-4Rα receptor resulted in significant attenuation of the EAR and LAR to inhaled allergen in mild allergic asthmatics, whereas no change in allergen inhalation challenge outcomes were observed by blocking IL-13 binding to IL-13R1 or IL-13R2. A few additional molecules targeting IL13 in asthma have been investigated. The most widely studied molecule was lebrikizumab. Initial studies generated positive data with respect to improving FEV_1 in asthmatics with high serum periostin and poor asthma control [43], but subsequent studies, both a dose-ranging study in steroid-naïve stable asthma [44] and an allergen challenge study in mild atopic asthma [45], failed. Additionally, two replicate study investigations were either canceled due to study drug impurity [46] or produced equivocal data [47]. Lebrikizumab development has been put on hold.

Tralokinumab has twice failed to meet the primary study objective in studies involving individuals with moderate to severe asthma [48, 49].

The level of fractional exhaled nitric oxide (FeNO) in a population of mild asthmatics was suppressed with anti-IL-13 treatment (GSK679586) [50]; however the molecule failed to show any benefit in severe refractory asthma [51].

In summary, monoclonal antibodies targeting IL-5 signaling have proven safe and beneficial in those who express an eosinophilic phenotype, but the cost-effectiveness of such treatment is not necessarily favorable. The benefits of targeting IL-4 and IL-13 with monoclonal antibodies for the treatment of asthma are difficult to define. While some studies have proven beneficial, others have failed to show clinical benefit or produced equivocal results in replicate studies. In these data together with the current socioeconomic burden of asthma and the projected increase in the prevalence of both asthma and allergy, there is a convincing need to pursue alternative therapies. RASONs are a promising therapeutic option for a number of reasons. These small inhalable molecules can be delivered directly to the lung and with high specificity prevent translation of key proteins (e.g. transcription factors, cytokines, receptors, intracellular signaling molecules) involved in the very complex inflammatory pathways of allergic asthma. RASONs can be developed rapidly, require less dosing frequency, have low systemic exposure, and can undergo prescreening for nonintended hybridization effects. The efficacy of these molecules in treating allergic asthma, both in preclinical and clinical investigations, has been encouraging.

14.3 Antisense Oligonucleotides in Animal Models

Novel therapeutics like antisense therapies [52–55] are first evaluated in animal models before moving into human trials [56]. Asthma is a disease with a high degree of genetic and phenotypic heterogeneity [57] and with therapeutics developed to control and manage the disease, but a cure or preventative measures have yet to be developed [58]. Since asthma is a multifaceted disease, there are an abundance of available targets for antisense therapy, all of which have the potential in mitigating the disease [59–61] (Table 14.1).

Although there is a wide range of antisense therapies tested in animal models of asthma, this chapter will focus on animal studies of oligonucleotides that have advanced to human clinical trials: 1018 immunostimulatory sequences (ISS), TPI ASM8, and IL-4Rα antisense oligonucleotides.

14.3.1 CpG Immunostimulatory Sequences

Toll-like receptor 9 (TLR-9) signaling is key for monocyte/macrophage/dendritic cell and B cell activity in the allergic cascade. There have been several

Table 14.1 Antisense therapeutics and their specific targets in animal models of asthma and the effects on asthma outcomes.

Drug	Target	Species	Route	Effect on asthma
C274 ISS	TLR-9	Rhesus macaques [62], BALB/c mice [63], C57BL/6 mice [63]	Intranasal	↑ DC activation ↑ expression of activation markers ↑ Th1 cytokines
1018 ISS	TLR-9	Rhesus macaques [63], BALB/c mice [63–65], C57BL/6 mice [63–66]	Intranasal	↑ DC activation ↑ expression of activation markers ↑ Th1 cytokines ↑ regulatory genes and proteins ↓ hyperresponsiveness of the airways ↓ airway eosinophilia, neutrophilia ↓ Th2 cytokines
1826G ISS	TLR-9	BALB/c mice [67, 68], C57BL/6 [69]	Intranasal, intraperitoneal	↑ Th1 cytokines ↑ regulatory cytokines ↑ neutralizing immunoglobulins ↓ hyperresponsiveness of the airways ↓ airway eosinophilia
1668 ISS	TLR-9	BALB/c mice [70]	Subcutaneous	↑ Th1 cytokines ↓ airway eosinophilia, neutrophilia ↓ Th2 cytokines and chemokines
AS107	Common β chain	Human cell lines (U937 and TF-1) [71]	*In vitro* culture	↓ β-chain protein ↓ eosinophil survival
AS143	Common β chain	BN rats [72, 73]	Intraperitoneal, intratracheal	↓ β-chain protein ↓ hyperresponsiveness of the airways ↓ airway eosinophilia

AS4	CCR3	BN rats [73, 74]	Intratracheal	↓ CCR3 protein ↓ hyperresponsiveness of the airways ↓ airway eosinophilia
ASM8	Common β chain and CCR3	Cynomolgus monkeys [73, 75]	Aerosol	↓ β-chain and CCR3 protein ↓ hyperresponsiveness of the airways ↓ airway eosinophilia
TPI 1100	PDE-4 and PDE-7	Cynomolgus monkeys [75]	Aerosol	↓ PDE-4 and PDE-7 protein
A1AS and B2AS	Adenosine A1 and B2 receptor	Rabbits [76, 77]	Aerosol	↓ A1 and B2 receptor protein ↓ hyperresponsiveness of the airways
IL-4Rα ASO	IL-4Rα	BALB/c mice [78, 79]	Aerosol	↓ IL-4Rα protein ↓ hyperresponsiveness of the airways ↓ airway eosinophilia, neutrophilia ↓ Th2 cytokines ↓ mucus production
IL-4 ASO	IL-4	BN rats [80, 81], BALB/c mice [82]	Adoptive transfer of ASO-treated CD4+ cells, Intranasal	↓ IL-4 protein ↓ hyperresponsiveness of the airways ↓ airway wall and smooth muscle area ↓ airway eosinophilia ↑ Th1 cytokines ↓ Th2 cytokines ↓ inflammatory immunoglobulins

(*Continued*)

Table 14.1 (Continued)

Drug	Target	Species	Route	Effect on asthma
IL-5 ASO	IL-5	BN rats [80, 83], BALB/c mice [84, 85]	Adoptive transfer of ASO-treated CD4+ cells	↓ IL-5 protein ↓ airway wall and smooth muscle area ↓ or no effect on hyperresponsiveness of the airways ↓ or no effect on airway eosinophilia
p38α ASO	p38α MAPK	BALB/c mice [86]	Aerosol	↓ p38α protein ↓ hyperresponsiveness of the airways ↓ mucus production ↓ airway eosinophilia ↑ Th1 cytokines ↓ Th2 cytokines
α_1 antitrypsin ASO	α_1 antitrypsin type 2	BALB/c mice [87], C57BL/6 mice [87]	Intranasal	↓ α_1 antitrypsin type 2 protein ↑ airway eosinophils
GATA-3 ASO	GATA-3	BALB/c [88]	Intranasal	↓ GATA-3 protein ↓ Th2 cytokines ↓ airway eosinophilia ↓ hyperresponsiveness of the airways
p65 ASO	NFκB	C57BL/6 [89]	Intravenous	↓ NFκB activation ↓ airway eosinophilia ↓ hyperresponsiveness of the airways ↓ Th2 cytokines ↓ inflammatory immunoglobulins

CCR3, C-C chemokine receptor type 3; DC, dendritic cells; IL, interleukin; NFkB, nuclear factor kappa light chain enhancer of activated B cells; PDE, phosphorodiesterase; Th1, T-helper cell type 1; Th2, T-helper cell type 2; TLR9, Toll-Like Receptor 9.

antisense therapies developed against TLR-9 that have been tested in animal models [62–70] including CpG ISS that increases Th1 responses through the production of interferons [62–64]. CpG ISS therapy is particularly important in treating Th2-mediated allergic diseases, due to the ability to tilt the balance between Th1 and Th2 immunity [65–70].

Campbell et al. [63] treated C57BL/6 and BALB/c mice with intranasal administration of 1018 ISS to investigate the cell types and cytokines implicated in ISS-induced toxicity in murine models of allergic asthma. After 1018 ISS treatment, mice had increased lung infiltrates, increased cytokine production (IFN-α, IFN-γ, TNF-α, IL-6, IL-12), and TNF-α induced weight loss, all of which was reversible several weeks after the 1018 ISS treatment. Although murine models showed that TNF-α was the key mediator for the toxicity after 1018 ISS treatment, when immune cells from human and nonhuman primates were treated *in vitro* with 1018 ISS, there was minimal to no TNF-α production. These findings suggest that 1018 ISS toxicity is TLR-9 dependent and presents mainly in murine models but has a minor impact in human and nonhuman primate models [63].

It was established by Campbell et al. [63] that 1018 ISS therapy potently inhibited Th2-mediated allergic disease by tipping the balance between Th1 and Th2 immunity toward Th1-mediated allergic disease. Since a single treatment with 1018 ISS demonstrated transient effects, a multitreatment regimen was tested that demonstrated sustained efficacy in the inhibition of Th2-mediated allergic asthma in a murine model [64]. BALB/c strain mice were intraperitoneally sensitized and intranasally challenged weekly with ragweed allergen in alum adjuvant; 1018 ISS was coadministered with the ragweed allergen in the treated mice. Mice that were chronically exposed to allergen alone had increased Th2-mediated allergic disease marked by lung inflammation, increase Th2 cytokine production (IL-4, IL-5, and IL-13), and increased eosinophilia. When mice were chronically exposed to allergen plus 1018 ISS, there was a marked reduction in allergic responses. A separate experiment in mice with established allergic airway inflammation then challenged with ragweed demonstrated that concurrent serial treatments with 1018 ISS suppressed measures of allergic asthma [64]. This study found that 5 weekly treatments with 1018 ISS were needed to reduce the Th2-mediated responses even with continual exposure to allergen posttreatment. Furthermore, concurrent exposure to ragweed plus 1018 ISS shifted the inflammatory response from a ragweed-specific Th2 response to a ragweed-specific Th1 response. The expression of Th1-associated genes (IFN-γ, CXCL9, and CXCL10) in the lungs increased after ragweed plus 1018 ISS exposure. Furthermore, ragweed-specific CD4+ T cells from the lungs were found to be Th1-like, and serum measurements showed reduced IgE and elevated IgG2a in mice treated with ragweed plus 1018 ISS. Dendritic cells from 1018 ISS-treated mice were exposed to ragweed in an attempt to restore allergic responses; however these antigen-presenting cells were unable to successfully activate Th2 cells. Microarray analysis demonstrated that there were genes differentially expressed between treatment types. Both regulatory genes and proteins for FoxP3, IL-10, and transforming growth factor β (TGF-β) and regulatory-associated proteins CD223, CD134, and CD25 were expressed at higher levels in ragweed plus 1018 ISS-treated mice compared with ragweed-only treated mice. A chronic treatment regimen with 1018 ISS in this murine model [64] skews the inflammatory response to Th1, with markedly reduced lung inflammation, reduced Th2 cytokine production, reduced IgE production, increased IgG2a production, and increased immunomodulation. Thus 1018 ISS therapies have been shown to be efficacious in controlling allergic asthma in mouse models of asthma.

14.3.2 Antisense to Receptors on Eosinophils

Granulocyte-macrophage colony-stimulating factor, interleukin-3, and interleukin-5 are important growth factors in the pathogenesis of allergic inflammation. Receptors for each of these cytokines are heterodimers composed of an α chain that binds with high affinity and specificity to its respective cytokine and a common β chain that is necessary for the transduction of cell signaling. In order to block the effects of these three cytokines in Th2-mediated responses, an antisense technology was developed to specifically inhibit the messenger ribonucleic acid of the GM-CSF, IL-3, and IL-5 receptor common β-chain(β_c).

Allam and Renzi [71] carried out a series of experiments to test beta receptor chain (βc) antisense oligodeoxynucleotide (ASO) therapy using human TF-1 and U937 cell lines. Of a test panel of 19-base ASO targeting the common βc, the nucleotide coding sequence 107–126 (5′-GGG TCT GCA GCG GGA TGG TGG T-3′) was found to effectively inhibit mRNA expression of the common β_c. The cell surface expression of the β_c was also reduced treatment in a dose-dependent and time-dependent manner.

In a cell proliferation assay, the addition of oligodeoxynucleotide AS107 to GM-CSF, IL-3, and IL-5 markedly reduced cell proliferation in comparison with the sense oligodeoxynucleotide [71]; however higher concentrations of AS107 induced toxicity in the cells. The effector role of eosinophils in the pathogenesis of allergy and allergic asthma together with their expression of GM-CSF, IL-3, and IL-5 receptors makes eosinophils a key cell for testing the efficacy of AS107. Eosinophils treated with AS107 for 15 h demonstrated a dose-dependent reduction of survival and viability.

Allakhverdi et al. [72] evaluated the effects of antisense therapy, corresponding to the coding nucleotide sequence 143–161 (AS143) in the Brown Norway rat model of asthma. Rats were sensitized and challenged subcutaneously with ovalbumin (OVA) and alum adjuvant. AS143 administered intraperitoneally (i.p.) and intratracheally (i.t.) before OVA challenge significantly reduced β_c mRNA and protein expression in the lungs compared with control rats treated with mismatched base pairs (MM143). After OVA challenge, eosinophilia in the bronchoalveolar lavage fluid was reduced by AS143, demonstrating overall inhibitory effects on cellular infiltration into the lung. Allakhverdi et al. [73] also evaluated the effect of AS143 on leukotriene $_D$4-induced bronchoconstriction in Brown Norway rats and reported a dose-dependent reduction in airway hyperresponsiveness after LT_D4 challenge. These findings suggest that AS143 is an effective treatment in the Brown Norway rat model of asthma not only by reducing eosinophilia in the lungs but also by reducing airway hyperresponsiveness [72].

Recruitment of eosinophils into the airways is driven by eotaxin and chemokines that bind to the CC chemokine receptor-3 (CCR3). CCR3 is a member of the G protein-coupled receptor family and binds to chemokines responsible for eosinophil migration. Chemokine that binds to CCR3 includes chemokine (C-C motif) ligand 5 (CCL5), CCL13, CCL24, and CCL26, which are also known as RANTES (regulated on activation, normal T cell expressed and secreted), MCP-4 (monocyte chemoattractant protein-4), eotaxin-2, and eotaxin-3, respectively.

Fortin et al. [74] tested the effects of antisense therapy against CCR3 using the ASO AS4 (5′-ACT CAT ATT CAT AGG GTG-3′) in the Brown Norway rat model of asthma. Rats sensitized and challenged with aerosolized OVA with alum adjuvant were first treated with either saline, AS4, or AS4 with mismatched base pair (MM4) through an intratracheal (i.t.) injection. These investigators described a reduced expression of CCR3 mRNA and CCR3 protein in the lungs of rats receiving AS4 treatment. This reduction translated into reduced levels of eosinophils in the bronchoalveolar lavage fluid of rats treated with AS4 when compared with control animals. There was no difference in eosinophil levels in the blood and bone marrow nor eotaxin levels in the bronchoalveolar lavage fluid and blood: thus, the inhibition of eosinophil infiltration in the lungs was not due to reduced cell migration rather than reduced circulating eosinophils or eotaxin. AS4 treatment reduced LT_D4-induced airway hyperresponsiveness in these rats.

A multitargeted ASO approach was evaluated by Allkhverdi et al. [73] by using a combination of AS4 and AS143 to target the common β_c and CCR3, respectively. Suboptimal doses of fluorescently labeled AS4 and AS143 were delivered in the lungs of Brown Norway rats by i.t. injection, and immunofluorescence confirmed specific localization of the probes in the lung tissues. Suboptimal doses of either ASO alone showed no inhibitory effects on eosinophilia or AHR; however, combination therapy demonstrated synergistic effects by significantly reducing AHR and lung eosinophilia. Additional experiments were conducted to confirm that humanized forms of AS4 and AS143 (TOP004 and TOP005, respectively) did not form complexes in solution and that the combination inhibited βc and CCR3 gene expression in peripheral blood mononuclear cells from cynomolgus monkey.

Very few studies have evaluated the effectiveness and toxicity of antisense oligonucleotide treatments after lung inhalation, which is the common administration route for asthma medications. A cynomolgus monkey model was used to test aerosolized ASM8 [75]. The inhalation of low-dose ASM8 was well tolerated; however, high-dose treatment of ASM8 resulted in some aberrant findings that were deemed unrelated to ASM8 and consistent between all treatment groups. Plasma pharmacokinetics and tissue distribution demonstrated that ASM8 was only detectable in animals treated with the highest doses, with no evidence of accumulation several weeks after administration. Furthermore, trachea tissue samples were collected from the high-dose treated monkeys, and a very low percentage of ASM8 based on the administered dose was measured. Together, these findings suggest that ASM8 inhalation at low doses does not cause toxicity in primates [75].

14.3.3 Antisense to IL-4 and IL-13 Receptors

As reviewed earlier, IL-4, IL-13, and IL-4Rα are prime targets for antisense therapy [76–82] as treatment for asthma. Evaluations of IL-4 ASO were found to be effective in reducing IL-4 levels as well as airway hyperresponsiveness and airway eosinophilia, and the level of Th1 cytokines increased and Th2 cytokines decreased after IL-4 ASO treatment [80–82]. Since IL-4 ASO treatment only blocks one Th2 cytokine, targeting the IL-4Rα receptor to block the effects of both IL-4 and IL-13 is a better approach for inhibiting this pathway.

Karras et al. [78] developed an IL-4Rα ASO to complex with the intermediary IL-4Rα and to theoretically block IL-4 and IL-13 signaling. Using a murine model of asthma to test the effects of this IL-4Rα ASO, BALB/c mice were sensitized and challenged with OVA and alum adjuvant and further administered aerosolized IL-4Rα ASO. Control groups were sensitized and challenged with saline (control group) and treated with dexamethasone (positive control group). To show that IL-4Rα ASO treatment specifically targeted IL-4Rα, cell surface expression of IL-4Rα was measured on lung eosinophils, macrophages, dendritic cells, and epithelial cells and shown to be reduced in these mice. In addition, mice treated with IL-4Rα ASO had restored airway hyperresponsiveness and airway inflammation as well as reduced Th2 cytokines (IL-5, IL-13, CXCL-8, and CCL-2) in bronchial alveolar lavage fluid. IL-4Rα ASO treatment also had an effect on mucus production as shown by reduced Muc-5A expression and mucus index.

14.3.4 Summary of Antisense Oligonucleotides in Animal Models

Antisense therapies in animal models of asthma show promising results, including regulation of eosinophilia, airway hyperresponsiveness, and Th2 cytokine production. Furthermore, toxicity has not been observed at the optimal effective doses tested in animal models, providing a strong rationale for their application in human clinical trials.

14.4 Clinical Data

14.4.1 Allergen Challenge: A Model of Asthma Exacerbation

Allergen inhalation challenge is a valuable research procedure that induces a controlled asthma exacerbation. Airway responses to inhaled allergens include the development of T-helper (Th) 2 airway inflammation, persistent bronchoconstriction, and sustained airway hyperresponsiveness [90–92]. The measurement of these allergen-induced responses permits the study of various features of asthma and the evaluation of asthma therapies targeting the critical mechanistic pathways [93]. Well-defined subject inclusion and exclusion criteria are required to ensure subjects' safety and data integrity [94]. Additional safety measures can be applied using allergen skin prick testing and airway hyperresponsiveness to predict the allergen PC_{20} [95, 96], and by stepwise challenging subjects with increasing allergen concentrations to avoid the possibility of overdose.

Assessments of bronchoconstriction, sputum eosinophils, and soluble mediators have shown consistent and reproducible increases following allergen challenges [97–101] thus allowing for investigation of drug intervention [102]. Care must be taken to ensure sufficient recovery from a previous allergen challenge before performing subsequent challenges, and a 14-day recovery between consecutive challenges has been successfully implemented [103–114]

14.4.2 Allergen Challenge for Evaluation of Efficacy

Generally, inhibition of the late asthmatic response and associated sequelae in early efficacy studies has proven to be an effective predictor of clinical efficacy [7, 115]. For most established asthma controller therapies, inhibition of allergen-induced increases in sputum eosinophils has been shown to be associated with inhibition of the late asthmatic response [103, 116, 117], and the ICS demonstrate superior clinical efficacy in parallel with the most complete inhibition of the allergen-induced late phase sequelae [103, 106, 116, 118–120]. New experimental therapies, such an anti-TSLP, are beginning to show similar efficacy [121]. Drugs acting at several points upstream of the inflammatory cascade are generally less specific, while targeted therapies generally need to show at least 50% protection against the LAR (and/or associated sequelae) to prove clinical efficacy. Although the magnitude of the actual clinical benefit cannot be extrapolated from its effects on allergen response, the model shows an excellent negative predictive value [7, 115] and provides proof-of-mechanism approach for asthma investigational therapies such as antisense. Delivery of antisense to the lung is also an ideal target because of the ease of direct access and the likelihood of low systemic toxicity from deposited aerosols [64, 116]. A limited number of antisense therapies have been tested for efficacy in asthma using the allergen challenge model.

14.4.3 1018 Immunostimulatory Sequence

1018 ISS is a CpG-containing oligonucleotide that exerts an immunomodulatory effect through activation of Toll-like receptor 9 (TLR-9) [122–125]. Studies in animal models for allergic asthma demonstrate that treatment with 1018 ISS leads to significant inhibition of the Th2 response as well as inhibition of many features of allergic asthma, such as AHR, eosinophilia, and the production of Th2 cytokines in the airways [52, 126–129].

14.4.3.1 Study Design for 1018 ISS

A multicentre, double-blind, randomized, placebo-controlled parallel group study evaluated for the first time in asthmatic subjects the effects of 1018 ISS on allergen-induced responses [114]. Forty subjects were randomized to either 1018 ISS ($n = 21$) or placebo ($n = 19$) in a 1 : 1 ratio. Allergen challenges were conducted 2 weeks and 4 weeks of post-dosing to evaluate efficacy across several time points because the onset of 1018 ISS activity was unknown (Figure 14.1).

Randomized subjects reported to the laboratory for four consecutive weeks for treatment with 36 mg 1018 ISS administered by inhalation. The dose of 1018 ISS was selected based on efficacy in preclinical experiments and tolerability in preclinical and phase 1 (unpublished data), and, notably, this dose was associated with changes in mRNA indicative of bioactivity of 1018 ISS. The 1018 ISS was delivered with a PARI LC STAR™ jet nebulizer filled with 2.5 ml of solution and inhaled over 7.0–8.5 min until the nebulizer cup was empty. An estimated 20% of the actual dose was delivered to the lung. Study medication was administered in the laboratory, and all subjects completing the study were 100% compliant with treatment. As 1018 ISS had not been used previously in asthmatic subjects, safety measures were put in place to determine the acute effects on lung function and systemic markers of inflammation. Spirometry, vital signs, and asthma symptomatology were evaluated at 30 and 60 min post-dose, and a safety visit was scheduled 2–6 days after inhalation of the first dose, with a physical exam and blood tests to assess tolerability. Subjects reported to the lab 1 month later for a follow-up visit including a physical exam, electrocardiogram, chest X-ray, pregnancy test and blood for pregnancy test, antinuclear antibodies, IgE levels, hematology, and chemistry.

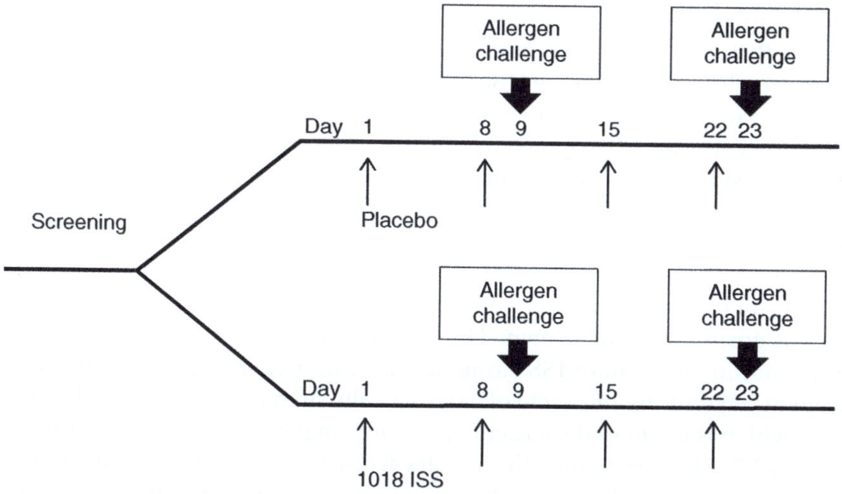

Figure 14.1 Parallel group design to study the effects of 1018 on allergen-induced responses. 1018 ISS or placebo was administered on days 1, 8, 15, and 22, and allergen challenges were conducted on days 9 and 23.

14.4.3.2 Results for 1018 ISS

There were no serious adverse events reported during this study, and no adverse event considered to be related to the study medication. The most common adverse event was headache, which was reported by six subjects in the placebo group and six subjects in the 1018 ISS group. Neither bronchoconstriction nor changes in vital signs were observed during or after the inhalation period.

Sputum cells and peripheral blood mononuclear cells were lysed for total RNA extraction. Expression of IFN-inducible genes was upregulated in peripheral blood mononuclear cells within 24 h and in sputum cells within 31 h of treatment with 1018 ISS but not placebo, indicating that 1018 ISS was pharmacologically active in the airways [114]. Changes of these IFN-inducible genes in peripheral blood mononuclear cells were smaller and more variable than in sputum, returned to baseline by 48 h of post-dosing, whereas IFN-inducible genes in sputum cells remained elevated 48 h of post-dosing. These differences likely reflect the size of oligonucleotides that are not readily passed into the blood when given as an aerosol.

There was no inhibition of allergen-induced Th2 mRNA measured from inflammatory cells measured in the airways, nor a reduction of allergen-induced bronchoconstriction or airway hyperresponsiveness. These results contrast those in murine models of asthma showing that dosing with 1018 ISS for 4 weeks inhibits allergen-induced airway hyperresponsiveness and airway eosinophilia [65, 126, 128, 129]. Results from animal models, however, are not necessarily predictive of results in humans; ISS specifically target TLR-9, which is preferentially activated by CpG motifs that are different between humans and mice [130, 131]. In addition, TLR-9 is expressed on a broader range of cells in mice, including plasmacytoid dendritic cells, B cells, mast cells, and basophils, as well as monocytes/macrophages. Such wide expression of TLR-9 on murine immune cells permits ISS to directly target several immune functions, including the induction of Th1 and modification of antigen presentation.

Dosing for this study was selected based on the highest amount of 1018 ISS for which safety data were available. The optimal dose and required length of dosing was not known. Therapeutic benefit in asthmatic subjects through administration of ISS alone may require extending the duration of treatment beyond the life cycle of existing cells such as memory T cells and B cells, which is estimated to be considerably longer than the 4-week dosing period [132, 133]. Although the results from the current study have not shown an effect on allergen-induced responses in this model of allergic asthma, we did observe a pharmacological effect through the induction of IFN-inducible genes. 1018 ISS has been successfully developed as an adjuvant for hepatitis B vaccination [134]; however the complex nature of ISS immune effects and optimal dosing leading to regulation of Th2 pathways in humans for treatment of allergic asthma is the focus of ongoing experimentation [64].

14.4.4 AIR645

The IL-4/IL-13 pathway has been shown to play a key role in asthma pathogenesis by promoting airway hyperresponsiveness and lung inflammation in preclinical models [135]. Protein inhibitors of IL-4 and IL-13 administered as intravenous and subcutaneous formulations were showing promise in early human clinical trials in asthma [38–51]. Inhaled antisense molecules provide an alternate approach for inhibition of the IL-4/IL-13 pathway through direct targeting of the lungs, and the pharmacokinetics, safety, and efficacy of inhaled antisense to IL-4Rα have been described [136, 137].

AIR645 is an ASO targeted to the IL-4Rα chain. This oligonucleotide is synthesized using second-generation chemistry to increase hybridization affinity and protection from nuclease-mediated degradation, which in turn enhances the stability of the oligonucleotide in tissue. As such, AIR645 has the potential for less frequent administration with improved local and systemic potency. When delivered topically to the airways, AIR645 is expected to inhibit expression of IL-4Rα and thereby inhibit downstream inflammatory events with potential for treatment of allergic asthma. Results from a phase 1 study of inhaled AIR645 demonstrated an excellent safety profile following single or multiple dose [137].

14.4.4.1 Study Design for AIR645

A multicenter, randomized, double-blind, placebo-controlled, parallel group bronchial allergen challenge study was conducted to test the efficacy of AIR645 on allergen-induced outcomes. Thirty subjects with well-controlled mild allergic asthma were randomized 1 : 1 to either 20 mg of AIR645 or placebo, and each subject received six doses of study drug over a period of 22 days. During the first week subjects were dosed three times (days 1, 3, 5) for loading and then again on study days 8, 15, and 22. Allergen challenges were carried out on study day 24 (Figure 14.2). The dose of 20 mg was selected because it was the highest dose given to subjects in the phase 1 study. AIR645 was diluted with 0.9% sodium chloride into a volume of 3.5 ml immediately before use and administered by nebulization using a PARI LC SPRINT® breath enhanced nebulizer, with slow deep breaths to inhale the aerosol medication slowly from the mouthpiece until the nebulizer was empty. The duration of drug administration was approximately 5–10 min.

14.4.4.2 Results for AIR645

There were no dose-limiting toxicities, serious or severe AEs following dosing in subjects with mild asthma. AEs were mostly mild and did not lead to study discontinuation. There were no clinically significant changes in laboratory safety tests or spirometry, and the drug was well tolerated.

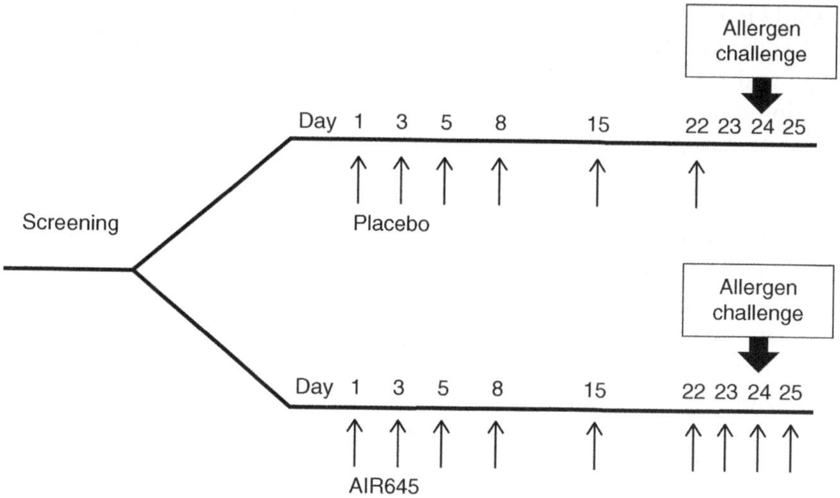

Figure 14.2 Parallel group design to study the effects of AIR645 on allergen-induced responses. AIR645 or placebo was administered on days 1, 3, 5, 8, 15, and 22. Allergen challenge was conducted on day 24.

A subset of subjects provided paired sputum samples on days 1 and 8 to examine changes in IL-4Rα and IL-4Rα-inducible proteins (VCAM-1) in permeabilized macrophages, using an assay developed for flow cytometry [138]. Following AIR645 treatment, the mean IL-4Rα protein MFI was reduced by 47% in AIR645-treated subjects (range: −74 to +1% increase) compared with 21% reduction in placebo subjects (range: −66 to +270%). Furthermore, the mean MFI of the IL-4-dependent VCAM-1 protein was reduced by 38% (range: −4 to −63%) in AIR645-treated subjects compared with an increase of 12% in the placebo group (range: −23 to +183%) (unpublished data). These differences between AIR645 and placebo treatment are suggestive of a pharmacologic effect of AIR645 in the airways; however they did not reach the level of statistical significance.

There was no effect of AIR645 on the allergen-induced late asthmatic response, nor on measurements of inflammation (sputum eosinophils, fractional exhaled oxide) or airway hyperresponsiveness. This is in contrast to studies of an IL-4 variant and IL-13 monoclonal antibody showing some attenuation of the allergen-induced responses [38] that indicate IL-4Rα is an effective target in the allergen challenge model. To date, no additional dose-optimization studies have been carried out with AIR645 in allergic asthma.

14.4.5 TPI ASM8

14.4.5.1 Mechanism of TPI ASM8

TPI ASM8 is a drug consisting of 1 : 1 ratio of two synthetic antisense oligonucleotides: the first is TOP 004, an ASO that targets the IL-3/5/GM-CSF common β_c, and the second is TOP 005, an ASO against the eotaxin receptor, CCR3. Inhibiting expression of β_c on inflammatory cells such as mast cells, eosinophils, basophils, neutrophils, and macrophages renders these cells unable to proliferate and survive in response to the growth factors IL-3, IL-5, and GM-CSF. The CCR3 receptor is particularly important for the recruitment of eosinophils, basophils, and their precursors to the inflamed lungs through binding with chemokines such as eotaxin, which is expressed at increased levels in the airways of patients with asthma [139]. Owing to the blockade of two critical pathways in the same cell, combining these two oligonucleotides results in higher blockade of eosinophils at lower doses compared with either oligonucleotide alone [72–75]. Thus TPI ASM8 was developed to potently inhibit CCR3 and β_c mRNA and protein expression [72–74] and thereby reduce the influx of eosinophils in the airways in allergic asthma [73, 75]. Two studies have been published examining TPI ASM8, each conducted for proof of concept of activity in the airways before and after allergen challenge. Importantly, this is the first antisense approach to be reported as efficacious in the airways of subjects with asthma.

14.4.5.2 Study #1 for TPI ASM8

14.4.5.2.1 First Study Design for TPI ASM8

The first study to examine the effect of TPI ASM8 on allergen-induced responses was designed as a placebo-controlled crossover trial, which compared 4 days treatment with a nominal dose of 1500 μg inhaled TPI ASM8, to placebo in 17 subjects [104]. TPI ASM8 was administered by inhaling 1 ml from an Aeroneb nebulizer with an output of 0.49 ml/min. Lung deposition was estimated to be 6–15%. The dose selected for this study was based on the results of the phase 1 trials demonstrating tolerability of TPI ASM8 at doses up to 6 mg, and this dose was then divided over 4 days. Each treatment period consisted of five consecutive morning visits with dosing on days 1–4. On day 3, the third dose was administered 30 min before an inhaled allergen challenge (Figure 14.3).

14.4.5.2.2 First Results for TPI ASM8

Activation of the alternative complement system was monitored by measurement of the plasma levels of Bb, safety labs, and immunogenicity skin test. Following dosing, hematology, chemistry, urinalysis, and complement pathway activation was similar across the two groups. Immunogenicity to TPI ASM8 was tested using both skin prick test and intradermal skin testing. A test was considered positive if subjects developed erythema with a wheal that increased by greater than 3 mm compared with PBS. Skin tests were positive in 1 of 17 subjects on TPI ASM8, and intradermal tests demonstrated similar low frequency of positive tests for both TPI ASM8 and placebo.

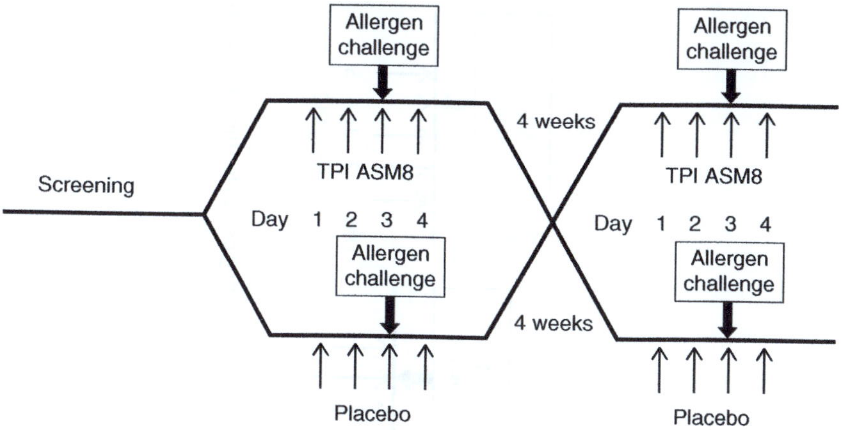

Figure 14.3 Crossover design to study the effects of TPI ASM8 on allergen-induced responses. TPI ASM8 or placebo was administered on days 1–4. Allergen challenge was conducted on day 3.

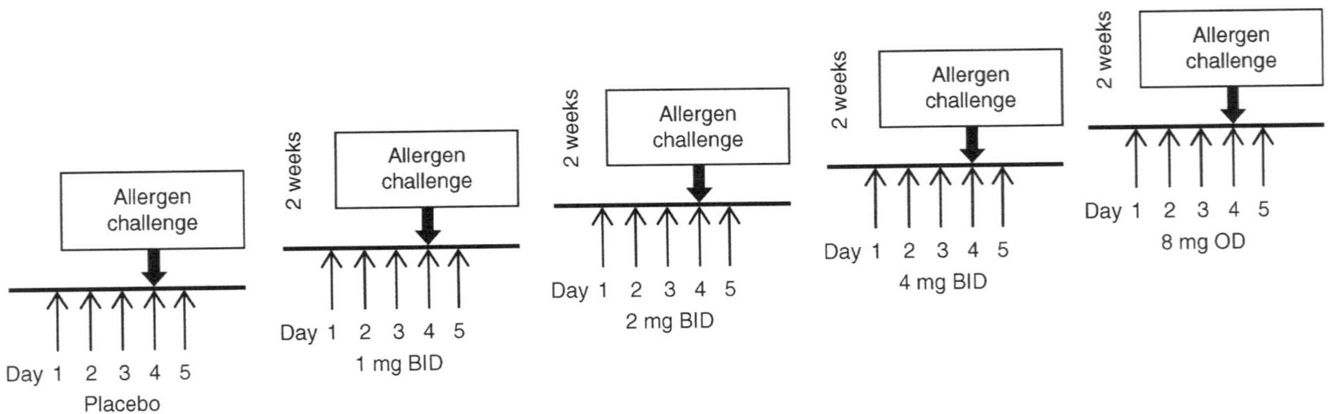

Figure 14.4 Dose-escalating design to study the effects of TPI ASM8 on allergen-induced responses.

Peak sputum drug concentrations were generally attained between 2.5- and 4.5-h post-dose, and these levels dropped steadily over the 24h throughout post-dosing period [107]. A subset of subjects provided sputum samples for assessment of mRNA levels. At 7h following allergen challenge, there was a 6.0-fold and 4.5-fold increase in β_c and CCR3 mRNA in sputum cells compared with preallergen; treatment with TPI ASM8 significantly inhibited the increase in β_c mRNA ($p = 0.039$) and attenuated the increase in CCR3 mRNA in six of the eight samples assessed [107]. TOP004 and TOP005 were not measurable in the plasma of any subjects at any time after administration.

TPI ASM8 attenuated the early and late asthmatic responses; this effect on the late asthmatic response was not statistically significant when evaluated across all study subjects, largely due to one subject with no detectable TPI ASM8 on all sputum PK measurements. With this subject removed TPI ASM8 significantly inhibited the late asthmatic response. The increase in sputum eosinophils that occurs 7h after allergen was also significantly inhibited by TPI ASM8 treatment [107], and this reduction in sputum eosinophils with TPI ASM8 treatment supports the concept of reduced trafficking of eosinophils to the airways.

14.4.5.3 Study #2 for TPI ASM8

14.4.5.3.1 Second Study Design for TPI ASM8

The second study to examine the effect of TPI ASM8 on allergen-induced responses was designed as single-center, open-label, dose-ranging study that evaluated stepwise increasing doses of TPI ASM8 (1mg BID, 2mg BID, 4mg BID, and 8mg OD) inhaled daily for 4 consecutive days [140]. TPI ASM8 was administered using an I-Neb™nebulizer that delivered a set dose of medication in 2–3min. Allergen challenges were conducted on the third day of dosing, with a 2–3 week washout between treatment periods (Figure 14.4).

14.4.5.3.2 Second Results for TPI ASM8

Complement pathway activation measured by Bb split product measurements remained within normal ranges during study, and lung function as measured by carbon dioxide diffusing capacity was not negatively affected by TPI ASM8. There were no serious adverse events, and none of the nonserious adverse events reported during the study were considered to be related to the drug. There were no increases in inflammatory cells in the airways following treatment.

The allergen-induced increase in mRNA levels for β_c and CCR3 was attenuated by TPI ASM8 (8mg OD) by 49 and 42%, respectively, demonstrating a clear effect of on target mRNA. TPI ASM8 significantly inhibited allergen-induced increase in sputum eosinophils, ECP, and total sputum cell numbers. The highest doses significantly attenuated the early and late asthmatic responses, and furthermore there was a steady decrease in airway hyperresponsiveness throughout the study [140]. We believe that the actions of TOP004 and TOP005 were indeed occurring in the airways, as drug levels were consistently not measurable in the plasma in both studies. In sputum, drug levels measured 7-h post-dosing were consistently higher than 24-h post-dosing, reflecting the short half-life of oligonucleotides. Despite this short half-life, the steadily improving airway hyperresponsiveness measured each time subjects returned for the next treatment period during the dose-ranging study suggests that the biological effects are more prolonged from this acute inhibition of the gene targets.

14.5 General Conclusion

Development of antisense therapy aimed at treatment of allergic asthma is an intriguing area of study and has the advantage of being administered directly to the lungs by inhalation. The studies described here were the first to deliver oligonucleotides to asthmatic subjects by inhalation. Pharmacological effects of these inhaled oligonucleotides were indeed observed in cells collected from airway secretions, including the induction of IFNs and IFN-inducible genes (by 1018 ISS), downregulation of IL-4Rα (by AIR645), and suppression of allergen-induced increases in β_c and CCR3 mRNA (TPI ASM8).

It is clear from these studies that inhaled oligonucleotides at the doses administered do not have harmful proinflammatory effects, as there was no increase in inflammatory cells or markers of activation. Even 1018 ISS, which was specifically designed to be immunostimulatory, had no negative impact on eosinophil influx or airway tone following allergen challenge.

For measures of efficacy there are several limitations associated with the selection of the drug dose. The doses administered in these studies were chosen based only on the highest doses that had been tested in phase 1 studies showing safety and tolerability in humans. No markers of efficacy in humans existed at the time of protocol development. Preclinical testing provides some evidence for efficacy; however models of asthma in mice and rats are not necessarily predictive of results in humans due to differences in immune activation, antigen presentation, and protein expression. Moving forward, the complex nature of oligonucleotide interactions in asthmatic airways for the purpose of therapeutic intervention will require further experimentation and refinement.

References

1 Masoli, M., Fabian, D., Holt, S. et al. (2004). The global burden of asthma: executive summary of the GINA Dissemination Committee report. *Allergy* 59 (5): 469–478.

2 Bousquet, J. and Khaltaev, N. (2007). *Global surveillance, prevention and control of chronic respiratory diseases. A comprehensive approach.* World Health Organization.

3 Hermus, G., Stonebridge, C., Goldfarb, D., et al. Cost Risk Analysis for Chronic Lung Disease in Canada. (2012). The conference board of Canada March 2012. p. 25.
4 National Heart, Lung and Blood Institute (2012). Fact book fiscal year. National Institutes of Health. p. 51. https://www.nhlbi.nih.gov/about/documents/factbook/2012 (accessed 13 April 2017).
5 Wenzel, S.E. (2012). Asthma phenotypes: the evolution from clinical to molecular approaches. *Nat. Med.* 18 (5): 716–725.
6 Bostock, J. (1819). Case of a periodical affection of the eyes and chest. *Medico Chirurgical Trans* 10: 161–164.
7 Boulet, L.P., Gauvreau, G., Boulay, M.E. et al. (2007). The allergen bronchoprovocation model: an important tool for the investigation of new asthma anti-inflammatory therapies. *Allergy* 62 (10): 1101–1110.
8 Cockcroft, D.W., Ruffin, R.E., Dolovich, J., and Hargreave, F.E. (1977). Allergen-induced increase in non-allergic bronchial reactivity. *Clin. Allergy* 7 (6): 503–513.
9 De Monchy, J.G.R., Kauffman, H.F., Venge, P. et al. (1985). Bronchoalveolar eosinophilia during allergen-induced late asthmatic reactions. *Am. Rev. Respir. Dis.* 131 (3): 373–376.
10 Pin, I., Freitag, A.P., O'Byrne, P.M. et al. (1992). Changes in the cellular profile of induced sputum after allergen-induced asthmatic responses. *Am. Rev. Respir. Dis.* 145 (6): 1265–1269.
11 Lougheed, M.D., Leniere, C., Ducharme, F.M. et al. (2012). Canadian Thoracic Society 2012 guideline update: diagnosis and management of asthma in preschoolers, children and adults: executive summary. *Can. Respir. J.* 19 (6): e81–e88.
12 O'Connor, B.J., Aikman, S.L., and Barnes, P.J. (1992). Tolerance to the nonbronchodilator effects of inhaled beta 2-agonists in asthma. *N. Engl. J. Med.* 327 (17): 1204–1208.
13 Cockcroft, D.W., McParland, C.P., Britto, S.A. et al. (1993). Regular inhaled salbutamol and airway responsiveness to allergen. *Lancet* 342 (8875): 833–837.
14 Gordon, J.R., Swystun, V.A., Li, F. et al. (2003). Regular salbutamol use increases CXCL8 responses in asthma: relationship to the eosinophil response. *Eur. Respir. J.* 22 (1): 118–126.
15 Dahl, R. (2006). Systemic side effects of inhaled corticosteroids in patients with asthma. *Respir. Med.* 100 (8): 1307–1317.
16 Leckie, M.J., ten Brinke, A., Khan, J. et al. (2000). Effects of an interleukin-5 blocking monoclonal antibody on eosinophils, airway hyper-responsiveness, and the late asthmatic response. *Lancet* 356: 2144–2148.
17 Kips, J.C., O'Connor, B.J., Langley, S.J. et al. (2003). Effect of SCH55700, a humanized anti-human interleukin-5 antibody, in severe persistent asthma. *Am. J. Respir. Crit. Care Med.* 167: 1655–1659.
18 Flood-Page, P.T., Menzies-Gow, A.N., Kay, A.B., and Robinson, D.S. (2003). Eosinophil's role remains uncertain as anti-interleukin-5 only partially depletes numbers in asthmatic airway. *Am. J. Respir. Crit. Care Med.* 167: 199–204.

19 Menzies-Gow, A., Flood-Page, P., Sehmi, R. et al. (2003). Anti-IL-5 (mepolizumab) therapy induces bone marrow eosinophil maturational arrest and decreases eosinophil progenitors in the bronchial mucosa of atopic asthmatics. *J. Allergy Clin. Immunol.* 111: 714–719.
20 Buttner, C., Lun, A., Splettstoesser, T. et al. (2003). Monoclonal anti-interleukin-5 treatment suppresses eosinophil but not T-cell functions. *Eur. Respir. J.* 21: 799–803.
21 Flood-Page, P., Menzies-Gow, A., Phipps, S. et al. (2003). Anti-IL-5 treatment reduces deposition of ECM proteins in the bronchial subepithelial basement membrane of mild atopic asthmatics. *J. Clin. Invest.* 112: 1029–1036.
22 Flood-Page, P., Swenson, C., Faiferman, I. et al. (2007). A study to evaluate safety and efficacy of mepolizumab in patients with moderate persistent asthma. *Am. J. Respir. Crit. Care Med.* 176: 1062–1071.
23 Halder, P., Brightling, C.E., Hargadon, B. et al. (2009). Mepolizumab and exacerbations of refractory eosinophilic asthma. *N. Engl. J. Med.* 360: 973–984.
24 Nair, P., Pizzichini, M.M.M., Kjarsgaard, M. et al. (2009). Mepolizumab for prednisone-dependent asthma with sputum eosinophilia. *N. Engl. J. Med.* 360: 985–993.
25 Pavord, I.D., Korn, S., Howarth, P. et al. (2012). Mepolizumab for severe eosinophilic asthma (DREAM): a multicenter, double-blind, placebo-controlled trial. *Lancet* 380: 61–69.
26 Bel, E.H., Wenzel, S.E., Thompson, P.J. et al. (2014). Oral glucocorticoid-sparing effect of mepolizumab in eosinophilic asthma. *Lancet* 371: 1189–1197.
27 Castro, M., Mathur, S., Hargreave, F.E. et al. (2011). Reslizumab for poorly controlled, eosinophilic asthma: a randomized, placebo-controlled study. *Am. J. Respir. Crit. Care Med.* 184: 1125–1132.
28 Corren, J., Weinstein, S., Janka, L. et al. (2016). Phase 3 study of reslizumab in patients with poorly controlled asthma. Effects across a broad range of eosinophil counts. *Chest* 150: 799–810.
29 Castro, M., Zangrilli, J., Wechsler, M.E. et al. Reslizumab for inadequately controlled asthma with elevated blood eosinophil counts: results from two multicenter, parallel, double-blind, randomized, placebo-controlled, phase 3 trials. *Lancet Respir. Med.* 3: 355–366.
30 Laviolette, M., Gossage, D., Gauvreau, G. et al. (2013). Effects of benralizumab on airway eosinophils in asthmatic patients with sputum eosinophilia. *J. Allergy Clin. Immunol.* 132: 1086–1096.
31 Castro, M., Wenzel, S.E., Bleecker, E.R. et al. (2014). Benralizumab, an anti-interleukin 5 receptor alpha monoclonal antibody, versus placebo for uncontrolled eosinophilic asthma: a phase 2b randomized dose-ranging study. *Lancet Respir. Med.* 2: 878–890.
32 Nowak, R.M., Parker, J.M., Silverman, R.A. et al. (2015). A randomized trial of benralizumab, an antiinterleukin 5 receptor alpha monoclonal antibody, after acute asthma. *Am. J. Emerg. Med.* 33: 14–20.

33 Haldar, P., Brightling, C.E., Singapuri, A. et al. (2014). Outcomes after cessation of mepolizumab therapy in severe eosinophilic asthma: a 12-month follow-up analysis. *J. Allergy Clin. Immunol.* 133: 921–922.

34 Whittington, M.D., McQueen, R.B., Ollendorf, D.A. et al. (2017). Assessing the value of mepolizumab for severe eosinophilic asthma: a cost-effectiveness analysis. *Ann Allergy Asthma Immunol* 118: 220–225.

35 Basu, A., Dalal, A., Canonica, G.W. et al. (2017). Economic analysis of the phase III MENSA study evaluating mepolizumab for severe asthma with eosinophilic phenotype. *Expert Rev. Pharmacoecon. Outcomes Res.* 17: 121–131.

36 Borish, L.C., Nelson, H.S., Lanz, M.J. et al. (1999). Interleukin-4 receptor in moderate atopic asthma. A phase I/II randomized, placebo-controlled trial. *Am. J. Respir. Crit. Care Med.* 160: 1816–1823.

37 Borish, L.C., Nelson, H.S., Corren, J. et al. (2001). Efficacy of soluble IL-4 receptor for the treatment of adults with asthma. *J. Allergy Clin. Immunol.* 107: 963–970.

38 Wenzel, S., Wilbraham, D., Fuller, R. et al. (2007). Effect of an interleukin-4 variant on late phase asthmatic response to allergen challenge in asthmatic patients: results of two phase 2a studies. *Lancet* 370 (9596): 1422–1431.

39 Corren, J., Busse, W., Meltzer, E.O. et al. (2010). A randomized, controlled, phase 2 study of AMG 317, an IL-4R alpha antagonist, in patients with asthma. *Am. J. Respir. Crit. Care Med.* 181: 788–796.

40 Wenzel, S., Ford, L., Pearlman, D. et al. (2013). Dupilumab in persistent asthma with elevated eosinophil levels. *N. Engl. J. Med.* 368: 2455–2466.

41 Wenzel, S., Castro, M., Corren, J. et al. (2016). Dupilumab efficacy and safety in adults with uncontrolled persistent asthma despite use of medium-to-high-dose inhaled corticosteroids plus a long-acting β_2 agonist: a randomized double-blind placebo-controlled pivotal phase 2b dose-ranging trial. *Lancet* 388: 31–44.

42 Gauvreau, G.M., Boulet, L.-P., Cockcroft, D.W. et al. (2011). Effects of interleukin-13 blockade on allergen-induced airway responses in mild atopic asthma. *Am. J. Respir. Crit. Care Med.* 183: 1007–1014.

43 Corren, J., Lemanske, R.F., Hanania, N.A. et al. (2011). Lebrikizumab treatment in adults with asthma. *N. Engl. J. Med.* 365: 1088–1098.

44 Noonan, M., Korenblast, P., Mosesova, S. et al. (2013). Dose-ranging study of lebrikizumab in asthmatic patients not receiving inhaled steroids. *J. Allergy Clin. Immunol.* 132: 567–574.

45 Scheerens, H., Arron, J.R., Zheng, Y. et al. (2013). The effects of lebrikizumab in patients with mild asthma following whole lung allergen challenge. *Clin. Exp. Allergy* 44: 38–46.

46 Hanania, N.A., Noonan, M., Corren, J. et al. (2015). Lebrikizumab in moderate-to-severe asthma: pooled data from two randomised placebo-controlled studies. *Thorax* 70: 748–756.

47 Hanania, N.A., Korenblat, P., Chapman, K.R. et al. (2016). Efficacy and safety of lebrikizumab in patients with uncontrolled asthma (LAVOLTA I and LAVOLTA II): replicate, phase 3, randomised, double-blind, placebo-controlled trials. *Lancet Respir. Med.* 4: 781–796.

48 Piper, E., Brightling, C., Niven, R. et al. (2013). A phase II placebo-controlled study of tralokinumab in moderate-to-severe asthma. *Eur. Respir. J.* 41: 330–338.

49 Brightling, C.E., Chanez, P., Leigh, R. et al. (2015). Efficacy and safety of tralokinumab in patients with severe uncontrolled asthma: a randomized, double-blind, placebo-controlled, phase 2b trial. *Lancet Respir. Med.* 3: 692–701.

50 Hodsman, P., Ashman, C., Cahn, A. et al. (2012). A phase 1, randomized, placebo-controlled, dose-escalation study of an anti-IL-13 monoclonal antibody in healthy subjects and mild asthmatics. *Br. J. Clin. Pharmacol.* 75: 118–128.

51 DeBoever, E.H., Ashman, C., Cahn, A.P. et al. (2014). Efficacy and safety of an anti-IL-13 mAb in patients with severe asthma: a randomized trial. *J. Allergy Clin. Immunol.* 133: 989–996.

52 Tanaka, M. and Nyce, J.W. (2001). Respirable antisense oligonucleotides: a new drug class for respiratory disease. *Respir. Res.* 2: 5–9.

53 Sandrasagra, A., Tang, L., Leonard, S.A. et al. (2001). RASONs: a novel antisense oligonucleotide therapeutic approach for asthma. *Expert. Opin. Biol. Ther.* 1: 979–983.

54 Nyce, J. (2002). Respirable antisense oligonucleotides: a new, third drug class targeting respiratory disease. *Curr. Opin. Allergy Clin. Immunol.* 2: 533–536.

55 Séguin, R.M. and Ferrari, N. (2009). Emerging oligonucleotide therapies for asthma and chronic obstructive pulmonary disease. *Expert Opin. Investig. Drugs* 18: 1505–1517.

56 Zosky, G.R. and Sly, P.D. (2007). Animal models of asthma. *Clin. Exp. Allergy* 37: 973–988.

57 Bush, A. and Menzies-Gow, A. (2009). Phenotypic differences between pediatric and adult asthma. *Proc. Am. Thorac. Soc.* 6: 712–719.

58 Anderson, S.D. (2008). Provocative challenges to help diagnose and monitor asthma: exercise, methacholine, adenosine, and mannitol. *Curr. Opin. Pulm. Med.* 14: 39–45.

59 Popescu, F.-D. and Popescu, F. (2007). A review of antisense therapeutic interventions for molecular biological targets in asthma. *Biologics* 1: 271–283.

60 Popescu, F.-D. (2005). Antisense- and RNA interference-based therapeutic strategies in allergy. *J. Cell. Mol. Med.* 9: 840–853.

61 Horner, A.A., Van Uden, J.H., Zubeldia, J.M. et al. (2001). DNA-based immunotherapeutics for the treatment of allergic disease. *Immunol. Rev.* 179: 102–118.

62 Teleshova, N., Kenney, J., Jones, J. et al. (2004). CpG-C immunostimulatory oligodeoxyribonucleotide activation of plasmacytoid dendritic cells in rhesus macaques to augment the activation of IFN-gamma-secreting simian immunodeficiency virus-specific T cells. *J. Immunol.* 173: 1647–1657.

63 Campbell, J.D., Cho, Y., Foster, M.L. et al. (2009). CpG-containing immunostimulatory DNA sequences elicit TNF-alpha-dependent toxicity in rodents but not in humans. *J. Clin. Invest.* 119: 2564–2576.

64 Campbell, J.D., Kell, S.A., Kozy, H.M. et al. (2014). A limited CpG-containing oligodeoxynucleotide therapy regimen induces sustained suppression of allergic airway inflammation in mice. *Thorax* 1.

65 Sur, S., Wild, J.S., Choudhury, B.K. et al. (1999). Long term prevention of allergic lung inflammation in a mouse model of asthma by CpG oligodeoxynucleotides. *J. Immunol.* 162: 6284–6293.

66 Kline, J.N., Kitagaki, K., Businga, T.R., and Jain, V.V. (2002). Treatment of established asthma in a murine model using CpG oligodeoxynucleotides. *Am. J. Phys. Lung Cell. Mol. Phys.* 283: L170–L179.

67 Mousavi, T., Salek Moghadam, A., Falak, R., and Tebyanian, M. (2008). Co-administration of CpG oligonucleotides and chenopodium album extract reverse IgG2a/IgG1 ratios and increase IFN-gamma and IL-10 productions in a murine model of asthma. *Iran. J. Allergy Asthma Immunol.* 7: 1–6.

68 Fedulov, A., Silverman, E., Xiang, Y. et al. (2005). Immunostimulatory CpG oligonucleotides abrogate allergic susceptibility in a murine model of maternal asthma transmission. *J. Immunol.* 175: 4292–4300.

69 Kitagaki, K., Jain, V.V., Businga, T.R. et al. (2002). Immunomodulatory effects of CpG oligodeoxynucleotides on established th2 responses. *Clin. Diagn. Lab. Immunol.* 9: 1260–1269.

70 Ashino, S., Wakita, D., Zhang, Y. et al. (2008). CpG-ODN inhibits airway inflammation at effector phase through down-regulation of antigen-specific Th2-cell migration into lung. *Int. Immunol.* 20: 259–266.

71 Allam, M. and Renzi, P.M. (2001). Inhibition of GM-CSF/IL-3/IL-5 signaling by antisense oligodeoxynucleotides targeting the common beta chain of their receptors. *Antisense Nucleic Acid Drug Dev.* 11: 289–300.

72 Allakhverdi, Z., Allam, M., and Renzi, P.M. (2002). Inhibition of antigen-induced eosinophilia and airway hyperresponsiveness by antisense oligonucleotides directed against the common beta chain of IL-3, IL-5, GM-CSF receptors in a rat model of allergic asthma. *Am. J. Respir. Crit. Care Med.* 165: 1015–1021.

73 Allakhverdi, Z., Allam, M., Guimond, A. et al. (2006). Multitargeted approach using antisense oligonucleotides for the treatment of asthma. *Ann. N. Y. Acad. Sci.* 1082: 62–73.

74 Fortin, M., Ferrari, N., Higgins, M.-E. et al. (2006). Effects of antisense oligodeoxynucleotides targeting CCR3 on the airway response to antigen in rats. *Oligonucleotides* 16: 203–212.

75 Guimond, A., Viau, E., Aubé, P. et al. (2008). Advantageous toxicity profile of inhaled antisense oligonucleotides following chronic dosing in non-human primates. *Pulm. Pharmacol. Ther.* 21: 845–854.

76 Nyce, J.W. and Metzger, W.J. (1997). DNA antisense therapy for asthma in an animal model. *Nature* 385: 721–725.
77 Ali, S., Leonard, S.A., Kukoly, C.A. et al. (2001). Absorption, distribution, metabolism, and excretion of a respirable antisense oligonucleotide for asthma. *Am. J. Respir. Crit. Care Med.* 163: 989–993.
78 Karras, J.G., Crosby, J.R., Guha, M. et al. (2007). Anti-inflammatory activity of inhaled IL-4 receptor-alpha antisense oligonucleotide in mice. *Am. J. Respir. Cell Mol. Biol.* 36: 276–285.
79 Ripple, M.J., You, D., Honnegowda, S. et al. (2010). Immunomodulation with IL-4R alpha antisense oligonucleotide prevents respiratory syncytial virus-mediated pulmonary disease. *J. Immunol.* 185: 4804–4811.
80 Molet, S., Ramos-Barbón, D., Martin, J.G., and Hamid, Q. (1999). Adoptively transferred late allergic response is inhibited by IL-4, but not IL-5, antisense oligonucleotide. *J. Allergy Clin. Immunol.* 104: 205–214.
81 Cao, Y., Zeng, D., Song, Q. et al. (2010). The effects of antisense interleukin-4 gene transferred by recombinant adeno-associated virus vector on the airway remodeling in allergic rats. *J. Asthma* 47: 951–958.
82 Seong, J.-H., Lee, K.-M., Kim, S.T. et al. (2006). Polyethylenimine-based antisense oligodeoxynucleotides of IL-4 suppress the production of IL-4 in a murine model of airway inflammation. *J. Gene. Med.* 8: 314–323.
83 Zeng, D., Cao, Y., Song, Q. et al. (2011). Recombinant adeno-associated virus vector-mediated delivery of antisense interleukin-5 gene attenuates airway remodeling in allergic rats. *Int. Arch. Allergy Immunol.* 154: 207–215.
84 Huang, F., Wang, C., Zhou, S. et al. (2009). Antisense interleukin-5 reduces eosinophil infiltration and hyperresponsiveness in an allergic asthma model. *Asian Pac. J. Allergy Immunol.* 27: 35–41.
85 Karras, J.G., McGraw, K., RA, M.K. et al. (2000). Inhibition of antigen-induced eosinophilia and late phase airway hyperresponsiveness by an IL-5 antisense oligonucleotide in mouse models of asthma. *J. Immunol.* 164: 5409–5415.
86 Duan, W., Chan, J.H.P., McKay, K. et al. (2005). Inhaled p38alpha mitogen-activated protein kinase antisense oligonucleotide attenuates asthma in mice. *Am. J. Respir. Crit. Care Med.* 171: 571–578.
87 Sung, H.-N., Jeon, C.-H., Gill, B.-C. et al. (2009). Enhancement of ovalbumin-induced pulmonary eosinophilia by intranasal administration of alpha1-proteinase inhibitor type 2 antisense oligonucleotides. *Immunol. Lett.* 122: 76–83.
88 Finotto, S., De Sanctis, G.T., Lehr, H.A. et al. (2001). Treatment of allergic airway inflammation and hyperresponsiveness by antisense-induced local blockade of GATA-3 expression. *J. Exp. Med.* 193: 1247–1260.
89 Choi, I.-W., Kim, D.-K., Ko, H.-M., and Lee, H.-K. (2004). Administration of antisense phosphorothioate oligonucleotide to the p65 subunit of NF-kappaB inhibits established asthmatic reaction in mice. *Int. Immunopharmacol.* 4: 1817–1828.

90 Gauvreau, G.M., Lee, J.M., Watson, R.M. et al. (2000). Increased numbers of both airway basophils and mast cells after allergen inhalation challenge in atopic asthmatics. *Am. J. Respir. Crit. Care Med.* 161: 1473–1478.

91 Kawayama, T., O'Byrne, P.M., Watson, R.M. et al. (2006). Effects of inhaled ciclesonide on circulating T-helper type 1/T-helper type 2 cells in atopic asthmatics after allergen challenge. *Clin. Exp. Allergy* 36: 1417–1424.

92 Cockcroft, D.W. and Murdock, K.Y. (1987). Changes in bronchial responsiveness to histamine at intervals after allergen challenge. *Thorax* 42: 302–308.

93 Gauvreau, G.M., Harris, J.M., Boulet, L.P. et al. (2014). Targeting membrane-expressed IgE B cell receptor with an antibody to the M1 prime epitope reduces IgE production. *Sci. Transl. Med.* 6 (243): 243ra85.

94 Sterk, P.J., Fabbri, L.M., Quanjer, P.H. et al. (1993). Airway responsiveness. Standardized challenge testing with pharmacological, physical and sensitizing stimuli in adults. Report working party standardization of lung function tests, European Community for Steel and Coal. Official statement of the European Respiratory Society. *Eur. Respir. J. (suppl)* 16: 53–83.

95 Cockcroft, D.W., Murdock, K.Y., Kirby, J., and Hargreave, F. (1987). Prediction of airway responsiveness to allergen from skin sensitivity to allergen and airway responsiveness to histamine. *Am. Rev. Respir. Dis.* 135 (1): 264–267.

96 Cockcroft, D.W., Davis, B.E., Boulet, L.P. et al. (2005). The links between allergen skin test sensitivity, airway responsiveness and airway response to allergen. *Allergy* 60 (1): 56–59.

97 Gauvreau, G.M., Watson, R.M., and O'Byrne, P.M. (1999). Kinetics of allergen-induced airway eosinophilic cytokine production and airway inflammation. *Am. J. Respir. Crit. Care Med.* 160: 640–647.

98 Boot, J.D., De Haas, S., Tarasevych, S. et al. (2007). Effect of an nk_1/nk_2 receptor antagonist on airway responses and inflammation to allergen in asthma. *Am. J. Respir. Crit. Care Med.* 175 (5): 450–457.

99 Gauvreau, G.M., Watson, R.M., Rerecich, T.J. et al. (1999). Repeatability of allergen-induced airway inflammation. *J. Allergy Clin. Immunol.* 104 (1): 66–71.

100 Kharitonov, S.A., O'Connor, B.J., Evans, D.J., and Barnes, P.J. (1995). Allergen-induced late asthmatic reactions are associated with elevation of exhaled nitric oxide. *Am. J. Respir. Crit. Care Med.* 151: 1894–1899.

101 Duong, M., Cockcroft, D., Boulet, L.P. et al. (2008). The effect of IVX-0142, a heparin-derived hypersulfated disaccharide, on the allergic airway responses in asthma. *Allergy* 63 (9): 1195–1201.

102 Diamant, Z., Boot, J.D., Kamerling, I., and Bjermer, L. (2008). Methods used in clinical development of novel anti-asthma therapies. *Respir. Med.* 102 (3): 332–338.

103 Gauvreau, G.M., Doctor, J., Watson, R.M. et al. (1996). Effects of inhaled budesonide on allergen-induced airway responses and airway inflammation. *Am. J. Respir. Crit. Care Med.* 154 (5): 1267–1271.

104 Gauvreau, G.M., Jordana, M., Watson, R.M. et al. (1997). Effect of regular inhaled albuterol on allergen-induced late responses and sputum eosinophils in asthmatic subjects. *Am. J. Respir. Crit. Care Med.* 156 (6): 1738–1745.

105 Gauvreau, G.M., Watson, R.M., and O'Byrne, P.M. (1999). Protective effects of inhaled PGE2 on allergen-induced airway responses and airway inflammation. *Am. J. Respir. Crit. Care Med.* 159 (1): 31–36.

106 Inman, M.D., Watson, R.M., Rerecich, T. et al. (2001). Dose-dependent effects of inhaled mometasone furoate on airway function and inflammation after allergen inhalation challenge. *Am. J. Respir. Crit. Care Med.* 164 (4): 569–574.

107 Gauvreau, G.M., Boulet, L.P., Cockcroft, D.W. et al. (2008). Antisense therapy against CCR3 and the common beta chain attenuates allergen-induced eosinophilic responses. *Am. J. Respir. Crit. Care Med.* 177 (9): 952–958.

108 Diamant, Z., Timmers, M.C., van der Veen, H. et al. (1995). The effect of MK-0591, a novel 5-lipoxygenase activating protein inhibitor, on leukotriene biosynthesis and allergen-induced airway responses in asthmatic subjects in vivo. *J. Allergy Clin. Immunol.* 95: 42–51.

109 Diamant, Z., Van der Veen, H., Kuijpers, E.A. et al. (1996). The effect of inhaled thiorphan on allergen-induced airway responses in asthmatic subjects. *Clin. Exp. Allergy* 26: 525–532.

110 Diamant, Z., Grootendorst, D.C., Veselic-Charvat, M. et al. (1999). The effect of montelukast (MK-0476), a cysteinyl leukotriene receptor antagonist, on allergen-induced airway responses and sputum cell counts in asthma. *Clin. Exp. Allergy* 29: 42–51.

111 Diamant, Z., Kuperus, J., Baan, R. et al. (2005). Effect of a very late antigen-4 antagonist on allergen-induced airway responses and inflammation in asthma. *Clin. Exp. Allergy* 35: 1080–1087.

112 Kent, S.E., Boyce, M., Diamant, Z. et al. (2013). The 5-lipoxygenase activating protein inhibitor, GSK2190915, attenuates the early and late responses to inhaled allergen in mild asthma. *Clin. Exp. Allergy* 43 (2): 177–186.

113 Fahy, J.V., Cockcroft, D.W., Boulet, L.P. et al. (1999). Effect of aerosolized anti-IgE (E25) on airway responses to inhaled allergen in asthmatic subjects. *Am. J. Respir. Crit. Care Med.* 160 (3): 1023–1027.

114 Gauvreau, G.M., Hessel, E.M., Boulet, L.P. et al. (2006). Immunostimulatory sequences regulate interferon-inducible genes but not allergic airway responses. *Am. J. Respir. Crit. Care Med.* 174 (1): 15–20.

115 O'Byrne, P.M. (2009). Allergen-induced airway inflammation and its therapeutic intervention. *Allergy, Asthma Immunol. Res.* 1 (1): 3–9.

116 Gauvreau, G.M., Boulet, L.P., Postma, D.S. et al. (2005). Effect of low-dose ciclesonide on allergen-induced responses in subjects with mild allergic asthma. *J. Allergy Clin. Immunol.* 116 (2): 285–291.

117 Ruddy, M., Zuiker, R., Morelli, N. et al. (2010). Detection of increased levels of Th2-derived cytokines in ultracentrifuged sputum following allergen challenge and their responsiveness to fluticasone. *Am. J. Respir. Crit. Care Med.* A4043.

118 Kidney, J.C., Boulet, L.P., Hargreave, F.E. et al. (1997). Evaluation of single-dose inhaled corticosteroid activity with an allergen challenge model. *J. Allergy Clin. Immunol.* 100 (1): 65–70.

119 Swystun, V.A., Bhagat, R., Kalra, S. et al. (1998). Comparison of 3 different doses of budesonide and placebo on the early asthmatic response to inhaled allergen. *J. Allergy Clin. Immunol.* 102 (3): 363–367.

120 Parameswaran, K., Inman, M.D., Watson, R.M. et al. (2000). Protective effects of fluticasone on allergen-induced airway responses and sputum inflammatory markers. *Can. Respir. J.* 7 (4): 313–319.

121 Gauvreau, G.M., O'Byrne, P.M., Boulet, L.-P. et al. (2014 May 29). Effects of an anti-TSLP antibody on allergen-induced asthmatic responses. *N. Engl. J. Med.* 370 (22): 2102–2110.

122 Templin, M., Levin, A., Graham, M. et al. (2000). Pharmacokinetic and toxicity profile of a phosphorothioate oligonucleotide following inhalation delivery to lung in mice. *Antisense Nucleic Acid Drug Dev.* 10 (5): 359–368.

123 Krieg, A.M. (2002). CpG motifs in bacterial DNA and their immune effects. *Annu. Rev. Immunol.* 20: 709.

124 Chuang, T.H., Lee, J., Kline, L. et al. (2002). Toll-like receptor 9 mediates CpG-DNA signaling. *J. Leukoc. Biol.* 71: 538.

125 Takeshita, F., Leifer, C.A., Gursel, I. et al. (2001). Cutting edge: role of toll-like receptor 9 in CpG DNA-induced activation of human cells. *J. Immunol.* 167: 3555.

126 Broide, D., Schwarze, J., Tighe, H. et al. (1998). Immunostimulatory DNA sequences inhibit IL-5, eosinophilic inflammation, and airway hyperresponsiveness in mice. *J. Immunol.* 161: 7054–7062.

127 Shirota, H., Sano, K., Kikuchi, T. et al. (2000). Regulation of T-helper type 2 cell and airway eosinophilia by transmucosal coadministration of antigen and oligodeoxynucleotides containing CpG motifs. *Am. J. Respir. Cell Mol. Biol.* 22: 176.

128 Broide, D.H., Stachnick, G., Castaneda, D. et al. (2001). Systemic administration of immunostimulatory DNA sequences mediates reversible inhibition of Th2 responses in a mouse model of asthma. *J. Clin. Immunol.* 21: 175–182.

129 Kline, J.N., Waldschmidt, T.J., Businga, T.R. et al. (1998). Modulation of airway inflammation by CpG oligodeoxynucleotides in a murine model of asthma. *J. Immunol.* 160: 2555–2559.

130 Roberts, T.L., Sweet, M.J., Hume, D.A., and Stacey, K.J. (2005). Cutting edge: species-specific TLR9-mediated recognition of CpG and non-CpG phosphorothioate-modified oligonucleotides. *J. Immunol.* 174: 605–608.

131 Bauer, S., Kirschning, C.J., Hacker, H. et al. (2001). Human TLR9 confers responsiveness to bacterial DNA via species-specific CpG motif recognition. *Proc. Natl. Acad. Sci. U. S. A.* 98: 9237–9242.

132 Pawelec, G., Mariani, E., Bradley, B., and Solana, R. (2000). Longevity in vitro of human CD4+ T helper cell clones derived from young donors and elderly donors, or from progenitor cells: age-associated differences in cell surface molecule expression and cytokine secretion. *Biogerontology* 1: 247–254.

133 McCarron, M., Osborne, Y., Story, C.J. et al. (1987). Effect of age on lymphocyte proliferation. *Mech. Ageing Dev.* 41: 211–218.

134 Janssen RS, Mangoo-Karim R, Pergola PE, Girndt M, Namini H, Rahman S, Bennett SR, Heyward WL, Martin JT. Immunogenicity and safety of an investigational hepatitis B vaccine with a toll-like receptor 9 agonist adjuvant (HBsAg-1018) compared with a licensed hepatitis B vaccine in patients with chronic kidney disease. *Vaccine* 4 November 2013;31(46):5306–5313.

135 Wills-Karp, M., Luyimbazi, J., Xu, X. et al. (1998). Interleukin-13: central mediator of allergic asthma. *Science* 282 (5397): 2258–2261.

136 Ball, H.A., Sandrasagra, A., Tang, L. et al. (2003). Clinical potential of respirable antisense oligonucleotides (RASONs) in asthma. *Am. J. Pharmacogenomics* 3: 97–106.

137 Hodges, M.R., Castelloe, E., Chen, A. et al. (2009). Randomized, double-blind, placebo controlled first in human study of inhaled AIR645, an IL-4R alpha oligonucleotide, in healthy volunteers. *Am. J. Respir. Crit. Care Med.* 179: A3640.

138 Beaudin, S.M., Karras, J.G., Strinich, T.X. et al. (2010). Detection of IL-4 receptor alpha in sputum eosinophils and Macrophages. *Am. J. Respir. Crit. Care Med.* 181: A1266.

139 Lamkhioued, B., Renzi, P.M., Abi-Younes, S. et al. (1997). Increased expression of eotaxin in bronchoalveolar lavage and airways of asthmatics contributes to the chemotaxis of eosinophils to the site of inflammation. *J. Immunol.* 159 (9): 4593–4601.

140 Gauvreau, G.M., Pageau, R., Séguin, R. et al. (2011). Dose response effects of TPI ASM8 in asthmatics after allergen. *Allergy* 66 (9): 1242–1248.

15

Antisense Oligonucleotides for Treatment of Neurological Diseases

Rosanne Seguin

Montreal Neurological Institute, McGill University, Montreal, Quebec, Canada

15.1 Introduction

An expanding area of interest is in the use of antisense oligonucleotides (ASOs) in the research of neurodegenerative disorders such as Alzheimer's disease (AD), Parkinson's disease, Huntington's disease (HD), spinal muscular atrophy, amyotrophic lateral sclerosis, and multiple sclerosis (MS). Gene silencing techniques of ASO and siRNA have been utilized *in vitro* and *in vivo* studies to assist in identifying potential therapeutic targets for these diseases. More recently for several neurodegenerative disorders, ASO therapy has moved from preclinical stage development to efficacy assessment in clinical trials.

15.1.1 Delivery of ASO to Central Nervous System

In the majority of clinical trials to date, treatment with ASO has been through systemic delivery, and other chapters in this book discuss the challenges with this route of delivery. A limitation in developing ASO as treatment for neurodegenerative diseases has been the delivery of ASO to the central nervous system (CNS) in concentrations sufficient to render a therapeutic effect. The blood–brain barrier (BBB) and the blood–spinal cord barrier (BSCB) remain formidable challenges for accessing cells in the CNS [1, 2]. In animal models of the diseases, both intraparenchymal and intracerebroventricular routes of administration have been used to directly deliver ASO to the CNS and spinal cord [2–4]. While intrathecal (IT) administration is possible in mice, it is far easier to achieve in larger species [2, 3]. In humans, IT injection circumvents the BBB/BSCB through direct entry into the cerebrospinal fluid (CSF), delivering ASO into the subarachnoid space of the spinal cord [5] where it will pass

Oligonucleotide-Based Drugs and Therapeutics: Preclinical and Clinical Considerations for Development, First Edition. Edited by Nicolay Ferrari and Rosanne Seguin.

the pia mater to enter the parenchyma, to be taken up by neurons and glial cells [6, 7]. Thus, the IT delivery route enables access throughout the entire CNS, distributing ASO to regions of the brain and spinal cord necessary to obtain a clinical effect. Furthermore, it is possible that the IT route would circumvent some of the challenges associated with systemic delivery including, but not limited to, binding of ASO to plasma proteins, clearance of ASO by the liver and kidney, and induction of proinflammatory effects in the periphery. IT delivery may enable lower doses of ASO to be administered as compared with systemic delivery yet with more effectiveness due to proximity of ASO to the target cells [8, 9]. For humans, IT administration is preferred route for clinical applications and has been shown to be well tolerated [10]. Delivering ASO by IT injection makes them realistic options for neurodegenerative diseases, which often have no therapies available and for which there is a high unmet need.

15.2 Potential ASO Therapies in Neurodegenerative Diseases

15.2.1 Spinal Muscular Atrophy (SMA)

The leading cause of hereditary infant mortality is the spinal muscular atrophy (SMA) [11], which also is the second most common autosomal recessive-inherited disorder in humans [12]. SMA affects an estimated 1 in 6000 to 1 in 10 000 births [11]. The symptoms of SMA are muscle atrophy and weakness resulting from motor neuron degeneration in the spinal cord and brain stem [11]. Paralysis, brain stem defects, and respiratory defects are the primary manifestations of this disease and ultimately lead to a shortened life span. SMA is classified into clinical subtypes based on the severity of the disease and the age of onset [11]. SMA is caused by mutations in or deletions of the survival motor neuron (*SMN*)1 gene, which in turn leads to insufficient levels of the SMN protein [13]. SMN protein is critical for motor neuron maintenance [14]. Humans, unlike mice, have a nearly identical second *SMN* gene copy, *SMN2* [13]. Both *SMN1* and *SMN2* genes encode SMN protein; however, while *SMN1* encodes a functional full-length protein, an estimated 90% of *SMN2* mRNA transcripts encode a truncated SMN protein that is rapidly degraded, with perhaps 10% of transcripts encoding full-length protein. The severity of SMA is partially determined by the number of copies of the *SMN2* gene [15]. *SMN2* differs from *SMN1* by 11 nucleotide substitutions: seven in intron 6, two in intron 7, and one in exon 7 [16, 17]. The truncation is due to the nucleotide substitution in exon 7 (a translationally silent C to T transition) that leads to a differential splicing of the *SMN2* pre-mRNA, producing an isoform lacking exon 7 [16, 17]. As the inclusion of exon 7 in the *SMN2* gene product results in a fully functional SMN protein, research has been devoted to identifying means of modifying this altered splicing to render exon 7 intact. A 15-nucleotide-long motif downstream of the 5'ss of exon 7, termed intronic splicing silencer N1 (ISS-N1), was identified whereby deletion of ISS-N1 fully restored inclusion of *SMN2* exon 7 [18]. Furthermore, when ASO targeting ISS-N1 was delivered *in vitro* to cultured fibroblasts from a SMA patient, exon 7 inclusion in *SMN2* was fully restored [18]. Numerous studies with a various modified ASO targeting ISS-N1 were effective in SMA mouse models to increase inclusion of exon 7 in mRNA transcripts, leading to the production of full-length SMN protein, increased body weight, improved neuromuscular junction maturation, protected motor neurons, improved motor neuron function, and increased median survival of the animals (reviewed in [19]).

Nusinersen (ISIS-SMN_{Rx} or ISIS 396443), a 18-mer-modified 2′-*O*-2-methoxyethyl (MOE) phosphorothioate (PS) ASO that binds to the specific sequence in the intron downstream of exon 7 on the *SMN2* pre-mRNA [20, 21] with the goal of blocking the exon skipping of exon 7 leading to the production of full-length SMN protein, was tested in an open-label, phase 1 study. A single dose of nusinersen (1, 3, 6, or 9 mg) was delivered IT via lumbar puncture (LP) into children aged 2–14 years with documented homozygous deletion of *SMN1* gene and exhibited clinical symptoms of SMA (clinical trials.gov NCT01494701) [22]. The goals of the study were to determine the safety, tolerability, pharmacokinetics, and preliminary clinical efficacy of nusinersen at four ascending single-dose levels [22]. The post-dose monitoring period was 29 days for the cohorts exposed to 1- or 3-mg dose and 85 days for the cohorts exposed to 6 and 9 mg. Patients were eligible to receive additional study drug by enrolling in an open-label extension study (clinicaltrials.gov NCT01780246) at 9–14 months after their initial dose of nusinersen. No serious adverse events (SAEs) were observed nor did any patient experience an adverse event (AE) that led to discontinuation in the study. Most common AEs were headache, post-LP headache, and back pain. Thus, a single dose of up to 9 mg of nusinersen was shown to be well tolerated and not associated with safety or tolerability concerns. Moreover, the pharmacokinetic profiles showed nusinersen was cleared from CSF into systemic circulation consistent with normal CSF turnover, with peak plasma concentrations at 1.7–6 h post-IT administration [22]. These plasma levels slowly declined over the next 20 h and then declined even further over the next 7 days [22]. Nusinersen CSF concentrations are still quantifiable 9–14 months after dosing, indicating prolonged CSF and CNS tissue exposure [22]. At 9–14 months' post-dosing, SMN protein levels in the CSF had a mean increase of 161% as compared with baseline for the 9-mg dose group ($p = 0.06$) [22]. In terms of clinical efficacy, the 9-mg dose increased motor function up to 9–14 months' post-treatment as assessed by the Hammersmith Functional Motor Scale Expanded (HFMSE) [22]. None of the other dose groups demonstrated clinically meaningful changes. Key finding from this study was the tolerability of children with SMA for LP administration of nusinersen and the favorable risk profile of this drug.

In the open-label dose-escalating phase 2 trial (NCT01839656), infants of between 3 weeks and 7 months of age, who had *SMN1* homozygous gene deletion and symptoms of SMA, received loading doses of either 6 or 12 mg on days 1, 15, and 85 and then 12-mg doses on day 253 and every 4 months thereafter [23]. Overall the repeated IT injections were well tolerated by the infants. The serious AEs were considered not related or unlikely related to the study drug. Most common were respiratory distresses or respiratory infections [23]. All 15 of 15 participants in the 12-mg group demonstrated improvements in developmental motor milestones, with a significant change in the Hammersmith Infant Neurological Exam part 2 (HINE-2) score from baseline to last visit ($p < 0.0001$) [23]. Patients in the 12-mg dose group demonstrated various motor milestones ranging from grasping, ability to kick, sitting, rolling, standing, crawling, and walking [23]. At the date of analysis, the median age of death or permanent ventilation had not been reached with most trial participants surviving without permanent ventilation [23]. In the 6-mg dose group, one patient died at 12.6 months from accidental asphyxia, whereas in the 12-mg dose group, one died from SMA disease progression at 6.9 months, and two died from pulmonary infection at 5.1 and 11.2 months. Autopsy data from three of the infants who died during the study showed drug concentrations in all areas of the spinal cord at pharmacologically relevant concentrations [23]. Immunohistochemical staining confirmed localization of nusinersen within neurons, glial cells, and vascular endothelial cells in the CNS. The clearance of nusinersen from the CSF into systemic circulation was demonstrated by the localization of nusinersen in peripheral tissue of the liver and kidney [23]. Primary route of elimination of nusinersen and its shortened metabolites is through urinary excretions. Analysis of the spinal cord tissue from these patients showed 50–69% of SMN2 transcripts containing exon 7, which corresponded to 2.6-fold increase in full-length SMN2 mRNA as compared with untreated infants with SMA [23]. This study demonstrated the tolerability of nusinersen with repeated IT injections as well as clinical efficacy.

The phase 3, placebo-controlled, double-blinded trials ENDEAR (NCT02193074) and CHERISH (NCT02292537) were both terminated early because the primary endpoints at interim analyses were met, and the patients rolled over into open-label extension studies. In 2016, nusinersen was approved in the United States for IT use in pediatric and adult patients with SMA [24].

The story of nusinersen in SMA demonstrated the successful transition from fundamental discovery of the ISS-N1 target through preclinical validations and finally independent clinical trials. However, the need for SMA therapies to be developed remains. There is the risk that nusinersen may not act with equal efficacy in all SMA patients either due to low number of *SMN2* copies or the delayed age of administration. Nevertheless, the door is open to the development of other ASO for SMA either by targeting other regions (e.g. intron 6 or ISS-N2 in intron 7) of *SMN2*.

15.2.2 Amyotrophic Lateral Sclerosis (ALS)

Amyotrophic lateral sclerosis (ALS) is the most common of the motor diseases, with an incidence of approximately 2/100 000 [25]. ALS is a rapidly progressive neurodegenerative disorder caused by the selective death of motor neurons within the brain, brain stem, and spinal cord [26]. Symptoms of ALS include progressive loss of muscle mass, strength, and function with all muscles under voluntary control, except those muscles responsible for eye movement and urinary sphincter function, which are relatively spared [26]. The disease is terminal, usually through respiratory failure with most individuals succumbing within 2–5 years after diagnosis [26]. Currently there are no curative therapies for ALS, and there are only two disease-modifying therapies that have been approved for its treatment. Riluzole®, approved since 1995 and delays the onset of ventilator dependences, has been associated with increasing survival of about 2–3 months [27, 28]. In 2017, edaravone (Radicava) was approved although the mechanism through which edaravone manifests its effect remains undefined. In post hoc analyses of a previous phase 3 study, edaravone was shown to significantly reduce the decline in the ALS Functional Rating Scale Revised (ALSFRS-R) score as compared with placebo in a small group of ALS patients [29]. Despite these two treatments, there remains a desperate need for effective ALS therapies.

Inherited (familial ALS, FALS) accounts for less than 10% of all ALS cases with the rest arising sporadically (sporadic ALS, SALS). Clinically, FALS and SALS cases are indistinguishable from each other; thus family history and genetics are both used to discriminate between the two forms. Over the past two decades, a number of ALS-causing mutations have been uncovered in studies [26]. Approximately 10–20% of autosomal dominant FALS cases have mutations occurring in the superoxide dismutase 1 (*SOD1*) gene [30–32]. As the *SOD1* gene knockout in mice does not produce motor neuron disease phenotype nor decrease the life span, this suggests that loss of the normal function of SOD1 is not the cause of ALS [33]. Studies of transgenic mice overexpressing mutated *SOD1* develop motor neuron difficulties providing evidence that *SOD1* mutations lead to a toxic "gain of function" pathology [34–36]. Mutations in *SOD1* can affect multiple aspects of *SOD1* including the enzymatic activity of SOD1 proteins, stability of SOD1, its ability to dimerize, its hydrophobicity, and its ability to chelate copper ions [37]. Taken together all evidence suggests that reducing levels of mutated SOD1 would be beneficial for ALS patients, hence the interest in developing ASO targeting this gene.

A 20-mer MOE-containing PS-ASO (ASO SOD^{r146192}) was identified following screening in tissue culture where transfection led to decreased *SOD1* mRNA levels. Normal rats injected intraperitoneally with $SOD^{r146192}$ had a 5% reduction in *SOD1* mRNA and 10% reduction in SOD1 protein levels [2] in the livers of these animals. However, in neither the brains nor the spinal cords of these animals, no change in *SOD1* mRNA or protein levels was observed [2]; a finding not unexpected as the intraperitoneal route of peripheral administration leads to minimal ASO concentrations in the CNS. To achieve concentrations of ASO in the CNS needed to demonstrate the potential of ASO efficacy, normal rats were infused with ASO $SOD^{r146192}$ into the right lateral ventricle via catheter that was surgically implanted through the skull and connected via an osmotic pump imbedded subcutaneously [2]. Through this delivery mechanism, concentrations of accumulated ASO were measured in the brains, brain stems, and throughout the spinal cords of the animals. Fourteen days after infusion, there was an approximate 40% decrease in *SOD1* mRNA levels, and after 28 days of infusion, a 50% decrease in SOD1 protein levels in the right frontal cortex and lumbar and the cervical spinal cord were observed [2]. Dosing with scrambled ASO that contained similar chemical modifications had no effect on either SOD1 mRNA or protein levels [2]. ASO complementary to human *SOD1* mRNA and selected for efficacy *in vitro* was tested in the transgenic rat model of ALS [2]. This model utilizes rats expressing the human ALS-linked mutant SOD1G93A, which leads to between 5 and 10 times increased SOD1 levels in the brains of these animals compared with wild-type rats, and the development of a progressive ALS-like motor neuron disease. The ASO complementary to human SOD1, $SOD^{r/h333611}$, when infused into the right ventricle of the transgenic rats for 1 month led to decreased levels of both SOD1 mRNA and protein. Treatment with ASO $SOD^{r/h333611}$ did slow disease progression and significantly increased survival of the rats compared with saline-infused rats or scrambled ASO infused rats [2]. Together these results demonstrated the efficacy of ASO $SOD^{r/h333611}$ to reduce the mRNA and protein of its target and to prolong survival in the treated animals, thereby paving the way for the first-in-human studies. The safety and tolerability of a single dose of ASO ISIS 333611 were assessed in patients with SOD1 FALS [38]. In this study, a single dose of ASO ISIS 333611 was delivered to patients by IT infusion using an external pump over 11.5 h [38]. The placebo-controlled, double-blind, randomized study assessed single-escalating doses of ISIS 33611 of either 0.15, 0.5, 1.5 , or 3.0 mg in cohorts of patients, randomized 3 : 1 to ISIS 33611 or placebo. In assessing safety and tolerability, 88% of placebo-treated patients and 83% of ISIS 333611-treated patients reported AEs, the most common of which were post-LP syndrome and back pain, which were associated with IT infusion and not to the drug dosed [38]. Furthermore, no increase in the number of AEs with increasing dose of ISIS 333611 was observed [38]. No related changes in vital signs, neurological or physical exams, clinical chemistry, coagulation parameters, complement, urinalysis, or ECG were reported in the ISIS 333611-treated cohorts. The ALSFRS-R and respiratory function (forced vital capacity) were generally stable throughout the study and did not differ between the treatment groups [38].

With regard to pharmacokinetics, ISIS 333611 was detected in the CSF of all treated participants and the levels increased with increasing doses. Although ASOs when administered systemically do not cross the BBB, following IT delivery of ASO is cleared from CSF into the plasma [2, 22, 23]. Plasma concentrations of ISIS 333611 for 1.5- and 3-mg dosing cohorts were shown to increase during the 12-h infusion period and then rapidly decrease over the next 12 h [38]. For the lower dose cohorts, ASO plasma concentrations were below the limit of detection. A single patient who participated in 1.5-mg dose cohort and then 13 weeks later participated in 3-mg dose cohort and then died 12.5 weeks after dosing provided an opportunity to evaluate ISIS 333611 concentrations in human tissue. In autopsy samples of the lumbar, thoracic, and cervical spinal cord, ISIS 333611 was detected at concentrations of 218, 122 and 39 ng/g, respectively [38]. Although a single-case phenomenon, these tissue levels were consistent with the expected levels based on monkey nonclinical studies [38]. SOD1 concentrations were assessed in the CSF for those patients who provided sufficient material for testing. There was relative lack of change in CSF SOD1 protein levels (12%), a finding not surprising following a single dose of ASO. Similar to the results from the initial SMA studies, this study demonstrated that IT delivery of the ASO appeared well tolerated in ALS patients and the pharmacokinetics fell within predictable parameters.

BIIB067, a 20-mer oligonucleotide complementary to a region of the 3′ UTR of mRNA for human SOD1, is in development by Biogen, in collaboration with Ionis Pharmaceuticals Inc. Chemically, BIIB067 contains a chimeric mixed backbone of PS and phosphodiester linkages and incorporates sugar residues of 2-deoxy-D-ribose and MOE. The phase 1 is a placebo-controlled, single, and multiple ascending dose study with the primary objective to evaluate the safety, tolerability, and PK of BIIB067 in adults with ALS. Secondary objective is to evaluate the effects of BIIB067 on levels of SOD1 protein in the CSF of patients. The study began in January 2016 and is expected to be completed by July 2018 (clinicaltrials.gov identifier NCT02623699).

Along with the goal of improving the lives of those ALS patients with SOD1 mutation, demonstrating that ASO methodology (technology and delivery) works in ALS patients would potentially open the door for other ASO targeting other genetic mutations in ALS. The most common cause of FALS and frontotemporal dementia (FTD) are repeats of the hexanucleotide (GGGGCC) in chromosome 9 open reading frame 72 (C9orf72) gene [39–42]. Healthy people may have up to 10 of these repeats, whereas FALS patients can have repeats that number in hundreds to thousands [43]. However, the role of these repeats in the pathology of the disease remains to be determined. Mouse-specific ASO targeting C9orf72 injected intracerebroventricularly into mice led to a reduction of c9orf72 RNA levels in the spinal cord and brain by 30 and 40%, respectively [44]. In mice expressing bacterial artificial chromosome C9orf72 RNA with 450 repeats, a single intraventricular injection of ASO targeting repeat-containing RNAs reduced these RNA levels in the cortex and spinal cord and alleviated the anxiety and cognitive impairments that were observed in the control treated animals [45]. As of the time of this publication, no clinical development of ASO targeting C9orf72 is underway; however, if ASO targeting SOD1 demonstrates efficacy in those trials, it is to be expected that the next potential therapeutic target would be C9orf72 both in ALS and perhaps FTD.

15.2.3 Huntington's Disease (HD)

HD is a rare disorder affecting approximately 5–10 in 100 000 people in American, Australian, and European populations [46, 47]. HD is progressive neurodegenerative disorder that typically manifests in patients between 30 and 45 years of age [46, 47]. Patients with HD exhibit progressive movement dysfunction including hyperkinetic involuntary movements (jerky movements), an abnormal gait and difficulty with speech and swallowing. There are personality changes as well as cognitive impairments [3]. HD is fatal and patients succumb to the disease 10–20 years following diagnosis. In 1872, Huntington identified the description of the disorder as well as the autosomal dominant inheritance pattern [48]. Over a century later, HD was the first disease mapped to a specific chromosome [49], followed a decade later by the identification of repeated CAG triplet mutation within the huntingtin (HTT) gene [50], which translates into an elongated glutamine tract in the N terminus of the huntingtin (htt) protein [3]. In a normal *HTT* gene, the CAG triplet may repeat between 10 and 36 in number, whereas in patients with HD, CAG repeats may be between 37 and 121 in number, suggesting a gain of function by the mutated gene [51, 52]. Although one of the first identified genetic disorders, no effective treatment or cure for HD currently exists.

Once the gene was identified, research began to investigate whether the reduction or elimination of the mutant protein would be a viable treatment option for HD. Complete suppression of wild-type htt protein may not be desirable as HTT is involved in numerous physiological activities in cells that are important for neuronal function [1]. Mice homozygous for HTT knockout (Hdh −/−) die between days 8.5 and 10 of gestation, which occurs before nervous system development, indicating htt protein is indispensable for survival [53]. Nevertheless, investigation of htt inhibition using ASO was enacted shortly after the identification of gene mutation. Initial work by Haque and Isacson [54] attempted to use an 18-mer fully PS-ASO targeting the HTT sequence. The ASO was also fluorescein labeled at the 5′ end to monitor uptake. CD-1 mice were treated daily with ASO for four consecutive days via direct injection into the striatum and then sacrificed the following day, and the brain tissue was assessed by immunostaining and Western blot. Upon viewing, fluorescent cells were observed in the tissue up to 100 μm from the needle tract. Repeated injections into the striatum showed no evidence of neuronal loss or gross damage nor was there a decrease in levels of htt protein. Western blotting demonstrated a modest reduction of 17% in levels of htt protein in the htt-targeted ASO-treated animals compared with those treated with either sense ASO or vehicle. The reason for the failure of the ASO to reduce htt protein levels has not been defined.

Meanwhile, the team of Gagnon et al. investigated a panel of modified ASO targeted to the CAG repeat and assessed efficacy of these ASOs to inhibit htt *in vitro* in patient-derived fibroblasts that harbored a normal HTT allele with 17 CAG repeats and a mutant allele with 69 repeats [55]. The proposed goal was for selective allele reduction of the mutant HTT RNA with limited effect on the normal HTT mRNA levels as the side effects of long-term inhibition of the normal htt protein remain to be determined. In this study, ASO with various modifications were identified that could achieve allele-selective inhibition through a steric block of translation rather than through RNase H activity [55]. In a follow-up study, the ASOs that were allele selective for HTT inhibition were also potent inhibitors of the CAG repeat gene ATX3, the mutant gene involved in spinocerebellar ataxia type 3 [56], but were not allele selective.

The team of Kordasiewicz et al. tested a 20-mer PS-ASO complementary to human HTT mRNA containing 2′-MOE modifications on 5 nucleotides on the 3′ and 5′ ends (HuASO). HuASO at doses of 10, 25, or 50 μg/day was continuously infused for 2 weeks into the right lateral ventricle of BACHD mice that harbor a full-length mutant human HTT gene with 97 CAG repeats [3]. These mice express 1.5 times more mutant human protein than endogenous mouse HTT [57]. A dose-dependent decrease in human HTT mRNA was observed following infusion with HuASO, such that the 50 μg/day dose was used in subsequent studies. Two weeks after continuous infusion, human HTT mRNA levels were reduced to 38% ($p < 0.001$) compared with vehicle-infused animals and the inhibition persisted for 12 weeks [3]. A 66% reduction in levels of mutant human htt protein was observed although the timing was delayed to 4 weeks post-termination of ASO infusion [3]. HuASO infusion did not alter the levels of murine HTT mRNA and protein [3].

To determine whether the suppression of the mutant HTT by ASO would affect the disease development, YAC128 mice, which develop motor and cognitive dysfunctions and brain atrophy [58], were evaluated. Three-month-old YAC128 mice were infused with HuASO for 2 weeks into the right lateral ventricle and then the osmotic pumps removed and animals allowed to recover for an additional 2 weeks [3]. HTT mRNA and htt protein levels were significantly reduced in these animals in 6 weeks posttreatment [3]. Motor deficits in the HuASO-treated animals were significantly different from the saline-treated animals [3].

To test whether there would be a therapeutic benefit in older animals, 6-month-old animals were then infused for 2 weeks with HuASO, and again reduction in HTT mRNA (42%, $p < 0.001$) and protein (44%, $p = 0.0057$) was observed 2.5 months after discontinuing treatment [3]. Motor function in the HuASO-treated animals improved, with treated animals no longer significantly different from nontransgenic animals [3].

Six-month-old BACHD mice that develop progressive motor incoordination, hypokinetic motor activity, and brain atrophy were infused for 2 weeks with HuASO or vehicle and then followed by 6 months, 3 months longer than the YAC128 animals [3]. Again, HuASO reduced human HTT mRNA, but the key in these animals was the alteration of the motor skills. BACHD mice develop a latency to fall in a rotarod task by 6 months of age. Eight weeks following the initiation of treatment, the motor skills of the HuASO-treated BACHD mice were improved compared with their pretreatment performance ($p = 0.002$) and with their control ASO-treated littermates [3]. This improvement in performance persisted for a further 6 months [3].

A third mouse model, the R6/2 mice, which express a fragment of human HTT gene with an expanded CAG repeat and develop a very rapidly progressing fatal disease, were evaluated. Eight-week-old R6/2 mice, at an age when they already developed symptoms and had gross loss of brain mass, were infused for 4 weeks with HuASO ex1, an ASO designed to target the mutant R6/2 transgene. Infusion with HuASO ex1 blocked further brain loss in these animals, such that the brain mass of 12-week-old treated animal was comparable to the 8-week untreated animals and significantly larger than vehicle-treated 12-week-old animals [3]. The life span of these animals was increased to a median of 136 days compared with vehicle-treated littermates (median survival 113 days, $p = 0.0498$) [3]. Suppression of human HTT in these murine models of HD demonstrated a reversal of existing behavioral phenotypes and prevented progressive loss of brain mass.

The earlier the treatment was initiated in the disease, the more robust the disease reversal. Transient ASO suppression of mHTT 4 months was sufficient to obtain a long-term benefit over 9 months. One can imagine a treatment where intermittent suppression of HTT or a "HTT holiday" is all that is required to provide continuous benefit and could be incorporated into the clinical trial design [59].

In preparation of drug development, the effectiveness of ASO delivery into nonhuman primates was assessed. ASO complementary to both monkey and human HTT mRNA (MkHuASO) was infused into the CSF for 21 days. ASO was distributed to neurons of the periventricular and lateral brain regions and accumulated in the cortex [3]. HTT mRNA levels were decreased in the anterior and posterior cortex of the ASO-treated animals by 47 and 63% immediately postinfusion and remained reduced for 4 weeks after the termination of treatment [3].

With these efforts to show that ASO targeting the human HTT mRNA can lead to reduction in the mutant mRNA, protein, and ameliorate the disease phenotype, as well as the demonstration of ASO infusion achieving a broad distribution in the nonhuman primate brain, the next step was to embark on a phase 1 clinical study. In July 2015, the first-in-man phase 1/2a trial of ASO IONIS-HTTrx was embarked in early manifest HD patients. The primary goal is to evaluate the safety, tolerability, and pharmacokinetics of IT administration of ASO. The rationale for selecting early manifest HD patients as the population was due to concerns that later manifest HD patients may not have been cognitively able to provide informed consent and the exclusion of prodromal patients was that they do not yet have motor symptoms and the risk to these patients was deemed inappropriate.

15.2.4 Muscular Sclerosis (MS)

The chronic inflammatory demyelinating disease of the CNS, MS, is the most common neurological disease of young adults [60]. Patients typically present initially as the relapsing–remitting form of MS (RR-MS) characterized by episodes of active disease (relapse) and periods of inactivity (remission) [60]. Over decades, the majority of RR-MS patients will convert to a secondary progressive form (SP-MS), which is characterized by persistent increase in disability [60]. In a small percentage of patients, the disease is progressive at onset and is classified as primary progressive (PP-MS) [60]. In MS, the lesions that arise in the CNS are distinctive and can be viewed by MRI [60]. These lesions are believed to arise as a consequence of demyelination, leading to a loss of oligodendrocytes, resulting in axonal loss and ultimately neurodegeneration. MS is believed to be an autoimmune disease involving autoreactive T and B cells, which attack myelin sheaths leading to demyelination and axonal damage [61]. Therapies for MS have therefore focused on modifying the immune responses, in particular, those that can reduce lymphocyte migration into the CNS. In the past two decades, there has been progress in developing treatments of MS with at least a dozen disease-modifying therapies on the market for the relapsing–remitting form of the disease. Significantly less progress has been made in treatment of the primary progressive and slow progressive forms of the disease, with only the anti-CD20 monoclonal antibody (mAb) ocrelizumab approved by the FDA for the treatment of PP-MS [62].

ATL1102, an ASO that targets the mRNA of CD49d, a subunit of the very late antigen-4 (VLA-4) integrin, is in development as a treatment for MS [63]. VLA-4 is expressed by activated T and B lymphocytes and is critical for their adherence to endothelial cells of the CNS as a precursor step for migration into the CNS. Natalizumab, an approved therapy for MS, targets the protein CD49d, is a mAb targeting CD49d, demonstrating the validity of CD49d as clinical target in MS [64]. Natalizumab treatment has been shown to significantly reduce brain lesions [65], relapse frequency, and the progression of disability in patients with RRMS [66]. Although highly efficacious as a treatment in RR-MS, natalizumab treatment has also been associated with a risk of the development of progressive multifocal leukoencephalopathy (PML) [67–69]. It has been shown that the binding of natalizumab to VLA-4 induces a direct signal and supports a proinflammatory phenotype. In the MS treatment field, development of a drug that possessed similar levels of efficacy of natalizumab without the risk of PML would be of great benefit. As an ASO, ATL1102 would not bind to cell surface VLA-4 but rather would function by binding to the CD49d mRNA, which may possibly avoid the induction of PML.

ALT1102 is 20-mer gapmer design with MOE and PS modifications [63]. In the phase 2 trial, RR-MS patients ($n = 77$) were randomized to either placebo or ATL1102 (200 mg/dose) given subcutaneously initially for three injections in the first week then twice a week for a further 7 weeks [63]. As an autoimmune disease, treatment for MS can be delivered systemically rather than through the IT route described in this chapter for the other neurodegenerative diseases. The efficacy endpoints focused on MRI measures between ALT1102-treated patients and placebo-treated patients including cumulative number of new active lesions (either new T1 gadolinium lesions or nonenhancing, new, or enlarging T2 lesions) and cumulative volume of all T1 Gd lesions [63]. Safety analyses included number of AEs, laboratory data, vital signs, MRI assessment for PML, physical examinations, 12-lead ECG, and local tolerance [63].

In terms of efficacy based on MRI measures, there were significant differences between the ALT1102-treated group and placebo group. There was an observed 54% reduction in the cumulative number of new lesions in the MRIs for weeks 4, 8, and 12 in the ALT1102-treated group than the placebo group [63]. ALT1102 reduced the cumulative number of new T1 Gd lesions for weeks 4, 8, and 12 by 68% compared with placebo and reduced the cumulative number of T1 Gd lesions for weeks 4, 8, and 12 by 67% compared with placebo [63]. At week 12 (4 weeks after treatment period), ATL1102 significantly reduced T1 Gd and new T1 Gd lesion numbers by 86 and 91%, respectively, compared with placebo [63]. The study was not powered to observe changes in number of relapses or change in EDSS scores [63], nor would changes be expected in the short period of the study.

With regard to safety, most common observed treatment-emergent events were mild to moderate injection site erythema (25% of patients) and thrombocytopenia (22% of patients) in the ALT1102-treated group only. Both treatment groups reported incidents of increase in alanine aminotransferase (ALT), increase in aspartate aminotransferase, MS relapses, and headaches [63]. The decrease in platelet in the ALT1102-treated group led to two patients prematurely discontinuing with the study medication. Platelet counts returned to normal range after dosing terminated and did not require any additional intervention [63]. Of key interest considering the VLA-4 target of the ASO, no PML was observed.

In terms of pharmacodynamics, the efficacy of ALT1102 could be assessed by a reduction in levels of CD49d RNA and/or changes in CD49d receptor levels or changes in numbers of peripheral blood lymphocytes. At week 8 of ATL1102 treatment, a significant reduction in the number of total blood leukocytes, granulocytes, T cells, and B cells was observed, while the numbers of monocytes and NK cells were unchanged. Flow cytometry methodology was used to quantify VLA-4 levels on the cell surface of blood cells and the percentage of cell subtypes expressing VLA-4. At week 8, the percentage of B cells expressing VLA-4 in the ATL1102 group compared was reduced compared with either baseline or placebo group [63]. Both CD3+ and CD4+ T cells had a reduction in the percentage of cells expressing VLA-4 at week 12 compared with baseline for the ATL1102 treatment group [63]. VLA-4 acts on B and T cells for their maturation, activation, adhesion, and migration [70–73]. ATL1102 appears to be reducing numbers of these cells in circulating blood, and perhaps the activities of these cells, leading to fewer cells migrating to CNS, which would result in a reduction in the number and volume of MS brain lesions as demonstrated in this study.

Although no PML cases were observed in this trial, the risk of ATL1102 with regard to PML also needs to be further studied. The mechanism of action through which ATL1102 functions to reduce VLA-4 expression may be beneficial as compared with the broader anti-VLA-4 effects demonstrated by natalizumab. Longer-term trials will aid to clarify these positions. In future longer-term studies, there is proposition of less frequent or lower dosing than the current 200-mg twice weekly dosing in this study. The goal for reduced dosing/decreased frequency is to retain the potential to reduce MRI brain lesions yet also minimizing side effects including platelet reductions. Preparations are underway for a phase 2b trial in MS patients as well as a small investigative study of ATL1102 in relapsing secondary progressive MS.

15.2.5 Alzheimer's Disease (AD)

AD is the most common form of dementia, accounting for 60–80% of cases [74]. Currently, drugs approved for the treatment of AD have at best a modest impact upon the disease, and there is little evidence that they modify the underlying pathological processes of the disease. Elevated levels of amyloid β (Aβ) peptide resulting in the formation of Aβ plaques are considered to be involved in the pathogenesis of these diseases [74]. Whether the elevated levels result from overexpression of amyloid precursor protein (APP), altered cleavage of APP or decreased clearance of Aβ or combinations thereof remains to be defined. To demonstrate that Aβ is a causal agent of dementia, Kumar and colleagues [75] delivered a 42-mer PS-modified ASO targeting amino acids 17–30 of Aβ intracerebroventricularly into a strain of mice with a spontaneous mutation to overexpress APP as they age. Multiple injections of the APP target ASO resulted in a reduction in the protein levels of APP as compared with animals treated with control ASO [75]; however, mRNA levels were not changed in the amygdala, septum, or hippocampi of these animals. Animals treated with multiple doses of the ASO showed a statistically significant improvement ($p < 0.001$) in acquisition in footshock avoidance when compared with vehicle-treated animals [75]. Another prominent feature of AD is tangles that form inside dying cells. These tangles are of tau protein. Tau is encoded by the microtubule-associated protein tau (MAPT) gene on chromosome 17q21 [76]. Tau protein undergoes post-translational modifications including phosphorylation. Hyperphosphorylated MAPT in AD patients results in tau protein that is unable to interact with microtubules and instead forms filaments and aggregates of neurofibrillary tangles [77]. Tau deposits appear to map closely to where atrophy occurs in the brain. While tau is not considered to be initiating AD, it is key as a secondary effector of the disease and remains a potential therapeutic target. AD mouse models when placed on a tau knockout (tau −/−) background show improved learning and memory [78]. Inhibiting tau by ASO, human-specific tau-targeted ASO (Tau^{Aso-12}) delivered intracerebroventrically to PS19 mice carrying human tau with the P301S mutation [79], led to reduced levels of human tau mRNA but not murine mRNA [80]. Human tau protein was reduced in the Tau^{ASO-12} group compared with the control group ($p = 0.011$) [80]. Tau^{ASO-12} treatment reduced the tau pathology deposition in the hippocampus ($p = 0003$) and whole brain sections ($p = 0.008$) [80]. Furthermore, Tau^{ASO-12} treatment in aged PS19 mice (9 months of age) reduced tau inclusions, prevented hippocampal volume decrease, and reduced hippocampal neuron loss [80]. Mice treated with TauASO-12 performed better in a behavior test of nestlet task, making better nests ($p = 0.040$) and leaving less untorn nestlet weight behind ($p = 0.014$). Devos and team proceeded to treat cynomologus monkeys with ASO complementary to monkey tau by IT bolus administration and demonstrated reduced tau mRNA and protein in the spinal cord and brain of the animals [80]. These results are encouraging for human trials.

IONIS-MAPTRx (ISIS 814907) is an 18-mer chimeric MOE PS/PO-modified ASO targeting human MAPT mRNA and is undergoing phase 1 testing in patients with mild AD (NCT03186989). This dose-escalating, placebo-controlled trail will assess the safety, tolerability, pharmacokinetics, and pharmacodynamics of IT-delivered ASO at 4-week intervals over 13 weeks. The study is expected to be completed by 2020. As mutations in MAPT have also been identified in other neurodegenerative disorders (FTD, ALS, PD) [81, 82]; if phase1 trial is successful, it is possible for IONIS-MAPTRx to be tested in these other diseases as well.

The most prevalent known genetic risk factor for late-onset AD is apolipoprotein E (ApoE) [83, 84]; however the mechanism by which ApoE modifies AD risk is not clear. ApoE receptor 2 mediates a signaling cascade that affects learning and memory [85–87]. ApoER2 signaling requires 59 amino acids encoded by exon 19 in mice, and those ApoER2 protein isoforms that lack exon 19 are associated with defects in long-term memory storage and spatial learning [87, 88]. Following screening *in vitro* in mouse kidney cell line and then in wild-type animals, ASO-21 was selected for its capacity to retain exon 19 in mRNA isoforms [89]. When tested in transgenic mouse model of human AD (TgCRND8 mice), a single dose of ASO increased exon 19 inclusion for up to 6 months and improved performance of the mice in the Morris water maze task as an assessment of spatial memory [89]. As yet, human studies have not been enacted.

15.3 Conclusion

With the ongoing clinical trials, it is becoming increasingly likely that ASO may be at the forefront in altering the course of neurodegenerative diseases [90]. ASOs are flexible as therapies as they can function to allow transcription of a full-length isoform protein (e.g. SMA), limit production of a mutant isoform (e.g. ALS, HD), or reduce protein levels of a wild-type protein that may not be beneficial in the disease state (e.g. MS).

ASOs as drug therapies have faced historical challenges with regard to target organ delivery but with the safety and tolerability of repeated IT injections shown in the clinical trials, it is possible that ASOs have found their niche as treatment for neurological diseases. As most neurological disease may require chronic dosing, there remains a need to assess whether repeated long-term infusions of ASO or continuous delivery through the use of implantable infusion would be more beneficial and desirous. The decades spent advancing the chemical modifications of ASO now enable ASO to have long half-lives in the CNS. However, the question is not just how long the drugs are stable but rather how long the effect can last. Reduction in both gene target mRNA and protein has been shown to persist for several months following termination of dosing of ASO into the brains of animals [3], which could lead to even less frequent dosing. However, as for any therapy, the risk of being able to rapidly terminate drug exposure is that there should be safety issues, whether due to on-target or nonspecific effects [1], also needs to be weighed and considered. Even with these considerations, this is clearly an exciting time for gene suppression therapies in neurodegenerative diseases.

References

1 Sah, D.W. and Aronin, N. (2011). *Oligonucleotide therapeutic approaches for Huntington disease. J. Clin. Invest.* 121 (2): 500–507.
2 Smith, R.A., Miller, T.M., Yamanaka, K. et al. (2006). *Antisense oligonucleotide therapy for neurodegenerative disease. J. Clin. Invest.* 116 (8): 2290–2296.

3 Kordasiewicz, H.B., Stanek, L.M., Wancewicz, E.V. et al. (2012). *Sustained therapeutic reversal of Huntington's disease by transient repression of huntingtin synthesis*. *Neuron* 74 (6): 1031–1044.

4 Passini, M.A., Bu, J., Richards, A.M. et al. (2011). *Antisense oligonucleotides delivered to the mouse CNS ameliorate symptoms of severe spinal muscular atrophy*. *Sci. Transl. Med.* 3 (72): 72ra18.

5 Evers, M.M., Toonen, L.J., and van Roon-Mom, W.M. (2015). *Antisense oligonucleotides in therapy for neurodegenerative disorders*. *Adv. Drug Deliv. Rev.* 87: 90–103.

6 Butler, M., Hayes, C.S., Chappell, A. et al. (2005). *Spinal distribution and metabolism of 2′-O-(2-methoxyethyl)-modified oligonucleotides after intrathecal administration in rats*. *Neuroscience* 131 (3): 705–715.

7 Whitesell, L., Geselowitz, D., Chavany, C. et al. (1993). *Stability, clearance, and disposition of intraventricularly administered oligodeoxynucleotides: implications for therapeutic application within the central nervous system*. *Proc. Natl. Acad. Sci. U. S. A.* 90 (10): 4665–4669.

8 Banks, W.A., Farr, S.A., Butt, W. et al. (2001). *Delivery across the blood-brain barrier of antisense directed against amyloid beta: reversal of learning and memory deficits in mice overexpressing amyloid precursor protein*. *J. Pharmacol. Exp. Ther.* 297 (3): 1113–1121.

9 Erickson, M.A., Niehoff, M.L., Farr, S.A. et al. (2012). *Peripheral administration of antisense oligonucleotides targeting the amyloid-beta protein precursor reverses AbetaPP and LRP-1 overexpression in the aged SAMP8 mouse brain*. *J. Alzheimers Dis.* 28 (4): 951–960.

10 Hache, M., Swoboda, K.J., Sethna, N. et al. (2016). *Intrathecal Injections in Children With Spinal Muscular Atrophy: Nusinersen Clinical Trial Experience*. *J. Child Neurol.* 31 (7): 899–906.

11 Lunn, M.R. and Wang, C.H. (2008). *Spinal muscular atrophy*. *Lancet* 371 (9630): 2120–2133.

12 Madocsai, C., Lim, S.R., Geib, T. et al. (2005). *Correction of SMN2 Pre-mRNA splicing by antisense U7 small nuclear RNAs*. *Mol. Ther.* 12 (6): 1013–1022.

13 McAndrew, P.E., Parsons, D.W., Simard, L.R. et al. (1997). *Identification of proximal spinal muscular atrophy carriers and patients by analysis of SMNT and SMNC gene copy number*. *Am. J. Hum. Genet.* 60 (6): 1411–1422.

14 Singh, R.N., Howell, M.D., Ottesen, E.W., and Singh, N.N. (2017). *Diverse role of survival motor neuron protein*. *Biochim. Biophys. Acta* 1860 (3): 299–315.

15 Monani, U.R., Sendtner, M., Coovert, D.D. et al. (2000). *The human centromeric survival motor neuron gene (SMN2) rescues embryonic lethality in Smn(−/−) mice and results in a mouse with spinal muscular atrophy*. *Hum. Mol. Genet.* 9 (3): 333–339.

16 Monani, U.R., Lorson, C.L., Parsons, D.W. et al. (1999). *A single nucleotide difference that alters splicing patterns distinguishes the SMA gene SMN1 from the copy gene SMN2*. *Hum. Mol. Genet.* 8 (7): 1177–1183.

17 Lorson, C.L., Hahnen, E., Androphy, E.J., and Wirth, B. (1999). *A single nucleotide in the SMN gene regulates splicing and is responsible for spinal muscular atrophy. Proc. Natl. Acad. Sci. U. S. A.* 96 (6311): 6307–6311.

18 Singh, N.K., Singh, N.N., Androphy, E.J., and Singh, R.N. (2006). *Splicing of a critical exon of human Survival Motor Neuron is regulated by a unique silencer element located in the last intron. Mol. Cell. Biol.* 26 (4): 1333–1346.

19 Singh, N.N., Howell, M.D., Androphy, E.J., and Singh, R.N. (2017). *How the discovery of ISS-N1 led to the first medical therapy for spinal muscular atrophy. Gene Ther.* 24 (9): 520–526.

20 Hua, Y., Vickers, T.A., Okunola, H.L. et al. (2008). *Antisense masking of an hnRNP A1/A2 intronic splicing silencer corrects SMN2 splicing in transgenic mice. Am. J. Hum. Genet.* 82 (4): 834–848.

21 Hua, Y., Sahashi, K., Hung, G. et al. (2010). *Antisense correction of SMN2 splicing in the CNS rescues necrosis in a type III SMA mouse model. Genes Dev.* 24 (15): 1634–1644.

22 Chiriboga, C.A., Swoboda, K.J., Darras, B.T. et al. (2016). *Results from a phase 1 study of nusinersen (ISIS-SMN(Rx)) in children with spinal muscular atrophy. Neurology* 86 (10): 890–897.

23 Finkel, R.S., Chiriboga, C.A., Vajsar, J. et al. (2016). *Treatment of infantile-onset spinal muscular atrophy with nusinersen: a phase 2, open-label, dose-escalation study. Lancet* 388 (10063): 3017–3026.

24 Hoy, S.M. (2017). *Nusinersen: First Global Approval. Drugs* 77 (4): 473–479.

25 Chio, A., Logroscino, G., Traynor, B.J. et al. (2013). *Global epidemiology of amyotrophic lateral sclerosis: a systematic review of the published literature. Neuroepidemiology* 41 (2): 118–130.

26 Valdmanis, P.N. and Rouleau, G.A. (2008). *Genetics of familial amyotrophic lateral sclerosis. Neurology* 70 (2): 144–152.

27 Lacomblez, L., Bensimon, G., Leigh, P.N. et al. (1996). *Dose-ranging study of riluzole in amyotrophic lateral sclerosis. Amyotrophic Lateral Sclerosis/Riluzole Study Group II. Lancet* 347 (9013): 1425–1431.

28 Zoccolella, S., Santamato, A., and Lamberti, P. (2009). *Current and emerging treatments for amyotrophic lateral sclerosis. Neuropsychiatr. Dis. Treat.* 5: 577–595.

29 Writing, G. and A.L.S.S.G. Edaravone (2017). *Safety and efficacy of edaravone in well defined patients with amyotrophic lateral sclerosis: a randomised, double-blind, placebo-controlled trial. Lancet Neurol.* 16 (7): 505–512.

30 Rosen, D.R., Siddique, T., Patterson, D. et al. (1993). *Mutations in Cu/Zn superoxide dismutase gene are associated with familial amyotrophic lateral sclerosis. Nature* 362 (6415): 59–62.

31 Pramatarova, A., Figlewicz, D.A., Krizus, A. et al. (1995). *Identification of new mutations in the Cu/Zn superoxide dismutase gene of patients with familial amyotrophic lateral sclerosis. Am. J. Hum. Genet.* 56 (3): 592–596.

32 Deng, H.X., Hentati, A., Tainer, J.A. et al. (1993). *Amyotrophic lateral sclerosis and structural defects in Cu,Zn superoxide dismutase. Science* 261 (5124): 1047–1051.

33 Reaume, A.G., Elliott, J.L., Hoffman, E.K. et al. (1996). *Motor neurons in Cu/Zn superoxide dismutase-deficient mice develop normally but exhibit enhanced cell death after axonal injury. Nat. Genet.* 13 (1): 43–47.

34 Gurney, M.E., Pu, H., Chiu, A.Y. et al. (1994). *Motor neuron degeneration in mice that express a human Cu,Zn superoxide dismutase mutation. Science* 264 (5166): 1772–1775.

35 Ripps, M.E., Huntley, G.W., Hof, P.R. et al. (1995). *Transgenic mice expressing an altered murine superoxide dismutase gene provide an animal model of amyotrophic lateral sclerosis. Proc. Natl. Acad. Sci. U. S. A.* 92 (3): 689–693.

36 Wong, P.C., Pardo, C.A., Borchelt, D.R. et al. (1995). *An adverse property of a familial ALS-linked SOD1 mutation causes motor neuron disease characterized by vacuolar degeneration of mitochondria. Neuron* 14 (6): 1105–1116.

37 Shaw, B.F. and Valentine, J.S. (2007). *How do ALS-associated mutations in superoxide dismutase 1 promote aggregation of the protein? Trends Biochem. Sci.* 32 (2): 78–85.

38 Miller, T.M., Pestronk, A., David, W. et al. (2013). *An antisense oligonucleotide against SOD1 delivered intrathecally for patients with SOD1 familial amyotrophic lateral sclerosis: a phase 1, randomised, first-in-man study. Lancet Neurol.* 12 (5): 435–442.

39 DeJesus-Hernandez, M., Mackenzie, I.R., Boeve, B.F. et al. (2011). *Expanded GGGGCC hexanucleotide repeat in noncoding region of C9ORF72 causes chromosome 9p-linked FTD and ALS. Neuron* 72 (2): 245–256.

40 Renton, A.E., Majounie, E., Waite, A. et al. (2011). *A hexanucleotide repeat expansion in C9ORF72 is the cause of chromosome 9p21-linked ALS-FTD. Neuron* 72 (2): 257–268.

41 van Blitterswijk, M., DeJesus-Hernandez, M., and Rademakers, R. (2012). *How do C9ORF72 repeat expansions cause amyotrophic lateral sclerosis and frontotemporal dementia: can we learn from other noncoding repeat expansion disorders? Curr. Opin. Neurol.* 25 (6): 689–700.

42 Ling, S.C., Polymenidou, M., and Cleveland, D.W. (2013). *Converging mechanisms in ALS and FTD: disrupted RNA and protein homeostasis. Neuron* 79 (3): 416–438.

43 Picher-Martel, V., Valdmanis, P.N., Gould, P.V. et al. (2016). *From animal models to human disease: a genetic approach for personalized medicine in ALS. Acta Neuropathol. Commun.* 4 (1): 70.

44 Lagier-Tourenne, C., Baughn, M., Rigo, F. et al. (2013). *Targeted degradation of sense and antisense C9orf72 RNA foci as therapy for ALS and frontotemporal degeneration. Proc. Natl. Acad. Sci. U. S. A.* 110 (47): E4530–E4539.

45 Jiang, J., Zhu, Q., Gendron, T.F. et al. (2016). *Gain of Toxicity from ALS/FTD-Linked Repeat Expansions in C9ORF72 Is Alleviated by Antisense Oligonucleotides Targeting GGGGCC-Containing RNAs*. *Neuron* 90 (3): 535–550.

46 Morrison, P.J. (2010). *Accurate prevalence and uptake of testing for Huntington's disease*. *Lancet Neurol.* 9 (12): 1147.

47 Pringsheim, T., Wiltshire, K., Day, L. et al. (2012). *The incidence and prevalence of Huntington's disease: a systematic review and meta-analysis*. *Mov. Disord.* 27 (9): 1083–1091.

48 Huntington, G. (1872). *On chorea*. *Med Surg Rep* 26: 320–321.

49 Gusella, J.F., Wexler, N.S., Conneally, P.M. et al. (1983). *A polymorphic DNA marker genetically linked to Huntington's disease*. *Nature* 306 (5940): 234–238.

50 MacDonald, M.E., Barnes, G., Srinidhi, J. et al. (1993). *Gametic but not somatic instability of CAG repeat length in Huntington's disease*. *J. Med. Genet.* 30 (12): 982–986.

51 Kremer, B., Weber, B., and Hayden, M.R. (1992). *New insights into the clinical features, pathogenesis and molecular genetics of Huntington disease*. *Brain Pathol.* 2 (4): 321–335.

52 Losekoot, M., van Belzen, M.J., Seneca, S. et al. (2013). *EMQN/CMGS best practice guidelines for the molecular genetic testing of Huntington disease*. *Eur. J. Hum. Genet.* 21 (5): 480–486.

53 Zeitlin, S., Liu, J.P., Chapman, D.L. et al. (1995). *Increased apoptosis and early embryonic lethality in mice nullizygous for the Huntington's disease gene homologue*. *Nat. Genet.* 11 (2): 155–163.

54 Haque, N. and Isacson, O. (1997). *Antisense gene therapy for neurodegenerative disease?* *Exp. Neurol.* 144 (1): 139–146.

55 Gagnon, K.T., Pendergraff, H.M., Deleavey, G.F. et al. (2010). *Allele-selective inhibition of mutant huntingtin expression with antisense oligonucleotides targeting the expanded CAG repeat*. *Biochemistry* 49 (47): 10166–10178.

56 Hu, J., Gagnon, K.T., Liu, J. et al. (2011). *Allele-selective inhibition of ataxin-3 (ATX3) expression by antisense oligomers and duplex RNAs*. *Biol. Chem.* 392 (4): 315–325.

57 Gray, M., Shirasaki, D.I., Cepeda, C. et al. (2008). *Full-length human mutant huntingtin with a stable polyglutamine repeat can elicit progressive and selective neuropathogenesis in BACHD mice*. *J. Neurosci.* 28 (24): 6182–6195.

58 Hodgson, J.G., Agopyan, N., Gutekunst, C.A. et al. (1999). *A YAC mouse model for Huntington's disease with full-length mutant huntingtin, cytoplasmic toxicity, and selective striatal neurodegeneration*. *Neuron* 23 (1): 181–192.

59 Lu, X.H. and Yang, X.W. (2012). *"Huntingtin holiday": progress toward an antisense therapy for Huntington's disease*. *Neuron* 74 (6): 964–966.

60 Palace, J. (2001). *Making the diagnosis of multiple sclerosis*. *J. Neurol. Neurosurg. Psychiatry* 71 (Suppl 2): ii3–ii8.

61 Frohman, E.M., Racke, M.K., and Raine, C.S. (2006). *Multiple sclerosis – the plaque and its pathogenesis. N. Engl. J. Med.* 354 (9): 942–955.

62 Frampton, J.E. (2017). *Ocrelizumab: first global approval. Drugs* 77 (9): 1035–1041.

63 Limmroth, V., Barkhof, F., Desem, N. et al. (2014). *CD49d antisense drug ATL1102 reduces disease activity in patients with relapsing-remitting MS. Neurology* 83 (20): 1780–1788.

64 O'Connor, P. (2007). *Natalizumab and the role of alpha 4-integrin antagonism in the treatment of multiple sclerosis. Expert. Opin. Biol. Ther.* 7 (1): 123–136.

65 Tubridy, N., Behan, P.O., Capildeo, R. et al. (1999). *The effect of anti-alpha4 integrin antibody on brain lesion activity in MS. The UK Antegren Study Group. Neurology* 53 (3): 466–472.

66 Polman, C.H., O'Connor, P.W., Havrdova, E. et al. (2006). *A randomized, placebo-controlled trial of natalizumab for relapsing multiple sclerosis. N. Engl. J. Med.* 354 (9): 899–910.

67 Kleinschmidt-DeMasters, B.K. and Tyler, K.L. (2005). *Progressive multifocal leukoencephalopathy complicating treatment with natalizumab and interferon beta-1a for multiple sclerosis. N. Engl. J. Med.* 353 (4): 369–374.

68 Langer-Gould, A., Atlas, S.W., Green, A.J. et al. (2005). *Progressive multifocal leukoencephalopathy in a patient treated with natalizumab. N. Engl. J. Med.* 353 (4): 375–381.

69 Van Assche, G., Van Ranst, M., Sciot, R. et al. (2005). *Progressive multifocal leukoencephalopathy after natalizumab therapy for Crohn's disease. N. Engl. J. Med.* 353 (4): 362–368.

70 Carrasco, Y.R. and Batista, F.D. (2006). *B-cell activation by membrane-bound antigens is facilitated by the interaction of VLA-4 with VCAM-1. EMBO J.* 25 (4): 889–899.

71 Lo, C.G., Lu, T.T., and Cyster, J.G. (2003). *Integrin-dependence of lymphocyte entry into the splenic white pulp. J. Exp. Med.* 197 (3): 353–361.

72 Alter, A., Duddy, M., Hebert, S. et al. (2003). *Determinants of human B cell migration across brain endothelial cells. J. Immunol.* 170 (9): 4497–4505.

73 Niino, M., Bodner, C., Simard, M.L. et al. (2006). *Natalizumab effects on immune cell responses in multiple sclerosis. Ann. Neurol.* 59 (5): 748–754.

74 Frisoni, G.B., Boccardi, M., Barkhof, F. et al. (2017). *Strategic roadmap for an early diagnosis of Alzheimer's disease based on biomarkers. Lancet Neurol.* 16 (8): 661–676.

75 Kumar, V.B., Farr, S.A., Flood, J.F. et al. (2000). *Site-directed antisense oligonucleotide decreases the expression of amyloid precursor protein and reverses deficits in learning and memory in aged SAMP8 mice. Peptides* 21 (12): 1769–1775.

76 Andreadis, A., Brown, W.M., and Kosik, K.S. (1992). *Structure and novel exons of the human tau gene. Biochemistry* 31 (43): 10626–10633.

77 Strong, M.J., Yang, W., Strong, W.L. et al. (2006). *Tau protein hyperphosphorylation in sporadic ALS with cognitive impairment. Neurology* 66 (11): 1770–1771.

78 Roberson, E.D., Scearce-Levie, K., Palop, J.J. et al. (2007). *Reducing endogenous tau ameliorates amyloid beta-induced deficits in an Alzheimer's disease mouse model. Science* 316 (5825): 750–754.

79 Yoshiyama, Y., Higuchi, M., Zhang, B. et al. (2007). *Synapse loss and microglial activation precede tangles in a P301S tauopathy mouse model. Neuron* 53 (3): 337–351.

80 DeVos, S.L., Miller, R.L., Schoch, K.M. et al. (2017). *Tau reduction prevents neuronal loss and reverses pathological tau deposition and seeding in mice with tauopathy. Sci. Transl. Med.* 9 (374).

81 Hutton, M., Lendon, C.L., Rizzu, P. et al. (1998). *Association of missense and 5′-splice-site mutations in tau with the inherited dementia FTDP-17. Nature* 393 (6686): 702–705.

82 Lynch, T., Sano, M., Marder, K.S. et al. (1994). *Clinical characteristics of a family with chromosome 17-linked disinhibition-dementia-parkinsonism-amyotrophy complex. Neurology* 44 (10): 1878–1884.

83 Rebeck, G.W., Reiter, J.S., Strickland, D.K., and Hyman, B.T. (1993). *Apolipoprotein E in sporadic Alzheimer's disease: allelic variation and receptor interactions. Neuron* 11 (4): 575–580.

84 Strittmatter, W.J. and Roses, A.D. (1996). *Apolipoprotein E and Alzheimer's disease. Annu. Rev. Neurosci.* 19: 53–77.

85 Herz, J. and Chen, Y. (2006). *Reelin, lipoprotein receptors and synaptic plasticity. Nat. Rev. Neurosci.* 7 (11): 850–859.

86 Telese, F., Ma, Q., Perez, P.M. et al. (2015). *LRP8-reelin-regulated neuronal enhancer signature underlying learning and memory formation. Neuron* 86 (3): 696–710.

87 Wasser, C.R., Masiulis, I., Durakoglugil, M.S. et al. (2014). *Differential splicing and glycosylation of Apoer2 alters synaptic plasticity and fear learning. Sci. Signal.* 7 (353): ra113.

88 Beffert, U., Weeber, E.J., Durudas, A. et al. (2005). *Modulation of synaptic plasticity and memory by Reelin involves differential splicing of the lipoprotein receptor Apoer2. Neuron* 47 (4): 567–579.

89 Hinrich, A.J., Jodelka, F.M., Chang, J.L. et al. (2016). *Therapeutic correction of ApoER2 splicing in Alzheimer's disease mice using antisense oligonucleotides. EMBO Mol. Med.* 8 (4): 328–345.

90 Keiser, M.S., Kordasiewicz, H.B., and McBride, J.L. (2016). *Gene suppression strategies for dominantly inherited neurodegenerative diseases: lessons from Huntington's disease and spinocerebellar ataxia. Hum. Mol. Genet.* 25 (R1): R53–R64.

16

Nucleic Acids as Adjuvants

Kevin Brown[1], *Montserrat Puig*[2], *Lydia Haile*[3], *Derek Ireland*[2], *John Martucci*[2], *and Daniela Verthelyi*[2]

[1] *Fluidigm Corporation, South San Francisco, CA, USA*
[2] *Centre for Drug Evaluation and Research, Office of Biotechnology Products, US FDA, Silver Spring, MD, USA*
[3] *Centre for Drug Evaluation and Research, Office of New Drugs, US FDA, Silver Spring, MD, USA*

16.1 Introduction

The twentieth century saw dramatic progress in the prevention and treatment of infectious diseases due to the improvement of sanitary conditions, development of antibiotics, and implementation of vaccine regimens. Vaccinations have been a critical tool used in the control or eradication of diseases such as yellow fever, small pox, and polio, as well as significant reductions in the incidence of high morbidity diseases such as measles and rubella. The number of vaccines available to the population has progressively increased, and current immunization schedules include vaccines for polio; tetanus; diphtheria; pertussis (Tdap); measles, mumps, and rubella (MMR); influenza; human papillomavirus (HPV); hepatitis B virus (HBV); hepatitis A virus (HAV); varicella; *meningococcus*; and *pneumococcus*. The majority of these vaccines mediate their effect by inducing protective titers of antigen-specific antibodies and rely on alum as the adjuvant to induce strong humoral and T_H2 immune responses [1]. However, the development of vaccines that require cell-mediated immune responses such as those against human immunodeficiency virus (HIV), malaria, and tuberculosis, as well as those aimed at breaking tolerance to oncologic antigens or allergens, has lagged. These failures underscore the need to produce safe and effective vaccines that induce both strong T-cell-mediated immunity through TH1 and antigen-specific cytotoxic T-cell and humoral responses in order to control or eliminate intracellular pathogens and malignancies.

Oligonucleotide-Based Drugs and Therapeutics: Preclinical and Clinical Considerations for Development, First Edition. Edited by Nicolay Ferrari and Rosanne Seguin.

New insights into the role that innate immune cells play in dictating the type and magnitude of immune responses suggest that innate immune response modulators (IIRM) may be used as adjuvants to not only enhance responses but also promote the type of immune response required. Cells are equipped with pathogen recognition receptors (PRR) such as Toll-like receptors (TLR), nucleotide oligomerization domain (NOD)-like receptors (NLR), retinoic acid-inducible gene I (RIG-I)-like receptors (RLR), C-type lectin receptors, and other nonclassified innate immune receptors. PRR sense small molecular motifs conserved among microbes named pathogen-associated molecular patterns (PAMP), as well as cell damage products known as damage-associated molecular patterns (DAMP). A new generation of synthetic ligands for PRRs, particularly TLR, is being developed as vaccine adjuvants in an effort to dictate the type and magnitude of immune responses. The recent licensure of the IIRM monophosphoryl lipid A (MPL), a TLR4 agonist derived from the endotoxin of Gram-negative bacteria and agonist of TLR4, as the adjuvant for the HPV vaccine Cervarix, confirms that the use of IIRM as adjuvants is a feasible approach. This chapter discusses recent advances in the use of nucleic acids or molecular mimics of nucleic acids for this purpose. Importantly, adjuvants are generally linked to the development of a specific vaccine and are usually not regulated as stand-alone products.

16.1.1 TLR as Nucleic Acid-Sensing Pathogen Recognition Receptors (PRR)

Within the human innate immune system, nucleic acids are recognized by four TLR that are localized in the endosomes of immune cells – thus preventing accidental recognition of host DNA or RNA species [2] – and by several nucleic acid sensors found in the cytosol such as NOD, RIG-I, melanoma differentiation-associated protein 5 (MDA-5), DNA-dependent activator of IFN regulatory factors (DAI), interferon-gamma-inducible protein 16 (IFI16), and absent in melanoma 2 (AIM2) that can also be found in non-immune cells such as fibroblasts and endothelial cells. The redundancy of nucleic acid sensors suggests that nucleic acid recognition is a biological priority (reviewed in Refs. [3] and [4]). The nucleic acid-sensing TLR: TLR-3, -7, -8, and -9, are selectively activated by dsRNA, ssRNA, and unmethylated deoxycytosine-deoxyguanosine (CpG)-containing DNA, respectively, facilitating the recognition of different microbial species. TLR-7, -8, or -9 engagement results in the recruitment of the adapter protein myeloid differentiation primary response gene 88 (MyD88). TIR domain-containing adapter-inducing IFN-β (TRIF) is recruited by TLR-3. This initiates downstream signaling pathways and subsequent activation of the transcription factors: nuclear factor-κB (NF-κB)/activator protein 1 (AP1) and interferon regulatory factors (IRF) 3 and 7 that promote the production of type I IFN and proinflammatory cytokines such as TNF-α, IL-1, and IL-6.

The nucleic acid-sensing TLR are capable of recognizing differences in nucleic acid sequence, structure, and base modifications. In humans, the expression of TLR3 is restricted to the endosomes of macrophages, myeloid dendritic cells (mDC), and BDCA3$^+$CD141$^+$ DC [5–7] and selectively expressed to high levels in murine CD8α$^+$ cross-presenting DC and CD103$^+$ dermal DC [8]. Ligation of human or mouse TLR-3 in mDC results in the production of numerous cytokines including IL-12. TLR-7 and -9 are expressed by human plasmacytoid dendritic cells (pDC) and B cells. pDC respond to TLR ligation by primarily producing not only IFN-α but also additional proinflammatory or antiviral cytokines, while B cells respond by polyclonal activation and accelerated antigen-specific antibody production. TLR-8 has a broader cellular distribution including monocytes, macrophages, and mDC [9]. Despite sharing signaling pathways, the engagement of nucleic acid-sensing TLR in specific cell types results in the production of distinct cytokine patterns that ultimately trigger specific adaptive immune responses [10, 11]. For example, although most ligands of nucleic acid-sensing TLR induce type I IFN, single TLR stimulation is associated with a different signature of IFN subtype gene expression ratios [11, 12]. Conversely, different oligonucleotide sequences induce proportionally similar IFN subtype transcripts when acting via the same receptor even if the magnitude and kinetics of the response are very different [11].

16.2 Categories of Nucleic Acid Adjuvants

Nucleic acids in the form of DNA, RNA, or synthetic molecules that trigger PRR are interesting candidates as vaccine adjuvants because they induce type I IFN and inflammatory mediators, activate the inflammasome, enhance antigen cross-presentation and antibody responses, and elicit strong TH1 and CTL responses. To date, studies have centered on three types of nucleic acid adjuvants: synthetic analogues of double-stranded RNA, molecular mimics of single-stranded RNA (imidazoquinolines), and synthetic analogues of single-stranded DNA mostly in the form of oligodeoxynucleotides (ODNs) containing one or more unmethylated CpG motifs. The following section will review the studies conducted on the three types of nucleic acid adjuvants, mainly focusing on the use of synthetic DNA analogues containing CpG motifs as vaccine adjuvants for infectious diseases and cancer.

Synthetic dsRNA analogues: Viral dsRNA and dsRNA analogues are recognized by two main receptors, endosomal TLR-3 mainly expressed by cross-presenting DC and the cytosolic ribonucleic acid helicase (RLR) MDA-5 found in nonhematopoietic cell types (e.g. stromal cells). Polyriboinosinic polyribocytidylic acid (poly I:C) was the first synthetic dsRNA analogue discovered and tested as a vaccine adjuvant [13–15]. The effects of poly I:C involve the distinct roles of both TLR-3 and MDA-5 [14–16]. Poly I:C-mediated activation of TLR-3 promotes antigen cross-presentation by DC to CD8$^+$ T cells and systemic release of IL-12 [17]. The systemic type I IFN released from stromal cells via poly I:C-mediated activation of MDA-5 enhances antigen cross-presentation, B-cell immunity, and the generation of cytotoxic and memory $CD8^+$ T cells [18, 19].

In mice, poly I:C is a strong IFN-α inducer providing antiviral and adjuvant activity [20, 21]. In addition, poly I:C works as a mucosal adjuvant to protein antigens (synthetic HIV Tat protein and anthrax protective antigen protein, Influenza B [H1N1] vaccines) for the induction of both humoral and cell-mediated immune responses [22–24]. In other preclinical studies, primary immunization with SIV Gag protein plus poly I:C resulted in higher frequencies of $CD4^+$ TH1 responses and $CD8^+$ T-cell responses after boosting with recombinant adenoviral vector with enhanced $CD4^+$ T-cell responses in bronchoalveolar lavage after viral challenge, suggesting poly I:C improves the magnitude and durability of $CD4^+$ T-cell responses [25]. However, stability and toxicity issues have hindered the clinical application of poly I:C [26–28].

The pursuit of less toxic synthetic dsRNA analogues capable of inducing IFN resulted in the development of two additional compounds: poly I:CLC (Hiltonol, Oncovir) and poly I:C_{12}U (Ampligen, Hemispherx Biopharma). Poly I:CLC is a complex of poly I:C stabilized with poly-L-lysine and carboxymethylcellulose and is 5–10 times more resistant to hydrolysis by RNase in primate serum than parental poly I:C [29]. In mice, poly I:CLC induced similar antigen-specific $CD4^+$ T-cell responses, antibodies, systemic release of inflammatory cytokines, and serum type I IFN as parental poly I:C [18]. Preclinical studies showed that poly I:CLC had potent adjuvant activity on the induction of humoral immune responses in nonhuman primate (NHP) models of Venezuelan equine encephalomyelitis virus and swine influenza virus [30, 31]. In rhesus macaques, poly I:CLC induced HPV16-specific TH1 immune responses and anti-HPV16 antibodies [32]. Interestingly, subcutaneous delivery of poly I:CLC induced a similar signature of innate immune gene expression as the yellow fever vaccine YFV-17D, one of the most successful vaccines ever developed, suggesting poly I:CLC can foster an immune response elicited by a microbial infection, *in vivo* [33]. Peptide vaccination with poly I:CLC as an adjuvant showed rapid induction of NY-ESO-1-specific immune responses in ovarian cancer patients [34] and high levels of IgG and long-term immune memory in patients with premalignant colon lesions [35]. Phase I and II studies conducted in children and adults showed that poly I:CLC exhibited dose-dependent mild to moderate side effects consisting of hypotension, fever, and anemia [36–42]. The use of poly I:CLC as an adjuvant in either prophylactic or therapeutic vaccination is currently being explored in clinical trials of HIV prevention and multiple types of cancer (www.clinicaltrials.gov).

Another dsRNA analogue being pursued is poly I:C_{12}U (Ampligen) [43, 44]. Although the action of poly I:C_{12}U is largely dependent upon TLR-3 and not MDA-5, intranasal immunization with a flu vaccine in mice offered protection against virus challenge similar to parental poly I:C [24, 45]. *In vitro*, studies using monocyte-derived DC from cancer patients and healthy donors showed that poly I:C_{12}U effectively induced mDC maturation and IL-12 production with significantly less production of the suppressive cytokine IL-10 compared with poly I:C [17, 46]. Initial clinical studies suggest that poly I:C_{12}U has less toxicity than poly I:C in humans [47]. Poly I:C_{12}U is currently under development for the treatment of chronic fatigue syndrome, myalgic encephalomyelitis ([48], and trial NCT00215813 [active]) and as part of the combined therapy for acquired immunodeficiency syndrome (AIDS) [49, 50]. Nevertheless, experience as a vaccine adjuvant is limited, and phase I/II trials have been initiated to assess the use of poly I:C_{12}U to improve immune responses to HER2/neu in breast cancer patients [51], as well as autologous tumor cell lysates in ovarian, fallopian tube, or peritoneal cancer and in combination with FluMist™ (terminated 2015, no published results) (www.clinicaltrials.gov).

New dsRNA analogues continue to be developed. Innate Pharma has developed and tested IPH 3102, in an effort to increase the specificity and safety profile of a dsRNA analogue targeting TLR3 in comparison with parental poly I:C. Recently, a novel TLR-3 agonist, RGC100, with a defined chemical structure and only 100 base pairs (bp) in length has been developed and shown to activate mDC and stimulate T-cell proliferation, *in vitro* [52]. Their clinical efficacy and safety is still unknown.

Imidazoquinolines: TLR-7 and TLR-8 recognize guanosine- and uridine-rich ssRNA [53–55]. In humans, B cells and pDC respond to TLR-7 agonists by secreting type I IFN, IL-12, and other proinflammatory cytokines. TLR-8 is mainly expressed by monocytes, mDC, and macrophages, which respond to TLR-8 agonists by producing proinflammatory cytokines and chemokines such as TNF-α, IL-12, and MIP-1α [56, 57]. Because ssRNA is rapidly degraded, fully synthetic small molecules such as imiquimod (R-837, TLR-7 agonist) and resiquimod (R-848, TLR-7/8 agonist) have been used to trigger TLR-7 and/or TLR-8. Studies in mice suggest that administration of antigen with TLR7 and TLR8 agonists results in stronger TH1 and $CD8^+$ T-cell responses as well as rapid antibody production [58–60]. The TLR-7 agonist imiquimod, formulated as a topical 5% cream (Aldara) (3M Company, Maplewood, MN, USA), was the first small molecule TLR agonist approved for the treatment of HPV-mediated external genital warts, superficial basal cell carcinoma, and actinic keratosis [61]. Topical application to skin resulted in localized immune stimulation with low levels of systemic drug uptake and limited systemic induction of cytokines [62]. Topical use of imiquimod as a vaccine adjuvant has been explored during prophylactic and therapeutic vaccination regimens in cancer and infectious diseases. Preconditioning of vaccination sites with imiquimod showed some improvement in responses to tumor antigens in patients with melanoma [63, 64]. In addition, topical imiquimod combined with HPV vaccination in a phase II trial of patients with vulval intraepithelial neoplasia correlated HPV-specific T-cell responses with clinical response [65]. However, in certain settings, topical use of imiquimod was reported not to enhance T-cell responses in healthy human subjects receiving a hepatitis C virus peptide vaccine [66] or humoral immune responses after HBV vaccination in healthy nonresponder individuals [67]. Nevertheless, topical imiquimod continues to be investigated as an adjuvant in vaccines targeting various cancers as well as a recent efficacy, randomized, double-blind clinical trial of intradermal trivalent influenza vaccination in patients with chronic illness (NCT01508884). Current trials are testing the safety and efficacy of combining a therapeutic DNA HPV vaccine with and without imiquimod given topically (NCT00788164).

While imiquimod is the most widely used and studied TLR agonist-based topical, resiquimod has also been explored as a vaccine adjuvant. In rhesus macaques, subcutaneous vaccination with resiquimod and purified recombinant HIV Gag protein showed evidence of enhanced T-cell activation in blood. Furthermore, direct conjugation of resiquimod with antigen enhanced multiple cytokine-producing T cells (IL-2/TNF-α/IFN-γ-producing CD4+ and CD8+ T cells) [68]. Current clinical studies are exploring the use of resiquimod 0.01% gel to improve the response to gp100 and MAGE-3 in patients with melanoma (NCT00960752). In addition, resiquimod was investigated in a safety/efficacy study as a topical adjuvant for subcutaneous NY-ESO-1 protein vaccination in patients with surgically resected stage IIB, stage IIC, stage III, or stage IV (AJCC criteria) melanoma (NCT00821652, completed 2014) and in a phase I trial for the ability to enhance influenza immunization (Intanza®, Sanofi Pasteur) in seniors 65–75 years of age (NCT01737580, completed 8/2013).

Despite the clinical success of topical TLR-7/8 agonists, the associated psoriatic-like skin lesions have raised concerns regarding the local inflammatory response elicited by these agonists. Development of TLR-7 and -8 agonists as vaccine adjuvants will require improvements in formulation and delivery strategy. Currently, new compounds are being developed that afford more predictable conjugation of TLR7/8 agonists to antigen [69–71] and improved safety profile [72].

Stabilized immune modulatory RNA (SIMRA): SIMRA are a pool of RNA compounds where two short (11 bp) phosphorothioate backbone RNA sequences are attached through their 3′ ends via a glycerol linker in order to protect the molecules from rapid exonuclease digestion while enabling delivery in the absence of lipid formulation or encapsulation. Depending on sequence composition and chemical modifications, SIMRA compounds activate immune responses through TLR7, TLR8, or both in cell-based assays [73]. Only 7-deazaguanosine-modified SIMRA activate both human TLR-7 and TLR-8 and induce pDC to produce IFN-α *in vitro* and after subcutaneous injection in monkeys [74]. Further chemical modifications and varying sequence compositions have led to identification of TLR7-specific SIMRA, capable of inducing higher levels of IFN-α and CXCL10 in human PBMC [75]. Lastly, SIMRA have been discussed as candidates for vaccine development, but results from clinical studies have not been published [76].

16.2.1 DNA-Based Adjuvants and Vaccine Studies in Mice

DNA has significant immunomodulatory properties. The immunomodulatory effect of dsDNA depends on its localization and conformation with right-handed B-form dsDNA strongly inducing type I IFN production [15]. The earliest research related to bacterial genomic DNA as enhancing immune responses dates back more than a hundred years ago when Bruns et al. investigated heat-killed Gram-negative or Gram-positive bacteria as an immunotherapeutic agent termed Coley's toxin for cancer [77]. Years later, Krieg and Klinman identified the DNA immunostimulatory motifs showing that sequences containing unmethylated CpG motifs stimulated B cells and NK cells and induced inflammatory cytokine production, while similar sequences with methylated cytosines did not [78, 79]. The immunomodulatory effect of DNA is largely sequence specific as ssDNA containing unmethylated CpG embedded within specific base sequences (e.g. AACGTT) activated the innate immune system via TLR-9, while strands of similar length and base composition that contain sequences such as TTAGGG suppressed the immune response partly by blocking AIM2 [80, 81].

The frequency of unmethylated CpG motifs is higher in microbial DNA than in eukaryotic DNA, and among microbes, the ratio of immunostimulatory to immunosuppressive DNA differs between pathogenic bacteria such as *Escherichia coli* and "protective" bacteria such as *lactobacilli* [82]. During infections, as microbes die and release nucleic acids, the immunostimulatory CpG motifs are recognized by TLR9 in the endosomes of immune cells to trigger a protective response by the host mediated by the recruitment of MyD88, IRAK, and TRAF-6 and resulting in the activation of NF-κB, API, and IRF-7. This response results in the production of proinflammatory cytokines, polyclonal B-cell activation, and the maturation and activation of pDC that enhances the innate immune response and skews immunity toward a $CD4^+$ TH1 adaptive immune response. TLR9-stimulated B cells produce increased levels of IgM and IgG, IL-6, and IL-12 and the CXCR3 chemokines IP-10, BAFF-R, and TACI [78, 83, 84]. Coactivation of TLR-9 and B cell receptor (BCR) in naïve B cells efficiently drives antigen-specific antibody responses; however vaccine efficacy largely depends upon DC targeting and activation. TLR-9-stimulated pDC upregulate the expression of CD40, CD54, CD80, CD86, and MHC class II molecules [80, 85–87] and directly or indirectly increase production of IL-1, IL-6, IL-12, IL-8, TNF-α, and IFN types I, II, and III. This results in increased antigen processing and presentation and leads to the maturation of mDC, which display enhanced CD80, CD86, CD40, and IL-12p40 expression, antigen cross-presentation, and activation of antigen-specific $CD8^+$T cells and γδ T cells [10].

Small synthetic ODNs used as vaccine adjuvants contain CpG motifs (CpG ODNs) mimicking the immunostimulatory properties of microbial DNA and devoid of suppressive motifs. In contrast to naturally occurring ODNs with phosphodiester backbones, synthetic CpG ODNs have a fully or partly phosphorothioate backbone to improved stability with a 10–100 fold increase in activity. CpG ODNs are the best studied nucleic acid immune modulators as vaccine adjuvants with several hundred studies in animals (some of which are listed in Tables 16.1 and 16.2). Studies in mice targeting a diverse group of pathogens, including anthrax [107], leishmania [108], influenza virus [109], lymphocytic choriomeningitis virus [110], orthopoxviruses [111], measles virus [112], HBV [113], and *Clostridium tetani* [114], demonstrated that CpG ODNs accelerated the induction of an immune response, reduced the number of inoculations, increased the overall antibody titers, and improved survival following challenge. The humoral immune response is accompanied by a strong TH1 response with increased production of IFN-γ and a corresponding shift in Ig isotype from IgG1 to IgG2a. More recently, addition of CpG ODN also increased the cellular response against pathogens such as leishmania and tuberculosis, where mice showed an increase in antigen-specific $CD4^+$ and $CD8^+$ T cells and higher levels of IFN-γ production upon challenge [115, 116]. Similarly, boosting bacillus Calmette–Guérin (BCG)-primed mice with mycobacterium tuberculosis proteins plus CpG ODN resulted in increased IFN-γ and IL-17 responses as well as improved protection compared with BCG vaccination alone [115]. Lastly, addition of CpG ODN accelerated and magnified the response to heat-killed Leishmania parasites in primates [117].

The induction of antigen-specific cytotoxic T cells and preferential expansion of effector T cells over regulatory T cells [118] led to numerous studies using CpG ODN as adjuvants for vaccines against several forms of cancer including melanoma [119, 120], prostate [121], and bladder and gastrointestinal cancers [122, 123] as well as studies aiming to redirect atopic immune responses including ragweed [124]. More recently CpG ODNs have been combined with other adjuvants such as ISCOMATRIX™, HSP65-MUC1, or alum to improve the response to antigens (see below). In Balb/c mice, animals that received antigen together with CpG ODNs and ISCOMATRIX showed increased antigen-specific IFN-γ secreting cells to a variety of antigens including hepatitis B surface antigen (HBsAg) [125], influenza A hemagglutinin [126], or tumor antigens [127]. Also, CpG has been used as an adjuvant in combined therapy with recombinant serine protease inhibitor (TgPI1) to protect highly susceptible mice from *Toxoplasma gondii* [128].

Table 16.1 Human experience with CpG ODN as vaccine adjuvants.

Pathogen	ODN	Ab/T-cell response[a]	Protection?	Vaccine	References
HBV	CpG 7909	Ab response in HIV+ patients	TBD	Engerix-B + CpG	[88]
HBV	1018 ISS	TBD	Significantly higher seroprotection than licensed Engerix-B	HEPLISAV™ + CpG	www.dynavax.com; [89]
HCV	Type CCpG 10101	Serum TH1 cytokines	Decreased HCV RNA in serum	CpG alone	[90]
Influenza	CpG 7909	Ab response	TBD	Fluarix® + CpG	[91]
Malaria	CpG 7909	Ab response	TBD	$MSP1_{42}$ + Alhydrogel® + CpG AMA1-C1 or $MSP1_{42}$-C1 + Alhydrogel + CpG	[92, 93]
Anthrax	CpG 7909	Ab response	TBD	Biothrax® + CpG	[94]
CMV	CpG 7909	TBD	TBD	Fusion peptide (universal T-helper cell epitope + viral CD8+ cytotoxic T-cell epitope) + CpG	www.clinicaltrials.gov (NCT00722839)
Pneumococcus	CpG 7909	Antigen-specific IgG TH1 cytokines	ND	Prevnar + CpG	[95, 96]
Melanoma	A-ODN (G10)	*Ex vivo* T-cell response	ND	Melan-A/Mart-1 peptide + VLP-CpG	[97, 98]

Unless noted otherwise, all ODN listed are B-type ODN

CPG 7909, a B-Class CpG ODN of sequence 5′-TCGTCGTTTTGTCGTTTTGTCGTT-3′ was synthesized with a wholly phosphorothioate backbone.

Ab, antibody; CMV, cytomegalovirus; HBV, hepatitis B virus; HCV, hepatitis C virus; ND, not determined; ODN, oligodeoxynucleotide; TBD, to be determined; VLP, virus-like nanoparticle.

[a] All reported responses are those found to be increased versus vaccination without CpG ODN.

Table16.2 Animal experience with CpG ODN as adjuvants.

Pathogen	ODN	Ab/T-cell response	Protection	Model/formulation	Route	Reference
Anthrax	CpG 1555	Ab (IgG2a)	50–90% survival	A/J mice, AVA + CpG ODN-PLG	i.p.	[99]
Influenza	B-ODN	Ab (IgG2a, IgG3)	100% protection	BALB/c and CD-1 mice, Hib-CHO conjugated to TT + CpG	i.d.	[100]
Influenza	CpG 1826	Systemic/mucosal Ab	Not challenged	BALB/c mice, formalin-inactivated virus + CpG	i.m., i.n.	[80]
Influenza	CpG 1826	Ab (IgG1, IgG2a)	100% protection	BALB/c mice, influenza NP and M2 fusion protein (NPM2e) + CpG	i.m.	[101]
Vaccinia	B-ODN	NAb Ab (IgG1, IgG2a)	100% protection	BALB/c mice, A33 and L1 proteins + CpG ODN	s.c.	[82]
HBV	CpG 1826	Ab (10 fold increase)	Protective anti-HBsAg antibody levels	BALB/c mice, HBsAg + CpG	i.m.	[102]
HSV-1	CpG 2007	NAb (IgA, IgG), virus-specific T cells TH1 cytokines	Not challenged	Rabbits, gD peptides + CpG	Ocular	[103]
HSV-2	B-ODN	Ab, virus-specific T-cells TH1 cytokines	Increased survival	C57BL/6 mice, gB + CpG	Intravaginal	[104]
Cancers						
Neuroblastoma	CpG 1826	Tumor-specific T cells	100% survival with low tumor dose	A/J mice: GM/Neuro-2a + CpG	s.c.	[105]
Cervical cancer (HPV)	CpG 1826	Antigen-specific $CD8^+$ T cells	80% survival and tumor shrinkage with CpG combined treatment	C57BL/6 mice, HPV peptides + CpG	s.c	[106]

TT, tetanus toxoid.

16.2.2 Classes of CpG ODN that Activate Human TLR9

Several studies have shown that recognition of CpG motifs is species specific. Mice and humans not only have different cellular distributions of TLR-9 but also respond to different immunostimulatory sequences [85, 129, 130]. CpG ODNs sequences that activate human TLR9 were initially classified as K-type (also known as B-type) or D-type (also known as A-type) depending on their sequence and structure. K-type CpG ODNs (K-ODNs) have phosphorothioate backbones and multiple CpG motifs usually separated by four to six bases with a T preceding the CG (TCG) and do not form secondary structures [85, 131, 132]. K-/B-type CpG ODNs are strong inducers of B-cell activation (IL-6) and pDC maturation and induce high levels of TNF-α but only low levels of type I IFN *in vitro* [80, 85]. Interestingly, some differences have been observed in the optimal immunostimulatory sequence activating peripheral blood mononuclear cells (PBMC) from individual subjects, and combinations of different K-ODNs may result in stronger responses in a population [133]. K-ODNs have been studied as vaccine adjuvants, monotherapy, or both in clinical trials for cancer, infectious diseases, allergy, and asthma [134].

D-and A-type CpG ODNs are structurally different from K-/B-type CpG ODNs [(130, 135)]. D/A-type CpG ODNs have a mixed backbone with a central phosphodiester region, phosphorothioate linkages on both ends to reduce degradation by DNases, and a single CpG motif (e.g. ATCGAT) surrounded by self-complementary bases. It is thought that the combination of phosphodiester backbone and self-complementary bases surrounding the CpG motif allows for the formation of hairpins exposing the CpG motif on the apex and favors the formation of dimers [130]. D-/A-ODNs also have terminal poly-G motifs that may be critical for intracellular localization to specific endosomes and the induction of high levels of IFN-α [136, 137]. Although D- and A-ODNs are grouped together, they differ in the number of terminal poly-G motifs. A-ODNs have poly-G strands on both ends, while D-ODNs only have one poly-G strand at the 3′ end. Importantly, poly-G strands favor the formation of tetrads and multimerization that can impact product stability during manufacturing and storage [136, 138]. *In vitro* and *in vivo*, D-/A-type CpG ODNs induce human pDC to secrete high and sustained levels of IFN type I, II, and III [12, 86]. As a result these ODNs induce monocytes to mature into functionally active DC, increase cross-presentation of antigen to $CD8^+$ T cells, and activate γδ T cells and NK cells to secrete IFN-γ [139]. In contrast to K-ODNs, D-ODNs do not induce pDC maturation, do not activate B cells directly, and have reduced proinflammatory effects partly due to inhibition of the AIM2 inflammasome and Caspase1 [138, 140].

Building on the knowledge that the formation of secondary structure determines the subcellular localization of CpG ODNs and the type of immune response elicited, a second generation of CpG ODNs was developed to induce both polyclonal B-cell stimulation and strong type I IFN production by pDC.

C- and P-type CpG ODNs are examples of second-generation ODNs, combining features of D- or K-ODNs. C-type ODNs (C-ODNs) have phosphorothioate backbones and a hexameric CpG motif positioned at or near the 5′ end linked by a T spacer to a GC-rich self-complementary sequence but do not have a poly-G string at either end. C-ODNs stimulate B-cell proliferation and IL-6 secretion, as well as pDC production of IFN-α. Interestingly, in recent studies assessing the distribution of type I IFN subtypes following stimulation with different TLR agonists, it was noted that while stimulation of pDC with TLR-3, TLR-4, TLR-7, and TLR-9 agonists elicited distinct IFN subtype patterns, activation of TLR-9 by K-, D-, or C-ODNs resulted in very similar IFN subtype gene expression signatures, albeit to varying levels [11, 12]. P-type ODNs (P-ODNs) also have phosphorothioate backbones and two palindromic regions, forming hairpins at their 5′ end and GC-rich 3′ end. P-ODNs activate B cells and act on pDC to increase the expression of CCR7, CD80, and CD83 and stimulate IFN-α production to at least as similar degree as C-ODNs [141, 142].

Third-generation CpG ODNs utilize more advanced chemistry to improve delivery and stability. Examples include the addition of thermolytic protective groups to existing CpG ODNs (pro-ODNs) to prolong their therapeutic window and improve stability [143–145]. The addition of thermolytic moieties to the poly-G tails of D-ODNs reduced G-tetrad formation in solution. Because these moieties do not interfere with the uptake of ODNs and degrade at 37°C, this chemistry allows the ODNs to fold within the cell and localize to the appropriate endosomes where it activates TLR9 signaling to induce high levels of IFN-α, IFN-γ, and CXCL10 [145]. *In vitro* pro-ODNs showed similar cytokine induction as the parental sequence, although immune activation was delayed due to the required deprotection of bases that occurs over time at physiologic temperature. Preclinical studies demonstrated that pro-D-ODNs were as effective as D-ODNs in reducing the severity of cutaneous lesions following a challenge with *Leishmania major* in rhesus macaques [145]. Interestingly, the engineering of thermolytic groups that deprotect with different half-lives at 37°C suggests that mixtures of thermolytic ODNs may prolong or stagger TLR-9 activation *in vivo* [143]. However, this delay may make them less viable as vaccine adjuvants unless a depot system is used to keep antigen and adjuvant together in the tissue.

Lastly, both K- and D-ODNs have been formulated into microparticles, micelles, lipids, and liposomes to improve delivery and stability *in vivo* (see below). In addition several studies have assessed the adjuvant effect of CpG ODNs conjugated to peptides or proteins.

16.2.3 Preclinical Studies with Human CpG ODN

CpG ODNs, particularly D-ODNs, capable of stimulating human cells are only weakly active in mice and rats. Therefore other species including gerbils, pigs, dogs, and NHPs have been used to model CpG ODN safety and efficacy *in vivo*.

The observed response in most of these models only partially mimics the human response; however some promising data has been generated. For example, studies in gerbils using B-ODN as adjuvant for the *Entamoeba histolytica* galactose/*N*-acetyl-D-galactosamine-inhibitable lectin (Gal-lectin) showed that addition of B-ODN increased the antigen-specific neutralizing antibody titer, accelerated parasite clearance, and protected 100% of the animals from intrahepatic challenge as compared with 50% efficacy by Gal-lectin plus control CpG ODNs [146]. Although not tested in combination with antigen, K-ODNs stimulated pig and dog PBMC to increase mRNA expression and protein production of IL-6 and IFN-γ, as well as several heat shock proteins (Hsp60, Hsp70, and Hsp90) but not IL-10 or IL-12 [147, 148]. Coadministration of B-ODNs 2007 with NS3 as a vaccine adjuvant for HCV in piglets increased antibody titers only weakly [149] but was effective in a vaccine for *Pasteurella multocida* where piglets showed higher levels of IFN-γ, TNF-α, and IL-6 [150].

The response to D- and A-ODN was also studied in pigs, where a screen of ODN identified optimal stimulation was achieved with ODN containing the ATCGAT motif [151]. *In vitro*, responses were heterogenic but generally characterized by increased proliferation and production of IL-6, IFN-α, IL-12, and TNF-α [151, 152]. Coinjection of CpG ODN containing the ATCGAT motif increased the immunogenicity of DNA vaccines against rabies virus in pigs [153].

Extensive studies show that NHPs have a similar cell type distribution of TLR-9 [154] and respond to the same CpG ODN sequences as humans [84, 85, 117, 155, 156], making NHP the optimal model to evaluate the safety and efficacy of CpG ODN prior to tests in human clinical trials. Initial proof-of-concept studies in rhesus macaques, delivering ovalbumin (OVA) antigen together with CpG ODN and alum, significantly increased antibody responses over alum alone [117]. Similar increases in antibody responses were evident when CpG ODN (particularly K-ODN) was added to hepatitis B surface antigen (HBsAg) or heat-inactivated anthrax [88, 157, 158]. When used together with *Plasmodium vivax* antigen merozoite surface protein 1 (His_6MSP1), K-ODNs were less effective at inducing an antibody response than incomplete Freund's adjuvant in marmosets [159]. However recent studies in squirrel monkeys showed that when K-ODN K3 was used as an adjuvant for the SE35 antigen, the vaccine effectively suppressed parasitemia and symptoms of malaria following challenge [91]. These results led to clinical studies in the development of a malaria vaccine with CpG 7909 (NCT00320658, NCT00414336, NCT00740090).

Most promising, the addition of CpG ODN to the licensed HBV vaccine Engerix-B resulted in accelerated seroconversion and protective antibody titers in three different primate models: orangutans, macaques, and chimpanzees [94]. Importantly, the addition of CpG ODN overcame hyporesponsiveness to the HBV vaccine Engerix-B in orangutans, with more than 85% of the animals reaching protective anti-HBsAg antibody titers [157]. Similarly, in SIV-infected macaques that fail to develop protective antibody titers following

Engerix-B vaccination, the addition of K- or D-ODN accelerated and significantly increased anti-HBsAg Ab protective antibody titers [88].

As expected, K-ODNs are more effective at inducing antibody responses, but D-ODNs appear more effective as an adjuvant for vaccinations aimed at pathogens requiring strong $CD4^+$ TH1 cell-mediated immune responses. In macaques immunized with heat-killed Leishmania in alum, animals that received D-ODN had higher numbers of antigen-specific TH1 cells producing IFN-γ than those receiving antigen with alum alone or alum plus K-ODN. Interestingly, D-ODN treated animals had less severe cutaneous lesions and reduced parasite burden after challenge compared with K-ODN. Therefore, D-ODN may be more effective than K-ODN at inducing protective responses for intracellular pathogens [117].

The ability of CpG ODN to enhance T cell responses has also been tested using second-generation C-ODN. Macaques primed with emulsified HIV Gag p41 antigen plus C-ODN emulsified in Montanide ISA 51 and boosted with a replication-defective adenovirus expressing HIV Gag (rAd-Gag) vector had higher levels of IFN-γ producing cells compared with Gag alone [92]. In addition, the frequency of IFN-γ producing cells (likely $CD8^+$ T cells) peaked 2 weeks after immunization, increased further with boosting, and lasted for at least 12 weeks. Moreover, the authors showed that the type of agonist used in the primary immunization dictated the short- and long-term quality of the response after rAd-Gag boosting, suggesting that CpG ODN impacts the generation of memory T-cell responses, an important consideration for vaccine design.

16.2.4 Safety Issues Raised in Animal Models

No safety concerns were reported in rodents or NHP (including squirrel monkeys, rhesus, and cynomolgus macaques) receiving adjuvant doses of CpG ODNs. At higher doses (consistent with those used for immunomodulation and immunoprotection), CpG ODNs have induced on-target toxicities. Indeed, in mice, daily intraperitoneal administration of 60 μg CpG ODNs was shown to induce lymphoid follicle destruction and immune suppression and could be associated with liver toxicity and hemorrhagic ascites if used long term [160]. Similarly, intravenous administration of high doses of phosphorothioate CpG ODN in bolus was associated with complement activation and death in a rhesus macaque (Dr. Arthur Krieg, personal communication). An additional safety concern was raised by studies in pregnant mice where repeated administration of high-dose CpG ODN was associated with fetal craniofacial/limb defects [93]. However, repeated subcutaneous administration of D-ODN or pro-D-ODN (1 mg/kg) in rhesus macaques did not result in changes in weight and behavior or altered CBC, the liver, or renal panels (Verthelyi et al., unpublished).

Other concerns regarding the use of CpG ODN have centered on the possibility of increasing the risk of autoimmune disease based on the ability of CpG ODN to induce polyclonal B-cell activation. However, this is a subject of controversy due to conflicting data regarding the association between repeated CpG ODN dosing and the emergence of anti-DNA antibodies. Newer studies suggested that triggering TLR9 would not foster autoimmune disease [95, 161]. Regarding the induction of organ-specific autoimmunity in models known to be driven by TH1 or TH17 responses (including experimental autoimmune encephalomyelitis [162], uveitis [163], and arthritis [164]), the administration of CpG ODN was not consistently associated with accelerated disease.

16.2.5 Clinical Trial Experience

K-ODNs have been tested extensively in clinical trials as adjuvants in vaccines targeting infection, allergies, and cancer with mixed results. The most advanced trials currently correspond to the HEPLISAV™ vaccine targeting HBV. HBV infects the liver and, upon chronic infection, can cause liver cirrhosis and hepatocellular carcinoma. The current hepatitis B vaccines require three immunizations over 6 months to achieve protective antibody titers (>10 mIU/ml), and several populations, including adults greater than 40 years of age, smokers, and HIV-infected individuals, fail to achieve long-lasting seroprotection [165]. Instead of alum, HEPLISAV (Dynavax) uses recombinant HBsAg with ODN 1018 as an adjuvant, a 22 base phosphorothioate B-ODN (sequence 5′-TGACTGTGAACGTTCGAGATGA-3′) reported to improve induction of humoral and cell-mediated immune responses against HBV. Consistent with findings in NHP, clinical studies show HEPLISAV accelerated seroconversion and sustained higher antibody titers with only two immunizations 4 weeks apart compared with the currently licensed alum-adjuvanted vaccines. Of particular interest, early results report HEPLISAV improved immune responses in older populations, as well as in HIV-infected patients compared with licensed vaccines.

Similarly, improved responses in HIV-infected patients were evident when Engerix-B was administered with the B-ODN CpG 7909. Those patients had protective titers after the second dose and maintained protective titers for at least 12 months [166]. Of some concern, during the phase III studies of HEPLISAV, one patient developed Wegener's granulomatosis, and several reports of autoimmune events were noted [97]. The US FDA is currently reviewing the biologic license application containing phase III clinical data for HEPLISAV, and the decision on licensure is awaiting the provision of additional safety data.

Other infectious diseases where K/B-ODN have been used to enhance antibody responses include influenza, malaria, and pneumonia. For influenza, a phase Ib trial (randomized, double-blind, controlled) using the B-ODN CpG

7909 as the adjuvant for the influenza vaccine Fluarix® showed that the addition of CpG 7909 resulted in antibody responses despite using 1/10th of the vaccine antigen concentration. However, patients receiving 1/10th Fluarix® dose with CpG 7909 did not develop the expected cellular immune responses [98]. More recently Dynavax Technologies has been working on a universal influenza vaccine that is composed of N8295, a proprietary fusion protein of two highly conserved influenza antigens, nucleoprotein (NP), and matrix protein 2 (M2e), covalently linked to a second-generation CpG ODN. Phase I studies were completed in February 2011 (www.dynavax.com).

CpG ODN may improve the efficacy of the anthrax vaccine as well. The licensed vaccine for anthrax currently requires immunizations spread over 18 months to induce a persistent protective response and its use is associated with painful adverse events. Since 2011, the need for immunizing defense personnel against anthrax has become a priority. The B-ODN CpG 7909 is being tested as an adjuvant to BioThrax based on animal data that suggests faster development of protective antibody titers than the current vaccine. A phase I study tested the vaccine on a regimen of days 0, 14, and 28 in 69 healthy subjects. Reported results showed that healthy volunteers receiving BioThrax plus CpG 7909 not only showed an accelerated antibody response but also experienced a six- to eightfold increase in antibody titers compared with those receiving BioThrax alone [167]. Phase II trials showed protective titers of neutralizing antibody achieved at 63 days after initial injection, when given in two doses (0 and 28 days) or three doses (0, 14, and 28 days) (NCT01770743). The combined vaccine was reasonably well tolerated in all studies.

For malaria, there are several ongoing programs assessing the safety and efficacy of B-ODN CpG 7909 as adjuvants for *Plasmodium falciparum* antigens. Healthy volunteers immunized with apical membrane Ag 1-combination 1 (AMA1-C1) and merozoite surface protein 142-combination 1 ($MSP1_{42}$-C1) formulated in Alhydrogel plus CpG 7909 showed enhanced kinetics, magnitude, and longevity of antigen-specific memory B cells [168–170]. New phase I studies are testing K-ODN K3 together with SE36 antigen in patients (K.J. Ishii, personal communication). Lastly, a phase Ib/IIa trial randomizing HIV-positive patients to receive 7vPnC (Prevnar®, Pfizer) at 0 and 3 months and one dose of 23-valent pneumococcal polysaccharide vaccine (PPV-23; Pneumo Novum) alone or with CpG 7909 showed that CpG 7909 increased the seroconversion frequency and antibody titers [171].

Perhaps the greatest expectation of success for CpG ODN as adjuvants involves vaccines that target cancer antigens. The rationale for these studies is that CpG ODN treatment improves antigen presentation and facilitates cross-presentation to generate effective $CD8^+$ T-cell responses and dampen regulatory T cells that may interfere with the development of effective immune responses. For example, K-ODN (alone or in association with other adjuvants) have been relatively successful at increasing the immunogenicity of melanoma-associated antigens such as MAGEA3 and NY-ESO-1 in clinical trials [172–176]. Other studies have assessed the safety and effectiveness of CpG ODN in vaccines for renal cancer, Willms' tumor, lung cancer, gliomas, and lymphoma [177]. D-ODNs have been studied less in clinical trials due to concerns of poly-G strand-mediated aggregation of the ODNs; however phase I/II studies used a vaccine that linked the melanoma-specific Melan-A/Mart-1 peptide to viruslike nanoparticles loaded with A-ODN (5′-GGGGGGGGGGACGATCGTCGGGGGG GGG-3′). The report states that vaccination induced *ex vivo* $CD8^+$ T-cell responses to antigen in 14 of 22 patients [178, 179] consistent with the improved CTL activity observed *in vitro* [180].

16.2.6 Safety Issues from Human Clinical Trials

The testing of CpG ODN in humans has been extensive and over 3000 patients have received the product in the context of different clinical trials. Studies have shown that administration of K- or C-ODN can lead to local injection site reactions (erythema, edema, inflammation, and pain) and systemic flu-like symptoms (headache, rigors, pyrexia, nausea, and fever). Symptoms typically developed within 24h of administration were mild to moderate and transient, usually lasting for less than 48h. There were no major adverse events (grade 3–4) reported when CpG ODNs were used as vaccine adjuvants in adults [98, 169, 181]. In subjects that received repeated ODN doses, the main complaints involved local inflammation and flu-like symptoms. In several studies patients were monitored for development of anti-DNA antibodies. Although some patients developed antibodies to ssDNA, the antibodies were transient and not associated with autoimmune disease. In the HEPLISAV study, one subject developed Wegener's granulomatosis, and a few others reported signs of autoimmune processes [97]. The nature of these associations is unclear.

Similar to K-ODNs, the toxicities observed in clinical studies using D-ODNs were not severe. Systemic adverse events linked to the vaccine included myalgia and arthralgia, followed by headache and flu-like symptoms. All adverse events were classified as mild to moderate and thought to result from the local immune activation induced by the product. Local immune activation effects were transient, usually disappearing within 24h and included edema, erythema, induration, and pain as well as lymphadenopathy of draining lymph nodes [178, 179].

16.2.7 Novel Delivery Systems for CpG ODN as Adjuvants

A practical aspect of vaccine design is whether ensuring both antigen and adjuvant are delivered to the same DC for optimal antigen presentation and effective priming of $CD4^+$ and $CD8^+$ T cells. Several studies have shown that

TLR activation in the same DC presenting the antigen is critical for $CD4^+$ T-cell activation and TH1 cell differentiation [140, 182–184]. The use of unmodified phosphodiester CpG ODN *in vivo* posed an immediate problem as exonucleases could degrade the adjuvant prior to reaching the target endosomes [99]. Phosphorothioate-modified CpG ODNs overcome this problem by being stable, inexpensive, easily manufactured, and thus the most widely used vaccine adjuvants [185]. However, the phosphorothioate chemistry induces nonspecific inflammation that is likely associated with some of the local adverse events and lymphadenopathies reported for CpG ODN adjuvants [186, 187]. A number of delivery systems are being developed to overcome these problems including covalently cross-linking CpG ODN with antigen to create bioconjugates. Early studies showed that cross-linking CpG ODN with OVA antigen increased antigen internalization by APC 40-fold, facilitating antigen cross-presentation and cross-priming of $CD8^+$ and $CD4^+$ T cells [188]. Importantly, an accessible 5′ end of CpG DNA is required for immunostimulatory activity suggesting bioconjugates; using the 5′ end of CpG may interfere with receptor binding resulting in diminished immune stimulation [89, 90, 96].

Encapsulation of CpG ODN into cationic liposomes facilitates electrostatic interaction with the cell membrane and codelivery with the antigen via clathrin-mediated endocytosis into the endosomes of APC resulting in activation of both innate and antigen-specific adaptive immune responses in murine models [100, 101]. Studies showed that encapsulation of CpG ODN in cationic liposomes enhanced uptake by macrophages and $CD11c^+$ DC resulting in higher levels of IL-12 and IFN-γ production compared with naked CpG ODN [102–104]. Recent studies from Gursel's group showed that the use of different lipid combinations determined release and delivery rates and ultimately impacted the profile of cell activation [105]. In their study, encapsulating CpG ODN in anionic and neutral lipids resulted in higher IFN-γ production and CD86 expression on APC than cationic lipids, while D-ODN encapsulated in stealth liposomes (cholesterol, dioleylphosphatidylethanolamine, and polyethylene glycol-phosphatidylethanolamine) resulted in higher levels of IFN-α [106].

CpG ODNs have also been encapsulated into microparticles made from biodegradable and biocompatible polymers such as polylactide co-glycolide (PLGA or PLG) [189]. In mice, PLG microparticles adsorbed with CpG ODN and *Neisseria* meningitides B protein showed significantly higher antibody production and T cell responses than soluble CpG ODN [190]. Similarly, coencapsulation of CpG ODN with the anthrax vaccine AVA induced stronger IgG responses than the same vaccine coadministered with free CpG ODN [191]. Other microparticles tested include self-assembling nanoparticles of cholesterol, which efficiently delivered antigen to APC [192, 193] and gelatin, a natural and nontoxic macromolecule found to be more stable than liposomes *in vivo* [194].

Lastly, systems are being developed for oncology applications that conjugate whole tumor cells with biodegradable PLA-PEG-microparticles entrapping

CpG ODN. The potential advantage of these microparticle hybrid systems is allowing co-delivery of irradiated tumor cells containing a full array of tumor antigens as well as cells modified to secrete cytokines such as GM-CSF. This approach may ultimately increase antitumor immunity; however, there is no evidence regarding the efficacy and safety of this particular delivery system *in vitro* or *in vivo* [189].

16.3 Conclusion

Among the pharmaceutical armamentaria, vaccines are the most efficient tool to combat specific infectious diseases, but they are also mostly used in healthy populations and thus require unparalleled guarantees of safety. Recent discoveries in innate and adaptive immune responses and mechanisms allow for the design of a new generation of vaccines based on highly purified recombinant proteins, synthetic peptides, or viral subunits rather than attenuated pathogens. This approach to vaccines improves manufacturing reproducibility, reduces the risk of infection in immunocompromised subjects (Marciano B.E., 2014 237 /id), minimizes interference between different vaccines, and lowers the risk of adverse reactions. However the more simplified vaccine structure of the subunits compared with whole pathogens can also reduce immunogenicity and thus requires the use of powerful adjuvants. The development of new adjuvants that activate the innate immune system by targeting a specific TLR allows for the design of vaccines that induce very specific, targeted, and effective responses.

As described above, nucleic acid-derived adjuvants combine improved antigen uptake and presentation with strong cellular and humoral responses in a variety of models and are excellent candidates for vaccine adjuvants. Of particular interest, these adjuvants promise to improve the response to vaccines in populations that do not mount protective responses readily, such as the elderly and the immunosuppressed. Studies transitioned rapidly from the laboratory to clinical trials and initial reports demonstrate that they can be used relatively safely in patients. The development of new chemistries, formulations, and delivery methods should further propel the clinical development of these adjuvants.

Acknowledgments

The assertions herein are the private ones from the authors and are not to be construed as official or as reflecting the views of the Food and Drug Administration. The authors acknowledge CDER Critical Path funding that supported this research project. This study was supported in part by Senior Postgraduate Research Fellowship Award to K.B., L.A.H. and J.M. from the

Oak Ridge Institute for Science and Education (ORISE) through an interagency agreement between the US Department of Energy and the US Food and Drug Administration. We thank Mr. Logan Kelley Baker for careful review and research help with the manuscript.

References

1 Schijns, V.E. and Lavelle, E.C. (2011). Trends in vaccine adjuvants. *Expert Rev. Vaccines* 10: 539–550.
2 Barton, G.M., Kagan, J.C., and Medzhitov, R. (2006). Intracellular localization of Toll-like receptor 9 prevents recognition of self DNA but facilitates access to viral DNA. *Nat. Immunol.* 7 (1): 49–56.
3 Desmet, C.J. and Ishii, K.J. (2012). Nucleic acid sensing at the interface between innate and adaptive immunity in vaccination. *Nat. Rev. Immunol.* 12 (7): 479–491.
4 Paludan, S.R. and Bowie, A.G. (2013). Immune sensing of DNA. *Immunity* 38 (5): 870–880.
5 Bachem, A., Guttler, S., Hartung, E. et al. (2010). Superior antigen cross-presentation and XCR1 expression define human CD11c+CD141+ cells as homologues of mouse CD8+ dendritic cells. *J. Exp. Med.* 207 (6): 1273–1281.
6 Jongbloed, S.L., Kassianos, A.J., McDonald, K.J. et al. (2010). Human CD141+ (BDCA-3)+ dendritic cells (DCs) represent a unique myeloid DC subset that cross-presents necrotic cell antigens. *J. Exp. Med.* 207 (6): 1247–1260.
7 Poulin, L.F., Salio, M., Griessinger, E. et al. (2010). Characterization of human DNGR-1+ BDCA3+ leukocytes as putative equivalents of mouse CD8alpha+ dendritic cells. *J. Exp. Med.* 207 (6): 1261–1271.
8 Jelinek, I., Leonard, J.N., Price, G.E. et al. (2011). TLR3-specific double-stranded RNA oligonucleotide adjuvants induce dendritic cell cross-presentation, CTL responses, and antiviral protection. *J. Immunol.* 186 (4): 2422–2429.
9 Sarvestani, S.T., Williams, B.R., and Gantier, M.P. (2012). Human Toll-like receptor 8 can be cool too: implications for foreign RNA sensing. *J. Interferon Cytokine Res.* 32 (8): 350–361.
10 Verthelyi, D. and Zeuner, R.A. (2003). Differential signaling by CpG DNA in DCs and B cells: not just TLR9. *Trends Immunol.* 10: 519–522.
11 Puig, M., Tosh, K.W., Schramm, L.M. et al. (2012). TLR9 and TLR7 agonists mediate distinct type I IFN responses in humans and nonhuman primates *in vitro* and *in vivo*. *J. Leukoc. Biol.* 91 (1): 147–158.
12 Hillyer, P., Mane, V.P., Schramm, L.M. et al. (2012). Expression profiles of human interferon-alpha and interferon-lambda subtypes are ligand- and cell-dependent. *Immunol. Cell Biol.*

13 Mathe, G., Kamel, M., Dezfulian, M. et al. (1973). An experimental screening for "systemic adjuvants of immunity" applicable in cancer immunotherapy. *Cancer Res.* 33 (9): 1987–1997.
14 Gitlin, L., Barchet, W., Gilfillan, S. et al. (2006). Essential role of mda-5 in type I IFN responses to polyriboinosinic:polyribocytidylic acid and encephalomyocarditis picornavirus. *Proc. Natl. Acad. Sci. U. S. A.* 103 (22): 8459–8464.
15 Kato, H., Takeuchi, O., Sato, S. et al. (2006). Differential roles of MDA5 and RIG-I helicases in the recognition of RNA viruses. *Nature* 441 (7089): 101–105.
16 Martins, K.A., Bavari, S., and Salazar, A.M. (2015). Vaccine adjuvant uses of poly-IC and derivatives. *Expert Rev. Vaccines* 14 (3): 447–459.
17 Jasani, B., Navabi, H., and Adams, M. (2009). Ampligen: a potential toll-like 3 receptor adjuvant for immunotherapy of cancer. *Vaccine* 27 (25–26): 3401–3404.
18 Longhi, M.P., Trumpfheller, C., Idoyaga, J. et al. (2009). Dendritic cells require a systemic type I interferon response to mature and induce CD4+ Th1 immunity with poly IC as adjuvant. *J. Exp. Med.* 206 (7): 1589–1602.
19 Wang, Y., Cella, M., Gilfillan, S., and Colonna, M. (2010). Cutting edge: polyinosinic:polycytidylic acid boosts the generation of memory CD8 T cells through melanoma differentiation-associated protein 5 expressed in stromal cells. *J. Immunol.* 184 (6): 2751–2755.
20 Richmond, J.Y. and Hamilton, L.D. (1969). Foot-and-mouth disease virus inhibition induced in mice by synthetic double-stranded RNA (polyriboinosinic and polyribocytidylic acids). *Proc. Natl. Acad. Sci. U. S. A.* 64 (1): 81–86.
21 Thomson, A.W. (1977). The influence of poly I: poly C on kinetics of the primary immune response to sheep erythrocytes in the mouse spleen. *Biomedicine* 26 (5): 319–326.
22 Partidos, C.D., Hoebeke, J., Moreau, E. et al. (2005). The binding affinity of double-stranded RNA motifs to HIV-1 Tat protein affects transactivation and the neutralizing capacity of anti-Tat antibodies elicited after intranasal immunization. *Eur. J. Immunol.* 35 (5): 1521–1529.
23 Sloat, B.R. and Cui, Z. (2006). Nasal immunization with anthrax protective antigen protein adjuvanted with polyriboinosinic-polyribocytidylic acid induced strong mucosal and systemic immunities. *Pharm. Res.* 23 (6): 1217–1226.
24 Ichinohe, T., Watanabe, I., Ito, S. et al. (2005). Synthetic double-stranded RNA poly(I:C) combined with mucosal vaccine protects against influenza virus infection. *J. Virol.* 79 (5): 2910–2919.
25 Park, H., Adamson, L., Ha, T. et al. (2013). Polyinosinic-polycytidylic acid is the most effective TLR adjuvant for SIV Gag protein-induced T cell responses in nonhuman primates. *J. Immunol.* 190 (8): 4103–4115.

26 Robinson, R.A., DeVita, V.T., Levy, H.B. et al. (1976). A phase I-II trial of multiple-dose polyriboinosic-polyribocytidylic acid in patieonts with leukemia or solid tumors. *J. Natl. Cancer Inst.* 57 (3): 599–602.

27 Absher, M. and Stinebring, W.R. (1969). Toxic properties of a synthetic double-stranded RNA. Endotoxin-like properties of poly I. poly C, an interferon stimulator. *Nature* 223 (5207): 715–717.

28 Apostolico, J.d.S., Boscardin, S.B., Yamamoto, M.M. et al. (2016). HIV envelope trimer specific immune response is influenced by different adjuvant formulations and heterologous prime-boost. *PLoS One* 11 (1): e0145637.

29 Levy, H.B., Baer, G., Baron, S. et al. (1975). A modified polyriboinosinic-polyribocytidylic acid complex that induces interferon in primates. *J Infect Dis* 132 (4): 434–439.

30 Houston, W.E., Crabbs, C.L., Stephen, E.L., and Levy, H.B. (1976). Modified polyriboinosinic-polyribocytidylic acid, an immunological adjuvant. *Infect. Immun.* 14 (1): 318–319.

31 Stephen, E.L., Hilmas, D.E., Mangiafico, J.A., and Levy, H.B. (1977). Swine influenza virus vaccine: potentiation of antibody responses in rhesus monkeys. *Science* 197 (4310): 1289–1290.

32 Stahl-Hennig, C., Eisenblatter, M., Jasny, E. et al. (2009). Synthetic double-stranded RNAs are adjuvants for the induction of T helper 1 and humoral immune responses to human papillomavirus in rhesus macaques. *PLoS Pathog.* 5 (4): e1000373.

33 Caskey, M., Lefebvre, F., Filali-Mouhim, A. et al. (2011). Synthetic double-stranded RNA induces innate immune responses similar to a live viral vaccine in humans. *J. Exp. Med.* 208 (12): 2357–2366.

34 Sabbatini, P., Tsuji, T., Ferran, L. et al. (2012). Phase I trial of overlapping long peptides from a tumor self-antigen and poly-ICLC shows rapid induction of integrated immune response in ovarian cancer patients. *Clin. Cancer Res.* 18 (23): 6497–6508.

35 Kimura, T., McKolanis, J.R., Dzubinski, L.A. et al. (2013). MUC1 vaccine for individuals with advanced adenoma of the colon: a cancer immunoprevention feasibility study. *Cancer Prev. Res. (Phila.)* 6 (1): 18–26.

36 Bever, C.T. Jr., Salazar, A.M., Neely, E. et al. (1986). Preliminary trial of poly ICLC in chronic progressive multiple sclerosis. *Neurology* 36 (4): 494–498.

37 Champney, K.J., Levine, D.P., Levy, H.B., and Lerner, A.M. (1979). Modified polyriboinosinic-polyribocytidylic acid complex: sustained interferonemia and its physiological associates in humans. *Infect. Immun.* 25 (3): 831–837.

38 Durie, B.G., Levy, H.B., Voakes, J. et al. (1985). Poly(I,C)-LC as an interferon inducer in refractory multiple myeloma. *J. Biol. Response Mod.* 4 (5): 518–524.

39 Krown, S.E., Kerr, D., Stewart, W.E. 2nd et al. (1985). Phase I trials of poly(I,C) complexes in advanced cancer. *J. Biol. Response Mod.* 4 (6): 640–649.

40 Lampkin, B.C., Levine, A.S., Levy, H. et al. (1985). Phase II trial of a complex polyriboinosinic-polyribocytidylic acid with poly-L-lysine and carboxymethyl

cellulose in the treatment of children with acute leukemia and neuroblastoma: a report from the Children's Cancer Study Group. *Cancer Res.* 45 (11 Pt 2): 5904–5909.

41 Salazar, A.M., Levy, H.B., Ondra, S. et al. (1996). Long-term treatment of malignant gliomas with intramuscularly administered polyinosinic-polycytidylic acid stabilized with polylysine and carboxymethylcellulose: an open pilot study. *Neurosurgery* 38 (6): 1096–1103 discussion 103-4.

42 Stevenson, H.C., Abrams, P.G., Schoenberger, C.S. et al. (1985). A phase I evaluation of poly(I,C)-LC in cancer patients. *J. Biol. Response Mod.* 4 (6): 650–655.

43 Gowen, B.B., Wong, M.H., Jung, K.H. et al. (2007). TLR3 is essential for the induction of protective immunity against Punta Toro Virus infection by the double-stranded RNA (dsRNA), poly(I:C12U), but not Poly(I:C): differential recognition of synthetic dsRNA molecules. *J. Immunol.* 178 (8): 5200–5208.

44 Trumpfheller, C., Caskey, M., Nchinda, G. et al. (2008). The microbial mimic poly IC induces durable and protective CD4+ T cell immunity together with a dendritic cell targeted vaccine. *Proc. Natl. Acad. Sci. U. S. A.* 105 (7): 2574–2579.

45 Ichinohe, T., Kawaguchi, A., Tamura, S. et al. (2007). Intranasal immunization with H5N1 vaccine plus Poly I:Poly C12U, a Toll-like receptor agonist, protects mice against homologous and heterologous virus challenge. *Microbes Infect.* 9 (11): 1333–1340.

46 Navabi, H., Jasani, B., Reece, A. et al. (2009). A clinical grade poly I:C-analogue (Ampligen) promotes optimal DC maturation and Th1-type T cell responses of healthy donors and cancer patients *in vitro*. *Vaccine* 27 (1): 107–115.

47 Thompson, K.A., Strayer, D.R., Salvato, P.D. et al. (1996). Results of a double-blind placebo-controlled study of the double-stranded RNA drug polyI:polyC12U in the treatment of HIV infection. *Eur. J. Clin. Microbiol. Infect. Dis.* 15 (7): 580–587.

48 Strayer, D.R., Carter, W.A., Stouch, B.C. et al. (2012). A double-blind, placebo-controlled, randomized, clinical trial of the TLR-3 agonist rintatolimod in severe cases of chronic fatigue syndrome. *PLoS One* 7 (3): e31334.

49 Carter, W.A., Ventura, D., Shapiro, D.E. et al. (1991). Mismatched double-stranded RNA, Ampligen (poly(I): poly(C12U)), demonstrates antiviral and immunostimulatory activities in HIV disease. *Int. J. Immunopharmacol.* 13 (Suppl 1): 69–76.

50 Strayer, D.R., Carter, W.A., Brodsky, I. et al. (1994). A controlled clinical trial with a specifically configured RNA drug, poly(I).poly(C12U), in chronic fatigue syndrome. *Clin. Infect. Dis.* 18 (Suppl 1): S88–S95.

51 Saito, S., Ainai, A., Suzuki, T. et al. (2016). The effect of mucoadhesive excipient on the nasal retention time of and the antibody responses induced by an intranasal influenza vaccine. *Vaccine* 34 (9): 1201–1207.

52 Naumann, K., Wehner, R., Schwarze, A. et al. (2013). Activation of dendritic cells by the novel Toll-like receptor 3 agonist RGC100. *Clin. Dev. Immunol.* 2013 (283649).

53 Diebold, S.S., Kaisho, T., Hemmi, H. et al. (2004). Innate antiviral responses by means of TLR7-mediated recognition of single-stranded RNA. *Science* 303 (5663): 1529–1531.
54 Heil, F., Hemmi, H., Hochrein, H. et al. (2004). Species-specific recognition of single-stranded RNA via toll-like receptor 7 and 8. *Science* 303 (5663): 1526–1529.
55 Lund, J.M., Alexopoulou, L., Sato, A. et al. (2004). Recognition of single-stranded RNA viruses by Toll-like receptor 7. *Proc. Natl. Acad. Sci. U. S. A.* 101 (15): 5598–5603.
56 Gorden, K.B., Gorski, K.S., Gibson, S.J. et al. (2005). Synthetic TLR agonists reveal functional differences between human TLR7 and TLR8. *J. Immunol.* 174 (3): 1259–1268.
57 Ito, T., Amakawa, R., Kaisho, T. et al. (2002). Interferon-alpha and interleukin-12 are induced differentially by Toll- like receptor 7 ligands in human blood dendritic cell subsets. *J. Exp. Med.* 195 (11): 1507–1512.
58 Hamm, S., Heit, A., Koffler, M. et al. (2007). Immunostimulatory RNA is a potent inducer of antigen-specific cytotoxic and humoral immune response *in vivo*. *Int. Immunol.* 19 (3): 297–304.
59 Rajagopal, D., Paturel, C., Morel, Y. et al. (2010). Plasmacytoid dendritic cell-derived type I interferon is crucial for the adjuvant activity of Toll-like receptor 7 agonists. *Blood* 115 (10): 1949–1957.
60 Zhang, W.W. and Matlashewski, G. (2008). Immunization with a Toll-like receptor 7 and/or 8 agonist vaccine adjuvant increases protective immunity against Leishmania major in BALB/c mice. *Infect. Immun.* 76 (8): 3777–3783.
61 Wagstaff, A.J. and Perry, C.M. (2007). Topical imiquimod: a review of its use in the management of anogenital warts, actinic keratoses, basal cell carcinoma and other skin lesions. *Drugs* 67 (15): 2187–2210.
62 Harrison, L.I., Skinner, S.L., Marbury, T.C. et al. (2004). Pharmacokinetics and safety of imiquimod 5% cream in the treatment of actinic keratoses of the face, scalp, or hands and arms. *Arch. Dermatol. Res.* 296 (1): 6–11.
63 Adams, S., O'Neill, D.W., Nonaka, D. et al. (2008). Immunization of malignant melanoma patients with full-length NY-ESO-1 protein using TLR7 agonist imiquimod as vaccine adjuvant. *J. Immunol.* 181 (1): 776–784.
64 Shackleton, M., Davis, I.D., Hopkins, W. et al. (2004). The impact of imiquimod, a Toll-like receptor-7 ligand (TLR7L), on the immunogenicity of melanoma peptide vaccination with adjuvant Flt3 ligand. *Cancer Immun.* 4: 9.
65 Daayana, S., Elkord, E., Winters, U. et al. (2010). Phase II trial of imiquimod and HPV therapeutic vaccination in patients with vulval intraepithelial neoplasia. *Br. J. Cancer* 102 (7): 1129–1136.
66 Firbas, C., Boehm, T., Buerger, V. et al. (2010). Immunogenicity and safety of different injection routes and schedules of IC41, a Hepatitis C virus (HCV) peptide vaccine. *Vaccine* 28 (12): 2397–2407.

67 Roukens, A.H., Vossen, A.C., Boland, G.J. et al. (2010). Intradermal hepatitis B vaccination in non-responders after topical application of imiquimod (Aldara). *Vaccine* 28 (26): 4288–4293.
68 Wille-Reece, U., Flynn, B.J., Lore, K. et al. (2005). HIV Gag protein conjugated to a Toll-like receptor 7/8 agonist improves the magnitude and quality of Th1 and CD8+ T cell responses in nonhuman primates. *Proc. Natl. Acad. Sci.* 102 (42): 15190–15194.
69 Shukla, N.M., Lewis, T.C., Day, T.P. et al. (2011). Toward self-adjuvanting subunit vaccines: model peptide and protein antigens incorporating covalently bound toll-like receptor-7 agonistic imidazoquinolines. *Bioorg. Med. Chem. Lett.* 21 (11): 3232–3236.
70 Shinchi, H., Crain, B., Yao, S. et al. (2015). Enhancement of the immunostimulatory activity of a TLR7 ligand by conjugation to polysaccharides. *Bioconjug. Chem.* 26 (8): 1713–1723.
71 Gao, D., Liu, Y., Diao, Y. et al. (2015). Synthesis and evaluation of conjugates of novel TLR7 inert ligands as self-adjuvanting immunopotentiators. *ACS Med. Chem. Lett.* 6 (3): 249–253.
72 Smirnov, D., Schmidt, J.J., Capecchi, J.T., and Wightman, P.D. (2011). Vaccine adjuvant activity of 3M-052: an imidazoquinoline designed for local activity without systemic cytokine induction. *Vaccine* 29 (33): 5434–5442.
73 Lan, T. and Kandimalla, E.R. (2011). Synthesis, purification, and characterization of oligoribonucleotides that act as agonists of TLR7 and/or TLR8. *Methods Mol. Biol.* 764: 249–261.
74 Lan, T., Kandimalla, E.R., Yu, D. et al. (2007). Stabilized immune modulatory RNA compounds as agonists of Toll-like receptors 7 and 8. *Proc. Natl. Acad. Sci. U. S. A.* 104 (34): 13750–13755.
75 Lan, T., Dai, M., Wang, D. et al. (2009). Toll-like receptor 7 selective synthetic oligoribonucleotide agonists: synthesis and structure-activity relationship studies. *J. Med. Chem.* 52 (21): 6871–6879.
76 Philbin, V.J. and Levy, O. (2009). Developmental biology of the innate immune response: implications for neonatal and infant vaccine development. *Pediatr. Res.* 65 (5 Pt 2): 98R–105R.
77 Coley, W.B. (1991). The treatment of malignant tumors by repeated inoculations of erysipelas. With a report of ten original cases. 1893. *Clin. Orthop. Relat. Res.* (262): 3–11.
78 Klinman, D.M., Yi, A., Beaucage, S.L. et al. (1996). CpG motifs expressed by bacterial DNA rapidly induce lymphocytes to secrete IL-6, IL-12 and IFNg. *Proc. Natl. Acad. Sci. U. S. A.* 93: 2879–2883.
79 Krieg, A.M. (1995). CpG DNA: a pathogenic factor in systemic lupus erythematosus? *J. Clin. Immunol.* 15 (6): 284–292.
80 Krieg, A.M., Yi, A., Matson, S. et al. (1995). CpG motifs in bacterial DNA trigger direct B-cell activation. *Nature* 374: 546–548.

81 Kaminski, J.J., Schattgen, S.A., Tzeng, T.C. et al. (2013). Synthetic oligodeoxynucleotides containing suppressive TTAGGG motifs inhibit AIM2 inflammasome activation. *J. Immunol.* 191 (7): 3876–3883.

82 Bouladoux, N., Hall, J.A., Grainger, J.R. et al. (2012). Regulatory role of suppressive motifs from commensal DNA. *Mucosal Immunol.* 5 (6): 623–634.

83 Katsenelson, N., Kanswal, S., Puig, M. et al. (2007). Synthetic CpG oligodeoxynucleotides augment BAFF- and APRIL-mediated immunoglobulin secretion. *Eur. J. Immunol.* 37 (7): 1785–1795.

84 Verthelyi, D. and Klinman, D.M. (2003). Immunoregulatory activity of CpG oligonucleotides in humans and nonhuman primates. *Clin. Immunol.* 109 (1): 64–71.

85 Hartmann, G., Weeratna, R.D., Ballas, Z.K. et al. (2000). Delineation of a CpG phosphorothioate oligodeoxinucleotide for activating primate immune responses *in vitro* and *in vivo*. *J. Immunol.* 164: 1617–1624.

86 Krug, A., Rothenfusser, S., Hornung, V. et al. (2001). Identification of CpG oligonucleotide sequences with high induction of IFNa/b in plasmacytoid dendritic cells. *Eur. J. Immunol.* 31: 2154–2163.

87 Kerkmann, M., Rothenfusser, S., Hornung, V. et al. (2003). Activation with CpG-A and CpG-B oligonucleotides reveals two distinct regulatory pathways of type I IFN synthesis in human plasmacytoid dendritic cells. *J. Immunol.* 170 (9): 4465–4474.

88 Verthelyi, D., Wang, V.W., Lifson, J.D., and Klinman, D.M. (2004). CpG oligodeoxynucleotides improve the response to hepatitis B immunization in healthy and SIV-infected rhesus macaques. *AIDS* 18 (7): 1003–1008.

89 Yu, D., Zhao, Q., Kandimalla, E.R., and Agrawal, S. (2000). Accessible 5′-end of CpG-containing phosphorothioate oligodeoxynucleotides is essential for immunostimulatory activity. *Bioorg. Med. Chem. Lett.* 10 (23): 2585–2588.

90 Karlsen, K., Korsholm, K.S., Mortensen, R. et al. (2014). A stable nanoparticulate DDA/MMG formulation acts synergistically with CpG ODN 1826 to enhance the CD4(+) T-cell response. *Nanomedicine (Lond.)* 9 (17): 2625–2638.

91 Tougan, T., Aoshi, T., Coban, C. et al. (2013). TLR9 adjuvants enhance immunogenicity and protective efficacy of the SE36/AHG malaria vaccine in nonhuman primate models. *Hum. Vaccin. Immunother.* 9 (2): 283–290.

92 Wille-Reece, U., Flynn, B.J., Lore, K. et al. (2006). Toll-like receptor agonists influence the magnitude and quality of memory T cell responses after prime-boost immunization in nonhuman primates. *J. Exp. Med.* 203 (5): 1249–1258.

93 Prater, M.R., Johnson, V.J., Germolec, D.R. et al. (2006). Maternal treatment with a high dose of CpG ODN during gestation alters fetal craniofacial and distal limb development in C57BL/6 mice. *Vaccine* 24 (3): 263–271.

94 Payette, P.J., Xiaoying, M., Risini, D. et al. (2006). Testing of CpG-optimized protein and DNA vaccines against the hepatitis B virus in Chimpanzees for immunogenicity and protection from challenge. *Intervirology* 49 (3): 144–151.

95 Su, S.B., Silver, P.B., Grajewski, R.S. et al. (2005). Essential role of the MyD88 pathway, but nonessential roles of TLRs 2, 4, and 9, in the adjuvant effect promoting Th1-mediated autoimmunity. *J. Immunol.* 175 (10): 6303–6310.
96 Kramer, K., Shields, N.J., Poppe, V. et al. (2017). Intracellular cleavable CpG oligodeoxynucleotide-antigen conjugate enhances anti-tumor immunity. *Mol. Ther.* 25 (1): 62–70.
97 Cooper, C. and Mackie, D. (2011). Hepatitis B surface antigen-1018 ISS adjuvant-containing vaccine: a review of HEPLISAV safety and efficacy. *Expert Rev. Vaccines* 10 (4): 417–427.
98 Cooper, C.L., Davis, H.L., Morris, M.L. et al. (2004). Safety and immunogenicity of CPG 7909 injection as an adjuvant to Fluarix influenza vaccine. *Vaccine* 22 (23–24): 3136–3143.
99 Mutwiri, G.K., Nichani, A.K., Babiuk, S., and Babiuk, L.A. (2004). Strategies for enhancing the immunostimulatory effects of CpG oligodeoxynucleotides. *J. Control. Release* 97 (1): 1–17.
100 Zaks, K., Jordan, M., Guth, A. et al. (2006). Efficient immunization and cross-priming by vaccine adjuvants containing TLR3 or TLR9 agonists complexed to cationic liposomes. *J. Immunol.* 176 (12): 7335–7345.
101 Wilson, K.D., Raney, S.G., Sekirov, L. et al. (2007). Effects of intravenous and subcutaneous administration on the pharmacokinetics, biodistribution, cellular uptake and immunostimulatory activity of CpG ODN encapsulated in liposomal nanoparticles. *Int. Immunopharmacol.* 7 (8): 1064–1075.
102 Mui, B., Raney, S.G., Semple, S.C., and Hope, M.J. (2001). Immune stimulation by a CpG-containing oligodeoxynucleotide is enhanced when encapsulated and delivered in lipid particles. *J. Pharmacol. Exp. Ther.* 298 (3): 1185–1892.
103 Suzuki, Y., Wakita, D., Chamoto, K. et al. (2004). Liposome-encapsulated CpG oligodeoxynucleotides as a potent adjuvant for inducing type 1 innate immunity. *Cancer Res.* 64 (23): 8754–8760.
104 Yu, C., An, M., Li, M., and Liu, H. (2017). Immunostimulatory properties of lipid modified CpG oligonucleotides. *Mol. Pharm.* 14: 2815–2823.
105 Gungor, B., Yagci, F.C., Gursel, I., and Gursel, M. (2014). Forging a potent vaccine adjuvant: CpG ODN/cationic peptide nanorings. *Oncoimmunology* 3 (7): e950166.
106 Erikci, E., Gursel, M., and Gursel, I. (2011). Differential immune activation following encapsulation of immunostimulatory CpG oligodeoxynucleotide in nanoliposomes. *Biomaterials* 32 (6): 1715–1723.
107 Klinman, D.M., Xie, H., and Ivins, B.E. (2006). CpG oligonucleotides improve the protective immune response induced by the licensed anthrax vaccine. *Ann. N. Y. Acad. Sci.* 1082: 137–150.
108 Walker, P.S., Scharton-Kersten, T., Krieg, A.M. et al. (1999). Immunostimulatory oligodeoxynucleotides promote protective immunity

and provide systemic therapy for leishmaniasis via IL-12- and IFN-gamma-dependent mechanisms. *Proc. Natl. Acad. Sci. U. S. A.* 96 (12): 6970–6975.
109 Moldoveanu, Z., Love-Homan, L., Huang, W.Q., and Krieg, A.M. (1998). CpG DNA, a novel immune enhancer for systemic and mucosal immunization with influenza virus. *Vaccine* 16: 1216–1224.
110 Rottembourg, D., Filippi, C.M., Bresson, D. et al. (2010). Essential role for TLR9 in prime but not prime-boost plasmid DNA vaccination to activate dendritic cells and protect from lethal viral infection. *J. Immunol.* 184 (12): 7100–7107.
111 Fogg, C.N., Americo, J.L., Lustig, S. et al. (2007). Adjuvant-enhanced antibody responses to recombinant proteins correlates with protection of mice and monkeys to orthopoxvirus challenges. *Vaccine* 25 (15): 2787–2799.
112 Olszewska, W., Partidos, C.D., and Steward, M.W. (2000). Antipeptide antibody responses following intranasal immunization: effectiveness of mucosal adjuvants. *Infect. Immun.* 68 (9): 4923–4929.
113 Chen, Z., Cao, J., Liao, X. et al. (2011). Plasmids enriched with CpG motifs activate human peripheral blood mononuclear cells *in vitro* and enhance th-1 immune responses to hepatitis B surface antigen in mice. *Viral Immunol.* 24 (3): 199–209.
114 Pihlgren, M., Tougne, C., Schallert, N. et al. (2003). CpG-motifs enhance initial and sustained primary tetanus-specific antibody secreting cell responses in spleen and bone marrow, but are more effective in adult than in neonatal mice. *Vaccine* 21 (19–20): 2492–2499.
115 da Fonseca, D.M., Silva, C.L., Wowk, P.F. et al. (2009). Mycobacterium tuberculosis culture filtrate proteins plus CpG Oligodeoxynucleotides confer protection to mycobacterium bovis BCG-primed mice by inhibiting interleukin-4 secretion. *Infect. Immun.* 77 (12): 5311–5321.
116 Rhee, E.G., Mendez, S., Shah, J.A. et al. (2002). Vaccination with heat-killed leishmania antigen or recombinant leishmanial protein and CpG oligodeoxynucleotides induces long-term memory CD4+ and CD8+ T cell responses and protection against leishmania major infection. *J. Exp. Med.* 195 (12): 1565–1573.
117 Verthelyi, D., Kenney, R.T., Seder, R.A. et al. (2002). CpG oligodeoxynucleotides as vaccine adjuvants in primates. *J. Immunol.* 168 (4): 1659–1663.
118 Perret, R., Sierro, S.R., Botelho, N.K. et al. (2013). Adjuvants that improve the ratio of antigen-specific effector to regulatory T cells enhance tumor immunity. *Cancer Res.* 73 (22): 6597–6608.
119 Miconnet, I., Koenig, S., Speiser, D. et al. (2002). CpG are efficient adjuvants for specific CTL induction against tumor antigen-derived peptide. *J. Immunol.* 168 (3): 1212–1218.
120 Switaj, T., Jalili, A., Jakubowska, A.B. et al. (2004). CpG immunostimulatory oligodeoxynucleotide 1826 enhances antitumor effect of interleukin 12

gene-modified tumor vaccine in a melanoma model in mice. *Clin. Cancer Res.* 10 (12 Pt 1): 4165–4175.

121 Lubaroff, D.M., Karan, D., Andrews, M.P. et al. (2006). Decreased cytotoxic T cell activity generated by co-administration of PSA vaccine and CpG ODN is associated with increased tumor protection in a mouse model of prostate cancer. *Vaccine* 24 (35–36): 6155–6162.

122 Jacobs, C., Duewell, P., Heckelsmiller, K. et al. (2011). An ISCOM vaccine combined with a TLR9 agonist breaks immune evasion mediated by regulatory T cells in an orthotopic model of pancreatic carcinoma. *Int. J. Cancer* 128 (4): 897–907.

123 Shi, R., Hong, L., Wu, D. et al. (2005). Enhanced immune response to gastric cancer specific antigen peptide by coencapsulation with CpG oligodeoxynucleotides in nanoemulsion. *Cancer Biol. Ther.* 4 (2): 218–224.

124 Creticos, P.S., Schroeder, J.T., Hamilton, R.G. et al. (2006). Immunotherapy with a ragweed-toll-like receptor 9 agonist vaccine for allergic rhinitis. *N. Engl. J. Med.* 355 (14): 1445–1455.

125 McCluskie, M.J., Weeratna, R.D., Evans, D.M. et al. (2013). CpG ODN and ISCOMATRIX adjuvant: a synergistic adjuvant combination inducing strong T-Cell IFN-gamma responses. *Biomed. Res. Int.* 2013: 636847.

126 Gong, M., Zhou, J., Yang, C. et al. (2012). Insect cell-expressed hemagglutinin with CpG oligodeoxynucleotides plus alum as an adjuvant is a potential pandemic influenza vaccine candidate. *Vaccine* 30 (52): 7498–7505.

127 Yang, M., Yan, Y., Fang, M. et al. (2012). MF59 formulated with CpG ODN as a potent adjuvant of recombinant HSP65-MUC1 for inducing anti-MUC1+ tumor immunity in mice. *Int. Immunopharmacol.* 13 (4): 408–416.

128 Sanchez, V.R., Fenoy, I.M., Picchio, M.S. et al. (2015). Homologous prime-boost strategy with TgPI-1 improves the immune response and protects highly susceptible mice against chronic Toxoplasma gondii infection. *Acta Trop.* 150: 159–165.

129 Verthelyi, D., Ishii, K.J., Takeshita, F., and Klinman, D.M. (1999). Relative potency of immunostimulatory DNA motifs in humans. *DNA Vaccines: Immune Responses, Mechanisms, and Manipulating Antigen Processing.* Keystone Symposia, Snowbird, UT.

130 Verthelyi, D., Ishii, K.J., Gursel, M. et al. (2001). Human peripheral blood cells differentially recognize and respond to two distinct CpG motifs. *J. Immunol.* 166: 2372–2377.

131 Bauer, S., Kirschning, C.J., Hacker, H. et al. (2001). Human TLR9 confers responsiveness to bacterial DNA via species specific CpG motif recognition. *Proc. Natl. Acad. Sci. U. S. A.* 98 (16): 9237–9242.

132 Hartmann, G. and Krieg, A.M. (2000). Mechanism and function of a newly identified CpG DNA motif in human primary B cells. *J. Immunol.* 164 (1): 944–952.

133 Leifer, C.A., Verthelyi, D., and Klinman, D.M. (2003). Heterogeneity in the human response to immunostimulatory CpG oligodeoxynucleotides. *J. Immunother.* 26 (4): 313–319.

134 Klinman, D.M. (2004). Immunotherapeutic uses of CpG oligodeoxynucleotides. *Nat. Rev. Immunol.* 4 (4): 249–258.

135 Gursel, M., Verthelyi, D., Gursel, I. et al. (2002). Differential and competitive activation of human immune cells by distinct classes of CpG oligodeoxynucleotide. *J. Leukoc. Biol.* 71 (5): 813–820.

136 Costa, L.T., Kerkmann, M., Hartmann, G. et al. (2004). Structural studies of oligonucleotides containing G-quadruplex motifs using AFM. *Biochem. Biophys. Res. Commun.* 313 (4): 1065–1072.

137 Wu, C.C.N., Lee, J., Raz, E. et al. (2004). Necessity of oligonucleotide aggregation for toll-like receptor 9 activation. *J. Biol. Chem.* 279 (32): 33071–33078.

138 Hanagata, N. (2012). Structure-dependent immunostimulatory effect of CpG oligodeoxynucleotides and their delivery system. *Int. J. Nanomedicine* 7: 2181–2195.

139 Gursel, M., Verthelyi, D., and Klinman, D.M. (2002). CpG oligodeoxynucleotides induce human monocytes to mature into functional dendritic cells. *Eur. J. Immunol.* 32 (9): 2617–2622.

140 Blander, J.M. and Medzhitov, R. (2006). Toll-dependent selection of microbial antigens for presentation by dendritic cells. *Nature* 440 (7085): 808–812.

141 Samulowitz, U., Weber, M., Weeratna, R. et al. (2010). A novel class of immune-stimulatory CpG oligodeoxynucleotides unifies high potency in type I interferon induction with preferred structural properties. *Oligonucleotides* 20 (2): 93–101.

142 Donhauser, N., Helm, M., Pritschet, K. et al. (2010). Differential effects of P-class versus other CpG oligodeoxynucleotide classes on the impaired innate immunity of plasmacytoid dendritic cells in HIV type 1 infection. *AIDS Res. Hum. Retroviruses* 26 (2): 161–171.

143 Grajkowski, A., Pedras-Vasconcelos, J., Wang, V. et al. (2005). Thermolytic CpG-containing DNA oligonucleotides as potential immunotherapeutic prodrugs. *Nucleic Acids Res.* 33 (11): 3550–3560.

144 Grajkowski, A., Pedras-Vasconcelos, J., Ausin, C. et al. (2005). Design and development of thermolytic DNA oligonucleotide prodrugs. *Ann. N. Y. Acad. Sci.* 1058: 26–38.

145 Puig, M., Grajkowski, A., Boczkowska, M. et al. (2006). Use of thermolytic protective groups to prevent G-tetrad formation in CpG ODN type D: structural studies and immunomodulatory activity in primates. *Nucleic Acids Res.* 34 (22): 6488–6495.

146 Ivory, C.P., Keller, K., and Chadee, K. (2006). CpG-oligodeoxynucleotide is a potent adjuvant with an *Entamoeba histolytica* Gal-inhibitable lectin vaccine against amoebic liver abscess in gerbils. *Infect. Immun.* 74 (1): 528–536.

147 Im Hof, M., Williamson, L., Summerfield, A. et al. (2008). Effect of synthetic agonists of toll-like receptor 9 on canine lymphocyte proliferation and cytokine production *in vitro*. *Vet. Immunol. Immunopathol.* 124 (1–2): 120–131.
148 Nagarajan, G., Kuo, C.C., Liang, C.M. et al. (2007). Effects of CpG-B ODN on the protein expression profile of swine PBMC. *Vet. Res.* 38 (6): 795–808.
149 Yu, H., Babiuk, L.A., and van Drunen Littel-van den Hurk, S. (2004). Priming with CpG-enriched plasmid and boosting with protein formulated with CpG oligodeoxynucleotides and Quil A induces strong cellular and humoral immune responses to hepatitis C virus NS3. *J. Gen. Virol.* 85 (Pt 6): 1533–1543.
150 Zhang, L., Tian, X., and Zhou, F. (2007). CpG oligodeoxynucleotides augment the immune responses of piglets to swine Pasteurella multocida living vaccine *in vivo*. *Res. Vet. Sci.* 83 (2): 171–181.
151 Kamstrup, S., Verthelyi, D., and Klinman, D.M. (2001). Response of porcine peripheral blood mononuclear cells to CpG- containing oligodeoxynucleotides. *Vet. Microbiol.* 78 (4): 353–362.
152 Guzylack-Piriou, L., Balmelli, C., McCullough, K.C., and Summerfield, A. (2004). Type-A CpG oligonucleotides activate exclusively porcine natural interferon-producing cells to secrete interferon-alpha, tumour necrosis factor-alpha and interleukin-12. *Immunology* 112 (1): 28–37.
153 Dory, D., Beven, V., Torche, A.M. et al. (2005). CpG motif in ATCGAT hexamer improves DNA-vaccine efficiency against lethal Pseudorabies virus infection in pigs. *Vaccine* 23 (36): 4532–4540.
154 Ketloy, C., Engering, A., Srichairatanakul, U. et al. (2008). Expression and function of Toll-like receptors on dendritic cells and other antigen presenting cells from non-human primates. *Vet. Immunol. Immunopathol.* 125 (1–2): 18–30.
155 Abel, K., Wang, Y., Fritts, L. et al. (2005). Deoxycytidyl-deoxyguanosine oligonucleotide classes A, B, and C induce distinct cytokine gene expression patterns in rhesus monkey peripheral blood mononuclear cells and distinct alpha interferon responses in TLR9-expressing rhesus monkey plasmacytoid dendritic cells. *Clin. Diagn. Lab. Immunol.* 12 (5): 606–621.
156 Verthelyi, D., Gursel, M., Kenney, R.T. et al. (2003). CpG Oligodeoxynucleotides protect normal and SIV infected macaques from Leishmania infection. *J. Immunol.* 170 (9): 4717–4723.
157 Davis, H.L., Suparto, I.I., Weeratna, R.R. et al. (2000). CpG DNA overcomes hyporesponsiveness to hepatitis B vaccine in orangutans. *Vaccine* 19 (4–5): 413–422.
158 Klinman, D.M., Xie, H., Little, S.F. et al. (2004). CpG oligonucleotides improve the protective immune response induced by the anthrax vaccination of rhesus macaques. *Vaccine* 22 (21–22): 2881–2886.
159 Rosa, D.S., Iwai, L.K., Tzelepis, F. et al. (2006). Immunogenicity of a recombinant protein containing the Plasmodium vivax vaccine candidate

MSP1(19) and two human CD4+ T-cell epitopes administered to non-human primates (Callithrix jacchus jacchus). *Microbes Infect.* 8 (8): 2130–2137.

160 Heikenwalder, M., Polymenidou, M., Junt, T. et al. (2004). Lymphoid follicle destruction and immunosuppression after repeated CpG oligodeoxynucleotide administration. *Nat. Med.* 10 (2): 187–192.

161 Fallarino, F., Volpi, C., Zelante, T. et al. (2009). IDO mediates TLR9-driven protection from experimental autoimmune diabetes. *J. Immunol.* 183 (10): 6303–6312.

162 Ichikawa, H.T., Williams, L.P., and Segal, B.M. (2002). Activation of APCs through CD40 or Toll-like receptor 9 overcomes tolerance and precipitates autoimmune disease. *J. Immunol.* 169 (5): 2781–2787.

163 Shi, G., Vistica, B.P., Nugent, L.F. et al. (2013). Differential involvement of Th1 and Th17 in pathogenic autoimmune processes triggered by different TLR ligands. *J. Immunol.* 191 (1): 415–423.

164 Deng, G.M. and Tarkowski, A. (2000). The features of arthritis induced by CpG motifs in bacterial DNA. *Arthritis Rheum.* 43: 356–364.

165 Eng, N.F., Bhardwaj, N., Mulligan, R., and Diaz-Mitoma, F. (2013). The potential of 1018 ISS adjuvant in hepatitis B vaccines: HEPLISAV review. *Hum. Vaccin. Immunother.* 9 (8): 1661–1672.

166 Cooper, C.L., Davis, H.L., Angel, J.B. et al. (2005). CPG 7909 adjuvant improves hepatitis B virus vaccine seroprotection in antiretroviral-treated HIV-infected adults. *AIDS* 19 (14): 1473–1479.

167 Rynkiewicz, D., Rathkopf, M., Sim, I. et al. (2011). Marked enhancement of the immune response to BioThrax(R) (Anthrax Vaccine Adsorbed) by the TLR9 agonist CPG 7909 in healthy volunteers. *Vaccine* 29 (37): 6313–6320.

168 Ellis, R.D., Martin, L.B., Shaffer, D. et al. (2010). Phase 1 trial of the Plasmodium falciparum blood stage vaccine MSP1(42)-C1/Alhydrogel with and without CPG 7909 in malaria naive adults. *PLoS One* 5 (1): e8787.

169 Mullen, G.E., Ellis, R.D., Miura, K. et al. (2008). Phase 1 trial of AMA1-C1/Alhydrogel plus CPG 7909: an asexual blood-stage vaccine for Plasmodium falciparum malaria. *PLoS One* 3 (8): e2940.

170 Crompton, P.D., Mircetic, M., Weiss, G. et al. (2009). The TLR9 ligand CpG promotes the acquisition of Plasmodium falciparum-specific memory B cells in malaria-naive individuals. *J. Immunol.* 182 (5): 3318–3326.

171 Sogaard, O.S., Lohse, N., Harboe, Z.B. et al. (2010). Improving the immunogenicity of pneumococcal conjugate vaccine in HIV-infected adults with a Toll-like receptor 9 agonist adjuvant: a randomized, controlled trial. *Clin. Infect. Dis.* 51 (1): 42–50.

172 Guha, M. (2012). Anticancer TLR agonists on the ropes. *Nat. Rev. Drug Discov.* 11 (7): 503–505.

173 Bioley, G., Guillaume, P., Luescher, I. et al. (2009). Vaccination with a recombinant protein encoding the tumor-specific antigen NY-ESO-1 elicits an A2/157-165-specific CTL repertoire structurally distinct and of reduced

tumor reactivity than that elicited by spontaneous immune responses to NY-ESO-1-expressing Tumors. *J. Immunother.* 32 (2): 161–168.

174 Fourcade, J., Kudela, P., Andrade Filho, P.A. et al. (2008). Immunization with analog peptide in combination with CpG and montanide expands tumor antigen-specific CD8+ T cells in melanoma patients. *J. Immunother.* 31 (8): 781–791.

175 Haining, W.N., Davies, J., Kanzler, H. et al. (2008). CpG oligodeoxynucleotides alter lymphocyte and dendritic cell trafficking in humans. *Clin. Cancer Res.* 14 (17): 5626–5634.

176 Valmori, D., Souleimanian, N.E., Tosello, V. et al. (2007). Vaccination with NY-ESO-1 protein and CpG in Montanide induces integrated antibody/Th1 responses and CD8 T cells through cross-priming. *Proc. Natl. Acad. Sci. U. S. A.* 104 (21): 8947–8952.

177 Ohno, S., Okuyama, R., Aruga, A. et al. (2012). Phase I trial of Wilms' Tumor 1 (WT1) peptide vaccine with GM-CSF or CpG in patients with solid malignancy. *Anticancer Res* 32 (6): 2263–2269.

178 Goldinger, S.M., Dummer, R., Baumgaertner, P. et al. (2012). Nano-particle vaccination combined with TLR-7 and -9 ligands triggers memory and effector CD8(+) T-cell responses in melanoma patients. *Eur. J. Immunol.* 42 (11): 3049–3061.

179 Speiser, D.E., Schwarz, K., Baumgaertner, P. et al. (2010). Memory and effector CD8 T-cell responses after nanoparticle vaccination of melanoma patients. *J. Immunother.* 33 (8): 848–858.

180 Katsuda, M., Iwahashi, M., Matsuda, K. et al. (2011). Comparison of different classes of CpG-ODN in augmenting the generation of human epitope peptide-specific CTLs. *Int. J. Oncol.* 39 (5): 1295–1302.

181 Halperin, S.A., Ward, B., Cooper, C. et al. (2012). Comparison of safety and immunogenicity of two doses of investigational hepatitis B virus surface antigen co-administered with an immunostimulatory phosphorothioate oligodeoxyribonucleotide and three doses of a licensed hepatitis B vaccine in healthy adults 18–55 years of age. *Vaccine* 30 (15): 2556–2563.

182 Joffre, O., Nolte, M.A., Sporri, R., and Reis e Sousa, C. (2009). Inflammatory signals in dendritic cell activation and the induction of adaptive immunity. *Immunol. Rev.* 227 (1): 234–247.

183 Sporri, R. and Reis e Sousa, C. (2005). Inflammatory mediators are insufficient for full dendritic cell activation and promote expansion of CD4+ T cell populations lacking helper function. *Nat. Immunol.* 6 (2): 163–170.

184 Tighe, H., Takabayashi, K., Schwartz, D. et al. (2000). Conjugation of immunostimulatory DNA to the short ragweed allergen amb a 1 enhances its immunogenicity and reduces its allergenicity. *J. Allergy Clin. Immunol.* 106: 124–134.

185 Kurreck, J. (2003). Nucleic acids chemistry and biology. *Angew. Chem. Int. Ed. Engl.* 42 (44): 5384–5385.

186 Brown, D.A., Kang, S.H., Gryaznov, S.M. et al. (1994). Effect of phosphorothioate modification of oligodeoxynucleotides on specific protein binding. *J. Biol. Chem.* 269 (43): 26801–26805.

187 Lipford, B.G., Sparwasser, T., Zimmermann, S. et al. (2000). CpG-DNA-mediated transient lymphadenopathy is associated with a state of Th1 predisposition to antigen-driven responses. *J. Immunol.* 165: 1228–1235.

188 Heit, A., Huster, K.M., Schmitz, F. et al. (2004). CpG-DNA aided cross-priming by cross-presenting B cells. *J. Immunol.* 172 (3): 1501–1507.

189 Krishnamachari, Y. and Salem, A.K. (2009). Innovative strategies for co-delivering antigens and CpG oligonucleotides. *Adv. Drug Deliv. Rev.* 61 (3): 205–217.

190 Malyala, P., Chesko, J., Ugozzoli, M. et al. (2008). The potency of the adjuvant, CpG oligos, is enhanced by encapsulation in PLG microparticles. *J. Pharm. Sci.* 97 (3): 1155–1164.

191 Xie, H., Gursel, I., Ivins, B.E. et al. (2005). CpG oligodeoxynucleotides adsorbed onto polylactide-co-glycolide microparticles improve the immunogenicity and protective activity of the licensed anthrax vaccine. *Infect. Immun.* 73 (2): 828–833.

192 Rattanakiat, S., Nishikawa, M., and Takakura, Y. (2012). Self-assembling CpG DNA nanoparticles for efficient antigen delivery and immunostimulation. *Eur. J. Pharm. Sci.* 47 (2): 352–358.

193 De Paula, D., Bentley, M.V., and Mahato, R.I. (2007). Hydrophobization and bioconjugation for enhanced siRNA delivery and targeting. *RNA* 13 (4): 431–456.

194 Zwiorek, K., Bourquin, C., Battiany, J. et al. (2008). Delivery by cationic gelatin nanoparticles strongly increases the immunostimulatory effects of CpG oligonucleotides. *Pharm. Res.* 25 (3): 551–562.

17

Splice-Switching Oligonucleotides

Isabella Gazzoli and Annemieke Aartsma-Rus

Department of Human Genetics, Leiden University Medical Center, Leiden, The Netherlands

17.1 Introduction of Splice Switching

Chemically modifying synthetic oligonucleotides to improve stability and alter pharmacokinetic properties allowed antisense-mediated manipulation of gene expression. This has had a large impact on basic biomedical research and has opened the way for the development of therapies for human diseases with unmet medical need. Antisense oligonucleotides (AONs) share the fundamental principle of binding and modulating target RNA function. This modulation can be classified in two types: oligonucleotides that bind the target and interfere with its function without RNA degradation [1, 2] and oligonucleotides that promote degradation of target RNA by endogenous enzymes or cleavage mechanisms [3]. The latter, which includes the use of small interfering RNAs (siRNAs) and RNase H oligonucleotides, has been developed to promote selective degradation of a single transcript, and several drugs are in clinical development, while one (mipomersen) has been approved by the Food and Drug Administration (FDA) as a treatment for familial hypercholesterolemia [4–7]. Currently, oligonucleotides are also recognized as a powerful tool to manipulate splicing of endogenous pre-messenger RNA (mRNA) in cultured cells, animal models, and humans [8, 9], and recently the first splice-modulating AON (eteplirsen) was approved by the FDA.

Normally, the RNA splicing complex engages the spliceosome and multiple trans-splicing factors to recognize and bind to specific sequences and remove introns from the nascent pre-mRNA, generating mature mRNA [10, 11]. After splicing is completed, spliced transcripts are exported into the cytoplasm where they are translated into protein. Alternative splicing – the phenomenon where multiple transcripts are generated by a single gene [12] by inclusion or

Oligonucleotide-Based Drugs and Therapeutics: Preclinical and Clinical Considerations for Development, First Edition. Edited by Nicolay Ferrari and Rosanne Seguin.

exclusion of (part of) an exon or intron, or by using different promoters or polyadenylation sites – has been observed in more than 90% of human genes [13]. Alternative splicing allows the generation of different protein isoforms, sometimes with differing or opposing functions (e.g. calcitonin and cGRP are both products of same calcitonin gene).

As without proper splicing, protein translation cannot occur, it is not surprising that an increasing number of diseases turn out to be caused by mis-splicing of pre-mRNA transcripts, due to inherited or sporadic changes in the DNA sequence. Recently mis-splicing events have been reported to underlie 60% of disease with pathogenic effects [14, 15]. Mutations have been reported in all possible sites that are normally involved in proper splicing: donor and acceptor splice sites, branch points, polypyrimidine tracts, and exonic or intronic splicing enhancers (ESEs and ISEs, respectively) or exonic or intronic splicing silencers (ESSs and ISSs, respectively). Alternatively, mutations can create "cryptic" splicing motifs, leading to the aberrant inclusion of intronic sequences.

AONs can be used to modulate splicing. AONs act by binding to their reverse complementary target sequences in pre-mRNA transcripts [16–19]. Through this binding they interfere with the splicing process, e.g. by interfering with the binding of splicing factors that are needed for inclusion of the targeted exon into the spliced mRNA (Figure 17.1). Antisense-mediated splicing modulation can be applied to correct cryptic splicing, restore the open reading frame to allow the production of a partially functional protein, alter levels of alternative splicing (isoform switching), and achieve exon inclusion or to induce knockdown on protein level by skipping an out-of-frame exon or skipping an exon that encodes a functional protein domain (Figure 17.2a–e). There are many diseases for which AON-mediated exon skipping is being investigated as a therapeutic option [8, 20–22]. Here we will provide some examples to illustrate the different ways exon skipping can be exploited, which are summarized in Table 17.1.

17.1.1 Correct Cryptic Splicing

The loss of normal splicing can be caused by mutations that create aberrant donor and acceptor splice sites, leading to the inclusion of intronic sequence, which generally disrupts the open reading frame and/or introduces a premature stop codon. In this situation, AONs can be engaged to restore normal splicing.

17.1.1.1 β-Thalassemia

Absence or reduction of the β-globin protein, a subunit of the hemoglobin, causes β-thalassemia, which is an inherited autosomal recessive blood disease that, if untreated, leads to death. Genetic alterations introducing splicing

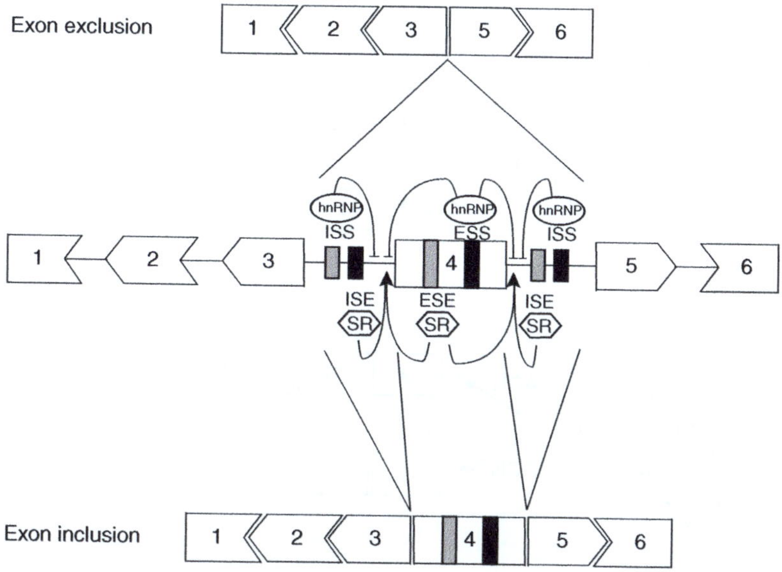

Figure 17.1 Alternative splicing. Exon exclusion (upper panel) can be regulated by splicing repressors proteins (such as heterogeneous nuclear ribonucleoproteins [hnRNPs]) that bind to silencing motifs (black boxes) in the exons (exonic splicing silencers [ESS]) or in the introns (intron splicing silencers [ISS]). Splicing activator proteins (such as serine–arginine [SR]-rich domain proteins) can influence the recognition and the splice of the exons by binding exonic or intronic splicing enhancers (ESE or ISE, respectively) (grey boxes), allowing the inclusion of the exon in the transcript (exon inclusion, lower panel). Mutations in these repressor or activator motifs, variation in levels of activating or repressing factors between tissues, or developmental stages can influence exon exclusion or inclusion. Exon inclusion can also be modulated by targeting the silencing or enhancing motifs with antisense oligonucleotides (AONs) to induce exon inclusion or exclusion, respectively.

defects in the β-globin gene are common in thalassemic patients [37–39]. The first correction of cryptic splicing was achieved on a mutated human β-globin transcript, restoring the use of the proper splice sites and expression of β-globin protein *in vitro*. This was accomplished in transcripts carrying two different mutations in intron 2 (at position 654 or 705, IVS2-654 and IVS2-705, respectively), which both lead to the creation of a new donor splice site and activation of an acceptor cryptic splice site (indicated as mut. and c.s.s., respectively, in Figure 17.2a). The resulting cryptic exons lead to an in-frame stop codon and no translated β-globin protein [40]. Incubation with AONs targeting the cryptic splice sites resulted in restoration of the normal transcript. Further studies involved patient-derived erythroid progenitor cells (with the IVS2-654 and IVS2-745 mutations) [41] and erythroid cells from transgenic mice (carrying IVS2-654) [42]. Treatment with AONs resulted in a correction of splicing and a significant increase of β-globin production. In thalassemic

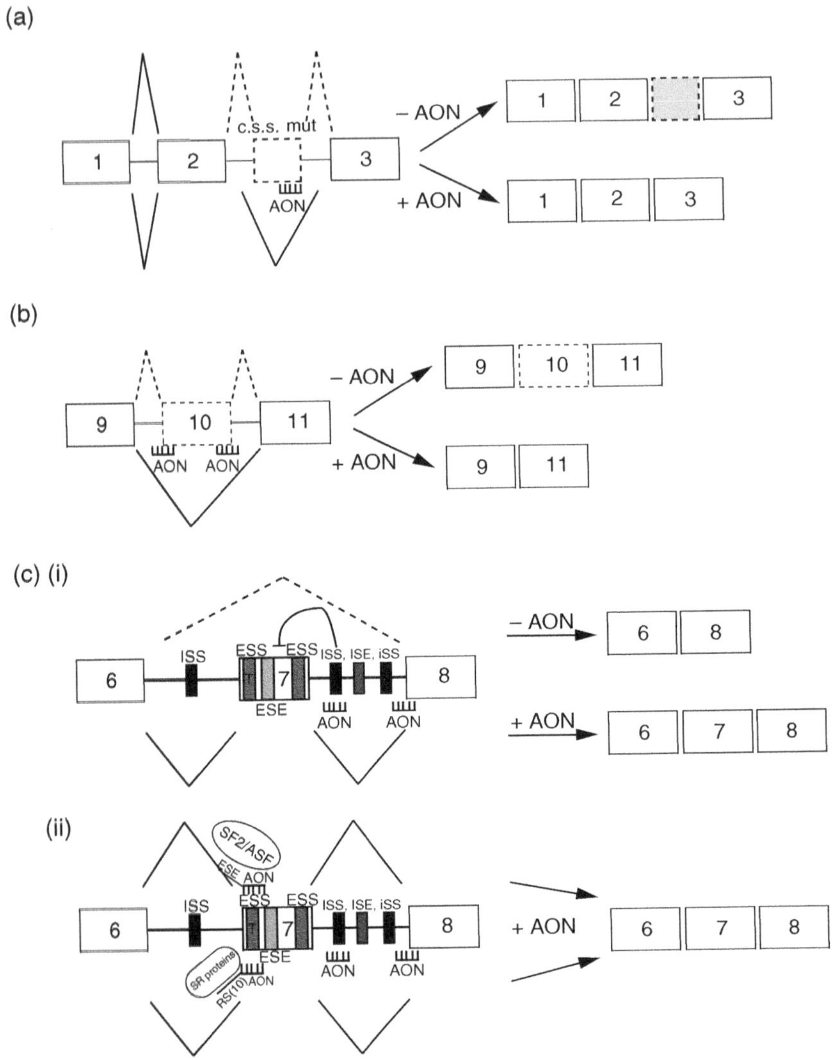

Figure 17.2 Antisense-mediated splicing modulation. (a) **Correct cryptic splicing**. A mutation (mut, a T>C transition at position 705 in intron 2) in the β-globin gene generates a new donor splice site and activates an acceptor cryptic splice site (c.s.s). Rather than splicing out intron 2, part of the intronic sequence (the cryptic exon [dashed grey box]) is restored in the mRNA (route –AON). Antisense oligonucleotides (AONs) targeting the cryptic exon restore normal splicing (route +AON). (b) **Isoform switching.** AONs can modulate alternative splicing by binding to the donor (start of intron) or the acceptor (end of intron) splice site of an alternatively spliced exon (dashed box), resulting in primarily transcripts where the exon is excluded. This mechanism switches from one isoform (route –AON) to another one (+AON).

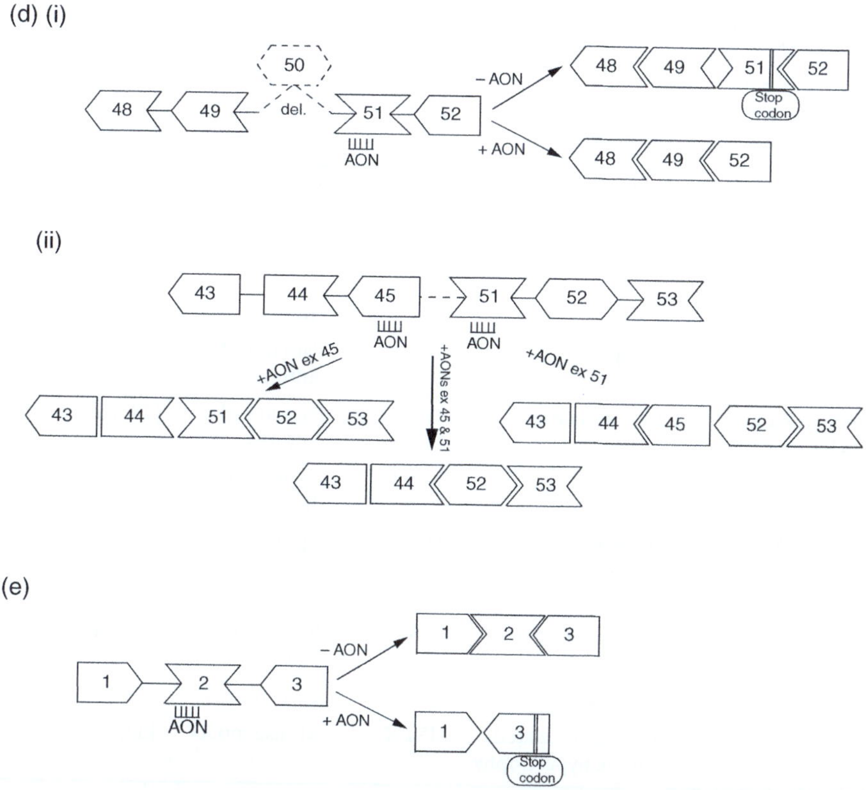

Figure 17.2 (Cont'd) (c) **Induce exon inclusion.** (i) In the *SMN2* gene, exon 7 is generally skipped (route –AON) and is due to a transition (C to T) that has changed an ESE site into an ESS site (light and dark grey boxes). Exon exclusion is further enhanced by an ISS present in intron 7. AON targeting ISS (black box) or the acceptor splice site of intron 7 can enhance the inclusion of exon 7 (route +AON). (ii) The use of bifunctional AONs, targeting the ESS (mutated T allele) and carrying a tail mimicking an ESE (ESE-AON, upper part of figure) or 10 RS repeats (RS10-AON, bottom part) can recruit splicing factors and correction of the spliced transcript. Regulatory sequences in exon 7 (ESS and ESE) and introns 6 and 7 (ISS and ISE [dark gray box]) can silence or enhance the exclusion or inclusion of an exon, respectively. (d) **Reading frame correction.** (i) In this example, a deletion of exon 50 (dashed exon) in the *DMD* gene disrupts the open reading frame, resulting in a premature stop codon in exon 51 (route –AON). AON can be used to hide exon 51 from the splicing machinery, leading to the skip of the exon and restoration of the reading frame allowing the production of a shorter, partially functional protein. (ii) Double exon skipping with a combination of two AONs (route +AONs ex45 and 51), targeting exons 45 and 51, restores the DMD reading frame in patient cell line carrying a 46–50 exon deletion (dashed exons). The use of single AON (route +AON ex45, route +AON ex51) provides out-of-frame transcripts. (e) **Knockdown.** An opposite scenario results in downregulation of a protein. Using AON targeting exon 2, one of the three exons of myostatin gene (*MSTN*), results in exon 2 exclusion and a transcript from which no functional protein can be produced (route +AON).

Table17.1 List of disorders and related splice-switching oligonucleotides discussed in the chapter.

Splice-switching model	Disorder	Target gene	Preclinical/ clinical study	References
Correct cryptic splicing	β-Thalassemia	β-Globin	Mouse model	[23]
	Cystic fibrosis	CTFR	Minigene	[24]
Isoform switching	Anticancer	BCL-X	Patient cell cultures	[25–27]
	Tauopathies	MAPT	Patient cell cultures	[28]
Induce exon inclusion	Tumorigenesis	FGFR1	Patient cell cultures	[29–31]
	Spinal muscular atrophy	SMN2	Approved drug	
Reading frame correction	Duchenne muscular vdystrophy	DMD	Approved drug	
	Dysferlinopathies	DYSF	Patient cell cultures	[32–34]
Knockdown	Atherosclerosis	ApoB	Cell cultures	[35]
	Myostatin-related muscle hypertrophy	MSTN	Mouse model	[36]

patients, the balance between α- and β-globin chains (both subunits of hemoglobin) is disrupted by the absence of β-globin, which results in free precipitated α-globin chains and iron accumulation in the liver in β-thalassemia patients. Recently, Xie et al. reported that the imbalance between α- and β-globin chains can be restored by combining AONs (targeting IVS2-654) and RNAi, achieving the correction of β-globin splicing and production of low levels of β-globin protein and the downregulation of α-globin transcripts in the IVS2-654 mouse model [23]. Even though AON-mediated splicing modulation was shown first for the β-globin gene, no trials are to our knowledge being conducted for β-thalassemia.

17.1.1.2 Cystic Fibrosis

Cystic fibrosis (CF) is a common autosomal recessive genetic disorder in the Caucasian population. Homozygous loss of the CF transmembrane conductance regulator (*CFTR*) gene results in the absence of protein involved in the movement and conductance of chloride and bicarbonate ions across to the apical membrane of epithelial cells. The dysfunction of the protein raises an ionic imbalance of epithelial secretions in several organ systems, such as the pancreas, gastrointestinal tract, liver, and the respiratory system, causing severe damage due to secretions of large amounts of thick mucus, which reduces the functional capacity of the affected organs and predisposes them to infections. Since the discovery of the gene (27 exons, 190 kb) in 1989 [43], studies have been focused on targeting mutations in order to restore functional CFTR transcript levels and reduce the severity of the disease. Around 15% of the reported mutations are mis-splicing events that either affect regulatory target motifs for the spliceosome or disrupt or create donor or acceptor splice sites [44–46]. One example is a common cryptic splice mutation in intron 19, which affects 2% of the population and has an even higher incidence in the Ashkenazi Jewish population. A construct carrying this C>T mutation at position 3849 of intron 19 that creates a new donor splice site was expressed in different cell lines (3T3 mouse fibroblasts, CFT1 human tracheal, and C127 mouse mammary epithelial cells). Treatment of the three cell lines with an AON targeting the aberrant donor and acceptor splice sites or the retained introns sequence restored normal CFTR transcript splicing and led to a dose- and sequence-dependent increase in protein expression [24].

17.1.2 Isoform Switching

An alteration of ratios of two alternative splicing transcripts underlies multiple diseases but also offers interesting opportunities for treatments.

17.1.2.1 Anticancer

The *BCL-X* gene is one of the members of the bcl protein family, which is involved in the regulation of programmed cell death (apoptosis). Alternative splicing, through two different 5′splice sites of BCL-X in intron 2, results in two different BCL-X mRNAs, having opposite functions. Both long (bcl-xL) and short (bcl-xS) isoforms are located at the outer mitochondrial membrane and regulate cell death by binding and changing the voltage-dependent anion channel (VDAC), but they act as apoptotic inhibitor and apoptotic activators, respectively [47]. In healthy tissue, both isoforms are expressed at similar levels. However, in some types of cancer, the long protein was found to be overexpressed [48–51]. Therefore, AON-mediated splicing modulation has been used to target the 5′splice site of the bcl-xL isoform with the intention to lower the bcl-xL/bcl-xS ratio and increase a proapoptotic response. In prostate cancer PC3 cells, AONs successfully enhanced the levels of the bcl-xS isoform [25]. Similar experiments have been performed in a lung epithelial cancer cell line, where the accomplished balance between the two isoforms and the resulted apoptotic effect was achieved by a combination of AONs and the *cis*-platinum [26]. Increased sensitivity to chemotherapeutic drugs after AON treatment, leading to apoptosis, was also confirmed in other cancer cell lines. In addition, the same study showed that the amount of the newly expressed bcl-xS protein correlated with apoptosis sensitivity of AON-treated cells [27].

17.1.2.2 Tauopathies

Six different tau isoforms are formed through alternative splicing of the human *MAPT* (microtubule-associated protein tau) gene on chromosome 17[52]. The protein is localized in neurons of the central nervous system (CNS) and interacts with tubulin to modulate the stability of microtubules in the distal part of the axons [53]. Tau protein can be phosphorylated by activated kinases; however, hyperphosphorylation can result in the disruption of microtubule organization, and that can lead to protein deposits, which have been identified in different neurodegenerative disorders, including Alzheimer [54]. In the human brain, alternative splicing results in three N-terminal and three C-terminal isoforms, which include exon 2 and/or exon 3 and/or exon 10. Normally, the pre-mRNA structure of the *MAPT* gene results in the skip of exon 10 due to an unrecognized splice site (Figure 17.2b). Four microtubule repeats (4R) are localized in the C-terminal part of the tau protein, and the absence of exon 10 results in the loss of one of them generating the 3R isoform. Studies showed that equal amounts of 4R and 3R isoforms are important for normal human neuronal function [55, 56]. Mis-splicing events involving missense mutations or deletions within exon 10 that disrupt the secondary RNA structure, leading to the inclusion of this exon, and a disturbed ratio of 3R and 4R tau proteins have been associated with frontotemporal dementia and parkinsonism linked to chromosome 17 (FTDP-17) [54, 57]. Correction of mis-splicing was achieved by AONs targeting 5′ or 3′ splice site of exon 10 in a minigene carrying an FTDP-17 mutation expressed by neuronal pheochromocytoma cells. This resulted in increased expression of the short 3R protein, leading to improved microtubule function and suggesting that this could be a putative genetic therapy for this disorder [28].

17.1.3 Induce Exon Inclusion

Exclusion of exons that are normally present in full-length mRNA transcripts has been described to occur, e.g. as a consequence of single point mutations in intronic sequences. This results in shorter mRNAs that often do not encode functional proteins. While from the examples given in 17.1.1 and 17.1.2 it can be concluded that inducing the skipping of (part of) an exon by AONs is rather straightforward, inducing the inclusion of a skipped exon is much more challenging and only applies to a few genes.

17.1.3.1 Tumorigenesis

Several types of cancers (glioblastoma [58], breast [59], and pancreatic cancers [60]) have been associated with aberrant splicing of fibroblast growth factor receptor 1 (*FGFR1*). This gene encodes a transmembrane tyrosine kinase receptor, which belongs to the highly conserved FGFR protein family and differs from other members by its ligand binding affinities and tissue specificity. Alternative splicing of *FGFR1* originates two receptor forms, containing either three (FGFR1α) or two (FGFR1β) extracellular immunoglobulin-like domains. The latter variant lacks exon 3 (α-exon), putatively as a result of abnormal expression of trans-regulator elements. The alternative exon contains two small ESEs and one ESS motif and is flanked by two ISS sequences (ISS1 and ISS2), which play an important role in the regulation of the excluded exon in glioblastoma cells [29, 30]. Using AONs targeting the ISS1 and ISS2 sequences to block the binding of transacting factors splicing of the FGFR1β transcript was redirected to the normal α-isoform in a glioblastoma cell line. Inclusion of the α-exon was achieved in 35–60% of transcripts. The presence of an inhibitor of exon recognition and other splicing inhibitors was suggested to contribute to regulating α-exon skipping [31]. These results show that AON-mediated exon inclusion of FGFR1 is a possible therapeutic approach for cancer cells in which the alternative splicing of FGFR1 is disrupted.

17.1.3.2 Spinal Muscular Atrophy (SMA)

Spinal muscular atrophy (SMA) is an autosomal recessive neurodegenerative disease characterized by the loss of alpha motor neurons, resulting in progressive muscle weakness, atrophy, and in severe cases paralysis and death. SMA occurs in about one in 6000–10 000 live births, and four different types have been classified, based on the age of onset, the severity of the clinical symptoms, and the achieved motor functions [61, 62]. The disease is caused by mutations in the *SMN1* gene that abolish SMN protein production. Lack of SMN protein is embryonically lethal. However, humans carry two copies of the survival motor neuron (*SMN*) gene: *SMN1* and a duplicated and inverted centromeric copy, *SMN2*. All patients retain one or more copies of the *SMN2* gene, but this does not fully compensate for the loss of *SMN1*. The two genes differ by only 11 nucleotides at the genomic level. However, *SMN2* contains a C to T transition at position 6 in exon 7 that is translationally silent but leads to exon 7 skipping during SMN2 splicing in the vast majority of transcripts. The resulting transcripts encode a truncated, unstable SMN protein that is rapidly degraded. The small proportion of SMN2 transcripts that retain exon 7 results in a normal full-length SMN protein that is apparently sufficient to prevent embryonic lethality, but insufficient to maintain motor neurons [63]. Since the severity of the SMA phenotype is inversely correlated with expression levels of SMN from *SMN2* copies, a therapeutic strategy focused on increasing SMN protein levels, by increasing the level of inclusion of exon 7 in SMN2-derived transcripts, may prove helpful. The mechanism for exon 7 inclusion has been extensively studied by different groups through targeting different sequences of intron 7 with AONs. Indeed, exon 7 inclusion has been achieved using an AON masking the acceptor splice site of intron 7 in cell cultures [64] or targeting an ISS in adult mice expressing a human *SMN2* transgene [65] (Figure 17.2c-i). This approach has been developed further and the lead compound (nusinersen) has now been approved for treatment of all SMA subtypes by the FDA and EMA (https://www.fda.gov/NewsEvents/Newsroom/PressAnnouncements/ucm534611.htm, http://www.ema.europa.eu/ema/index.jsp?curl=pages/news_and_events/news/2017/04/news_detail_002732.jsp&mid=WC0b01ac058004d5c1), and an expanded access program has been initiated for the most severe form (NCT02865109). Additionally, bifunctional AONs targeting partial complementary sequence and carrying a recruiting splice factor domain have been tested *in vivo* but not *in vitro* (Figure 17.2c-ii) (see 17.2.7.1 for more details).

17.1.4 Reading Frame Correction

Mutations can generate out-of-frame transcripts, leading to premature truncation of protein translation, often resulting in a nonfunctional protein. For some proteins, AON-mediated exon skipping can be engaged to restore the open reading frame, resulting in a shorter, partly functional protein and converting a severe disease into a milder form. Notably, this primarily applies to long modular proteins; for most genes lacking even a single exon renders the resulting protein nonfunctional.

17.1.4.1 Duchenne Muscular Dystrophy

The most prominent example of AON-mediated exon skipping is Duchenne muscular dystrophy (DMD). This lethal, X-linked muscle-wasting disease arises from large deletions or duplications, nonsense, and small frameshifting mutations in the *DMD* gene, which lead to out-of-frame dystrophin transcripts and absence of functional dystrophin protein in muscle [66]. Normally, in muscle fibers, dystrophin connects the actin cytoskeleton to the extracellular matrix, absorbing force generated during muscle fiber contractions. In DMD patients, this connection is lost, due to the absence of the protein, resulting in chronic inflammation and damage and loss of muscle fibers due to normal exercise, which eventually leads to loss of muscle function. Affected boys are wheelchair bound by the age of 12, gradually lose the ability for any voluntary movement, and suffer from respiratory and cardiac weakness, leading to an early death in the third decade of life in the Western world [67]. In contrast, Becker muscular dystrophy (BMD) is a milder allelic form of this disease. BMD is caused by deletions that maintain the reading frame of the *DMD* gene, allowing the translation of a shorter but partially functional dystrophin protein that can fulfill its function of connecting actin to the extracellular matrix. Generally, BMD is a milder disease, with a slower disease progression. Patients remain ambulant for longer and have normal or near-normal life span [68]. Lacking a cure for this disease, a conversion of a DMD into a BMD phenotype has been suggested as a potential therapeutic approach. First, pioneer exon skipping experiments showed proof of principle (exon skipping and dystrophin restoration after AON treatment) in patient-derived cell cultures [69, 70], in mouse models [71], and in a dog model [72]. Considering that DMD exon skipping is mutation specific but that most of DMD boys are affected by deletions that cluster in a hotspot region in the *DMD* gene from exons 43 to 55, skipping of some exons applies to larger groups of patients [73]. Exon 51 skipping, which applies to the largest group of patients, has been developed furthest [74] (Figure 17.2d-i). Achievements of preclinical studies using exon 51 AONs have led to two European clinical programs with very different backbone chemistries using local and systemic administration [75–78]. One chemistry (drisapersen) was tested in multiple trials, including a phase 3 placebo-controlled trial, in >300 DMD patients worldwide. While phase 2 trials in young patients showed encouraging trends in muscle function, the primary endpoint in the phase 3 trial was not met, and the FDA ruled that drisapersen was not ready for approval (http://www.medscape.com/viewarticle/857406). The other chemistry (eteplirsen) recently received accelerated approval from the FDA based on increases in dystrophin levels in some treated patients (http://www.fda.gov/NewsEvents/Newsroom/PressAnnouncements/ucm521263.htm). Skipping of multiple exons simultaneously (multiexon skipping) has also been tested on different patient cell lines carrying mutations that need skipping of two or more exons to restore the reading frame [79] (Figure 17.2d-ii).

17.1.4.2 Dysferlinopathies

Mutations in the dysferlin (*DYSF*) gene have been identified in different dysferlinopathies, such as limb girdle muscle dystrophy type 2B (LGMD2B), Miyoshi myopathy (MM), and distal myopathy with anterior tibial onset (DMAT) [80–83]. This disease belongs to the limb girdle muscular dystrophy (LGMD) group, which contains autosomal inherited heterogeneous disorders characterized by a progressive muscle weakness and wasting of the shoulder and pelvic girdles [84, 85]. The human *DYSF* gene is mapped to chromosome 2p13 and contains 55 exons, and alternative splicing results in at least 14 isoforms. Dysferlin protein is localized at the sarcolemma in the adult skeletal muscle [86, 87] and belongs to the ferlin family. It contains six or seven calcium-dependent C2 lipid-binding (C2) domains, with partial redundant function, a transmembrane domain and multiple "fer" and "dysf" domains. Known dysferlin functions involve calcium-dependent vesicle fusion with the plasma membrane through C2 domains and anchoring the protein to the plasma membrane through the transmembrane domain [88]. The functions of the fer and dysf domains are as yet unknown. Due to the modular build of the dysferlin protein, internal deletions can lead to partly functional proteins, as exemplified by a case report of a mother of two LGMD2B patients, who turned out to be compound heterozygous for a disease-causing mutation as well as a mutation in intron 31, subsequent skipping of exon 32, leading to the production of an internally deleted dysferlin protein at 10% of wild-type levels [87]. This resulted in only very mild proximal muscle weakness and high creatine kinase (CK) levels and normal ambulation. By contrast, the two daughters, who were homozygous for the null mutation, had a severe phenotype. An additional case of a mildly affected patient ambulant until midlife who expressed a truncated dysferlin (containing the last two C2 and the transmembrane domains) has been reported [89]. Thus skipping of in-frame exons that carry mutations offers therapeutic potential for dysferlinopathies. AON-mediated exon skipping has been reported for exons 17, 24, 30, 32, 34, 35, 36, and 42, suggesting that exon skipping is feasible for the *DYSF* gene [32–34]. However, not all exons may be suitable targets for exon skipping as some are essential for the protein function or stability, e.g. a patient homozygous for an exon 25 skipping mutation in the *DYSF* gene turned out to have no dysferlin at all, suggesting that exon 25 is not an applicable target for exon skipping therapy [90]. In addition, exon 49 seems to be an inappropriate target, since a patient with a mutation in a splice site, causing the skipping of exon 49, showed the clinical features of LGMD2B [91, 92].

17.1.5 Knockdown

Downregulation of protein expression can be achieved using AON-mediated splice modulation. This can be either achieved by inducing the skipping of an out-of-frame exon (anticipated to result in protein knockdown) or by skipping an in-frame exon encoding a functional domain (anticipated to result in a nonfunctional protein).

17.1.5.1 Atherosclerosis

Apolipoprotein B (ApoB) is the primary structural apolipoprotein of chylomicron particles and low-density lipoproteins (LDL), very (V)LDL, intermediate-density lipoprotein (IDL), and lipoprotein(a) (Lp(a)). APOB protein acts as a ligand for LDL receptors in various cells by delivering cholesterol to the tissue. High levels of ApoB can lead to plaques that cause a vascular pathology, known as atherosclerosis, which causes disease of the coronary, cerebral, and peripheral arteries. The *ApoB* gene codes for two natural protein isoforms – the full-length APOB100, which is synthesized in the liver, and the C-terminal truncated APOB48, which is a small intestinal tissue-specific protein. The truncated isoform (APOB48) is the result of RNA editing, which takes place within exon 26 (at residue 2153 a CAA codon is changed into a premature termination codon UAA) in the intestine [93]. The presence of the ApoB100's C-terminal LDL receptor binding region confers a central role in the association with atherogenesis to this isoform, which results in high levels of the APOB100-containing particles LDL and Lp(a). Therefore, downregulation of the protein level with AONs has been attempted to decrease cholesterol levels. Different AONs targeting the donor and acceptor splice sites of exon 27, the branch-point sequence of introns 26–27, have been tested in HepG2 cells, which naturally express APOB100 [35]. Skipping of exon 27 was only achieved by the combination of two AONs, more efficiently with AONs targeting the acceptor splice site and branch-point sequence. The skip of exon 27 results in a new secreted isoform (APOB87^{SKIP27}), carrying a premature termination codon within the exon 28 sequence, downregulating APOB100 expression while not affecting the intestinal Apo48 expression.

17.1.5.2 Myostatin-Related Muscle Hypertrophy

Genetic alterations that reduce the function of the protein myostatin, which is encoded by the *MSTN* gene, lead to an overgrowth of muscle tissue. Myostatin is a negative regulator of skeletal muscle mass and is a member of a transforming growth factor-β family. Several muscle-wasting disease conditions could benefit from knockdown of this protein, which would be anticipated to lead to an increase in skeletal muscle growth. Therefore, different groups have tried myostatin knockdown using AONs in combination with other therapeutic approaches for DMD [94, 95]. Two different types of AONs have been tested, leading to both successful skipping of exon 2 and downregulation of the MSTN transcript level in DMD culture cells or in a mouse model (Figure 17.2e). In the animal study this was accompanied by increased muscle size when AONs were delivered systemically [36].

17.2 Preclinical and Clinical Development of Splice-switching Oligos

17.2.1 Introduction to Different Chemistries to be Used for Splice Switching

AONs are synthetic DNAs or RNAs with variable length (11–50 nt) that are complementary to their target sequences. The possibility to use synthetic AONs to modulate gene expression has been described for the first time in 1978 by Zamecnik. He reported efficient inhibition of protein expression using a tridecamer oligodeoxynucleotide (i.e. unmodified DNA) with sequence complementary to the Rous sarcoma virus 35S RNA [96]. However, in the last three decades, alternative oligonucleotide chemistries have been generated with different characteristics. Chemical modifications have been made to generate AONs that are more stable, have higher affinity for their target sequence, and are resistant to endo- and exonuclease and RNase H cleavage. Furthermore, modifications also improved pharmacokinetic properties by preventing renal clearance and improving cellular uptake. A first biochemical modification to be reported was the replacement of a phosphodiester backbone with a phosphorothioate (PS) backbone [97]. This modification increases nuclease resistance, enhances transport over cell membranes, and prevents clearance of AONs by the kidney through low-affinity binding to serum proteins, but it also reduces the affinity for target RNA and DNA [98]. PS-modified AONs activate RNAse H cleavage of targeted RNA. Modification of the 2′-hydroxyl group of the RNA ribose renders AONs RNase H resistant and leads to increased affinity. Examples of 2′ modifications are 2′-*O*-methyl (2OMe), 2′-fluro (2F), and 2′-*O*-methoxyethyl (MOE) [99–101] (Figure 17.3). Further improvement of binding affinity can be achieved by fixing the oligonucleotides in the "endo" conformation by "locking" the ribose moiety, e.g. with a methylene bridge connecting the 2′ oxygen and 4′ carbon (locked nucleic acid [LNA] [or BNA]) or by an ethylene-bridged nucleic acids (ENA) [102–105]. These bridged nucleic acids do not activate RNase H and therefore can be used to modulate splicing. However, already relative short LNA AONs (over 15 mers) can form self-annealing structures and result in off-target effects due to high affinity for

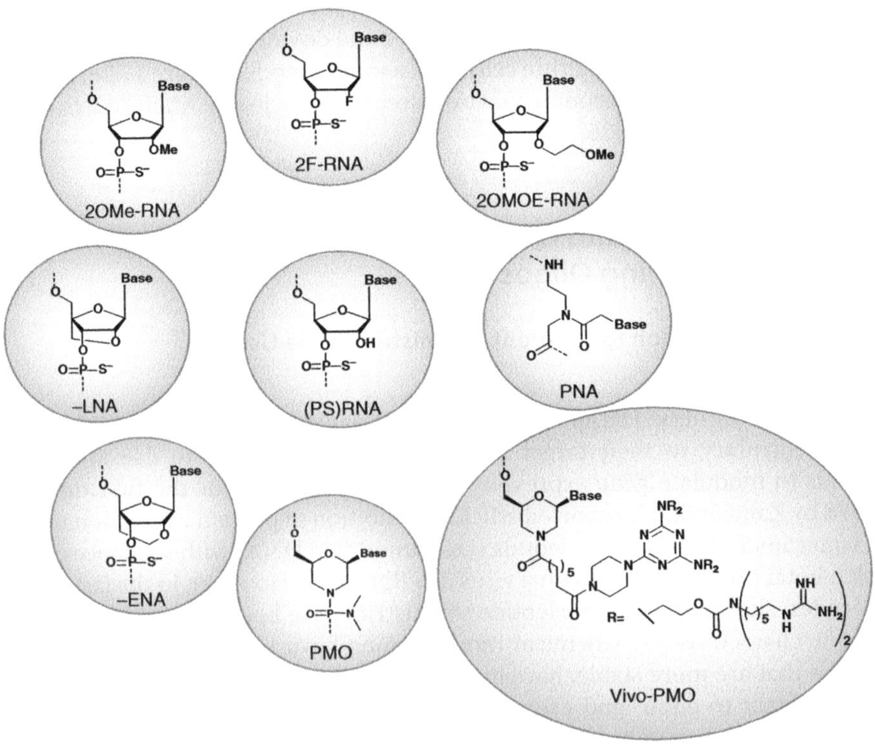

Figure 17.3 Antisense oligonucleotide backbone structures of the most in use compounds (2OMe-RNA, 2F-RNA, 2OMOE-RNA, PNA, LNA, ENA, PMO, vivo-PMO, and phosphorothioate (PS)-RNA).

unwanted target sequences [79]. Therefore often mixmers (or chimeric AONs) are used that contain bridged nucleotides as well as 2OMe or MOE nucleotides. Besides these sugar modifications, the phosphate backbone can be replaced by peptide bonds linking repeat units of *N*-(2-aminoethyl)-glycine to form a peptide nucleic acid (PNA) [106]. These changes result in an AON with low toxicity, high affinity for RNA target, and good sequence specificity that can be used to modulate splicing. However, the lack of a charged phosphate group causes a hydrophobic status, which requires the attachment of a short peptide in order to cross the cell membranes and be delivered inside the cells [107, 108]. Furthermore, PNAs do not contain a PS backbone, meaning that they are nonmetabolized, resulting in quickly and efficiently filtered by the kidney.

Finally, phosphorodiamidate morpholino oligomers (PMOs) are used for splice modulation (see Section 17.2.5) [9]. PMOs are uncharged due to the replacement of an anionic phosphate with a phosphorodiamidate group. The nucleobases are bound to morpholino rings instead of ribose sugar. PMOs are nontoxic and RNase H and nuclease resistant. As for the PNA, *in vitro* cellular uptake is negatively affected by the neutral charge of the molecule, while due to the lack of serum protein binding, PMOs are quickly cleared by the kidney after systemic treatment *in vivo*. This limitation can be avoided by conjugation of dendrimeric octaguanidine polymer to PMOs (vivo-morpholino) [109]. An alternative way to improve uptake by tissues is to conjugate arginine-rich cell-penetrating peptides (CPPs) to peptide-conjugated PMO (PMO) (pPMO). Some pPMOs have resulted in toxicity in nonhuman primates and a possible immune response [110, 111]. However, there are many different peptide conjugates in development [112–114] that may have a better safety profile. PMOs appear to be very well tolerated even at high doses in mouse, monkey, and humans [115–117].

17.2.2 AON Targets

In order for AONs to be effective and to induce exon skipping or exon inclusion, an AON has to bind stably to its target, and AON efficiency can often be improved by selecting AONs that adhere to certain criteria, such as good pre-mRNA target accessibility in secondary RNA structures, good thermodynamic properties, and targeting a motif that is essential for splicing that exon. Part of these criteria can be predicted by software programs during the development design stage of AON. Notably retrospective analysis of two reported large series of effective and ineffective DMD AONs from different groups [118, 119] has resulted in guidelines [19, 120].

Regarding target accessibility, pre-mRNA can form, by itself, different secondary structures containing open and close loops, which may be accessible or not to an AON. Retrospective analysis revealed that in general more effective

AONs contain more Gs and Cs, have higher melting temperature (Tms), have a higher binding energy, and target sequences closer to the acceptor splice sites than ineffective AONs. Furthermore, AONs targeting exon-internal sequences generally outperform AONs targeting the splice sites. This latter finding may be because splice sites are better defined than ESE motifs, making it easier to disrupt the binding of splicing factors to ESEs. In addition to these guidelines, different types of exon skipping approaches may require different target sites. For instance, AONs correcting a cryptic splice site could target the specific mutation and/or cryptic splice site. For exon skipping to induce knockdown, isoform switching, or restoring the reading frame, exon-internal sites may be better targets, although splice sites and the branch-point site are also possible targets. Finally, to achieve exon inclusion, one would have to target ISS or ESS motifs (if they are present). In some cases, in-frame transcript or transcriptional expression levels can be restored or changed using a mutation-specific AON targeting, e.g. single nucleotide polymorphism on the mutated allele [121].

17.2.3 AON Development for DMD

As the therapeutic development of antisense-mediated modulation of splicing is most advanced for DMD (see 17.1.4.1 for the rationale of this approach), we will here describe the preclinical and clinical development of AON-mediated exon skipping for DMD in detail and will then briefly summarize the current state of the art for other AON-mediated splicing modulation approaches.

Two animal models have been used toward the clinical development of AONs for DMD: a mouse and a dog model. The *mdx* mouse carries a stop mutation in exon 23 of the mouse dystrophin gene and consequently does not produce dystrophin. *Mdx* mice are less severely affected than DMD patients, having a mild nonfatal phenotype. However, they do suffer from a dystrophic phenotype, have reduced muscle function, and can be used for preclinical development of 2OMePS and PMOs [122]. Since the mutation is located in an in-frame exon, exon 23 skipping can bypass the mutation without disrupting the reading frame.

The best known canine model is the golden retriever muscular dystrophy (GRMD) dog, which carries a mutation in the 3′ splice site of intron 6, resulting in a skip of the exon 7 and truncated out-of-frame transcript [123]. The lack of dystrophin results in a severe phenotype affecting the muscle functions and reducing the life span to a few years. At pathological level, the skeletal muscles are replaced by fibrosis and necrosis as in the DMD patients [124].

The aim of AON-mediated exon skipping for DMD is to reframe the dystrophin transcript. The *DMD* gene contains 79 exons spanning around 2.3 Mb, is transcribed in 16 h, and is spliced in an mRNA of 14 kb, coding for a muscle isoform of 427 kDa. The size and complexity of the gene and the inaccessibility and abundance of skeletal muscle have seriously hampered the development of other forms of therapy, such as gene [125] and cell therapy [126], leaving an open field to the manipulation of splicing. Exon skipping for DMD was first shown in the early 1990s in human control lymphoblastoid cells [127]. The proof of principle of this approach, showing both exon skipping and dystrophin restoration after transfection with a 2OMePS AON, was reported in the same year in cultured cells from DMD patients [69] and in cultured *mdx* mouse cells [71]. Exon 23 skipping and local dystrophin restoration were confirmed *in vivo* intramuscular (i.m.) injection of 2OMePS AONs [128, 129]. Dystrophin was restored in cultured cells from patients carrying various deletions, point mutations, or duplications [130, 131]. Notably, skipping of multiple exons (2 or more) and dystrophin restoration have been achieved as well using a combination of AONs targeting different exons [79, 132, 133].

However, the DMD exon skipping approach is not applicable to patients carrying mutations that involve the functional N- and C-terminal domains of the dystrophin protein. Fortunately, these mutations are rare (<5% of all mutations [74]). In addition, the mutation-specific nature requires extensive optimization since different AONs are needed to skip different exons.

Parallel studies in cultured cells have been performed using PMO, resulting in successful murine exon skipping of exon 23 *in vitro* [134] and human exons 51 and 53 in control and patient-derived cell cultures [135, 136].

17.2.4 2′-*O*-Methyl Phosphorothioate AONs

17.2.4.1 Animal Studies

Initial *in vivo* tests were performed with 2OMePS AONs targeting exon 23 by i.m. injections in the *mdx* mouse model. This resulted in exon 23 skipping and restored dystrophin expression to near-normal levels in a large numbers of fibers that was accompanied by functional improvement [129]. This study used the block copolymer F127 to improve delivery of AONs. However, later studies revealed that equal levels of exon skipping and dystrophin restoration could be achieved by 2OMePS injections in saline [137]. Considering that all skeletal (limb and respiratory) muscles and the heart are affected in DMD patients, delivery and uptake of AONs in the whole skeletal and cardiac musculature is required. Thus systemic treatment is needed. In a follow-up study of the Lu et al. i.m. treatment with repeated intravenous (i.v.) injections of this AON in saline or combined with block copolymer F127 was tested in *mdx* mice. Both induced dystrophin expression to up to 5% of the normal levels of skeletal muscles, but not in the heart. No toxicity was detected in the liver, kidney, lung, and muscles of treated *mdx* mice [138]. While 2OMePS AONs are not filtered out by the kidney due to their PS backbone, the majority of 2OMePS AONs are normally taken up by the kidney and liver, and uptake in muscles is low. Thus the observed effects in skeletal muscles after i.v. were unexpected; this effect was hoped for even before i.v. injection. However, a later study revealed that actually the dystrophic phenotype facilitates AON uptake by muscle, leading to up to 10-fold higher 2OMePS levels in dystrophic muscle compared with healthy muscle and thus higher exon skipping levels [137]. The half-life of the 2OMePS in muscle was ~4 weeks, while it was longer (~6 weeks) in cardiac muscle. Comparison of subcutaneous (s.c.), i.v., and intraperitoneal (i.p.) (5 daily injections of 250 mg kg^{-1}) delivery revealed a higher clearance level from the plasma for i.v. than s.c. and i.p. delivery and two fold higher 2OMePS levels in the liver and kidney and ~20% higher levels in the muscle. Exon skipping levels were also highest after i.v. delivery, but the ratio of exon skipping levels and 2OMePS ending up in the liver and kidney was more favorable for s.c. and i.p. delivery. A last experiment using s.c. administration and increasing doses (25, 50, or 100 mg kg^{-1} twice weekly for 4 or 8 weeks) of 23 AON showed a dose-dependent effect where higher doses or longer treatment resulted in higher AON levels in the muscles and higher exon skipping and dystrophin levels. In addition, Heemskerk et al. [137] suggested the arrector pili smooth muscle (attached to hair follicles) as a potential method to assess treatment efficacy that can be performed using a skin biopsy rather than a more invasive muscle biopsy. In recent study, different dosing regimens of 2OMePS have been tested in 4-week-old *mdx* mice for 8 weeks comparing s.c. injections of the same total dose but with different numbers of administrations (either once 200 mg kg^{-1}, twice 100 mg kg^{-1}, or 7 times 28.6 mg kg^{-1} per week). This revealed that exon skipping levels in the heart and diaphragm were higher with daily administration than single and double weekly doses, while no differences were observed for other muscles. In the same study different maintenance regimens were compared as well. After an initial loading dose in the first week involving five daily injections of 250 mg kg^{-1}, mice received one or two weekly, one or two monthly, or no further treatments of 100 mg kg^{-1}. This revealed that AON, exon skipping, and protein levels were highest in the animals receiving two weekly injections, followed by those receiving one weekly dose. Notably exon skipping could be observed in all groups (even those without any further treatments) [139].

The *mdx* mouse model is primarily used to study DMD pathology and therapy. However, in this model severe pathology is prevented through efficient regeneration of muscle and upregulation of the dystrophin homologue utrophin, which can partly take over dystrophin function. A mouse model in which both dystrophin and utrophin (*mxd/utrn*$^{-/-}$) are knocked out leads to a severe phenotype [140]. Animals generally die around 7–8 weeks of age and show severe muscle impairment, joint contracture, growth retardation, respiratory problems, and kyphosis, representing not ideal mouse model. Notably, *mdx* mice carrying only one utrophin allele (*mxd/utrn*$^{+/-}$) are less severely affected than *mdx/utrn* double knockout mice but more severely affected than regular *mdx* mice. Indeed, skeletal muscle inflammation and fibrosis are more severe in *mxd/utrn*$^{+/-}$ mice than in *mdx* mice, and functional performance is more impaired [141] (also Ref. [142]). A 6-month s.c. treatment of 200 mg kg^{-1} per week in *mdx* and *mdx/utrn*$^{+/-}$ mice revealed comparable levels of exon skipping and dystrophin restoration for both models. Muscle function was comparable after treatment, but because function is more impaired in the *mdx* mouse with one utrophin allele, this model facilitates picking up of functional improvements over the regular *mdx* mouse where muscle function is only slightly impaired [143, 144]. Long-term treatment with 2OMePS AONs was well tolerated. CK is a muscle enzyme that leaks into the bloodstream upon muscle damage. Thus, plasma CK levels are elevated in DMD patients and *mdx* mice. However, over time CK levels were significantly lower in *mdx* mice and *mdx* mice with one utrophin allele treated for 6 months than in saline-treated mice.

Animal models provide a good framework for preclinical studies to test proof of concept and pharmacokinetic and pharmacodynamic characteristics of AONs. However, for translation into the clinic, clinical trials have to be performed in patients, requiring human-specific AONs, since due to sequence mismatches AONs are generally species specific. Human AONs have been optimized in cultured cells from healthy individuals and patients. To enable *in vivo* confirmation of human-specific AONs, a mouse model carrying the human *DMD* gene integrated into chromosome 5 of the mouse genome has been generated [145]. The presence of the integrated human wild-type *DMD* gene in *mxd/utrn*$^{-/-}$ led to the production of full-length human dystrophin, compensating for the lack of mouse dystrophin and rescuing the dystrophic phenotype. Thus 2OMePS or PMO AONs uptake and skipping in the hDMD mouse were lower than those previously observed in mdx mouse. [142, 146, 147] Notably, vivo-morpholinos did result in high levels of exon skipping [148]. (See below for more details.)

17.2.4.2 Human Studies

Exon skipping is mutation specific. Different mutations require the skipping of different exons (and thus different AONs) to restore the reading frame. Each AON is recognized as unique drug, and currently independent clinical trials are needed to obtain regulatory approval (from the European Medicines Agency (EMA) in Europe or the FDA in the United States). Since exon 51 skipping applies to the largest group of DMD patients, clinical development has mainly focused on this exon. As exon 51 skipping would disrupt the open reading frame in normal individuals, studies in healthy volunteers have not been performed. The first proof-of-concept study, sponsored by Prosensa Therapeutics B.V., involved a single unilateral local i.m. administration of 0.8 mg PRO051 (a 2OMePS targeting exon 51, currently called drisapersen) in the tibialis anterior muscle of four Dutch DMD boys. The treatment was tolerated well and resulted in a successful skip of exon 51 and dystrophin restoration in the muscle biopsy of each of the treated boys, taken 4 weeks after the AON injection. Dystrophin expression was found at levels between 17 and 35% of wild type in 64–94% of fibers. Notably the oldest patient in whom the disease had progressed furthest had the lowest levels of dystrophin, probably due to the poor condition of the patients' muscle tissue at the time of injection. Considering that dystrophin is only expressed in muscle fibers, which are replaced by fibrotic and adipose tissues during the progression of the disease that do not express dystrophin, it can be concluded that treatment should be initiated as early as possible [75]. Subsequently a trial testing systemic administration (phase 1/2) has been performed in 12 DMD boys (three boys per group), who received five weekly abdominal s.c. injections of drisapersen at four different doses (0.5, 2, 4, or 6 $mg\,kg^{-1}$). Successful treatment was established by a dose-dependent increase of dystrophin expression in 60–90% of muscle fibers in 10 out of 12 patients. After complete analysis and an interval of 6–15 months, all 12 patients were enrolled in an open-label extension study, receiving weekly injections of the highest dose (6 $mg\,kg^{-1}$). After 12 weeks this resulted in an increase in the distance walked in 6 min (6 MWD) in the 10 patients who were ambulant at the start of the extension study [77]. The 6-min walk test is a validated outcome measure that is accepted by EMA and FDA as a primary endpoint in (young) patients with neuromuscular disorders. The open-label extension study has been ongoing for almost 4 years, and all patients are still participating [149]. After 72 weeks the dosing regimen changed from weekly dosing to cycles of weekly dosing for 8 weeks followed by a drug-free period of 4 weeks. Side effects, such as proteinuria (reversible during off-treatment periods), injection side reactions (a known side effect of s.c. delivery of 2OMePS AONs), and thrombocytopenia, for this long treatment have been reported. Notably, 8 of the 10 boys who were ambulant at the start of the study maintained their baseline 6 MWD for close to 4 years, while two boys have lost ambulation. These results are unprecedented and encouraging. However, given that there is no placebo group, the results have to be interpreted with caution [150]. Prosensa then licensed the development of exon 51 skipping 2OMePS AONs (PRO051/GSK2402968) to GlaxoSmithKline (GSK), who initiated phase 2 and phase 3 trials. A phase 2 placebo-controlled study involved young boys (aged 6–8 years) who were in the early stage of their disease (had to be able to get up from the floor in <7 s). The study involved 53 patients who received weekly s.c. treatment with 6 $mg\,kg^{-1}$ drisapersen (18), intermittent treatment with drisapersen (17), or placebo (18) for 24 weeks and maintained to 48 weeks [151]. The primary objective of the study was to observe a clinically meaningful improvement in 6 MWD for any of the treated groups over placebo after 24 weeks. For the weekly dosed patients, the 6 MWD was significantly higher than placebo after 24 weeks, while there was a trend for improvement after 48 weeks. For intermittently dosed patients, there was a trend for improvement after 48 weeks. Trends for improvement were also observed for both treated groups in other muscle function tests, while muscle strength did not differ significantly from the placebo-treated group [151]. The treatment was well tolerated, but side effects (injection side reactions and proteinuria and thrombocytopenia for some patients) were observed and more frequently in treated groups than in the placebo group. Dystrophin analysis in muscle biopsies obtained from these patients revealed increased dystrophin in 13/18 patients on continuous treatment, in 10/17 patients on intermittent treatment, and in 1/18 placebo-treated patient.

Based on the above encouraging results, the pivotal double-blind, placebo-controlled phase 3 trial of drisapersen was initiated in December 2010, which involved 186 ambulant patients between 5 and 16 years old (http://clinicaltrials.gov/ct2/showNCT01254019). These patients and those who had participated in the phase 2 trial could voluntarily be enrolled in an open-label study of 220 boys after completion of their respective trials (http://clinicaltrials.gov/ct2/show/NCT01480245).

Of the 186 DMD randomized trial participants, 125 boys received a weekly s.c. drisapersen injection at a dose level of $6\,mg\,kg^{-1}$, whereas 61 received weekly placebo injections for 48 weeks. The functional outcome of this trial has recently been publicly disclosed, reporting no statistical difference in the 6MWD between drisapersen and placebo groups (showing a decline of 42 and 53 m, respectively, treatment difference 10.3 m, P-value 0.415). Neither were differences observed for secondary functional outcome measures such as North Star Ambulatory Assessment (NSAA) or timed function tests. As reported for the phase 2 trials, proteinuria and injection side reactions were seen more frequently in the drisapersen-treated group than the placebo-treated group. Further analysis (e.g. dystrophin quantification, the analysis of subgroups as well as longer-term data of patients enrolled in the open-label study) is currently ongoing. Pending these results, treatment of patients involved in open-label studies was suspended, but later reinitiated. GSK gave the rights and data for drisapersen back to Prosensa, which was subsequently acquired by BioMarin Pharmaceuticals. Using post hoc analysis on the pooled data of the different trials, it was revealed that the 6MWD was significantly better in drisapersen-treated patients than placebo for a subset of patients. BioMarin applied for marketing authorization with the FDA and EMA. However, the FDA ruled that drisapersen was not yet ready for approval, while the company withdrew the application with the EMA. At the same time BioMarin indicated to stop the clinical development of drisapersen and to focus on better compounds for future clinical application (http://investors.bmrn.com/releasedetail.cfm?ReleaseID=973536).

In addition to exon 51 skipping, Prosensa Therapeutics had initiated dose-escalating clinical trials to evaluate dosing, safety, and tolerability for 2OMePS AONs targeting exon 44 (http://clinicaltrials.gov/ct2/show/NCT01037309), exon 53 (http://clinicaltrials.gov/ct2/show/NCT01957059), and exon 45 (http://clinicaltrials.gov/ct2/show/NCT01826474). Some of these studies included an i.v. dosing arm in addition to s.c. dosing. The clinical development of these compounds stopped as well when drisapersen development was stopped.

17.2.5 Phosphorodiamidate Morpholino Oligos

17.2.5.1 Animal Studies

The preclinical and clinical development of PMOs was done in parallel with the 2OMePS compound. As described in Section 17.2.1, PMOs are uncharged and do not bind serum proteins, nor do they interact with other cellular components, resulting in a high excretion by the kidney, a low toxicity level, and a short serum half-life (1.5–3 h) [152]. The first reported exon skipping of murine exon 23 *in vivo* in *mdx* mice involved i.m. injection of PMOs with a DNA leash that was complexed to lipoplexes [134]. Subsequently, systemic i.v. injections of PMOs without a leash showed dose-dependent dystrophin restoration, up to 60% of wild-type levels in 6-week-old *mdx* mice. After seven weekly treatments with a high dose of 2 mg PMO, dystrophin restoration was confirmed for all skeletal muscles, but not the heart [153]. Considering that DMD patients are affected by myocardial dysfunction, which with respiratory problems are the main cause of death, a lot of effort has been put in achieving dystrophin restoration in the cardiac muscle. Recently, posttreatment dystrophin restoration in levels up to 30% of normal was detected in hearts from the *mdx* mouse model using single i.v. injections of PMOs, albeit at heroic doses (ranging from 0.03 to $3\,g\,kg^{-1}$). No toxic effects were observed, while the higher doses resulted in a drop in plasma CK levels [154]. By contrast, when targeting the skeletal muscle, it has also been reported that multiple i.v. injections of low doses of PMO are more beneficial than a single high dose [155]. However, the improved muscle functionality observed after restoring dystrophin in the skeletal muscle (but not in the heart) with this dosing regimen worsened the cardiac phenotype in *mdx* mice.

PMOs have also been tested in a dog model. Breeding the *GRMD* dog with a smaller beagle gave rise to the canine X-linked muscular dystrophy (CXMDj) model. CXMDj dogs have a less severe phenotype than GRMD dogs, and because they are smaller, they are a cheaper alternative to test potential therapies [123]. Initial tests had already shown dystrophin restoration *in vitro* in cultured myoblasts of GRMD dog with different AONs targeting exons 6 and 8 [156]. In further experiments, a cocktail of three different AONs (1 for exons 8 and 2 for exon 6) was tested in isolated primary myotubes from CXMDj dog. This resulted in skipping of not only exons 6 and 8 but also exon 9 (which is occasionally skipped due to alternative splicing). Similar multiexon skipping was also obtained by i.v. injection with a combination of these PMOs in two dogs, which was accompanied by high levels of dystrophin restoration [157]. While this dog model is a valuable addition to available animal models, its size, long breeding time, and the large amount of compounds required (due to animal size and the fact that reframing the transcripts needs skipping of two exons) have so far hampered performing studies in large groups.

Many independent pharmacokinetic and pharmacodynamic studies have been done with PMO and 2OMePS AONs. However, due to differences in delivery routes, dosing and dosing regimen comparison of these studies is difficult. Heemskerk et al. reported a direct comparison *in vivo* study in *mdx* and hDMD mice using equal molar amounts of PMO (25-mer) and 2OMePS (17-, 20-, and 25-mer) via i.m. and i.v. administrations. For mouse exon 23 skipping, after i.m. administration, the PMO clearly outperformed the 2OMePS AON, regardless of its length. However, for human exons (44, 45, 46, and 51), 2OMePS and PMOs performed equally well regarding exon skipping levels, while 2OMePS appeared to be more sequence specific. After i.v. injections of equal molar amounts of PMO and 2OMePS targeting mouse exon 23, exon skipping and dystrophin restoration were again better for the PMO in the skeletal muscle, while similar low levels of dystrophin were observed in the heart. Biodistribution analysis revealed that the vast majority of PMO ended up in the kidney, while 2OMePS was found in both the kidney and liver and to a lesser extent in the muscle [146].

17.2.5.2 Human Studies

To optimize exon 51 PMOs, Arechavala-Gomeza et al. investigated eight different 2OMePS AONs targeting exon 51 *in vitro* using cell transfection. The best AONs were synthesized as PMOs and injected into hDMD gastrocnemius muscle, confirming their ability to induce exon 51 skipping *in vivo*. The most optimal PMO was then developed further for clinical trial in the United Kingdom [135].

The first phase 1 clinical trial using PMO was coordinated by the UK MDEX Consortium and Sarepta Therapeutics and involved seven DMD patients who received a single i.m. injection of 0.09 or 0.9 mg of AVI-4658 (currently called eteplirsen) in the extensor digitorum brevis (EDB) muscle and a contralateral muscle saline injection (http://clinicaltrials.gov/ct2/show/NCT00159250). Muscle biopsies of pre- and posttreatments were analyzed, and dystrophin expression was detected at the sarcolemma in 44–79% of fibers at 22–32% of wild-type levels in the five patients treated with high dose. Exon 51 skipping was observed for all seven treated biopsies [76]. A subsequent systemic study (phase 1/2) involving weekly i.v. infusions for 12 weeks at escalating doses of eteplirsen (0.5, 1, 2, 4, 10, and 20 mg kg^{-1}) took place in 19 ambulatory DMD patients [78] (http://clinicaltrials.gov/ct2/show/NCT00844597). Posttreatment biopsies, obtained 2 weeks after the last administration, showed exon skipping in all samples. However, dystrophin restoration was only observed in 7 of the 19 patients in up to 55% of fibers, at levels between 8 and 16% of wild type by fluorescent staining [78]. Following this study, a new phase 2 trial was conducted in the United States (Columbus, OH) using i.v. administration for 60 min of higher doses (30 or 50 mg kg^{-1}) or a placebo, once per week for 24 weeks in 12 ambulatory DMD patients (4 in each group) between 7 and

13 years old [115]. After 24 weeks the four placebo-treated patients were distributed over the 30 and 50 mg kg^{-1} groups (2 patients each). Biopsies were obtained before treatment, after 12 weeks for the 50 mg kg^{-1} group and two placebo-treated patients, after 24 weeks for the 30 mg kg^{-1} group and two placebo-treated patients, and after 48 weeks for all patients. No dystrophin-positive fibers were observed for the 50 mg kg^{-1} group after 12 weeks. However, after 24 weeks patients from the 30 mg kg^{-1} group showed a significant increase in the percentage of dystrophin-positive fibers compared with placebo-treated patients. Up to 47% of dystrophin-positive fibers ($p \leq 0.001$) were observed in the 48 weeks biopsies, and a significant increase in the percentage of positive fibers was found for all patients (i.e. also the patients receiving placebo for the first 24 weeks, up to 38.3% of normal) [115]. These patients were followed with the 6-min walk test as well. For the 30 mg kg^{-1} group two patients lost ambulation in the early phase of the study (week 24) and were removed from further analysis. In patients receiving the 50 mg kg^{-1} dose, the 6 MWD stabilized for up to 72 weeks, while the 6 MWD decreased for the group receiving placebo (24 weeks) and decreased further for the first 12 weeks of treatment (until week 36) and then stabilized up to 188 weeks. Lacking a placebo group, the 6 MWD was compared with natural history data in Belgium and Italy. This suggested a slower disease progression for the treated patients [158]. Sarepta applied for marketing authorization with the FDA based on the dystrophin restoration. The FDA requested quantification of the dystrophin levels using a validated method, which was done using biopsies obtained after 188 weeks of treatment. This revealed an increase of 0.9% of dystrophin. The FDA then requested for further confirmation of dystrophin restoration in additional patients. Using 13 biopsies from patients who were treated with eteplirsen for at least 48 weeks in a second phase 2 trial (https://clinicaltrials.gov/ct2/show/NCT02255552), Sarepta was able to show an increase in dystrophin levels of ~0.4% in some patients. Based on this data FDA granted eteplirsen accelerated approval, outlining that functional effects have not yet been confirmed and that the company will need to perform confirmatory studies focusing on functional effects (http://www.fda.gov/NewsEvents/Newsroom/PressAnnouncements/ucm521263.htm).

17.2.6 Other Chemistries

17.2.6.1 Peptide-Conjugated PMOs

To improve the uptake of AONs by the skeletal and especially heart muscle, short peptides have been conjugated to PMOs to increase uptake by tissues [109, 114, 159–161].

CPPs are short peptides that are able to transport compounds (such as AONs) across the membrane. The first peptide conjugated to a PNA AON was a shorter version (46–60 amino acids (AA) of the HIV-derived TAT peptide (86 AA). A comparison between PNA–peptide conjugates and unmodified

PNA showed improved exon 23 skipping *in vitro* in *mdx* cell cultures but not *in vivo*[162]. Further studies involved synthetic CPPs consisting of arginine-rich peptides, mainly studying two variants: R-PMO and B-PMO, containing $(RXR)_4XB$ and $(RXRRBR)_2XB$, respectively, where R is arginine, X is 6-aminohexanoic acid, and B is β-alanine) [79, 163–165]. For all R-PMO and B-PMO studies, body-wide dystrophin protein restoration was achieved in *mdx* mice compared with the unconjugated PMO. For the first compound, escalating doses (1–25 mg kg^{-1}) via i.p. administration resulted in dose-dependent dystrophin restoration in particular in the diaphragm. A follow-up study of a selected dose of 5 mg kg^{-1} for 4 weeks confirmed high dystrophin restoration in the diaphragm, but low levels in other skeletal muscles [166, 167]. Additional studies showed dystrophin restoration at high levels in the skeletal muscle (up to 72%) and notably the heart (up to 30%) using various doses (12–30 mg kg^{-1}) of R-PMO and B-PMO via i.v. administration [161, 168, 169]. This also improved survival in a dobutamine stress test [161]. Using 25 mg kg^{-1} per week, dystrophin restoration at high levels was also observed in mice lacking both utrophin and dystrophin, where treatment also resulted in survival beyond 6 months, while normally these animals die at ~2–3 months of age [170]. Notably, using a lower dose did not result in improved survival. No toxicity or immune activation was reported for any of these studies. In an attempt to improve the uptake by muscle, two conjugated chimeric muscle-specific heptapeptide (MSP) peptides (B-MSP-PMO and MSP-B-PMO) and B-PMO were tested in *mdx* mice. Systemic administration of B-MSP-PMO and B-PMO resulted in dystrophin protein restoration up to 25 and 10% of the normal level in the skeletal muscle, respectively. In addition, exon skipping was observed after treatment with the B-MSP-PMO but not the MSP-B-PMO variant, revealing that the relative MSP position is important for exon skipping efficiency [171].

Based on these promising results, a clinical development path was initiated for a B-PMO compound targeting exon 50 (AVI-5038). However, after i.v. administration of different doses (1.5, 6, and 15 mg kg^{-1}), a dose-dependent kidney toxicity was observed in four monkeys (25 March 2010 press release AVI BioPharma) [111, 172]. In contrast with the previous study, no toxicity was observed in *mdx* mice at doses of B-PMO of 20 or 30 mg kg^{-1} for six weekly or 3 months biweekly injections, respectively [173]. Further study of an increased single dose of B-PMO of 60, 90, and 120 mg kg^{-1} in *mdx* mice resulted in reduced activity or death, suggesting that the LD_{50} of the B-PMO is ~85 mg kg^{-1} and as previously reported [111] that detection of toxicity is strongly dose and frequency dependent. In addition, Wu et al. [173] assessed the half-life of dystrophin in *mdx* mice. Animals were sacrificed at different time points after a single i.v. injection of 30 mg kg^{-1} of B-PMO (2 days, 2 weeks, and 1, 1.5, 2.5, 4, and 5 months). This revealed that dystrophin expression was maintained to similar level up to 1.5 months and then decreased to 50% (2.5 months), 30% (4 months), and 10% (5 months) of normal levels. In cardiac muscle, dystrophin was only detected up to 2.5 months with efficiencies from 70 to 25% of normal levels. In an additional experiment it was shown that biweekly i.v. administration of B-PMO in *mdx* mice of 1.5 mg kg^{-1} for 1 year showed no significant dystrophin restoration, whereas this same regimen using a high dose (6 mg kg^{-1}) exhibited more than 20% dystrophin expression in skeletal muscles and around 5% in the heart, which was accompanied by reduced plasma CK levels and improved muscle function. Finally, monthly i.v. injections of 30 mg kg^{-1} for 1 year resulted in high levels of dystrophin in most of the skeletal muscles, as well as in cardiac muscle albeit at lower levels. No pathological changes were observed for this treatment regimen.

Further studies have been conducted to identify peptides that improve uptake without toxic effects. Pip peptides are based on R6-penetratin internalization peptides (originally derived from the homeobox peptide of *Drosophila* Antennapedia) and have been conjugated to PNA or PMO [174]. The first generation of Pip2(a/b)-PNA showed *in vitro* and *in vivo* skipping of murine exon 23 [112]. Further optimization resulted in new series of Pip peptides that were conjugated to PMOs. Single i.v. administration of 25 mg kg^{-1} of a selected Pip5e-PMO compound resulted in a high restoration of dystrophin protein in the skeletal muscle and the heart [113]. In addition, higher skipping activity was observed using Pip5e-PMO than Pip5e-PNA. Further characterization of new Pip peptides led to the discovery of a new series of Pip6-PMO conjugates, which were generated from Pip5e and with the idea of improving efficiency in the heart. Despite being a more advanced series, Pip6-PMO compounds did not perform better than Pip5-PMOs [114]. However, clinical development of Pips will of course be based on both efficacy and safety.

The delivery of the AON can be also improved by the addition of a dendrimeric octaguanidine group to PMOs (vivo-morpholino). Wu et al. [109] examined vivo-morpholinos targeting mouse exon 23 in *mdx* mice after i.p. and i.v. administration. Whereas single or repeated i.p. injection of 30 mg kg^{-1} resulted in exon skipping and dystrophin restoration mainly at the diaphragm and abdominal muscles, single and multiple i.v. administrations of 6 mg kg^{-1} resulted in dystrophin restoration to 50% of normal level in the skeletal muscles and 10% in the heart. Recently, Wu et al. [148] studied vivo-morpholinos targeting exon 50 *in vitro* in human cell cultures and *in vivo* in the hDMD/*mdx* mouse and reported the combination of both experiments as a possible approach for selecting AONs for further clinical development.

17.2.7 Preclinical and Clinical Studies for Other Diseases

17.2.7.1 Spinal Muscular Atrophy (SMA)

Strategies aiming to correct the splicing of the *SMN2* gene and increase the expression of SMN protein have made it to the clinical trial phase as well after extensive preclinical optimization (see Section 17.1.3 for more background on this approach) [175–177]. Different AON-mediated approaches have been compared with increase SMN protein levels. This characterization included many sites in the flanking regions of exon 7 (ISS motifs in introns 6 and 7 to enhance exon 7 inclusion) and exon 8 splice sites (to inhibit the spliceosome joining exons 6 and 8, to allow joining of exons 6 and 7). Since exon 7 contains the stop codon, SMN can be produced from an SMN transcript that lacks exon 8 [64, 178, 179]. An MOE PS AON, targeting an ISS in intron 7 (ISS-N1), showed exon 7 inclusion in control tissue cultures and patient fibroblasts. In the latter, exon 7 skipping was accompanied by increased protein levels [180–182]. Mice do not carry an *SMN2* gene, and mice lacking *SMN1* are embryonic lethal. Transgenic mice in which *SMN1* is deleted but carrying copies of human *SMN2* recapitulate different types of human SMA to some extent (a severe and milder phenotype for mice with two or four copies of *SMN2*, respectively). These models have allowed preclinical development involving testing the *in vivo* efficacy of AONs, optimizing dosing and dosing regimens, and performing pharmacokinetic and pharmacodynamic studies. After systemic administration of MOE PS AONs, exon 7 inclusion and SMN production were found in the skeletal muscle, kidney, and liver, but not in CNS tissue [181]. Considering that these AONs do not cross the blood–brain barrier (BBB), Hua et al. performed intracerebroventricular infusion using a micro-osmotic pump to deliver the AON into the cerebrospinal fluid in adult mice expressing a human *SMN2* transgene in the absence of mouse *SMN1*. A dose-dependent effect was observed with 90% of exon 7 inclusion and increased levels of hSMN protein achieved at a daily dose of 50 μg per 7 days [65]. During the 6 months after the infusion, animals were sacrificed at different times, and high efficiency of exon inclusion was observed for up to half a year after the treatment, suggesting a long half-life and stability of the AON in the CNS. In addition, authors observed that after a single intracerebroventricular injection, the rescue of the SMA phenotype was more effective in embryonic than in neonatal mice.

Efficient restoration of SMN protein has also been achieved using PMO targeting ISS-N1 by intracerebroventricular injection in postnatal SMA mice [183]. Based on this PMO sequence, a set of 14 overlapping PMOs (leash lipoplexes) targeting the same region have been tested in fibroblasts derived from an SMA type I patient. Two of these PMOs showed high exon 7 inclusion efficiency [184]. These were injected intracerebroventricular into SMA mice at different doses. Increased SMN protein expression was detected for both PMOs, although the long PMO outperformed the short PMO. Exon 7 inclusion and improved survival were dose dependent. These findings were confirmed by another study [185] using the same PMO. Conjugating a dendrimer to this PMO improved survival in neonatal type I SMA mice, although toxicity was observed for this compound. A different approach, the so-called "bifunctional" strategy, also showed increased levels of SMN protein after the reinclusion of exon 7. In both types of bifunctional chimera, AON splicing factors are specifically recruited to SMN exon 7 to effect exon inclusion. In SMN2 exon 7 the C>T transition leads to a switch from an ESE into an ESS that binds the heterogeneous ribonucleoprotein hnRNPA1, which combined with other hnRNPA1 molecules (binding to the ISS-N1 sequence) leads to the skip of exon 7. Cartegni and Krainer reported a bifunctional AON called ESSENCE (exon-specific splicing enhancement by small chimeric effectors) that targeted SMN2 exon 7 and also contained an RS domain to recruit SR protein to the exon. In a similar approach the AON targeted the 5′-end of exon 7 (complementary to the allele carrying T) and contained a tail sequence binding motif for splice regulating factors (e.g. ESE) [179], which recruits SF2/ASF splicing factor to further facilitate exon 7 inclusion (Figure 17.2c-ii).

Based on preclinical studies and encouraging results, a phase 1 trial in type 2/3 SMA patients was initiated (http://www.clinicaltrials.gov/ct2/show/NCT01494701) to test the safety, tolerability, and PK of escalating doses of nusinersen (an MOE PS AON targeting intron 7) administered into the spinal fluid as a single injection [186]. The trial investigated the safety and pharmacokinetics of a single intrathecal injection of four different doses of ISIS-SMNRx in 28 type 2 and 3 SMA patients (1, 3, 6, and 9 mg in 6, 6, 6, and 10 patients, respectively). The treatment and delivery method were well tolerated by all patients. A dose-dependent increase in compound levels could be observed in cerebrospinal fluid and plasma. An improvement in the Hammersmith function scale was observed for patients in the two highest dose groups, which is encouraging, but given the open-label nature of this study has to be interpreted with caution. Importantly, the half-life of the AON in the CNS appears to be 6–9 months. Following this trial, phase 2 trials in SMA type 1 and type 2/3 were initiated (http://www.clinicaltrials.gov/ct2/show/NCT01703988,http://www.clinicaltrials.gov/ct2/show/NCT01839656). Both trials have met their primary endpoints (http://ir.ionispharma.com/phoenix.zhtml?c=222170&p=irol-newsArticle&ID=2220037, http://ir.ionispharma.com/phoenix.zhtml?c=222170&p=irol-newsArticle&ID=2191319), and Ionis has received marketing authorization for nusinersen for type 1 SMA with both FDA and EMA (https://www.fda.gov/NewsEvents/Newsroom/PressAnnouncements/ucm534611.htm and http://www.ema.europa.eu/ema/index.jsp?curl=pages/news_and_events/news/2017/04/news_detail_002732.jsp&mid=WC0b01ac058004d5c1).

There are many other ways antisense-mediated splicing can be exploited as a therapeutic tool that are currently in the preclinical phase (see Refs. [8, 187]).

17.2.8 Biomarkers

Biomarkers are biological molecules that can be measured in blood, urine, or tissues. Generally, they are defined as molecular, cellular, or biochemical measurements by which normal and abnormal processes, as well as particular treatment response for a disease or condition, can be detected and/or monitored. For therapy development pharmacodynamic biomarkers can expedite the clinical trial process, e.g. when a change in a certain biomarker after short treatment is predicted for a functional change after long-term treatment. Pharmacodynamic biomarker can also confirm the biological response to a particular therapy in treated patients and can be used to provide further insight in dosing and dosing regimens in preclinical and clinical programs. For the two diseases for which antisense-mediated splicing modulation has reach the clinical trial phase, several instrumental and molecular biomarkers are under investigation. In SMA, the instrumental markers include compound motor action potential (CMAP) and the motor unit number estimation (MUNE), which detect skeletal muscle innervation status by monitoring the loss of motor neurons [188], whereas dual-energy X-ray absorptiometry (DXA) measures the reduction of bone mineral density (BMD), which is observed in patients with limited motility [189]. Given that AON treatment for SMA is done intrathecally, one anticipates that the main pharmacodynamic effects will be on the CNS. For obvious reasons it is not feasible to assess SMN transfect or protein levels after treatment.

Serum CK levels are elevated for DMD patients compared with controls, and thus CK would make an obvious biomarker. However, during disease progression, CK levels gradually decline for DMD patients since they lose muscle mass. This impedes using CK as a serum biomarker for therapeutic monitoring, since it will be difficult to interpret results, e.g. lower CK levels can be indicative of both improved muscle quality (and less CK leakage) or decreased muscle quality (and less CK to leak into blood). However, for DMD the target tissue (skeletal muscle) is more accessible, and muscle biopsies have been taken to confirm a pharmacodynamic action of the AON (i.e. dystrophin restoration). Thus far there appears to be no correlation between dystrophin levels measured after treatment and the functional performance of DMD patients in the 6-min walk test [190], while correlation with or prediction of future functional improvement is a prerequisite of regulators for biomarker validation. Another consideration is that muscle biopsies are invasive and it is not feasible to take repeated biopsies over time, especially for patients with impaired muscle quality to begin with. One should also bear in mind that a biopsy from one skeletal muscle may not be representative for the whole musculature due to, e.g. variations in disease progression in different muscles or different areas of one muscle. Magnetic resonance imaging can be used to provide a more global picture of muscle quality, and the possibility of using magnetic resonance imaging as a clinical outcome measure in trials is currently being explored. Finally, serum biomarkers are under investigation based on the premise that >30% of the human body consists of muscle and that processes of muscle regeneration and remodeling will be reflected in the circulation. Several candidate serum biomarkers have been reported [191, 192], but further validation is required.

17.3 Future Directions

As can be gathered from this chapter, AON-mediated modulation of splicing holds much promise to be a therapy for a plethora of genetic and acquired diseases. Especially for genetic diseases this is an unprecedented opportunity, as often these are diseases with unmet medical need. Therapeutic development has been advanced furthest for DMD and SMA, where drugs (eteplirsen and nusinersen) have now been approved. It is anticipated that this will lead to an acceleration of the preclinical and clinical development of AONs to induce skipping of additional exons and to manipulate splicing of other genes for other genetic diseases.

For most genetic therapy approaches, delivery is the main challenge. It appears that for muscular dystrophy and neurological diseases, this may not be the case. In muscular dystrophy there is preferential uptake by the target tissue due to leaky muscle fibers, and while AONs do not cross the BBB, once injected into the CNS, they are readily taken up by neurons and seem to have a half-life of months. Whether delivery is a challenge for other genetic diseases for which AON-mediated modulation of splicing might be beneficial depends on the target tissue(s) and will have to be assessed on a case-by-case basis.

Due to AON, transcript, and protein turnover, repeated treatment will be needed, probably every (few) week(s) for muscular dystrophy and every few months for neuron diseases like SMA. As these genetic diseases present in early age, lifelong treatment is likely needed. At the moment it is not known whether long-term treatment with AONs is safe. Nevertheless, using systemic delivery, the liver and kidney are the organs where the majority of the injected AONs end up for most chemistries, and therefore careful monitoring of liver and kidney function and damage is warranted.

Most splicing modulation approaches apply to (subsets of) rare diseases. This poses challenges for the clinical development of these compounds. The regulatory system was put in place primarily with common diseases in mind, and for rare diseases doing large placebo-controlled trials may not be feasible to the scarcity of trial subjects. Furthermore, treatment development for rare diseases is often a pioneering effort. Often good insight in natural history and validated clinical outcome measures are lacking, and without a proven therapy in place, it is difficult to assess whether a compound leads to a clinical benefit. Nevertheless, through a combined effort of academics, clinicians, patients, regulators, and industry, it should be possible to overcome these issues and allow exploitation of this approach for as many patients as possible [193].

Conflict of Interest

IG discloses being employed by LUMC and being coinventor of a patent application. Should this patent be granted, she is entitled to a share of any future royalties or license fees paid to LUMC for this patent. AAR reports being employed by LUMC and being coinventor of patent applications for antisense sequences and the exon skipping technology. LUMC has licensed the rights for some of these patents exclusively to Prosensa Therapeutics (now BioMarin). The inventors specified on the patents (including the author) are jointly entitled to a share of any future royalties paid to LUMC, should the therapy eventually be brought to the market. AAR also reports being on the scientific advisory boards of ProQR and Philae Pharmaceuticals and being ad hoc consultant for Grünenthal, BioMarin, PTC Therapeutics, Summit, BioClinica, GLG consulting, Deerfield consulting, and Global Guidepoint. Remuneration for these activities and speaker fees at satellite symposia (organized by PTC and BioMarin) go to her employer.

Acknowledgments

The authors thank Giles Campion and Mark Einerhand for reading the manuscript and useful suggestions.

References

1 Kole, R., Vacek, M., and Williams, T. (2004). Modification of alternative splicing by antisense therapeutics. *Oligonucleotides* 14 (1): 65–74.
2 Kole, R. (1997). Modification of pre-mRNA splicing by antisense oligonucleotides. *Acta Biochim. Pol.* 44 (2): 231–237.
3 Sioud, M. (2004). Therapeutic siRNAs. *Trends Pharmacol. Sci.* 25 (1): 22–28.
4 de Fougerolles, A., Vornlocher, H.P., Maraganore, J., and Lieberman, J. (2007). Interfering with disease: a progress report on siRNA-based therapeutics. *Nat. Rev. Drug Discov.* 6 (6): 443–453.
5 Nguyen, Q.D., Schachar, R.A., Nduaka, C.I. et al. (2012). Evaluation of the siRNA PF-04523655 versus ranibizumab for the treatment of neovascular age-related macular degeneration (MONET Study). *Ophthalmology* 119 (9): 1867–1873.
6 Nguyen, Q.D., Schachar, R.A., Nduaka, C.I. et al. (2012). Dose-ranging evaluation of intravitreal siRNA PF-04523655 for diabetic macular edema (the DEGAS study). *Invest. Ophthalmol. Vis. Sci.* 53 (12): 7666–7674.
7 Gauvreau, G.M., Pageau, R., Seguin, R. et al. (2011). Dose-response effects of TPI ASM8 in asthmatics after allergen. *Allergy* 66 (9): 1242–1248.

8 van Roon-Mom, W.M. and Aartsma-Rus, A. (2012). Overview on applications of antisense-mediated exon skipping. *Methods Mol. Biol.* 867: 79–96.
9 Hammond, S.M. and Wood, M.J. (2011). Genetic therapies for RNA mis-splicing diseases. *Trends Genet.* 27 (5): 196–205.
10 Will, C.L. and Luhrmann, R. (2001). Spliceosomal UsnRNP biogenesis, structure and function. *Curr. Opin. Cell Biol.* 13 (3): 290–301.
11 Cartegni, L., Chew, S.L., and Krainer, A.R. (2002). Listening to silence and understanding nonsense: exonic mutations that affect splicing. *Nat. Rev. Genet.* 3 (4): 285–298.
12 Keren, H., Lev-Maor, G., and Ast, G. (2010). Alternative splicing and evolution: diversification, exon definition and function. *Nat. Rev. Genet.* 11 (5): 345–355.
13 Matlin, A.J., Clark, F., and Smith, C.W. (2005). Understanding alternative splicing: towards a cellular code. *Nat. Rev. Mol. Cell Biol.* 6 (5): 386–398.
14 Jensen, C.J., Oldfield, B.J., and Rubio, J.P. (2009). Splicing, cis genetic variation and disease. *Biochem. Soc. Trans.* 37 (Pt 6): 1311–1315.
15 Lopez-Bigas, N., Audit, B., Ouzounis, C. et al. (2005). Are splicing mutations the most frequent cause of hereditary disease? *FEBS Lett.* 579 (9): 1900–1903.
16 Garcia-Blanco, M.A., Baraniak, A.P., and Lasda, E.L. (2004). Alternative splicing in disease and therapy. *Nat. Biotechnol.* 22 (5): 535–546.
17 Kole, R., Krainer, A.R., and Altman, S. (2012). RNA therapeutics: beyond RNA interference and antisense oligonucleotides. *Nat. Rev. Drug Discov.* 11 (2): 125–140.
18 Wood, M.J., Gait, M.J., and Yin, H. (2010). RNA-targeted splice-correction therapy for neuromuscular disease. *Brain* 133 (Pt 4): 957–972.
19 Aartsma-Rus, A., van Vliet, L., Hirschi, M. et al. (2009). Guidelines for antisense oligonucleotide design and insight into splice-modulating mechanisms. *Mol. Ther.* 17 (3): 548–553.
20 Bauman, J., Jearawiriyapaisarn, N., and Kole, R. (2009). Therapeutic potential of splice-switching oligonucleotides. *Oligonucleotides* 19 (1): 1–13.
21 Spitali, P. and Aartsma-Rus, A. (2012). Splice modulating therapies for human disease. *Cell* 148 (6): 1085–1088.
22 Aartsma-Rus, A. (2012). Overview on DMD exon skipping. *Methods Mol. Biol.* 867: 97–116.
23 Xie, S.Y., Li, W., Ren, Z.R. et al. (2011). Correction of beta654-thalassaemia mice using direct intravenous injection of siRNA and antisense RNA vectors. *Int. J. Hematol.* 93 (3): 301–310.
24 Friedman, K.J., Kole, J., Cohn, J.A. et al. (1999). Correction of aberrant splicing of the cystic fibrosis transmembrane conductance regulator (CFTR) gene by antisense oligonucleotides. *J. Biol. Chem.* 274 (51): 36193–36199.
25 Mercatante, D.R., Bortner, C.D., Cidlowski, J.A., and Kole, R. (2001). Modification of alternative splicing of Bcl-x pre-mRNA in prostate and breast cancer cells. analysis of apoptosis and cell death. *J. Biol. Chem.* 276 (19): 16411–16417.

26 Taylor, J.K., Zhang, Q.Q., Wyatt, J.R., and Dean, N.M. (1999). Induction of endogenous Bcl-xS through the control of Bcl-x pre-mRNA splicing by antisense oligonucleotides. *Nat. Biotechnol.* 17 (11): 1097–1100.

27 Mercatante, D.R., Mohler, J.L., and Kole, R. (2002). Cellular response to an antisense-mediated shift of Bcl-x pre-mRNA splicing and antineoplastic agents. *J. Biol. Chem.* 277 (51): 49374–49382.

28 Kalbfuss, B., Mabon, S.A., and Misteli, T. (2001). Correction of alternative splicing of tau in frontotemporal dementia and parkinsonism linked to chromosome 17. *J. Biol. Chem.* 276 (46): 42986–42993.

29 Jin, W. and Cote, G.J. (2004). Enhancer-dependent splicing of FGFR1 alpha-exon is repressed by RNA interference-mediated down-regulation of SRp55. *Cancer Res.* 64 (24): 8901–8905.

30 Jin, W., Huang, E.S., Bi, W., and Cote, G.J. (1999). Redundant intronic repressors function to inhibit fibroblast growth factor receptor-1 alpha-exon recognition in glioblastoma cells. *J. Biol. Chem.* 274 (39): 28035–28041.

31 Bruno, I.G., Jin, W., and Cote, G.J. (2004). Correction of aberrant FGFR1 alternative RNA splicing through targeting of intronic regulatory elements. *Hum. Mol. Genet.* 13 (20): 2409–2420.

32 Aartsma-Rus, A., Singh, K.H., Fokkema, I.F. et al. (2010). Therapeutic exon skipping for dysferlinopathies? *Eur. J. Hum. Genet.* 18 (8): 889–894.

33 Levy, N., Wein, N., Barthelemy, F. et al. (2010). Therapeutic exon “switching” for dysferlinopathies? *Eur. J. Hum. Genet.* 18 (9): 969–970 author reply 971.

34 Wein, N., Avril, A., Bartoli, M. et al. (2010). Efficient bypass of mutations in dysferlin deficient patient cells by antisense-induced exon skipping. *Hum. Mutat.* 31 (2): 136–142.

35 Khoo, B., Roca, X., Chew, S.L., and Krainer, A.R. (2007). Antisense oligonucleotide-induced alternative splicing of the APOB mRNA generates a novel isoform of APOB. *BMC Mol. Biol.* 8: 3.

36 Kang, J.K., Malerba, A., Popplewell, L. et al. (2011). Antisense-induced myostatin exon skipping leads to muscle hypertrophy in mice following octa-guanidine morpholino oligomer treatment. *Mol. Ther.* 19 (1): 159–164.

37 Sierakowska, H., Sambade, M.J., Agrawal, S., and Kole, R. (1996). Repair of thalassemic human beta-globin mRNA in mammalian cells by antisense oligonucleotides. *Proc. Natl. Acad. Sci. U. S. A.* 93 (23): 12840–12844.

38 Tan, J.A., George, E., Tan, K.L. et al. (2004). Molecular defects in the beta-globin gene identified in different ethnic groups/populations during prenatal diagnosis for beta-thalassemia: a Malaysian experience. *Clin. Exp. Med.* 4 (3): 142–147.

39 Donaldson, A., Old, J., Fisher, C. et al. (2000). Jamaican Sbeta+-thalassaemia: mutations and haematology. *Br. J. Haematol.* 108 (2): 290–294.

40 Dominski, Z. and Kole, R. (1993). Restoration of correct splicing in thalassemic pre-mRNA by antisense oligonucleotides. *Proc. Natl. Acad. Sci. U. S. A.* 90 (18): 8673–8677.

41 Lacerra, G., Sierakowska, H., Carestia, C. et al. (2000). Restoration of hemoglobin A synthesis in erythroid cells from peripheral blood of thalassemic patients. *Proc. Natl. Acad. Sci. U. S. A.* 97 (17): 9591–9596.

42 Suwanmanee, T., Sierakowska, H., Lacerra, G. et al. (2002). Restoration of human beta-globin gene expression in murine and human IVS2-654 thalassemic erythroid cells by free uptake of antisense oligonucleotides. *Mol. Pharmacol.* 62 (3): 545–553.

43 Riordan, J.R., Rommens, J.M., Kerem, B. et al. (1989). Identification of the cystic fibrosis gene: cloning and characterization of complementary DNA. *Science* 245 (4922): 1066–1073.

44 Tsui, L.C. (1992). The spectrum of cystic fibrosis mutations. *Trends Genet.* 8 (11): 392–398.

45 Tsui, L.C. and Dorfman, R. (2013). The cystic fibrosis gene: a molecular genetic perspective. *Cold Spring Harb. Perspect. Med.* 3 (2): a009472.

46 Zielenski, J., Bozon, D., Kerem, B. et al. (1991). Identification of mutations in exons 1 through 8 of the cystic fibrosis transmembrane conductance regulator (CFTR) gene. *Genomics* 10 (1): 229–235.

47 Boise, L.H., Gonzalez-Garcia, M., Postema, C.E. et al. (1993). bcl-x, a bcl-2-related gene that functions as a dominant regulator of apoptotic cell death. *Cell* 74 (4): 597–608.

48 Krajewska, M., Krajewski, S., Epstein, J.I. et al. (1996). Immunohistochemical analysis of bcl-2, bax, bcl-X, and mcl-1 expression in prostate cancers. *Am. J. Pathol.* 148 (5): 1567–1576.

49 Lee, T.L., Yeh, J., Friedman, J. et al. (2008). A signal network involving coactivated NF-kappaB and STAT3 and altered p53 modulates BAX/BCL-XL expression and promotes cell survival of head and neck squamous cell carcinomas. *Int. J. Cancer* 122 (9): 1987–1998.

50 Krajewski, S., Krajewska, M., Turner, B.C. et al. (1999). Prognostic significance of apoptosis regulators in breast cancer. *Endocr. Relat. Cancer* 6 (1): 29–40.

51 Tang, L., Tron, V.A., Reed, J.C. et al. (1998). Expression of apoptosis regulators in cutaneous malignant melanoma. *Clin. Cancer Res.* 4 (8): 1865–1871.

52 Goedert, M., Spillantini, M.G., Potier, M.C. et al. (1989). Cloning and sequencing of the cDNA encoding an isoform of microtubule-associated protein tau containing four tandem repeats: differential expression of tau protein mRNAs in human brain. *EMBO J.* 8 (2): 393–399.

53 Van Deerlin, V.M., Gill, L.H., Farmer, J.M. et al. (2003). Familial frontotemporal dementia: from gene discovery to clinical molecular diagnostics. *Clin. Chem.* 49 (10): 1717–1725.

54 Goedert, M., Ghetti, B., and Spillantini, M.G. (2000). Tau gene mutations in frontotemporal dementia and parkinsonism linked to chromosome 17 (FTDP-17). Their relevance for understanding the neurogenerative process. *Ann. N. Y. Acad. Sci.* 920: 74–83.

55 Goedert, M. and Jakes, R. (1990). Expression of separate isoforms of human tau protein: correlation with the tau pattern in brain and effects on tubulin polymerization. *EMBO J.* 9 (13): 4225–4230.

56 Hutton, M., Lendon, C.L., Rizzu, P. et al. (1998). Association of missense and 5′-splice-site mutations in tau with the inherited dementia FTDP-17. *Nature* 393 (6686): 702–705.

57 Hong, M., Zhukareva, V., Vogelsberg-Ragaglia, V. et al. (1998). Mutation-specific functional impairments in distinct tau isoforms of hereditary FTDP-17. *Science* 282 (5395): 1914–1917.

58 Yamaguchi, F., Saya, H., Bruner, J.M., and Morrison, R.S. (1994). Differential expression of two fibroblast growth factor-receptor genes is associated with malignant progression in human astrocytomas. *Proc. Natl. Acad. Sci. U. S. A.* 91 (2): 484–488.

59 Luqmani, Y.A., Mortimer, C., Yiangou, C. et al. (1995). Expression of 2 variant forms of fibroblast growth factor receptor 1 in human breast. *Int. J. Cancer* 64 (4): 274–279.

60 Vickers, S.M., Huang, Z.Q., MacMillan-Crow, L. et al. (2002). Ligand activation of alternatively spliced fibroblast growth factor receptor-1 modulates pancreatic adenocarcinoma cell malignancy. *J. Gastrointest. Surg.* 6 (4): 546–553.

61 Zerres, K. and Rudnik-Schoneborn, S. (1995). Natural history in proximal spinal muscular atrophy. Clinical analysis of 445 patients and suggestions for a modification of existing classifications. *Arch. Neurol.* 52 (5): 518–523.

62 Zerres, K., Wirth, B., and Rudnik-Schoneborn, S. (1997). Spinal muscular atrophy – clinical and genetic correlations. *Neuromuscul. Disord.* 7 (3): 202–207.

63 Lorson, C.L., Hahnen, E., Androphy, E.J., and Wirth, B. (1999). A single nucleotide in the SMN gene regulates splicing and is responsible for spinal muscular atrophy. *Proc. Natl. Acad. Sci. U. S. A.* 96 (11): 6307–6311.

64 Lim, S.R. and Hertel, K.J. (2001). Modulation of survival motor neuron pre-mRNA splicing by inhibition of alternative 3′ splice site pairing. *J. Biol. Chem.* 276 (48): 45476–45483.

65 Hua, Y., Sahashi, K., Hung, G. et al. (2010). Antisense correction of SMN2 splicing in the CNS rescues necrosis in a type III SMA mouse model. *Genes Dev.* 24 (15): 1634–1644.

66 Hoffman, E.P., Brown, R.H. Jr., and Kunkel, L.M. (1987). Dystrophin: the protein product of the Duchenne muscular dystrophy locus. *Cell* 51 (6): 919–928.

67 Emery, A.E. (2002). The muscular dystrophies. *Lancet* 359 (9307): 687–695.

68 Bushby, K.M., Gardner-Medwin, D., Nicholson, L.V. et al. (1993). The clinical, genetic and dystrophin characteristics of Becker muscular dystrophy. II. Correlation of phenotype with genetic and protein abnormalities. *J. Neurol.* 240 (2): 105–112.

69 van Deutekom, J.C., Bremmer-Bout, M., Janson, A.A. et al. (2001). Antisense-induced exon skipping restores dystrophin expression in DMD patient derived muscle cells. *Hum. Mol. Genet.* 10 (15): 1547–1554.
70 Takeshima, Y., Wada, H., Yagi, M. et al. (2001). Oligonucleotides against a splicing enhancer sequence led to dystrophin production in muscle cells from a Duchenne muscular dystrophy patient. *Brain and Dev.* 23 (8): 788–790.
71 Mann, C.J., Honeyman, K., Cheng, A.J. et al. (2001). Antisense-induced exon skipping and synthesis of dystrophin in the mdx mouse. *Proc. Natl. Acad. Sci. U. S. A.* 98 (1): 42–47.
72 Takeda, S. (2009). Exon skipping therapy for Duchenne muscular dystrophy by using antisense Morpholino. *Rinsho Shinkeigaku* 49 (11): 856–858.
73 Aartsma-Rus, A., Van Deutekom, J.C., Fokkema, I.F. et al. (2006). Entries in the Leiden Duchenne muscular dystrophy mutation database: an overview of mutation types and paradoxical cases that confirm the reading-frame rule. *Muscle Nerve* 34 (2): 135–144.
74 Aartsma-Rus, A., Fokkema, I., Verschuuren, J. et al. (2009). Theoretic applicability of antisense-mediated exon skipping for Duchenne muscular dystrophy mutations. *Hum. Mutat.* 30 (3): 293–299.
75 van Deutekom, J.C., Janson, A.A., Ginjaar, I.B. et al. (2007). Local dystrophin restoration with antisense oligonucleotide PRO051. *N. Engl. J. Med.* 357 (26): 2677–2686.
76 Kinali, M., Arechavala-Gomeza, V., Feng, L. et al. (2009). Local restoration of dystrophin expression with the morpholino oligomer AVI-4658 in Duchenne muscular dystrophy: a single-blind, placebo-controlled, dose-escalation, proof-of-concept study. *Lancet Neurol.* 8 (10): 918–928.
77 Goemans, N.M., Tulinius, M., van den Akker, J.T. et al. (2011). Systemic administration of PRO051 in Duchenne's muscular dystrophy. *N. Engl. J. Med.* 364 (16): 1513–1522.
78 Cirak, S., Arechavala-Gomeza, V., Guglieri, M. et al. (2011). Exon skipping and dystrophin restoration in patients with Duchenne muscular dystrophy after systemic phosphorodiamidate morpholino oligomer treatment: an open-label, phase 2, dose-escalation study. *Lancet* 378 (9791): 595–605.
79 Aartsma-Rus, A., Janson, A.A., Kaman, W.E. et al. (2004). Antisense-induced multiexon skipping for Duchenne muscular dystrophy makes more sense. *Am. J. Hum. Genet.* 74 (1): 83–92.
80 Liu, J., Aoki, M., Illa, I. et al. (1998). Dysferlin, a novel skeletal muscle gene, is mutated in Miyoshi myopathy and limb girdle muscular dystrophy. *Nat. Genet.* 20 (1): 31–36.
81 Bashir, R., Britton, S., Strachan, T. et al. (1998). A gene related to Caenorhabditis elegans spermatogenesis factor fer-1 is mutated in limb-girdle muscular dystrophy type 2B. *Nat. Genet.* 20 (1): 37–42.
82 Illa, I., Serrano-Munuera, C., Gallardo, E. et al. (2001). Distal anterior compartment myopathy: a dysferlin mutation causing a new muscular dystrophy phenotype. *Ann. Neurol.* 49 (1): 130–134.

83 van der Kooi, A.J., Frankhuizen, W.S., Barth, P.G. et al. (2007). Limb-girdle muscular dystrophy in the Netherlands: gene defect identified in half the families. *Neurology* 68 (24): 2125–2128.

84 Bushby, K.M. (1999). The limb-girdle muscular dystrophies: diagnostic guidelines. *Eur. J. Paediatr. Neurol.* 3 (2): 53–58.

85 Amato, A.A. and Brown, R.H. Jr. (2011). Dysferlinopathies. *Handb. Clin. Neurol.* 101: 111–118.

86 Bansal, D., Miyake, K., Vogel, S.S. et al. (2003). Defective membrane repair in dysferlin-deficient muscular dystrophy. *Nature* 423 (6936): 168–172.

87 Lennon, N.J., Kho, A., Bacskai, B.J. et al. (2003). Dysferlin interacts with annexins A1 and A2 and mediates sarcolemmal wound-healing. *J. Biol. Chem.* 278 (50): 50466–50473.

88 Bansal, D. and Campbell, K.P. (2004). Dysferlin and the plasma membrane repair in muscular dystrophy. *Trends Cell Biol.* 14 (4): 206–213.

89 Krahn, M., Wein, N., and Lostal, W. (2008). Partial functionality of a mini-dysferlin molecule identified in a patient affected with moderately severe primary dysferlinopathy. *Neuromuscul. Disord.* 18: 781–785.

90 Nalini, A. and Gayathri, N. (2008). Dysferlinopathy: a clinical and histopathological study of 28 patients from India. *Neurol. India* 56 (3): 379–385 discussion 386-7.

91 Santos, R., Oliveira, J., Vieira, E. et al. (2010). Private dysferlin exon skipping mutation (c.5492G>A) with a founder effect reveals further alternative splicing involving exons 49–51. *J. Hum. Genet.* 55 (8): 546–549.

92 De Luna, N., Freixas, A., Gallano, P. et al. (2007). Dysferlin expression in monocytes: a source of mRNA for mutation analysis. *Neuromuscul. Disord.* 17 (1): 69–76.

93 Chester, A., Scott, J., Anant, S., and Navaratnam, N. (2000). RNA editing: cytidine to uridine conversion in apolipoprotein B mRNA. *Biochim. Biophys. Acta* 1494 (1–2): 1–13.

94 Kemaladewi, D.U., Hoogaars, W.M., van Heiningen, S.H. et al. (2011). Dual exon skipping in myostatin and dystrophin for Duchenne muscular dystrophy. *BMC Med. Genet.* 4: 36.

95 Malerba, A., Kang, J.K., McClorey, G. et al. (2012). Dual myostatin and dystrophin exon skipping by morpholino nucleic acid oligomers conjugated to a cell-penetrating peptide is a promising therapeutic strategy for the treatment of Duchenne muscular dystrophy. *Mol. Ther. Nucleic Acids* 1: e62.

96 Stephenson, M.L. and Zamecnik, P.C. (1978). Inhibition of Rous sarcoma viral RNA translation by a specific oligodeoxyribonucleotide. *Proc. Natl. Acad. Sci. U. S. A.* 75 (1): 285–288.

97 Matsukura, M., Shinozuka, K., Zon, G. et al. (1987). Phosphorothioate analogs of oligodeoxynucleotides: inhibitors of replication and cytopathic effects of human immunodeficiency virus. *Proc. Natl. Acad. Sci. U. S. A.* 84 (21): 7706–7710.

98 Watanabe, T.A., Geary, R.S., and Levin, A.A. (2006). Plasma protein binding of an antisense oligonucleotide targeting human ICAM-1 (ISIS 2302). *Oligonucleotides* 16 (2): 169–180.

99 Kawasaki, A.M., Casper, M.D., Freier, S.M. et al. (1993). Uniformly modified 2′-deoxy-2′-fluoro phosphorothioate oligonucleotides as nuclease-resistant antisense compounds with high affinity and specificity for RNA targets. *J. Med. Chem.* 36 (7): 831–841.

100 Teplova, M., Minasov, G., Tereshko, V. et al. (1999). Crystal structure and improved antisense properties of 2′-*O*-(2-methoxyethyl)-RNA. *Nat. Struct. Biol.* 6 (6): 535–539.

101 Kurreck, J. (2003). Antisense technologies. Improvement through novel chemical modifications. *Eur. J. Biochem.* 270 (8): 1628–1644.

102 Surono, A., Van Khanh, T., Takeshima, Y. et al. (2004). Chimeric RNA/ethylene-bridged nucleic acids promote dystrophin expression in myocytes of duchenne muscular dystrophy by inducing skipping of the nonsense mutation-encoding exon. *Hum. Gene Ther.* 15 (8): 749–757.

103 Petersen, M., Nielsen, C.B., Nielsen, K.E. et al. (2000). The conformations of locked nucleic acids (LNA). *J. Mol. Recognit.* 13 (1): 44–53.

104 Hari, Y., Kodama, T., Imanishi, T., and Obika, S. (2011). 2′-*O*,4′-*C*-methyleneoxymethylene bridged nucleic acids (2′,4′-BNA(COC)). *Methods Mol. Biol.* 764: 31–57.

105 Yamamoto, T., Yasuhara, H., Wada, F. et al. (2012). Superior silencing by 2′,4′-BNA(NC)-based short antisense oligonucleotides compared to 2′,4′-BNA/LNA-based apolipoprotein B antisense inhibitors. *J. Nucleic Acids* 2012: 707323.

106 Nielsen, P.E., Egholm, M., Berg, R.H., and Buchardt, O. (1991). Sequence-selective recognition of DNA by strand displacement with a thymine-substituted polyamide. *Science* 254 (5037): 1497–1500.

107 Abes, S., Turner, J.J., Ivanova, G.D. et al. (2007). Efficient splicing correction by PNA conjugation to an R6-Penetratin delivery peptide. *Nucleic Acids Res.* 35 (13): 4495–4502.

108 Maier, M.A., Esau, C.C., Siwkowski, A.M. et al. (2006). Evaluation of basic amphipathic peptides for cellular delivery of antisense peptide nucleic acids. *J. Med. Chem.* 49 (8): 2534–2542.

109 Wu, B., Li, Y., Morcos, P.A. et al. (2009). Octa-guanidine morpholino restores dystrophin expression in cardiac and skeletal muscles and ameliorates pathology in dystrophic mdx mice. *Mol. Ther.* 17 (5): 864–871.

110 Moulton, H.M. and Moulton, J.D. (2004). Arginine-rich cell-penetrating peptides with uncharged antisense oligomers. *Drug Discov. Today* 9 (20): 870.

111 Moulton, H.M. and Moulton, J.D. (2010). Morpholinos and their peptide conjugates: therapeutic promise and challenge for Duchenne muscular dystrophy. *Biochim. Biophys. Acta* 1798 (12): 2296–2303.

112 Ivanova, G.D., Arzumanov, A., Abes, R. et al. (2008). Improved cell-penetrating peptide-PNA conjugates for splicing redirection in HeLa cells and exon skipping in mdx mouse muscle. *Nucleic Acids Res.* 36 (20): 6418–6428.

113 Yin, H., Saleh, A.F., Betts, C. et al. (2011). Pip5 transduction peptides direct high efficiency oligonucleotide-mediated dystrophin exon skipping in heart and phenotypic correction in mdx mice. *Mol. Ther.* 19 (7): 1295–1303.

114 Betts, C., Saleh, A.F., Arzumanov, A.A. et al. (2012). Pip6-PMO, A new generation of Peptide-oligonucleotide conjugates with improved cardiac exon skipping activity for DMD treatment. *Mol. Ther. Nucleic Acids* 1: e38.

115 Mendell, J.R., Rodino-Klapac, L.R., Sahenk, Z. et al. (2013). Eteplirsen for the treatment of Duchenne muscular dystrophy. *Ann. Neurol.* 74 (5): 637–647.

116 Wu, B., Xiao, B., Cloer, C. et al. (2011). One-year treatment of morpholino antisense oligomer improves skeletal and cardiac muscle functions in dystrophic mdx mice. *Mol. Ther.* 19 (3): 576–583.

117 Sazani, P., Ness, K.P., Weller, D.L. et al. (2011). Repeat-dose toxicology evaluation in cynomolgus monkeys of AVI-4658, a phosphorodiamidate morpholino oligomer (PMO) drug for the treatment of duchenne muscular dystrophy. *Int. J. Toxicol.* 30 (3): 313–321.

118 Aartsma-Rus, A., De Winter, C.L., Janson, A.A. et al. (2005). Functional analysis of 114 exon-internal AONs for targeted DMD exon skipping: indication for steric hindrance of SR protein binding sites. *Oligonucleotides* 15 (4): 284–297.

119 Wilton, S.D., Fletcher, S., McClorey, G. (2005). Antisense oligonucleotides for inducing exon skipping and methods of use thereof. Patent WO2006000057 A1.

120 Aartsma-Rus, A., Houlleberghs, H., van Deutekom, J.C. et al. (2010). Exonic sequences provide better targets for antisense oligonucleotides than splice site sequences in the modulation of Duchenne muscular dystrophy splicing. *Oligonucleotides* 20 (2): 69–77.

121 Gualandi, F., Manzati, E., Sabatelli, P. et al. (2012). Antisense-induced messenger depletion corrects a COL6A2 dominant mutation in Ullrich myopathy. *Hum. Gene Ther.* 23 (12): 1313–1318.

122 Chamberlain, J.S., Metzger, J., Reyes, M. et al. (2007). Dystrophin-deficient mdx mice display a reduced life span and are susceptible to spontaneous rhabdomyosarcoma. *FASEB J.* 21 (9): 2195–2204.

123 Sharp, N.J., Kornegay, J.N., Van Camp, S.D. et al. (1992). An error in dystrophin mRNA processing in golden retriever muscular dystrophy, an animal homologue of Duchenne muscular dystrophy. *Genomics* 13 (1): 115–221.

124 Valentine, B.A., Cooper, B.J., Cummings, J.F., and de Lahunta, A. (1990). Canine X-linked muscular dystrophy: morphologic lesions. *J. Neurol. Sci.* 97 (1): 1–23.

125 Wang, Z., Zhu, T., Qiao, C. et al. (2005). Adeno-associated virus serotype 8 efficiently delivers genes to muscle and heart. *Nat. Biotechnol.* 23 (3): 321–328.

126 Trollet, C., Athanasopoulos, T., Popplewell, L. et al. (2009). Gene therapy for muscular dystrophy: current progress and future prospects. *Expert. Opin. Biol. Ther.* 9 (7): 849–866.

127 Pramono, Z.A., Takeshima, Y., Alimsardjono, H. et al. (1996). Induction of exon skipping of the dystrophin transcript in lymphoblastoid cells by transfecting an antisense oligodeoxynucleotide complementary to an exon recognition sequence. *Biochem. Biophys. Res. Commun.* 226 (2): 445–449.

128 Mann, C.J., Honeyman, K., McClorey, G. et al. (2002). Improved antisense oligonucleotide induced exon skipping in the mdx mouse model of muscular dystrophy. *J Gene Med* 4 (6): 644–654.

129 Lu, Q.L., Mann, C.J., Lou, F. et al. (2003). Functional amounts of dystrophin produced by skipping the mutated exon in the mdx dystrophic mouse. *Nat. Med.* 9 (8): 1009–1014.

130 Aartsma-Rus, A., Janson, A.A., Kaman, W.E. et al. (2003). Therapeutic antisense-induced exon skipping in cultured muscle cells from six different DMD patients. *Hum. Mol. Genet.* 12 (8): 907–914.

131 Aartsma-Rus, A., Janson, A.A., van Ommen, G.J., and van Deutekom, J.C. (2007). Antisense-induced exon skipping for duplications in Duchenne muscular dystrophy. *BMC Med. Genet.* 8: 43.

132 Fall, A.M., Johnsen, R., Honeyman, K. et al. (2006). Induction of revertant fibres in the mdx mouse using antisense oligonucleotides. *Genet. Vaccines Ther.* 4: 3.

133 Aoki, Y., Yokota, T., Nagata, T. et al. (2012). Bodywide skipping of exons 45–55 in dystrophic mdx52 mice by systemic antisense delivery. *Proc. Natl. Acad. Sci. U. S. A.* 109 (34): 13763–13768.

134 Gebski, B.L., Mann, C.J., Fletcher, S., and Wilton, S.D. (2003). Morpholino antisense oligonucleotide induced dystrophin exon 23 skipping in mdx mouse muscle. *Hum. Mol. Genet.* 12 (15): 1801–1811.

135 Arechavala-Gomeza, V., Graham, I.R., Popplewell, L.J. et al. (2007). Comparative analysis of antisense oligonucleotide sequences for targeted skipping of exon 51 during dystrophin pre-mRNA splicing in human muscle. *Hum. Gene Ther.* 18 (9): 798–810.

136 Popplewell, L.J., Adkin, C., Arechavala-Gomeza, V. et al. (2010). Comparative analysis of antisense oligonucleotide sequences targeting exon 53 of the human DMD gene: Implications for future clinical trials. *Neuromuscul. Disord.* 20 (2): 102–110.

137 Heemskerk, H., de Winter, C., van Kuik, P. et al. (2010). Preclinical PK and PD studies on 2′-*O*-methyl-phosphorothioate RNA antisense oligonucleotides in the mdx mouse model. *Mol. Ther.* 18 (6): 1210–1217.

138 Lu, Q.L., Rabinowitz, A., Chen, Y.C. et al. (2005). Systemic delivery of antisense oligoribonucleotide restores dystrophin expression in body-wide skeletal muscles. *Proc. Natl. Acad. Sci. U. S. A.* 102 (1): 198–203.

139 Verhaart, I.E., Tanganyika-de Winter, C.L., Karnaoukh, T.G. et al. (2013). Dose-dependent pharmacokinetic profiles of 2′-*O*-methyl phosphorothioate antisense oligonucleotides in mdx mice. *Nucleic Acid Ther.* 23 (3): 228–237.

140 Deconinck, A.E., Rafael, J.A., Skinner, J.A. et al. (1997). Utrophin-dystrophin-deficient mice as a model for Duchenne muscular dystrophy. *Cell* 90 (4): 717–727.

141 Zhou, L., Rafael-Fortney, J.A., Huang, P. et al. (2008). Haploinsufficiency of utrophin gene worsens skeletal muscle inflammation and fibrosis in mdx mice. *J. Neurol. Sci.* 264 (1–2): 106–111.

142 Bremmer-Bout, M., Aartsma-Rus, A., de Meijer, E.J. et al. (2004). Targeted exon skipping in transgenic hDMD mice: a model for direct preclinical screening of human-specific antisense oligonucleotides. *Mol. Ther.* 10 (2): 232–240.

143 Tanganyika-de Winter, C.L., Heemskerk, H., Karnaoukh, T.G. et al. (2012). Long-term exon skipping studies with 2′-*O*-methyl phosphorothioate antisense oligonucleotides in dystrophic mouse models. *Mol. Ther. Nucleic Acids.* 1: e44.

144 van Putten, M., Kumar, D., Hulsker, M., Hoogaars, W.M., Plomp, J.J., van Opstal, A., van Iterson, M., Admiraal, P., van Ommen, G.J., t Hoen, P.A., and Aartsma-Rus, A. (2012). Comparison of skeletal muscle pathology and motor function of dystrophin and utrophin deficient mouse strains. *Neuromuscul. Disord.* 22 (5): 406–417.

145 t Hoen, P.A., de Meijer, E.J., Boer, J.M. et al. (2008). Generation and characterization of transgenic mice with the full-length human DMD gene. *J. Biol. Chem.* 283 (9): 5899–5907.

146 Heemskerk, H.A., de Winter, C.L., de Kimpe, S.J. et al. (2009). In vivo comparison of 2′-*O*-methyl phosphorothioate and morpholino antisense oligonucleotides for Duchenne muscular dystrophy exon skipping. *J. Gene Med.* 11 (3): 257–266.

147 Heemskerk, H., de Winter, C.L., van Ommen, G.J. et al. (2009). Development of antisense-mediated exon skipping as a treatment for Duchenne muscular dystrophy. *Ann. N. Y. Acad. Sci.* 1175: 71–79.

148 Wu, B., Benrashid, E., Lu, P. et al. (2011). Targeted skipping of human dystrophin exons in transgenic mouse model systemically for antisense drug development. *PLoS One* 6 (5): e19906.

149 Goemans, N.M., Tulinius, M., van den Hauwe, M. et al. (2016). Long-term efficacy, safety, and pharmacokinetics of drisapersen in Duchenne muscular dystrophy: results from an open-label extension study. *PLoS One* 11 (9): e0161955.

150 Goemans, N., Klingels, K., van den Hauwe, M. et al. (2013). Test-retest reliability and developmental evolution of the 6-min walk test in Caucasian boys aged 5–12 years. *Neuromuscul. Disord.* 23 (1): 19–24.

151 Voit, T., Topaloglu, H., Straub, V. et al. (2014). Safety and efficacy of drisapersen for the treatment of Duchenne muscular dystrophy (DEMAND II): an exploratory, randomised, placebo-controlled phase 2 study. *Lancet Neurol.* 13 (10): 987–996.

152 Amantana, A. and Iversen, P.L. (2005). Pharmacokinetics and biodistribution of phosphorodiamidate morpholino antisense oligomers. *Curr. Opin. Pharmacol.* 5 (5): 550–555.

153 Alter, J., Lou, F., Rabinowitz, A. et al. (2006). Systemic delivery of morpholino oligonucleotide restores dystrophin expression bodywide and improves dystrophic pathology. *Nat. Med.* 12 (2): 175–177.

154 Wu, B., Lu, P., Benrashid, E. et al. (2010). Dose-dependent restoration of dystrophin expression in cardiac muscle of dystrophic mice by systemically delivered morpholino. *Gene Ther.* 17 (1): 132–140.

155 Malerba, A., Thorogood, F.C., Dickson, G., and Graham, I.R. (2009). Dosing regimen has a significant impact on the efficiency of morpholino oligomer-induced exon skipping in mdx mice. *Hum. Gene Ther.* 20 (9): 955–965.

156 McClorey, G., Moulton, H.M., Iversen, P.L. et al. (2006). Antisense oligonucleotide-induced exon skipping restores dystrophin expression in vitro in a canine model of DMD. *Gene Ther.* 13 (19): 1373–1381.

157 Yokota, T., Lu, Q.L., Partridge, T. et al. (2009). Efficacy of systemic morpholino exon-skipping in Duchenne dystrophy dogs. *Ann. Neurol.* 65 (6): 667–676.

158 Mendell, J.R., Goemans, N., Lowes, L.P. et al. (2016). Longitudinal effect of eteplirsen versus historical control on ambulation in Duchenne muscular dystrophy. *Ann. Neurol.* 79 (2): 257–271.

159 Amantana, A., Moulton, H.M., Cate, M.L. et al. (2007). Pharmacokinetics, biodistribution, stability and toxicity of a cell-penetrating peptide-morpholino oligomer conjugate. *Bioconjug. Chem.* 18 (4): 1325–1331.

160 Yin, H., Betts, C., Saleh, A.F. et al. (2010). Optimization of peptide nucleic acid antisense oligonucleotides for local and systemic dystrophin splice correction in the mdx mouse. *Mol. Ther.* 18 (4): 819–827.

161 Wu, B., Moulton, H.M., Iversen, P.L. et al. (2008). Effective rescue of dystrophin improves cardiac function in dystrophin-deficient mice by a modified morpholino oligomer. *Proc. Natl. Acad. Sci. U. S. A.* 105 (39): 14814–14819.

162 Yin, H., Lu, Q., and Wood, M. (2008). Effective exon skipping and restoration of dystrophin expression by peptide nucleic acid antisense oligonucleotides in mdx mice. *Mol. Ther.* 16 (1): 38–45.

163 Nelson, M.H., Stein, D.A., Kroeker, A.D. et al. (2005). Arginine-rich peptide conjugation to morpholino oligomers: effects on antisense activity and specificity. *Bioconjug. Chem.* 16 (4): 959–966.

164 Abes, R., Moulton, H.M., Clair, P. et al. (2008). Delivery of steric block morpholino oligomers by (R-X-R)4 peptides: structure-activity studies. *Nucleic Acids Res.* 36 (20): 6343–6354.

165 Wu, R.P., Youngblood, D.S., Hassinger, J.N. et al. (2007). Cell-penetrating peptides as transporters for morpholino oligomers: effects of amino acid composition on intracellular delivery and cytotoxicity. *Nucleic Acids Res.* 35 (15): 5182–5191.
166 Fletcher, S., Honeyman, K., Fall, A.M. et al. (2007). Morpholino oligomer-mediated exon skipping averts the onset of dystrophic pathology in the mdx mouse. *Mol. Ther.* 15 (9): 1587–1592.
167 Moulton, H.M., Fletcher, S., Neuman, B.W. et al. (2007). Cell-penetrating peptide-morpholino conjugates alter pre-mRNA splicing of DMD (Duchenne muscular dystrophy) and inhibit murine coronavirus replication in vivo. *Biochem. Soc. Trans.* 35 (Pt 4): 826–828.
168 Jearawiriyapaisarn, N., Moulton, H.M., Buckley, B. et al. (2008). Sustained dystrophin expression induced by peptide-conjugated morpholino oligomers in the muscles of mdx mice. *Mol. Ther.* 16 (9): 1624–1629.
169 Yin, H., Moulton, H.M., Seow, Y. et al. (2008). Cell-penetrating peptide-conjugated antisense oligonucleotides restore systemic muscle and cardiac dystrophin expression and function. *Hum. Mol. Genet.* 17 (24): 3909–3918.
170 Goyenvalle, A., Babbs, A., Powell, D. et al. (2010). Prevention of dystrophic pathology in severely affected dystrophin/utrophin-deficient mice by morpholino-oligomer-mediated exon-skipping. *Mol. Ther.* 18 (1): 198–205.
171 Yin, H., Moulton, H.M., Betts, C. et al. (2009). A fusion peptide directs enhanced systemic dystrophin exon skipping and functional restoration in dystrophin-deficient mdx mice. *Hum. Mol. Genet.* 18 (22): 4405–4414.
172 Sazani, P. (2009). Treat-NMD/NIH Conference, Brussel, Belgium (17–19 November 2009).
173 Wu, B., Lu, P., Cloer, C. et al. (2012). Long-term rescue of dystrophin expression and improvement in muscle pathology and function in dystrophic mdx mice by peptide-conjugated morpholino. *Am. J. Pathol.* 181 (2): 392–400.
174 Ivanova, G.D., Fabani, M.M., Arzumanov, A.A. et al. (2008). PNA-peptide conjugates as intracellular gene control agents. *Nucleic Acids Symp. Ser. (Oxf)* (52): 31–32.
175 Douglas, A.G. and Wood, M.J. (2013). Splicing therapy for neuromuscular disease. *Mol. Cell. Neurosci.* 56: 169–185.
176 Hamilton, G. and Gillingwater, T.H. (2013). Spinal muscular atrophy: going beyond the motor neuron. *Trends Mol. Med.* 19 (1): 40–50.
177 Shababi, M., Mattis, V.B., and Lorson, C.L. (2010). Therapeutics that directly increase SMN expression to treat spinal muscular atrophy. *Drug News Perspect.* 23 (8): 475–482.
178 Hua, Y., Vickers, T.A., Baker, B.F. et al. (2007). Enhancement of SMN2 exon 7 inclusion by antisense oligonucleotides targeting the exon. *PLoS Biol.* 5 (4): e73.

179 Skordis, L.A., Dunckley, M.G., Yue, B. et al. (2003). Bifunctional antisense oligonucleotides provide a trans-acting splicing enhancer that stimulates SMN2 gene expression in patient fibroblasts. *Proc. Natl. Acad. Sci. U. S. A.* 100 (7): 4114–4119.

180 Singh, N.K., Singh, N.N., Androphy, E.J., and Singh, R.N. (2006). Splicing of a critical exon of human survival motor neuron is regulated by a unique silencer element located in the last intron. *Mol. Cell. Biol.* 26 (4): 1333–1346.

181 Hua, Y., Vickers, T.A., Okunola, H.L. et al. (2008). Antisense masking of an hnRNP A1/A2 intronic splicing silencer corrects SMN2 splicing in transgenic mice. *Am. J. Hum. Genet.* 82 (4): 834–848.

182 Singh, N.N., Shishimorova, M., Cao, L.C. et al. (2009). A short antisense oligonucleotide masking a unique intronic motif prevents skipping of a critical exon in spinal muscular atrophy. *RNA Biol.* 6 (3): 341–350.

183 Porensky, P.N., Mitrpant, C., McGovern, V.L. et al. (2012). A single administration of morpholino antisense oligomer rescues spinal muscular atrophy in mouse. *Hum. Mol. Genet.* 21 (7): 1625–1638.

184 Mitrpant, C., Porensky, P., Zhou, H. et al. (2013). Improved antisense oligonucleotide design to suppress aberrant SMN2 gene transcript processing: towards a treatment for spinal muscular atrophy. *PLoS One* 8 (4): e62114.

185 Zhou, H., Janghra, N., Mitrpant, C. et al. (2013). A novel morpholino oligomer targeting ISS-N1 improves rescue of severe spinal muscular atrophy transgenic mice. *Hum. Gene Ther.* 24 (3): 331–342.

186 Chiriboga, C.A., Swoboda, K.J., Darras, B.T. et al. (2016). Results from a phase 1 study of nusinersen (ISIS-SMN(Rx)) in children with spinal muscular atrophy. *Neurology* 86 (10): 890–897.

187 Aartsma-Rus, A., den Dunnen, J.T., and van Ommen, G.J. (2010). New insights in gene-derived therapy: the example of Duchenne muscular dystrophy. *Ann. N. Y. Acad. Sci.* 1214: 199–212.

188 Swoboda, K.J., Prior, T.W., Scott, C.B. et al. (2005). Natural history of denervation in SMA: relation to age, SMN2 copy number, and function. *Ann. Neurol.* 57 (5): 704–712.

189 Khatri, I.A., Chaudhry, U.S., Seikaly, M.G. et al. (2008). Low bone mineral density in spinal muscular atrophy. *J. Clin. Neuromuscul. Dis.* 10 (1): 11–17.

190 Lourbakos, A., Beekman, C., Holling, T. et al. (2013). Bioanalysis of a double blind, placebo-controlled clinical phase 2 study of drisapersen for the treatment of boys suffering from Duchenne muscular dystrophy and comparison to clinical outcome results. *Neuromuscul. Disord.* 23 (9): 805.

191 Nadarajah, V.D., van Putten, M., Chaouch, A. et al. (2011). Serum matrix metalloproteinase-9 (MMP-9) as a biomarker for monitoring disease progression in Duchenne muscular dystrophy (DMD). *Neuromuscul. Disord.* 21 (8): 569–578.

192 Zaharieva, I.T., Calissano, M., Scoto, M. et al. (2013). Dystromirs as serum biomarkers for monitoring the disease severity in duchenne muscular dystrophy. *PLoS One* 8 (11): e80263.

193 Straub, V., Balabanov, P., Bushby, K. et al. (2016). Stakeholder cooperation to overcome challenges in orphan medicine development: the example of Duchenne muscular dystrophy. *Lancet Neurol.* 15 (8): 882–890.

18

CMC Aspects for the Clinical Development of Spiegelmers

Stefan Vonhoff

NOXXON Pharma AG, Berlin, Germany

18.1 Introduction

The concept of Spiegelmers, mirror-imaged, L-configured aptamers, as new class of oligonucleotide therapeutics with enhanced biostability was conceived in the mid-1990s at the Free University of Berlin. Unlike oligonucleotides targeting intracellular RNA, such as antisense oligonucleotides or short interfering RNA, Spiegelmers, like aptamers, can be designed to bind with high affinity and specificity to almost any molecular entity ranging from small molecules to large macromolecules such as nucleic acids or proteins. For therapeutic or diagnostic applications, target structures can be in the intra- or extracellular compartment and are typically unbound to cell surfaces. From the list of Spiegelmers that have been generated against a variety of biologically active molecules, three compounds targeting the chemokines CCL2/MCP-1, CXCL12/SDF-1, and the extracellular iron regulatory peptide hepcidin were selected for pharmaceutical development and have entered clinical phase 2 for the treatment of diabetic nephropathy, hematological cancers, and anemia of cancer, respectively. Using the three Spiegelmer® lead candidates as examples, this chapter will present an update on the status of preclinical and clinical development and the accompanying chemistry, manufacturing, and controls (CMC) activities for this class of therapeutic oligonucleotides.

Oligonucleotide-Based Drugs and Therapeutics: Preclinical and Clinical Considerations for Development, First Edition. Edited by Nicolay Ferrari and Rosanne Seguin.

18.2 Technology (Mirror-imaged SELEX Process) Selected Pharmaceutical Properties

Since the discovery of the genetic code, the understanding of the functional role of DNA and RNA oligonucleotides in controlling biochemical processes has progressed fundamentally and is still evolving as illustrated by the discovery of siRNA and miRNA in the last decade. While some functional activities rely solely on sequence information of the oligonucleotide strand, others also require a three-dimensional structure to elicit their activity. This feature was first described for the ribosome and later also for ribozymes and riboswitches [1–4]. The ribozymes Angiozyme™ and Herzyme™ were the first RNA nucleic acids pursued for therapeutic applications by Ribozyme Pharmaceuticals in the 1990s [5]. Around the same time, another class of oligonucleotides acting through their three-dimensional structure, the aptamers, were isolated by Tuerk and Gold [6] using the *in vitro* selection of target-binding nucleic acids, also known as SELEX [7, 8]. The classic definition of an aptamer (from *aptus* Greek, to fit) is a single-stranded RNA or DNA oligonucleotide having the ability to form unique secondary and tertiary structures that confer high binding affinity and specificity for its target molecule. In this function, aptamers are the nucleic acid equivalent to protein-based antibodies, and thus it is not surprising that aptamers have found applications in the analytical, diagnostic, and therapeutic arenas [9, 10].

From the multitude of aptamers that have been generated by employing the SELEX process, about a dozen have entered clinical development in recent years, and one – Macugen™ – reached market approval in 2005 [10–12]. One reason for the slow progress and low success rate is the required differentiation from the competing and well-established field of antibodies. In this context it is interesting to note that the VEGF-binding aptamer Macugen™ was developed and approved for the treatment of age-related macular degeneration 2 years ahead of the competing antibody ranibizumab, Lucentis™, which together with other antibodies has taken the majority of the market share for the indication [13].

Another reason for the slow advancement of aptamers in drug development can be found in the technological challenges of identifying an aptamer against a given target that additionally needs to possess suitable therapeutic properties for the specific indication. In this context, one parameter, the stabilization of the D-configured aptamer against enzymatic degradation by nucleases, is pivotal for any further pharmaceutical development. Stabilization is achieved either by applying 2′-*O*-methyl- or 2′-fluoro-modified NTPs in the SELEX process [14, 15] or by point-mutating the aptamer sequence identified in the SELEX process by introducing enzymatically stable 2′-*O*-methyl nucleosides or 2′-fluoro pyrimidines [16]. Other post-SELEX modifications conveying enzymatic stability are the introduction of phosphorothioates or the capping of the aptamer strand by inverted thymidine or polyethylene glycol (PEG), which is also the commonly chosen modification for prolonging plasma half-life of the systemically active therapeutic aptamer.

An elegant way to introduce stability against enzymatic degradation is the Spiegelmer technology discovered in the mid-1990s at the Free University of Berlin [17–19]. Here, a D-configured aptamer is selected against the chemically synthesized enantiomer of the natural target, usually a D-peptide or protein. This D-configured pair of aptamer and protein exhibits identical binding properties as their L-configured enantiomers, which are the natural peptide/protein, and the L-aptamer, which is called a Spiegelmer (Figure 18.1).

Apart from increasing the enzymatic stability, the use of L-nucleic acids leads to some additional favorable therapeutic properties for Spiegelmers. Since complementary strands of D- and L-nucleic acids show no or greatly reduced binding affinity [20], the risk of side reactions, triggered by the Spiegelmer binding to natural nucleic acid sequences, is minimized. Also, the native immune system that recognizes exogenous D-nucleic acids through Toll-like receptors like TLR3, 8, and 9 is not activated by any of the Spiegelmers tested

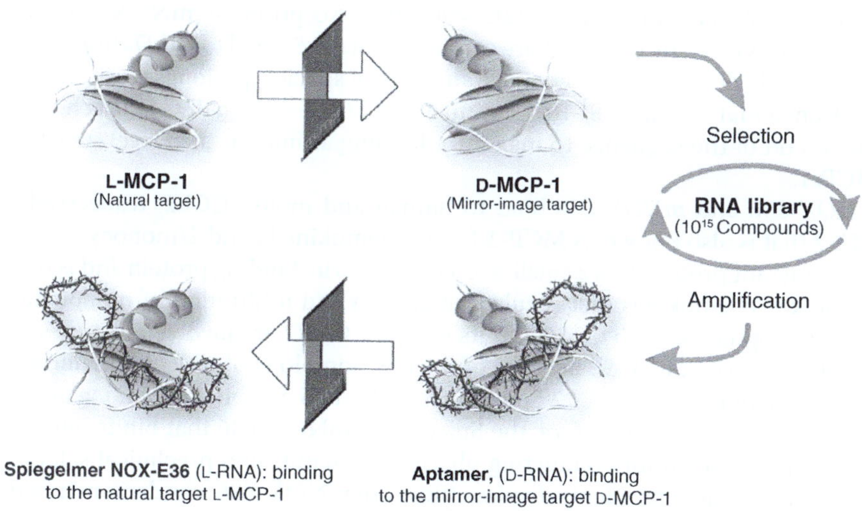

Figure 18.1 Illustration of the mirror-imaged SELEX process using the selection against the chemokine MCP-1 as example. Using a randomly synthesized RNA library of 10^{15} different sequences, or three-dimensional shapes, the selection is conducted against the chemically synthesized enantiomer of the natural target, i.e. D-MCP-1. Binding sequences are isolated and amplified for a new round of selection. After several rounds of selection during which the stringency is gradually increased, the sequences of the binding aptamers are identified and synthesized as L-enantiomers or Spiegelmers. The Spiegelmer sequences are optimized and characterized for their targeted therapeutic application by truncation and point mutations. At the end of the optimization process, the MCP-1 binding Spiegelmer NOX-E36 is identified.

so far [21]. Moreover, antibody formation against Spiegelmers has only been achieved in a rabbit model under challenging conditions of KHL conjugation demonstrating the low immunogenicity of Spiegelmers. This low immunogenicity of L-nucleic acids has been confirmed during all of the various clinical studies and is a differentiating feature of Spiegelmers from that of pharmaceutical antibodies.

18.3 Preclinical Efficacy Data for Spiegelmers

Table 18.1 lists a selection of Spiegelmers for which *in vivo* efficacy has been demonstrated in animal models. While the Spiegelmers in clinical development, i.e. NOX-E36, NOX-A12, and NOX-H94, are L-ribonucleic acids, some of the more recent preclinical candidates such as NOX-D19 or NOX-S93 contain point mutations of deoxynucleotides in the L-RNA sequence or, as in the case of the glucagon-binding Spiegelmer NOX-G15, consist entirely of L-DNA nucleotides. The choice of ribo- versus deoxynucleotide is driven by the optimization of binding affinity and can additionally lead to a shortening of the sequence of the Spiegelmer [22]. With the exception of mNOX-E36, all Spiegelmers listed in Table 18.1 are conjugated at the 5′- end to 40 kDa branched PEG, which, depending on sequence and animal species, prolongs plasma half-life from about 30 min to 8–10 h. For mNOX-E36 the 40 kDa PEG is attached at the 3′-end of the sequence to maximize binding affinity to the target, murine MCP-1.

NOX-E36 and *mNOX-E36* bind to human and mouse CCL2, respectively. CCL2 that is also known as MCP-1 (C—C chemokine ligand 2/monocyte chemoattractant protein 1) is a small, secreted, heparin-binding protein and is one of the key chemokines that regulate migration and infiltration of monocytes and macrophages. Both CCL2 and its receptor CCR2 are involved in a variety of inflammatory diseases. Because of the low homology of rodent and human CCL2, the human anti-CCL2 Spiegelmer NOX-E36 exhibits no binding affinity to mouse CCL2. Therefore, the Spiegelmer mNOX-E36 that binds murine CCL2 was selected for conducting pharmacodynamic and preclinical safety–toxicology studies in rodents. *In vivo* efficacy of mNOX-E36 was demonstrated in models of lupus nephritis, diabetic nephropathy, and chronic hepatic injury (Table 18.1).

In a lupus nephritis model with autoimmune-prone MRL$^{lpr/lpr}$ mice that were treated with mNOX-E36 from 14 to 24 weeks of age, prolonged survival associated with a robust improvement of lupus nephritis, peribronchial inflammation, and lupus-like inflammatory skin lesions was observed [21]. In a follow-up study using the same animal model, a 75% dose reduction of cyclophosphamide for controlling diffuse proliferative lupus nephritis and pneumonitis was observed when using mNOX-E36 [23]. Thus, inhibition of

Table 18.1 Summary of preclinical efficacy data for various Spiegelmers.

Spiegelmer	Target	Chemistry	Binding affinity[a] or activity	Demonstrated efficacy	References
mNOX-E36	Mouse CCL2/MCP-1	L-RNA	NOX-E36 to human MCP1: $K_D = 1.4\,nM$	Robust improvement of *lupus nephritis*, peribronchial inflammation, and lupus-like inflammatory skin lesions in a lupus mouse model with autoimmune MRLlpr/lpr mice	[21]
				75% dose reduction of cyclophosphamide for controlling diffuse proliferative *lupus nephritis* and pneumonitis in a lupus mouse model with autoimmune MRLlpr/lpr mice	[23]
				Reduction of glomerular macrophages, protection from diffuse glomerulosclerosis, and improvement of glomerular filtration rate in a model for *diabetic nephropathy* using uninephrectomized (1 K) db/db mice	[24]
				Acceleration of liver fibrosis regression and reduction of liver macrophage infiltration and steatohepatitis in mouse models of *chronic hepatic injury*. Inhibition of CCL2-dependent monocytes promoting inflammation-associated angiogenesis during fibrosis progression	[25–27]
				Inhibition of CCL2 reduces monocyte infiltration in acetaminophen-induced acute liver injury in mice	[28]
				Increase of graft survival of allogeneic heart transplants from BALB/c to C57BL/6 mice through a synergistic effect of mNOX-E36 with sub-therapeutic doses of cyclosporine A	[29]

(Continued)

Table 18.1 (Continued)

Spiegelmer	Target	Chemistry	Binding affinity[a] or activity	Demonstrated efficacy	References
NOX-A12	CXCL12/SDF-1	L-RNA	Human, mouse, and rat SDF-1α and SDF-1β: $K_D = 250\,pM$	***Mobilization of white blood cells*** **and hematopoietic stem and progenitor cells into peripheral blood of mice**	[30]
				In vitro mobilization and chemosensitization of chronic lymphocytic leukemia (CLL) cells	[31, 32]
				In vivo enhancement of multimyeloma (MM) cell sensitivity to bortezomib in a mouse model Induction of MM cell mobilization from the bone marrow to the peripheral blood	[33]
				Increase of T-cell and NK cell infiltration and synergistic effect with immune checkpoint blockade and NK cell-mediated antibody-dependent cellular cytotoxicity in tumor–stroma spheroids modeling tumor microenvironment	[34]
				Prolongation of life span and regression of tumor in an *autochthonous brain tumor model* in rats after irradiation followed by NOX-A12 treatment	[35]
				Reduction of mast cells, CD68(+) macrophages, and CD3(+) T cells in a *pulmonary arterial hypotension* model using human lung tissue	[36]
				Reduction of glomerulosclerosis, increase of podocyte numbers, and prevention of onset of albuminuria in a model of *diabetic nephropathy* using uninephrectomized db/db mice	[37]
mNOX-E36 and NOX-A12	Mouse CCL2 CXCL12	L-RNA	See above	Additive effect of CCL2 and CXCL12 blockage through mNOX-E36 and NOX-A12 in uninephrectomized db/db mice leading to suppressed proteinuria and increased glomerular filtration in a model of *diabetic nephropathy*	[38]

				Dual blockade of the proinflammatory chemokine CCl2 by mNOX-E36 and of the homeostatic chemokine CXCL12 by NOX-A12 in a murine proliferative lupus nephritis model is as effective as cyclophosphamide	[39]
NOX-H94	Hepcidin	L-RNA	$K_D = 0.65\,nM$	Prevention of serum iron reduction by NOX-H94 in an acute cynomolgus monkey model of *IL-6 -induced hypoferremia*	[40]
				Inhibition of decrease in Hb concentration in a subchronic model of *IL-6-induced anemia* in cynomolgus monkeys	
NOX-S93	S1P (sphingosine-1-phosphate)	L-RNA spiked with L-DNA	$K_D = 5\,nM$	Inhibits S1P, VEGF, and FGF 2- and IGF-induced angiogenesis in an assay employing human endothelial cells. In a murine xenograft model reduction of homing of human cancer cells to prometastatic bone marrow niches induced by whole body irradiation	[42, 43]
NOX-G15	Glucagon	L-DNA spiked with L-RNA	$K_D = 3.0\,nM$	Reduction of hyperglycemia after glucose challenge in *mouse models of type 1 and type 2 diabetes*	[44]
NOX-D19	C5a (complement component 5a)	L-RNA	K_D (huC5a) = 1.4 nM K_D (mC5a) = 51 pM	Reduced rejection of allograft airway transplants in mouse model	[45]
NOX-D20	C5a	L-RNA spiked with L-DNA	K_D (huC5a) = 0.3 nM K_D (mC5a) = 19 pM	Prolonged survival and reduction of liver and kidney damage in an experimental *murine model of polymicrobial septic shock*	[46]
NOX-L41	CGRP	L-RNA spiked with L-DNA	IC_{50} (huCGRP) 0.39 nM	Inhibition of neurogenic meningeal plasma protein extravasation in rats	[48]
			IC_{50} (ratCGRP) 3.3 nM	Increase of meningeal blood flow for a less active CGRP-binding Spiegelmer NOX-C89 in *a mouse model of migraine*	[49]

[a] As determined by binding competition assay.

CCL2 with the Spiegelmer NOX-E36 suggests a novel strategy for the treatment of autoimmune tissue injury, such as lupus nephritis, and could also reduce the side effects of the standard therapy using cyclophosphamide, which is known to produce myelosuppression and T-cell depletion.

In a model for diabetic nephropathy using uninephrectomized (1 K) db/db mice with advanced glomerulopathy, the subcutaneous dosing of mNOX-E36 (50 mg/kg three times per week) reduced the number of glomerular macrophages by 40% when compared with the control group [24]. This finding was associated with protection from diffuse glomerulosclerosis and significantly improved glomerular filtration rate (GFR) and prompted the clinical profiling of NOX-E36 for the therapeutic indication of diabetic nephropathy.

Finally, mNOX-E36 was shown to accelerate liver fibrosis regression and to diminish liver macrophage infiltration and steatohepatitis in mouse models for chronic hepatic injury [25, 26]. Employing microcomputed tomography, it was demonstrated that mNOX-E36 inhibits inflammation-associated angiogenesis that is promoted by CCL2-dependent monocytes during liver fibrosis progression [27]. In a model of acetaminophen-induced acute liver injury (AILI) in mice, early inhibition of CCL2 by mNOX-E36 reduced monocyte infiltration and did not interfere with any repair processes during resolution from injury, thereby pointing to the potential to reduce the inflammatory reaction during the early phase of AILI [28].

mNOX-E36 has also been investigated in other murine disease models such as adjunct immunosuppressant in transplantation [29].

NOX-A12 binds to the chemokine CXCL12 (stromal cell-derived factor-1 [SDF-1]) with picomolar affinity, thereby inhibiting CXCL12 signaling on both of its receptors, CXCR4 and CXCR7. The chemokine ligand CXCL12 is a key regulatory element in the homing and retention of hematopoietic stem cells in the bone marrow. Additionally, the CXCL12/CXCR4/CXCR7 axis has been shown to play a role in neo-vasculogenesis, tumor growth, metastasis, and chemo-desensitization of tumor cells [30].

In a first efficacy model in mice, NOX-A12 was shown to mobilize white blood cells (WBCs) and hematopoietic stem and progenitor cells (HSCs) into the peripheral blood [30]. The positive efficacy data prompted the evaluation of NOX-A12 in a clinical phase 1 study where long-lasting and dose-dependent WBC and HSC mobilization was observed in healthy volunteers (HV) (cf. to Table 18.2).

With the successful mobilization of white blood and hematopoietic stem cells, the effect of NOX-A12 on chronic lymphocytic leukemia (CLL) and multiple myeloma (MM) cell migration and drug sensitivity was explored [31–33]. An *in vitro* model demonstrated that NOX-A12 efficiently inhibited CXCL12-induced chemotaxis of CLL cells and interfered with stromal cell-mediated resistance to chemotherapeutic agents such as bendamustine or fludarabine, thus providing a rationale for the clinical development of NOX-A12 in

Table 18.2 Clinical studies conducted with Spiegelmers.

Spiegelmer	Clinical phase/ indication	Study objective	Study design	Dosage regiment	HV or patients	Route of administration	ID[a] on ClinicalTrials.gov
Emapticap pegol (NOX-E36)	Phase 1	First-in-human safety, PK, PD	Randomized placebo controlled	Dose-escalation single dose	Healthy volunteers (HV)	s.c. or i.v.	NCT00976729
	Phase 1	Safety in diabetic patients; multiple-dose PK; multiple-dose PD	Randomized placebo controlled	Dose-escalation Q2D for 27 days	HV and type 2 diabetes	i.v.	NCT01085292
	Phase 1	Evaluation of PK in renally impaired patients	Open label	Single dose	Various grades of renally insufficient patients	s.c.	NCT01372124
	Phase 2/diabetic nephropathy	12-week efficacy and safety; PoC in albuminuric patients	Randomized placebo controlled	Two fixed doses of 0.5 mg/kg per week for 12 weeks	Type 2 diabetes	s.c.	NCT01547897
Olaptesed pegol (NOX-A12)	Phase 1	First-in-human safety, PK, PD	Open label	Dose-escalation single dose	HV	i.v.	NCT00976378
	Phase 1	Safety in HV; multiple-dose PK; multiple-dose PD	Open label	2.0 mg/kg and 4.0 mg/kg every day for 5 days	HV	i.v.	NCT01194934

(Continued)

Table 18.2 (Continued)

Spiegelmer	Clinical phase/ indication	Study objective	Study design	Dosage regiment	HV or patients	Route of administration	ID[a] on ClinicalTrials.gov
	Phase 2/ hematological tumors, CLL	Safety/ tolerability; overall response rate	Open label	Six cycles of max. 4 mg/kg in combination with bendamustine and rituximab	Relapsed CLL patients	i.v.	NCT01486797
	Phase 2/ hematological tumors, MM	Safety/ tolerability; overall response rate	Open label	Eight cycles of max. 4 mg/kg in combination with bortezomib and dexamethasone	Relapsed MM patients	i.v.	NCT01521533
	Phase 2 combination study with pembrolizumabin solid tumors, metastatic colorectal and pancreatic cancer	Safety/ tolerability; disease control rate	Open label	300 mg olaptesed pegol in combination with 200 mg pembrolizumab every 3 weeks for up to 24 months	Patients with metastatic stage IV colorectal or pancreatic cancer with liver metastases after two or more lines of prior treatment	i.v.	NCT03168139
Lexaptepid pegol (NOX-H94)	Phase 1	First-in-human safety, PK, PD	Randomized, placebo controlled	Dose-escalation single dose and repeated dose (q2d)	HV	s.c. or i.v.	NCT01372137

Phase 1b	Proof-of-mechanism study in the human endotoxemia model; safety, PK, PD	Open label, placebo controlled	Single dose	HV	i.v.	NCT01522794
Phase 2a/anemia in cancer patients	Proof-of-concept study; safety, PK, PD	Open label	Two doses per week	Anemic patients with hematological or solid tumors	i.v.	NCT01691040
Phase 2a/anemia in dialysis patients	Proof-of-concept study; safety, PK, PD	Randomized, placebo controlled	Two doses per week	ESA-hyporesponsive dialysis patients	i.v.	NCT02079896

[a] Detailed information on all clinical studies can be retrieved under the identifier at ClinicalTrials.gov.

combination with conventional therapy in CLL. The analogous rational for the combination of NOX-A12 with standard chemotherapy as treatment for MM was provided by the enhancement of MM cell sensitivity to bortezomib in a mouse model. In this model NOX-A12 induced MM cell mobilization from the bone marrow to the peripheral blood [33].

Inhibition of CXCL12 by NOX-A12 also effects tumor microenvironment. In tumor–stroma spheroids, NOX-A12 increased T-cell and NK cell infiltration and showed a synergistic effect with checkpoint inhibitors or CAR-T cells as well as NK cell-based therapeutics [34].

The effect of NOX-A12 on inhibiting neovascularization by blocking the interaction of CXCL12 with its receptor CXCR4 was investigated in an autochthonous brain tumor model in rats after irradiation [35]. In this model, NOX-A12 inhibited tumor recurrences and increased the life span in irradiated rats when compared with irradiation alone or with the addition of temozolomide, thus pointing toward the application of NOX-A12 as new treatment of glioblastoma.

Inhibiting the CXCL12 signaling pathway, NOX-A12 is also a useful tool for elucidating the role of the chemokine in immune and inflammatory processes such as the pathogenesis of idiopathic pulmonary arterial hypotension (PAH) [36]. In a model with lung tissue sections from patients suffering from idiopathic PAH and healthy donors, it was shown that blocking CXCL12 with NOX-A12 reduced the infiltration of mast cells, CD68(+) macrophages, and CD3(+) T cells when compared with the control group. The reduction was associated with an improvement of hemodynamics and pulmonary vascular remodeling, which suggests a novel approach for the treatment of PAH.

In a model of diabetic nephropathy using uninephrectomized db/db mice, Sayyed et al. demonstrated that blocking CXCL12 with the Spiegelmer NOX-A12 reduced the degree of glomerulosclerosis, increased the number of podocytes, and prevented the onset of albuminuria [37]. Since both chemokines, CCL2 and CXCL12, contribute to glomerulosclerosis in mice with type 2 diabetes mellitus, it was subsequently investigated whether a combination of the CCL2- and CXCL12-binding Spiegelmers mNOX-E36 and NOX-A12 would have a synergistic protective effect on the progression of diabetic nephropathy [38]. Using the aforementioned mouse model, it was demonstrated that the effect of CCL2 blockage on reduction of glomerular leukocyte count and the effect of CXCL12 inhibition on maintaining podocyte numbers and renal nephrin expression were increased when compared with the group treated with the individual Spiegelmers. In addition, proteinuria was suppressed and glomerular filtration was increased in the dual-dose group.

The dual blockade of CCL2 and CXCL12 by mNOX-E36 and NOX-A12, respectively, was also studied in a murine proliferative lupus nephritis model where the synergistic effect of both compounds suppressed disease progression as efficiently as cyclophosphamide [39].

NOX-H94 binds hepcidin, the main regulator of iron metabolism via its effect on ferroportin. Hepcidin is upregulated by acute and chronic inflammatory reactions resulting in "iron restriction," which means that iron is blocked inside cellular stores and not available for hemoglobin synthesis. High hepcidin levels causing anemia of chronic disease are found mainly in infections, cancer, autoimmune diseases, and chronic kidney disease. Due to the low homology of rodent and human hepcidin, the Spiegelmer NOX-H94 binds only the human and not the murine variant with a K_D of 0.65 nM. Therefore, the pharmacodynamic effect of NOX-H94 was assessed in an acute cynomolgus monkey model of interleukin 6 (IL-6)-induced hypoferremia, where NOX-H94 inhibited serum iron reduction completely [40]. In a subchronic model of IL-6-induced anemia in cynomolgus monkeys, NOX-H94 inhibited the decrease in hemoglobin concentration, suggesting that NOX-H94 protects ferroportin from hepcidin-induced degradation.

Apart from the clinical stage candidates, a pharmacodynamic effect has been demonstrated for several additional Spiegelmers including NOX-S93, NOX-G15, NOX-D19, and NOX-L41 in the fields of inflammation, diabetes, oncology, and pain, respectively.

The Spiegelmer *NOX-S93* binds to the small molecular lipid S1P with low nanomolar molar affinity, demonstrating that hydrophobic targets can be addressed by Spiegelmers. Another example is the ghrelin-binding Spiegelmer NOX-B11 [41]. In a murine xenograft model, NOX-S93 inhibited S1P, VEGF, and FGF-2- and IGF-induced angiogenesis *in vitro* and reduced the homing of human cancer cells to prometastatic bone marrow niches induced by whole body irradiation [42]. The inhibition of S1P and aforementioned growth factors by NOX-S93 was also demonstrated in a cellular angiogenesis assay employing primary human endothelial cells [43], suggesting S1P neutralization by the Spiegelmer NOX-S93 as alternative antiangiogenesis treatment approach.

The peptide glucagon is selectively bound with nanomolar affinity by the L-DNA Spiegelmer *NOX-G15*, which contains six L-ribonucleotide positions. In mouse models of type 1 and type 2 diabetes mellitus, a single dose of NOX-G15 reduced hyperglycemia after an intraperitoneal glucose challenge [44].

The L-RNA Spiegelmer *NOX-D19* and its optimized L-DNA-spiked variant *NOX-D20* inhibit mouse and human complement component 5a (C5a)-induced chemotaxis in a C5a-receptor expressing cell line at low nanomolar to picomolar concentration. The anaphylatoxin C5a is a potent inflammatory mediator that has been implicated with organ damage. In a mouse model, NOX-D19 reduced the rejection of allograft airway transplants by preserving microvascular function and airway architecture [45]. The Spiegelmer NOX-D20 was shown to prolong survival and reduce liver and kidney damage in an experimental murine model of polymicrobial septic shock [46]. The interaction of NOX-D20 with mouse C5a has been elucidated by crystal structure analysis of the drug-target complex [47].

As a final example, the Spiegelmer *NOX-L41* binds with picomolar affinity and high specificity to calcitonin gene-related peptide (CGRP), which is a neuroinflammatory peptide involved in pain perception and migraine [22]. In rats, NOX-L41 extravasates from blood vessels in the dura mater and inhibits meningeal plasma protein extravasation for at least 18 h after single dosing [48]. In a mouse model of migraine, the Spiegelmer NOX-C89 that binds to the same target with less affinity was shown to increase meningeal blood flow, which is believed to be cause of migraine [49].

18.4 Clinical Development

From the aforementioned preclinical candidates, three Spiegelmers, emapticap pegol (NOX-E36), olaptesed pegol (NOX-A12), and lexaptepid pegol (NOX-H94), have entered clinical development (Table 18.2). All clinical development candidates are L-RNA nucleic acids with 40–45 nucleotides in length to which a branched 40 kDa PEG is attached via a hexylamino linker at the 5′-end of the oligonucleotide. The PEG, which is denoted in the INN name as pegol, is attached solely for prolonging plasma half-life by reducing renal elimination of the Spiegelmer.

In the preclinical safety–toxicology program, all Spiegelmers showed good safety and tolerability in murine and nonhuman primate models. So far, the no-adverse-effect level (NoAEL) has been assessed for chronic dosing for up to 1 month (NOX-H94) and for up to 3 months (NOX-E36 and NOX-A12) in nonhuman primates, which allows respective dosing in the clinical studies.

18.4.1 Emapticap Pegol: NOX-E36

As described in the preclinical section, Spiegelmer NOX-E36, also known under its INN name emapticap pegol, binds to the chemokine CCL2, which is involved in a variety of inflammatory diseases. There are additional human CCLx family members with high homology, and thus it is not surprising that NOX-E36 also binds to CCL8, CCL13, and CCL11. However, these other chemokines appear to play redundant or complementary roles in comparison with CCL2 so that synergistic activity can be anticipated. The interaction between NOX-E36 and its target CCL2 has been elucidated by analyzing the crystal structure of the drug-target complex, which also revealed insight into the selectivity of the Spiegelmer for related chemokines of similar amino acid sequence [50]. As illustrated by the demonstrated pharmacodynamic effects in a variety of animal models (Table 18.1), interrupting the CCL2 signaling pathway with NOX-E36 has wide therapeutic applications for the treatment of inflammatory complications. Emapticap pegol is currently being developed for the indication of diabetic nephropathy, an inflammatory condition developed by about one in three patients with diabetes mellitus. Diabetic nephropathy is a progressive kidney disease that is one of the main causes of end-stage renal disease and the need for dialysis. The recruitment of immune cells by CCL2 into the kidney along with direct effects of CCL2 on cells of the kidney is believed to be important in the progression of this disease.

Being the first Spiegelmer to enter a clinical study, the objective of the first-in-human (FIH) study with emapticap pegol was to obtain safety and tolerability data after single intravenous (i.v.) and subcutaneous (s.c.) doses in HV using a carefully controlled step-wise dose-escalation process from 0.03 to 2.0 mg/kg for i.v. administration and from 0.25 to 0.5 mg/kg for s.c. dosing. Together with the pharmacokinetic and pharmacodynamic data, e.g. changes in the CCL2-level, the safety and tolerability information was used to establish doses and dosage regimen for the subsequent multidose study in HV and patients with type 2 diabetes mellitus (cf. Table 18.2, study NCT01085292). In this study, i.v. administration of 0.0625–0.25 mg/kg NOX-E36 every other day for 27 days was well tolerated.

Patients in the emapticap lead indication of diabetic nephropathy have moderately to severely impaired kidney function, which could affect excretion and plasma levels of the Spiegelmer. Therefore, a pharmacokinetic study was conducted in patients with different grades of renal impairment demonstrating that no dose adjustment is necessary in this population. With this finding a randomized, double-blind, placebo-controlled clinical phase 2a study (Table 18.2, study NCT01547897) was conducted in 75 patients with type 2 diabetes mellitus and albuminuria receiving a twice weekly s.c. dose of 0.5 mg/kg NOX-E36 or placebo over 12 weeks. The study allowed inclusion of patients on stable antidiabetic therapy and RAS blockade who presented severely reduced kidney function (GFR down to 25 ml/min and an albumin–creatinine ratio [ACR] greater than 100 mg/g) [51]. During the course of the study and the 3-month follow-up period, the effect of emapticap pegol on ACR and glycosylated hemoglobin fraction (HbA1c) was measured among other parameters. At the end of treatment, ACR was reduced by 29% versus baseline ($p < 0.05$) and by 15% versus placebo ($p = 0.221$). During the 3-month follow-up, the therapeutic effect of emapticap was maintained. The maximum effect on mean ACR was observed 8 weeks after cessation of treatment (26% reduction versus placebo, $p = 0.064$), suggesting that emapticap interferes with the underlying pathophysiology of the disease.

The secondary efficacy parameter HbA1c showed a sustained steady decrease of HbA1c until the end of treatment, whereas the placebo group exhibited a slight initial decrease until day 57 with subsequent increase to and above baseline. The maximum difference for HbA1c was observed 4 weeks after the last dose with −0.35% for emapticap versus +0.12% for placebo ($p = 0.026$).

No relevant difference in blood pressure or GFR was observed between the treatment groups throughout the study.

The study demonstrated that the prolonged twice weekly subcutaneous treatment with emapticap was generally safe and well tolerated. It also indicated the renoprotective and antidiabetic potential of emapticap in type 2 diabetic patients with albuminuria [51].

18.4.2 Olaptesed Pegol: NOX-A12

Based on the mobilization of WBCs and HSCs into peripheral blood in mice, a clinical phase 1 study was conducted in HV investigating the safety and activity of NOX-A12 after i.v. administration (Table 18.2, NCT00976378). Single ascending doses of 0.05–10.8 mg/kg of NOX-A12 had a benign safety profile and also dose-dependently mobilized WBCs and HSCs into peripheral blood. HSC peak mobilization reached a plateau at five times the baseline level at a dose of 5.4 mg/kg and could only be increased moderately in a multidose study (Table 18.2, NCT01194934). In accordance with the plasma half-life of 38 h, the duration of the WBC and HSC mobilization was long lasting and increased dose dependently to more than 4 days at the highest dose of 10.8 mg/kg [31].

The combination of the observed long-lasting mobilization of stem cells and the inhibition of the CXCL12/CXCR4/CXCR7 signaling pathway, which sensitizes tumor cells to chemotherapy, suggests that olaptesed pegol may improve standard chemotherapy by mobilizing malignant cells from protective tissue microenvironments to the blood where they can be reached more effectively by standard therapeutics such as cytotoxic agents or antibody drugs. To this extent, two clinical phase 2a studies in different hematological oncology indications, previously treated CLL and previously treated MM, have been conducted (Table 18.2). Patients with relapsed CLL were treated with NOX-A12 in combination with the background therapy of bendamustine and rituximab and received NOX-A12 during six cycles of treatment at a dose between 1 and 4 mg/kg, which had been individually assessed for safety and tolerability. In the second study, patients with relapsed MM were receiving eight cycles of NOX-A12 at the same dose level in combination with the standard therapy using bortezomib and dexamethasone. The purpose of both studies was to evaluate the safety and efficacy of NOX-A12 in combination with the respective background therapy. Final data from both studies showed that the highest olaptesed pegol dose of 4 mg/kg was safe and well tolerated in combination with the individual background therapy. In both studies malignant cells were efficiently mobilized throughout the observational period of 3 days [52–54]. In the MM study, an overall response rate (ORR) of 68% was achieved [53]. In the CLL study, an ORR of 82% was achieved with 14% of the patients showing complete remission [54].

The preclinical finding that NOX-A12 modulates tumor microenvironment enabling T-cell infiltration into the cancer tissue prompted the investigation of the Spiegelmer in the clinical phase 2a combination study with the PD-1 checkpoint inhibitor pembrolizumab (Keytruda®) (Table 18.2, study NCT03168139). Patients with metastatic colorectal and pancreatic cancer are treated with 200 mg of pembrolizumab and 300 mg of olaptesed pegol every 3 weeks. The study has commenced in May 2017 and will evaluate changes in tumor microenvironment, safety and tolerability of the drug combination, and disease control rate.

18.4.3 Lexaptepid Pegol: NOX-H94

Some chronic diseases, e.g. tumors, inflammation, and renal disease, are associated with high hepcidin concentrations in the blood, leading to a reduction in iron concentrations in the blood and subsequent impairment of formation of red blood cells. The aforementioned preclinical efficacy data in cynomolgus monkey models suggest that NOX-H94 efficiently inhibits the hepcidin-induced iron reduction and thus may represent a broad application in the treatment of anemia associated with chronic disease. To evaluate this further, a clinical phase 1 study was conducted in HV investigating the safety, tolerability, pharmacokinetics, and pharmacodynamics of lexaptepid pegol after single and multiple i.v. or s.c. administration (Table 18.2, NCT01372137). The study demonstrated that NOX-H94 was safe and well tolerated over all dose levels and established an efficacious treatment regimen for subsequent clinical studies in patients with anemia of chronic disease. In the treated healthy population, a mild dose-dependent increase in serum iron, and transferrin saturation was observed [55].

In a subsequent clinical study, the pharmacodynamic effect of NOX-H94 inactivation of hepcidin was investigated in an acute model of systemic endotoxin-induced inflammation. In this model, intravenously administered lipopolysaccharide elicits an inflammatory response with release of proinflammatory cytokines, such as IL-6 and TNF-alfa, with subsequent induction of hepcidin. As a consequence of hepcidin induction, serum iron concentrations decrease. The randomized, double-blind, placebo-controlled trial was carried out in 24 healthy male volunteers (Table 18.2, NCT01522794) [56, 57]. At $T = 0$ h, 2 ng/kg *Escherichia coli* endotoxin was administered i.v., followed by i.v. dosing of 1.2 mg/kg NOX-H94 or placebo at $T = 0.5$ h. In both treatment groups, endotoxin administration led to flu-like symptoms peaking at $T = 1.5$ h. NOX-H94 effectively prevented the inflammation-induced drop in serum iron concentrations 9 h after endotoxin administration in over 90% of the subjects studied. In the lexaptepid-treated subjects, the serum iron had increased by 15.9 ± 9.8 μmol/l from baseline, while it was decreased by 8.3 ± 9.0 μmol/l in the control group ($P < 0.0001$). This endotoxemia study delivered the first clinical evidence that lexaptepid pegol is capable of neutralizing high levels of hepcidin in humans and of maintaining higher serum iron concentrations relative to subjects receiving placebo.

Subsequently, the safety and efficacy of hepcidin inhibition by NOX-H94 as treatment of anemia in chronic disease in patients with MM, low grade lymphoma, and CLL was investigated in a clinical phase 2 pilot study (Table 18.2, NCT 01691040, [58]). In responding patients hemoglobin increased to the targeted ≥1 g/dl within the treatment period of 4 weeks. This increase of hemoglobin was associated with a decrease in median serum ferritin from 253 to 203 μg/l and an increase in reticulocyte hemoglobin from 22.0 to 25.2 pg. Treatment with lexaptepid pegol was well tolerated without major adverse reactions. Currently, the effect of NOX-H94 on serum iron levels is being evaluated in a phase 2a study in dialysis patients with ESA-hyporesponsive anemia (Table 18.2, NCT02079896).

18.5 CMC Aspects for the Development of Spiegelmers

The following section highlights various CMC activities accompanying the development stages of a Spiegelmer from discovery to clinical phases 1 and 2.

18.5.1 Discovery and Early Preclinical Stage

During the discovery and early preclinical stage, standard processes are applied for the synthesis and quality control of the L-RNA and L-DNA Spiegelmers [59]. Since about 200–300 different Spiegelmers are required during the optimization phase in less than 1 mg quantity, the emphasis for the generic manufacturing process is placed on throughput and quality. The nucleic acids are produced on oligonucleotide synthesizers by standard coupling procedures using controlled pore glass (CPG) as solid support and commercially available L-DNA or *tert*-butyl-dimethylsilyl (TBDMS)-protected L-RNA amidites (ChemGenes, MA). After deprotection they are purified by PAGE or HPLC. The quality control is conducted by LC-MS and anion exchange (AX)-HPLC using elevated temperature to ensure denaturing of any secondary structure. Apart from testing the identity and purity of the Spiegelmers, the biological activity and/or binding activity is tested in cell culture, ELISA-based binding inhibition, and/or Biacore™ assays. Once short and high-affinity binding Spiegelmers have been identified, they are conjugated to 40 kDa PEG to ensure prolonged plasma half-life for subsequent preclinical and clinical development. Since solid-phase synthesis is conducted from the 3′-end of the sequence to its 5′-end, the preferred site for attaching the PEG from a manufacturing perspective is the 5′-terminus, so that only the full-length Spiegelmer carrying the terminally attached reactive group and not its capped failure sequences can be conjugated. The PEG is typically attached to a 5′-aminohexyl group via NHS-ester chemistry or to a 5′-thiohexyl group via a maleimide functionality. The

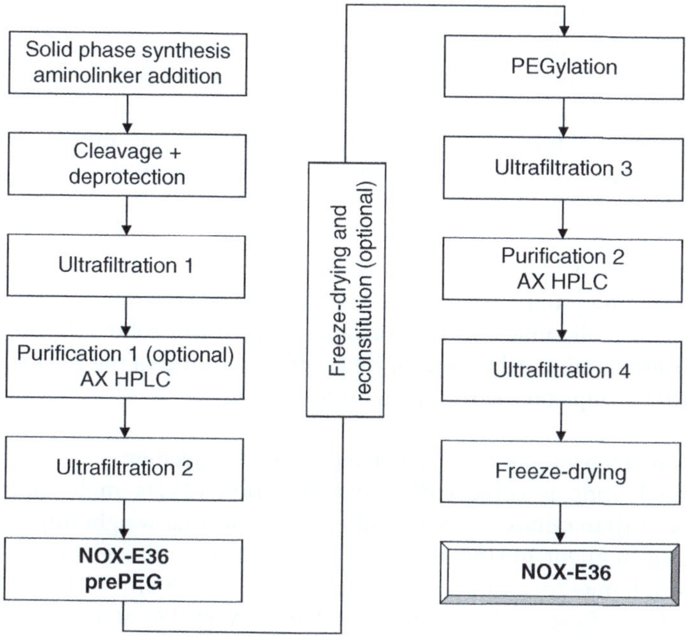

Figure 18.2 Manufacturing process flow diagram for Spiegelmers using NOX-E36 as example.

PEG–Spiegelmer conjugate can then be easily purified from its nonpegylated failure sequences by reversed phase (RP)-HPLC or AX-HPLC. Typically, no loss in binding affinity or biological activity is observed when attaching the PEG to the 5′-end, but in some cases, as for the Spiegelmer mNOX-E36, an attachment to the 3′-end led to improved affinity. At the end of the Spiegelmer sequence optimization phase, one Spiegelmer is selected to demonstrate *in vivo* efficacy in an established animal model. For this Spiegelmer an ELISA-based sandwich hybridization assay or a hybridization assay on the Biacore system is established to allow quantification of the Spiegelmer in biological samples [46]. In parallel, a less than 1 g batch of the Spiegelmer–PEG conjugate and of an inactive pegylated control Spiegelmer, which is commonly the reversed sequence of the active compound with identical base composition, are synthesized by a generic large-scale process, which flow scheme is shown in Figure 18.2 [59].

18.5.2 Generic Manufacturing Process

The process steps of this generic manufacturing process are outlined in the following section.

18.5.2.1 Solid-phase Synthesis

The process of chemically synthesizing L-RNA Spiegelmers, such as NOX-E36, is based on conventional solid-phase oligonucleotide synthesis using TBDMS-phosphoramidite chemistry on a solid support such as controlled pore glass (CPG) or polystyrene [60]. Once the target sequence has been constructed, the amino- or thiol-modified linker, utilized for later attachment of the PEG, is coupled.

18.5.2.2 Deprotection

After synthesis, the oligonucleotide is cleaved from the support, and the phosphate and base protecting groups are removed from the crude oligonucleotide with methylamine. The oligonucleotide is separated from the solid support and treated with triethylamine trihydrofluoride to effect removal of the 2′-silyl (TBDMS) protecting groups on the ribonucleotides [61].

18.5.2.3 Purification of the Intermediate Spiegelmer Prior to Pegylation

The crude oligonucleotide is subjected to ultrafiltration (2 kDa molecular weight cutoff [MWCO]) to remove salts and other low molecular weight impurities. In case no subsequent purification of the intermediate Spiegelmer is conducted, a salt exchange to sodium is performed during the ultrafiltration step. The optional purification is accomplished by AX-HPLC at elevated temperature. The purified Spiegelmer is subsequently desalted by ultrafiltration (2 kDa MWCO). This material can be freeze-dried and reconstituted or processed as solution.

18.5.2.4 Pegylation

The desalted amino- or thiol-modified Spiegelmer is conjugated to the *N*-hydroxysuccinimide (NHS)- or maleimide-activated 40 kDa PEG at ambient temperature [59]. The progress of the conjugation reaction is monitored by RP- or AX-HPLC methods, and additional PEG reagent may be added to ensure complete reaction.

18.5.2.5 Purification of the Pegylated Spiegelmer

Prior to purification, the crude conjugated Spiegelmer is subjected to ultrafiltration (30 kDa MWCO) to remove salts, organic solvents, other low molecular weight impurities, and nonpegylated oligonucleotides. Purification is accomplished by preparative AX-HPLC at neutral pH and elevated temperature to remove any remaining nonpegylated oligonucleotides and unreacted PEG reagent. Fractions containing the product are combined to provide a product pool of appropriate purity and yield. The resulting purified pegylated Spiegelmer is subsequently desalted by ultrafiltration (30 kDa MWCO) and can be stored frozen in solution or lyophilized.

For preclinical applications, the quality of the Spiegelmers is controlled by LC-MS and AX-HPLC analysis of the purified oligonucleotide prior to pegylation. The quality of the pegylated Spiegelmer is assessed by AX-HPLC, size-exclusion (SEC or GPC)-HPLC, bioactivity/binding activity measurement, and determination of the endotoxin burden. For use in animal studies, the Spiegelmer is dissolved to the desired concentration in water containing a tonicity adjuster such as glucose or mannitol. The concentration is determined by a UV assay using a previously established extinction coefficient specific for every sequence. Prior to administration, the solution is sterile filtered through a 0.2 μm filter.

18.5.3 CMC Aspects for the Selection of Development Candidates

The demonstration of *in vivo* efficacy in a suitable animal model is an essential requirement to select a Spiegelmer for clinical development. From the CMC perspective, additional criteria are the economy of the production process to control cost of goods (COG), the quality and stability of the Spiegelmer drug substance, and more importantly the stability of the Spiegelmer drug product. The evaluation of the production process can be based on data obtained during the synthesis of the preclinical material. For L-RNA Spiegelmers of 40–45 nucleotides in length, the crude product quality after deprotection is typically in the range of 50–55% full-length product (FLP) by AX-HPLC, yielding about 220–240 OD of oligonucleotide per micromole scale of synthesis. Longer sequences or sequences with long purine stretches are lower yielding and are not pursued for development. Another important factor in assessing the ability to manufacture the Spiegelmer is the stability of the specific secondary structure required to bind to its target. Depending on sequence, concentration, and salt conditions, alternative structures and/or aggregates could form in a reversible or irreversible manner. This is illustrated using the example of the two hepcidin-binding Spiegelmers NOX-H91 and the later chosen candidate NOX-H94 (Figure 18.3).

Under strong denaturing conditions using elevated temperature, NOX-H91 shows a second, later eluting peak by the AX-HPLC method that is characterized by the same molecular weight as NOX-H91. This putative aggregate, or differently folded structural isomer, could not be equilibrated to the main peak of the monomer and led to a broad product elution in preparative AX-HPLC purification. The additional complication for the process and for the product control strategy justified the change to the alternative lead candidate NOX-H94, which exhibits only one product peak under denaturing conditions. Thus, the early evaluation of the chromatographic behavior of a potential lead compound by generic RP-, AX-, and GPC-HPLC methods provides important information on the stability of the secondary structure and possible aggregation phenomena.

While the stability of the lyophilized Spiegelmers generally supports storage at −20 °C and 2–8 °C for several years, the quality and stability of the Spiegelmer

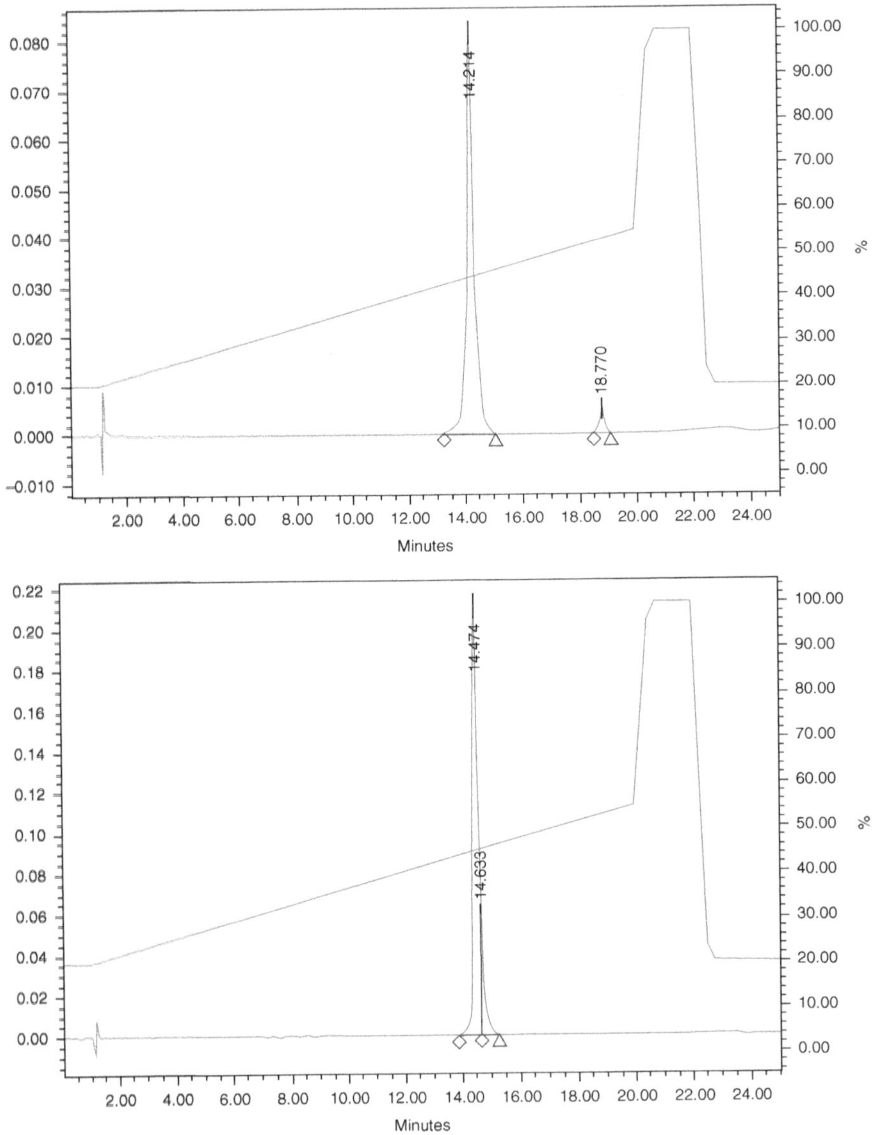

NOX-H91: 5′-GGCCGGACAGCCGGGGGACACCAUAUACAGACUACGAUACGGCC-3′
NOX-H94: 5′-GCGCCGUAUGGGAUUAAGUAAAUGAGGAGUUGGAGGAAGGGCGC-3′

Figure 18.3 Sequences and AX-HPLC chromatograms of the two hepcidin-binding Spiegelmers NOX-H91 (top chromatogram) and NOX-H94 (bottom chromatogram).

drug product solution is evaluated in formulation development and accelerated stress-stability studies. Adding the information on compatible buffers and salts from the discovery process to the knowledge on developing parenteral drug product solutions, various buffer and tonicity adjuster combinations are tested at 0.1 and 1 mM Spiegelmer concentration. The solutions are placed on accelerated stability conditions. The quality of the formulated Spiegelmer is assessed for purity and activity by AX-HPLC and cell culture or biacore assay, respectively. Incompatibility of the Spiegelmer with the excipients is usually first detected by a decrease in biological activity or binding affinity. An alternative high throughput analytical method is differential scanning calorimetry (DSC), which reliably detects any changes to the secondary structure of the Spiegelmer that might be caused by the added excipients (Figure 18.4).

The two μDSC diagrams in the top row of Figure 18.4 show the first and second scan of a 0.2 mM NOX-A12 solution in 200 mM glucose. Both scans

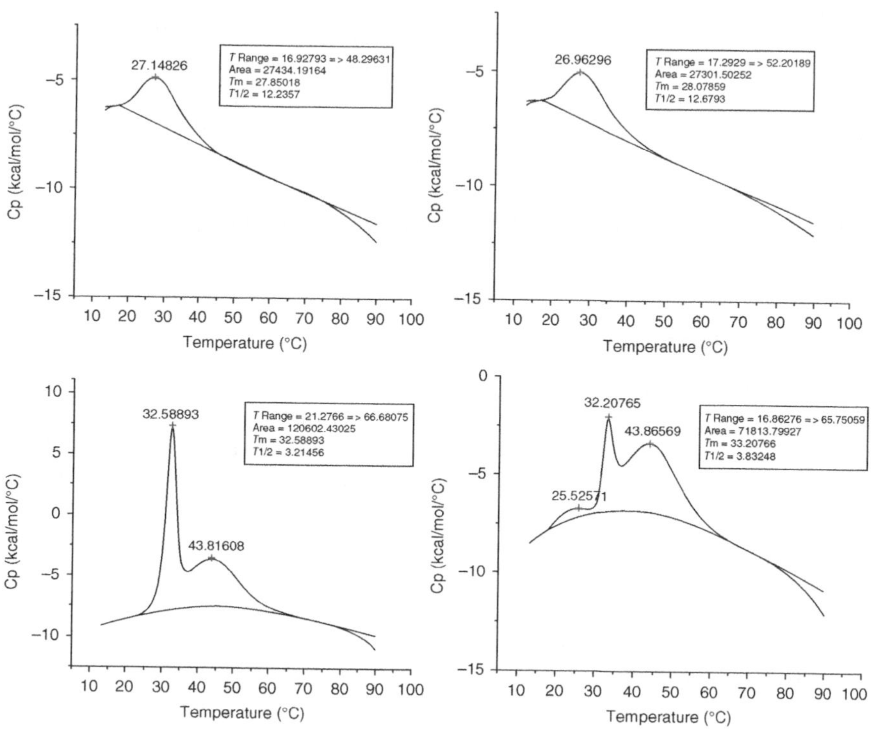

Figure 18.4 μDSC scans of 0.2 mM NOX-A12 in 200 mM glucose. Top row left: first scan with Tm 27.1°C, calorimetric enthalpy Δ*H* 27.4 kcal/mol; top row right: second scan with Tm 27.0°C, Δ*H* 27.3 kcal/mol, and in 20 mM sodium phosphate, 180 mM glucose, 1 mM EDTA; bottom row left: first scan with Tm 32.6 and 43.8°C, Δ*H* 120.6 kcal/mol; bottom row right: second scan with Tm 25.5, 32.2, and 43.9°C, Δ*H* 71.8 kcal/mol).

illustrate similar transition temperatures and calorimetric enthalpies indicating the reversibility of the unfolding and folding process upon heating, cooling, and repeated heating. In contrast to the reversible change in secondary structure, the 0.2 mM NOX-A12 solution dissolved in NOX-E36 formulation buffer, i.e. 20 mM sodium phosphate, 180 mM glucose, and 1 mM ethylenediaminetetraacetate (EDTA), exhibits an additional transition for the second scan at 25.5°C and a reduced calorimetric enthalpy from ΔH 120.9 kcal/mol to ΔH 71.8 kcal/mol. This irreversible change in secondary structure folding was accompanied with a loss in binding affinity to CXCL12 and reduction of biological activity in the chemotaxis assay.

The election of the Spiegelmer development candidate triggers process development for the subsequent GMP production of about 200 g of Spiegelmer, which is sufficient to support preclinical GLP safety–toxicology studies and the clinical phase 1 study. In parallel, candidate-specific analytical method development and validation is initiated to enable determination of the identity, quantity, purity, and binding affinity of the Spiegelmer in QC release and stability samples. It is also desirable to produce a smaller non-GMP batch of the Spiegelmer early on. Using the results from the formulation development experiments, this material can be processed in a non-GMP test fill to the drug product. Apart from evaluating the suitability of the drug product manufacturing process, the stability of the drug product solution can be examined for some time before the decision for the GMP manufacturing of the drug product will be made. Also, the data can be used to support the shelf life of the clinical drug product.

18.5.4 GMP Production of Spiegelmers

18.5.4.1 Starting Materials

Essential starting materials for all current Spiegelmers in clinical development are the L-configured 5′-dimethoxytrityl (DMT)-2′-TBDMS-protected nucleoside phosphoramidites for which manufacturing processes have been established at 50 kg scale. The scheme in Figure 18.5 illustrates the manufacturing process for the benzoyl-protected L-adenosine phosphoramidite as an example.

The common starting material for all L-nucleosides is 1-*O*-acetyl-2,3,5-tri-*O*-benzoyl-L-ribose (L-ATBR), which is readily obtained from L-ribose. L-Ribose is produced at metric ton scale either by epimerization of the 2′-hydroxyl-group of L-arabinose or by an enzymatic process starting from D-glucose [62]. The L-nucleosides are produced by trimethylsilyl triflate-assisted reaction of L-ATBR with the respective in situ silylated nucleobase [63] followed by deacylation with methylamine. To enable regioselective protection of the 2′-OH-position as TBDMS ether, the 3′- and 5′-OH groups are transiently protected by reacting the nucleoside with di-*tert*-butylsilyl bis-triflate [64, 65].

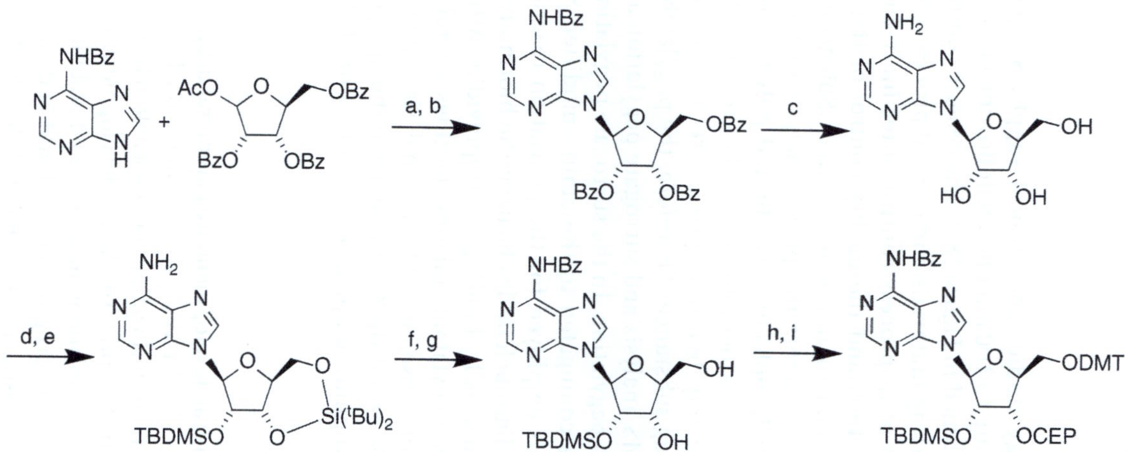

Figure 18.5 Synthesis scheme for *N*-benzoyl-2′-*tert*-butyldimethylsilyl-5′-(4,4′-dimethoxytrityl)-L-adenosine-3′-(2-cyanoethyl-*N*,*N*-diisopropyl) phosphoramidite. (a) Bis-*N*,*O*-trimethylsilyl-acetamide (BSA); (b) trimethylsilyl triflate (TMSOTf); (c) methylamine; (d) di-*tert*-butyl-silyl bis-triflate, DMF; (e) *tert*-butyl-dimethylsilyl chloride, imidazole; (f) benzoic anhydride; (g) HF-pyridine, THF; (h) dimethoxytrityl chloride (DMT-Cl), pyridine; (i) cyanoethyl *N*,*N*,*N*′,*N*′-tetra-isopropyl phosphane, benzylthiotetrazole (BTT). CEP, cyanoethyl *N*,*N*-di-isopropyl phosphoramidite.

After protection of the exocyclic amine group and selective removal of the transient protecting group with pyridine-HF, the 5′-DMT group is installed, and the phosphoramidite is produced. The manufacturing process is devoid of any intermediate chromatographic purification steps.

L-2′-Deoxy-nucleosides are produced by reductive 2′-deoxygenation of the appropriately protected corresponding L-nucleoside or by nucleosidation of protected L-2′-deoxy-ribosyl chloride [66]. Alternatively, the L-configured pyrimidines can be readily obtained from L-arabinose [67].

18.5.4.2 Drug Substance

For GLP safety–toxicology studies and clinical supply, Spiegelmers are chemically synthesized at up to 100 mmol scale following GMP guidelines using the process steps outlined in Figure 18.2. The generic control strategy entails the control of starting materials and critical process parameters, as well as the application of in-process controls after individual process steps, as shown in Table 18.3, and the quality control of the final drug substance. The in-process control of the intermediate Spiegelmer prior to pegylation is important to support the purity analysis of the pegylated Spiegelmer. Due to the high molecular weight and polydispersity of the PEG, the resolution of chromatographic methods such as AX-HPLC is often insufficient to adequately determine process-related oligonucleotide impurities in the final drug substance. Therefore, the pegylatable impurities are assessed by LC-MS analysis and surrogate pegylation at the stage of the Spiegelmer prior to pegylation. In the surrogate pegylation, a cleavable biotin affinity tag is conjugated to the amino-modified Spiegelmer using the same conditions employed for the pegylation reaction (Figure 18.6). The conjugatable oligonucleotides bearing the biotin are separated from the unreactive sequences by binding to streptavidin. After cleavage of the disulfide group, the pegylatable oligonucleotides are released from the streptavidin matrix for subsequent LC-MS analysis. The sequence of the Spiegelmer can also be established at the stage of the Spiegelmer prior to pegylation [68]. The lyophilized Spiegelmer can be stored at 2–8 °C for at least 2 years [69].

18.5.4.3 Drug Product

Current Spiegelmers in clinical development are parenteral drug product solutions manufactured by a sterile filtration process. Due to the heat sensitivity of the drug product solution, final sterilization is not feasible. The composition of the drug product is defined after preformulation development and evaluation of one to three candidate formulations in accelerated stability studies. The Spiegelmer concentration in the drug product is 1 mM and can be increased to 2–2.5 mM when the viscosity reaches about 100 cPs. Even at higher

Table 18.3 In-process controls for the manufacturing of Spiegelmer drug substance.

	Process step	Test	Purpose
Pre-pegylation	After cleavage + deprotection	UV	Determine yield
		AX-HPLC	Determine amount of full-length oligonucleotide and overall quality of crude
	After ultrafiltration 1	MALDI-TOF	Confirm correct mass of oligonucleotide
	During purification 1	UV	Determine yield
		AX-HPLC	Analyze fraction pools
	After ultrafiltration 2	LC-MS	Confirm correct mass of oligonucleotide
		AX-HPLC	Identify and quantify pegylatable impurities
	Prior to pegylation	UV	Determine concentration of oligonucleotide
Post-pegylation	Pegylation	AX-HPLC	Determine progress of reaction
	Ultrafiltration 3	UV	Determine yield
	During purification 2	UV	Determine yield
		AX-HPLC	Analyze fraction pools
	Ultrafiltration 4	UV	Determine yield

concentration, no aggregation phenomena have been observed in the current clinical development candidates. Other constituents of the drug product solution are a carbohydrate, such as glucose, sucrose, or mannitol as tonicity adjuster and optionally also buffering agents such as sodium phosphate or citrate. Additionally, the emapticap formulation consists of EDTA as a complexing agent for bivalent cations to reduce the risk of metal-catalyzed phosphodiester cleavage. All drug product solutions have been shown to be stable for at least 3 years at a conservative storage temperature of −20 °C or at 2–8 °C. The excellent stability of olaptesed drug product solution, which is a 1 mM solution of the drug substance in 200 mM glucose, demonstrates that the phosphate-containing oligonucleotide on its own exhibits sufficient buffering capacity to maintain a stable pH.

18.5.5 Analytical Methods for the Quality Control of Spiegelmers

The overall quality of the Spiegelmer drug product is assured through the control of starting materials, in-process controls, and release and stability

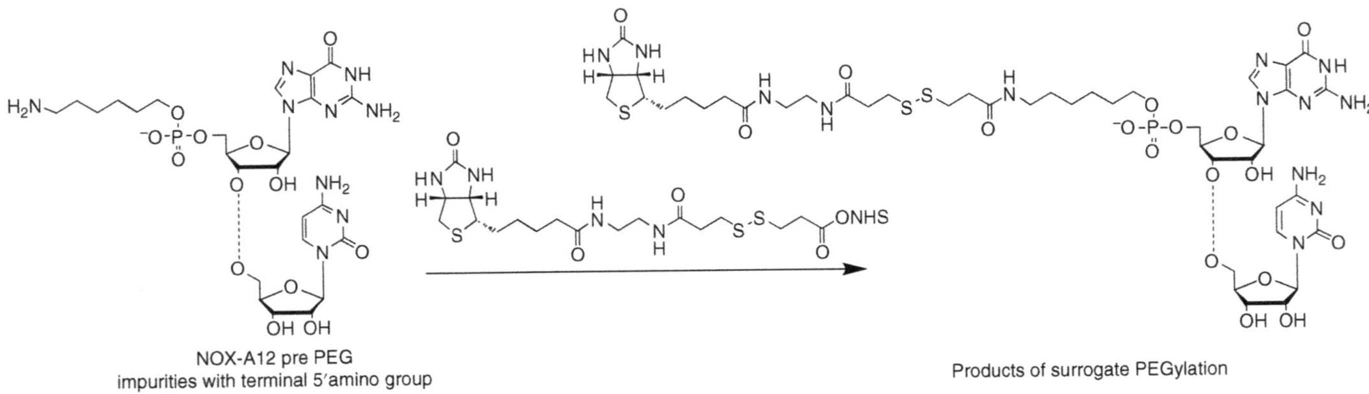

Figure 18.6 Surrogate pegylation of NOX-A12 pre-PEG with a cleavable biotin affinity label.

testing of the drug substance and the final drug product. As described in the quality guideline on good manufacturing practice, ICH Q7A, additional attention is paid to other factors such as manufacturing facilities, equipment, personnel, and systems that may affect final product quality [60, 70].

Table 18.4 lists the analytical test methods utilized for the quality control of Spiegelmer drug substance and drug product. For the compendial methods, equivalent USP versions might be used. Spiegelmer-specific test methods, i.e. the UV and activity assay and the AX- and GPC-HPLC method, have been phase-appropriately validated in line with the quality guideline ICH Q2 and are applied to both the drug substance and the drug product.

Specific challenges for the development of AX-HPLC methods for pegylated Spiegelmers are the requirement to resolve the secondary structure of the oligonucleotide and the polydisperse nature of the conjugated PEG. The former requires strong denaturing conditions, which are usually achieved at elevated temperature close to the specified operating range of the AX-HPLC column. Thus, testing the robustness of the methods by repeated analysis of samples of various purities followed by trend analysis of changes in retention time and purity values is an integral part of the method development activity. Robustness is usually tested on two columns from different lot numbers. The other challenge is caused by the polydispersity of the PEG, which leads to peak broadening, reducing the sensitivity of the method and complicating integration. While Gaussian skim is the best representation of the distribution of the polydisperse PEG, the reproduction of this integration practice between operators and labs is challenging and requires very specific instructions. In comparison with tangential and Gaussian skim, the drop-down integration is relatively easy to perform, but there might be difficulties in detecting the formation of degradation products that initially appear as slight shoulders of the main peak (Figure 18.7). The decision regarding the most appropriate integration method is based on the trend analysis of purity values obtained for a set of accelerated stability samples integrated by different routines.

Another quality control test method is the potency or activity assay, which in the case of emapticap and olaptesed is a biological chemotaxis assay. For the hepcidin-binding Spiegelmer lexaptepid, the biological assay proved to be highly variable and not sufficiently sensitive. This assay, which is based on the hepcidin affected expression of ferritin, was not suitable for the routine GMP analysis of QC release and stability samples, so that an alternative ELISA-based binding competition assay was developed. Equivalence of the biological ferritin assay and the binding competition assay was demonstrated by the analysis of lexaptepid samples of various purity and potency.

Table 18.4 Analytical test methods for the quality control of Spiegelmer drug substance and drug product.

Test	Analytical procedure	Test applicable for	
		Drug substance	Drug product
Identification tests			
By molecular weight (M_P)	GPC-HPLC	+	+
By HPLC retention time	AX-HPLC	+	+
Assay			
Oligonucleotide content	UV	+	+
Purity tests			
Purity	AX-HPLC	+	+
Impurity profile			
Purity	GPC-HPLC	+	+
Impurity profile			
Metals	ICP-MS	+	
Residual solvents[a]	GC, GC–MS or equivalent	+	
Endotoxins	Ph.Eur. 2.6.14 (test method C)	+	
Bioburden a. Total aerobic plate counts b. Total yeast and molds	Generic method	+	
Other tests			
Biological/binding activity	Chemotaxis/binding competition	+	+
Physical appearance	Visual	+	+
Sodium content (%w/w on anhydrous basis)	ICP-OES	+	
Moisture content	Karl Fischer	+	
pH	Ph. Eur. 2.2.3	+[b]	+
Osmolality	Ph. Eur. 2.2.35		+
Endotoxins	Ph. Eur. 2.6.14		+
Sterility	Ph. Eur. 2.6.1		+
Subvisible particles	Ph. Eur. 2.9.19		+

[a] Only testing for solvents that are used in the process and likely to be carried through to the final product.
[b] of a 1% solution.

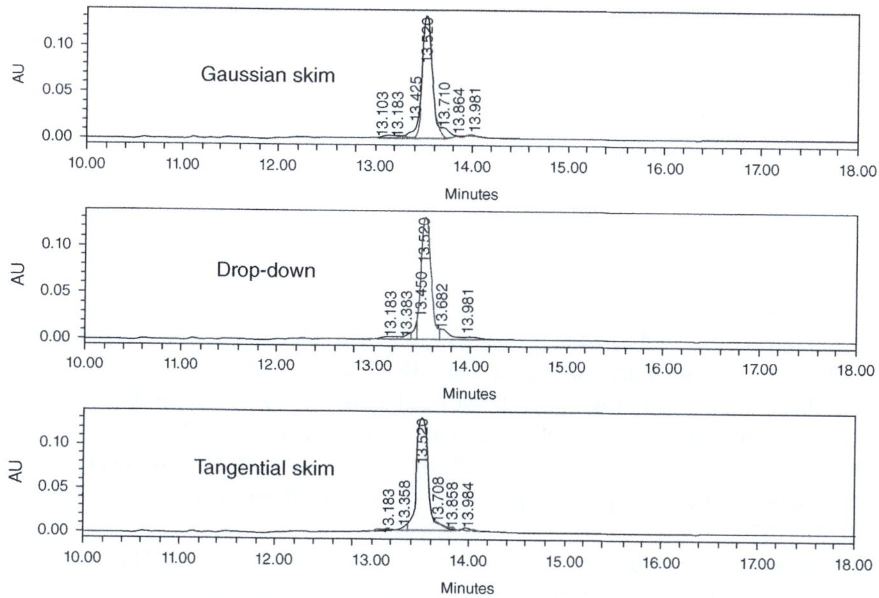

Figure 18.7 Integration of the same AX-HPLC of a 40 kDa pegylated Spiegelmer by Gaussian skim, drop-down, and Tangential skim.

18.6 Future Prospects for Spiegelmer Therapeutics

With several preclinical candidates with proven *in vivo* efficacy, additional Spiegelmers in a variety of indications ranging from diabetes to inflammation and oncology are likely to enter clinical studies. The existing development candidates have completed proof-of-concept studies and have displayed good safety, tolerability, and promising signs of efficacy. Additional cancer indications like glioblastoma and other solid tumors will be investigated with the Spiegelmer olaptesed. From a CMC perspective, the control of COGs by improved manufacturing processes and the continued optimization of the quality control methods and strategy will support the entry of Spiegelmers in later clinical phases.

References

1 Cech, T.R., Zaug, A.J., and Grabowski, P.J. (1981). In vitro splicing of the ribosomal RNA precursor of Tetrahymena: involvement of a guanosine nucleotide in the excision of the intervening sequence. *Cell* 27: 487–496.
2 Guerrier-Takada, C., Gardiner, K., Marsh, T. et al. (1983). The RNA moiety of ribonuclease P is the catalytic subunit of the enzyme. *Cell* 35: 849–857.

3 Winkler, W.C., Nahvi, A., Roth, A. et al. (2004). Control of gene expression by a natural metabolite-responsive ribozyme. *Nature* 428: 281–286.
4 Lafontaine, D. and Dubé, A. ed. (2014). Therapeutic applications of ribozymes and riboswitches: methods and protocols. In: *Springer Protocols*, vol. 1103. New York: Springer Science + Business Media.
5 Weng, D.E. and Usman, N. (2001). Angiozyme: a novel angiogenesis inhibitor. *Curr. Oncol. Rep.* 3: 141–146.
6 Tuerk, C. and Gold, L. (1990). Systematic evolution of ligands by exponential enrichment: RNA ligands to bacteriophage T4 DNA polymerase. *Science* 249: 505–510.
7 Ellington, A.D. and Szostak, J.W. (1990). *In vitro* selection of RNA molecules that bind specific ligands. *Nature* 346: 818–822.
8 Robertson, D.L. and Joyce, G.F. (1990). Selection *in vitro* of an RNA enzyme that specifically cleaves single-stranded DNA. *Nature* 344: 467–468.
9 Bouchard, P.R., Hutabarat, R.M., and Thompson, K.M. (2010). Discovery and development of therapeutic Aptamers. *Annu. Rev. Pharmacol. Toxicol.* 50: 237–257.
10 Sundaram, P., Kurniawan, H., Byrne, M.E., and Wower, J. (2013). Therapeutic RNA aptamers in clinical trials. *Eur. J. Pharm. Sci.* 48: 259–271.
11 Realini, T., Ng, E.W.M., and Adamis, A.P. (2006). Application in the clinic: the anti-VEGF aptamer. In: *Aptamer Handbook* (ed. S. Klussmann), 417–442. Weinheim: Wiley-VCH.
12 Missailidis, S. and Hardy, A. (2009). Aptamers as inhibitors of target proteins. *Expert Opin. Ther. Pat.* 19: 1–10.
13 Kovach, J.L., Schwartz, S.G., Flynn, H.W. Jr., and Scott, I.U. (2012). Anti-VEGF treatment strategies for wet AMD. *J. Ophthalmol.* 2012: 1–7.
14 Burmeister, P.E., Lewis, S.D., Silva, R.F. et al. (2005). Direct in vitro selection of a 2′-*O*-methyl aptamer to VEGF. *Chem. Biol.* 12: 25–33.
15 Yang, X., Li, N., and Gorenstein, D.G. (2011). Strategies for the discovery of therapeutic Aptamers. *Expert Opin. Drug Discovery* 6: 75–87.
16 Keefe, A.D. and Cload, S.T. (2008). SELEX with modified nucleotides. *Curr. Opin. Chem. Biol.* 12: 448–456.
17 Klussmann, S., Nolte, A., Bald, R. et al. (1996). Mirror-image RNA that binds d-adenosine. *Nat. Biotechnol.* 14: 1112–1115.
18 Eulberg, D., Jarosch, F., Vonhoff, S., and Klussmann, S. (2006). Spiegelmers for therapeutic applications – use of chiral principles in evolutionary selection techniques. In: *Aptamer Handbook* (ed. S. Klussmann), 417–442. Weinheim: Wiley-VCH.
19 Vater, A. and Klussmann, S. (2015). Turning mirror-image oligonucleotides into drugs: the evolution of Spiegelmer therapeutics. *Drug Discov. Today* 20: 147–155.
20 Garbesi, A., Capobianco, M.L., Colonna, F.P. et al. (1993). L-DNA as potential antimessenger oligonucleotide: a reassessment. *Nucleic Acids Res.* 21: 4159–4165.

21 Kulkarni, O., Pawar, R.D., Purschke, W. et al. (2007). Spiegelmer inhibition of CCL2/MCP-1 ameliorates lupus nephritis in MRL-(Fas)lpr mice. *J. Am. Soc. Nephrol.* 18: 2350–2358.

22 Jarosch, F., Klussmann, S., Sell, S., et al. (2012). Nucleic acid molecule having binding affinity to a target molecule and a method for generating the same. Patent WO 2012095303, 19 July 2012.

23 Kulkarni, O., Eulberg, D., Selve, N. et al. (2009). Anti-Ccl2 Spiegelmer permits 75% dose reduction of cyclophosphamide to control diffuse proliferative lupus nephritis and pneumonitis in MRL-Fas(lpr) mice. *J. Pharmacol. Exp. Ther.* 328: 371–377.

24 Ninichuk, V., Clauss, S., Kulkarni, O. et al. (2008). Late onset of CCL2 blockade with the Spiegelmer mNOX-E36-3'PEG prevents glomerulosclerosis and improves glomerular filtration rate in db/db mice. *Am. J. Pathol.* 172: 628–637.

25 Baeck, C., Wehr, A., Karlmark, K.R. et al. (2012). Pharmacological inhibition of the chemokine CCL2 (MCP-1) diminishes liver macrophage infiltration and steatohepatitis in chronic hepatic injury. *Gut* 61: 416–426.

26 Baeck, C., Wei, X., Bartneck, M. et al. (2014). Pharmacological inhibition of the chemokine C-C motif chemokine ligand 2 (monocyte chemoattractant protein 1) accelerates liver fibrosis regression by suppressing Ly-6C^{+} macrophage infiltration in mice. *Hepatology* 59: 1060–1072.

27 Ehling, J., Bartneck, M., Wei, X. et al. (2014). CCL2-dependent infiltrating macrophages promote angiogenesis in progressive liver fibrosis. *Gut* 63: 1960–1971.

28 Mossanen, J.C., Krenkel, O., Ergen, C. et al. (2016). Chemokine (C-C motif) receptor 2-positive monocytes aggravate the early phase of acetaminophen-induced acute liver injury. *Hepatology* 64: 1667–1682.

29 Kalnins, A., Thomas, M.N., Andrassy, M. et al. (2015). Spiegelmer inhibition of MCP-1/CCR2—potential as an adjunct immunosuppressive therapy in transplantation. *Scand. J. Immunol.* 82: 102–109.

30 Vater, A., Sahlmann, J., Kroeger, N. et al. (2013). Hematopoietic stem and progenitor cell mobilization in mice and humans by a first-in-class mirror-image oligonucleotide inhibitor of CXCL12. *Clin. Pharmacol. Ther.* 94: 150–157.

31 Hoellenriegel, J., Zboralski, D., Maasch, C. et al. (2014). The Spiegelmer® NOX-A12, a novel CXCL12 inhibitor, interferes with chronic lymphocytic leukemia cell motility and causes chemo-sensitization. *Blood* 123: 1032–1039.

32 Marasca, R. and Maffei, R. (2014). Comment on Hoellenriegel et al., NOX-A12: mobilizing CLL away from home. *Blood* 123: 952–953.

33 Roccaro, A., Sacco, A., Purschke, W.G. et al. (2014). SDF-1 inhibition targets the bone marrow niche for cancer therapy. *Cell Rep.* 9: 118–128.

34 Zboralski, D., Kruschinski, A., Eulberg, D., and Vater, A. (2016). CXCL12 inhibition with NOX-A12 (Olaptesed Pegol) increases T and NK cell infiltration and synergizes with immune checkpoint blockade and ADCC in tumor-Stroma spheroids. *ESMO Congress 2016*, Copenhagen, Denmark, Abstract 1083P (7–11 October 2016).

35 Liu, S.C., Alomran, R., Chernikova, S.B. et al. (2014). Blockade of SDF-1 after irradiation inhibits tumor recurrences of autochthonous brain tumors in rats. *Neuro Oncol.* 16: 21–28.

36 Savai, R., Pullamsetti, S.S., Kolbe, J. et al. (2012). Immune and inflammatory cell involvement in the pathology of idiopathic pulmonary arterial hypertension. *Am. J. Respir. Crit. Care Med.* 186: 897–908.

37 Sayyed, S.G., Haegele, H., Kulkarni, O.P. et al. (2009). Podocytes productive homeostatic chemokine stromal cell-derived factor-1/CXCL12, which contributes to glomerulosclerosis, podocyte loss and albuminuria in a mouse model of type 2 diabetes. *Diabetologia* 52: 2445–2454.

38 Darisipudi, M.N., Kulkarni, O.P., Sayyed, S.G. et al. (2011). Dual blockade of the homeostatic chemokine CXCL12 and the proinflammatory chemokine CCL2 has additive protective effects on diabetic kidney disease. *Am. J. Pathol.* 179: 116–124.

39 Davarapu, S.K., Kumar, V.S., Rupanagudi, K.V. et al. (2016). Dual blockade of the pro-inflammatory chemokine CCL2 and the homeostatic chemokine CXCL12 is as effective as high dose cyclophosphamide in murine proliferative lupus nephritis. *Clin. Immunol.* 169: 139–147.

40 Schwoebel, F., van Eijk, L.T., Zboralski, D. et al. (2013). The effects of the anti-hepcidin Spiegelmer NOX-H94 on inflammation-induced anemia in cynomolgus monkeys. *Blood* 121: 2311–2315.

41 Kobelt, P., Helmling, S., Stengel, A. et al. (2006). Anti-ghrelin Spiegelmer NOX-B11 inhibits neurostimulatory and orexigenic effects of peripheral ghrelin in rats. *Gut* 55: 788–792.

42 Schneider, G., Bryndza, E., Abdel-Latif, A. et al. (2013). Bioactive lipids S1P and C1P are pro-metastatic factors in human rhabdomyosarcoma, and their tissue levels increase in response to radio/chemotherapy. *Mol. Cancer Res.* 11: 793–807.

43 Purschke, W.G., Hoehlig, K., Buchner, K. et al. (2014). Identification and characterization of a mirror-image oligonucleotide that binds and neutralizes Sphingosine-1-phosphate, a central mediator of angiogenesis. *Biochem. J.* 462: 153–162.

44 Vater, A., Sell, S., Kaczmarek, P. et al. (2013). A mixed mirror-image DNA/RNA aptamer inhibits glucagon and acutely improves glucose tolerance in models of type 1 and type 2 diabetes. *J. Biol. Chem.* 288: 21136–21147.

45 Khan, M.A., Maasch, C., Vater, A. et al. (2013). Targeting complement component 5a promotes vascular integrity and limits airway remodeling. *PNAS* 110: 6061–6066.

46 Hoehlig, K., Maasch, C., Shushakova, N. et al. (2013). A novel C5a-neutralizing mirror-image (L)aptamer prevents organ failure and improves survival in experimental sepsis. *Mol. Ther.* 21: 2236–2246.

47 Yatime, L., Maasch, C., Hoehlig, K. et al. (2015). Structural basis for the targeting of complement anaphylatoxin C5a using a mixed L-RNA/L-DNA aptamer. *Nat. Commun.* 6: 6481–6483.
48 Hoehlig, K., Johnson, K.W., Pryazhnikov, E. et al. (2015). A novel CGRP-neutralizing Spiegelmer attenuates neurogenic plasma protein extravasation. *Br. J. Pharmacol.* 172: 3086–3098.
49 Denekas, T., Troeltzsch, M., Vater, A. et al. (2006). Inhibition of stimulated meningeal blood flow by a calcitonin gene-related peptide binding mirror-image RNA oligonucleotide. *Br. J. Pharmacol.* 148: 536–543.
50 Oberthuer, D., Achenbach, J., Gabdulkhakov, A. et al. (2015). Crystal structure of a mirror-image L-RNA aptamer (Spiegelmer) in complex with the natural target L-protein target CCL2. *Nat. Commun.* 6: 6923–6933.
51 Menne, J., Eulberg, D., Beyer, D. et al. (2017). C-C motif-ligand 2 inhibition with Emapticap Pegol (NOX-E36) in type 2 diabetic patients with albuminuria. *Nephrol. Dial. Transplant.* 32: 307–315.
52 Gobbi, M., Steurer, M., Caligaris-Cappio, F., Montillo, M., Janssens, A., Trentin, L., Duemmler, T., Zoellner, S., Zeitler, S., Riecke, K., Kruschinski, A. (2013). Anti-CXCL12/SDF-1 Spiegelmer NOX-A12 alone and in combination with bendamustine and rituximab in patients with relapsed chronic lymphocytic leukemia (CLL): results from a phase IIa study. *Blood (ASH Annual Meeting Abstracts)* 121, New Orleans, LA, Abstract 1635 (7–10 December 2013).
53 Ludwig, H., Weisel, K., Petrucci, M.T. et al. (2017). Olaptesed pegol, an anti-CXCL12/SDF-1 Spiegelmer, alone and with bortezomib-dexamethasone in relapsed/refractory multiple myeloma: a phase IIa study. *Leukemia* 31: 997–1000.
54 Steurer, M., Montillo, M., Scarfò, L., Mauro, F.R., Andel, J., Wildner, S., Trentin, L., Janssens, A., Burgstaller, S., Kruschinski, A., Duemmler, T., Riecke, K., Ghia, P., Caligaris-Cappio, F., Gobbi, M. (2014). Results from a phase IIa study of the anti-CXCL12 Spiegelmer Olaptesed Pegol (NOX-A12) in combination with Bendamustine/Rituximab in patients with chronic lymphocytic leukemia. *Poster #1996 at Session 642.* CLL: Therapy, excluding transplantation, ASH 2014, San Francisco, CA.
55 Boyce, M., Warrington, S., Cortezi, B. et al. (2016). Safety, pharmacokinetics and pharmacodynamics of the anti-hepcidin Spiegelmer lexaptepid pegol in healthy subjects. *Br. J. Pharmacol.* 173: 1580–1588.
56 van Eijk, L.T., John, A.S.E., Schwoebel, F. et al. (2014). Effect of the antihepcidin Spiegelmer lexaptepid on inflammation-induced decrease in serum iron in humans. *Blood* 124: 2643–2646.
57 Steinbicker, A.U. (2014). A novel treatment of anemia of inflammation. *Blood* 124: 2618–2619.
58 Georgiev, P., Lazaroiu, M., Ocroteala, L. et al. (2014). The anti-hepcidin Spiegelmer Lexaptepid Pegol (NOX-H94) as treatment of anemia of chronic disease in patients with multiple myeloma, low grade lymphoma, and CLL: a phase II pilot study. *Cancer Res.* 74: 3847.

59 Hoffmann, S., Hoos, J., Klussmann, S., and Vonhoff, S. (2011). RNA Aptamers and Spiegelmers: synthesis, purification, and post-synthetic PEG conjugation. *Curr. Protoc. Nucleic Acid Chem.* 4: 4.46.1–4.46.30.

60 Sanghvi, Y.S. (2011). A status update of modified oligonucleotides for chemotherapeutic applications. *Curr. Protoc. Nucleic Acid Chem.* 46: 4.1.1–4.1.22.

61 Wincott, F., DiRenzo, A., Shaffer, C. et al. (1995). Synthesis, deprotection, analysis and purification of RNA and ribozymes. *Nucleic Acids Res.* 23: 2677–2684.

62 Okano, K. and Ueda, M. (2005). L-ribose, a new chiral building block for nucleoside analogues. *Spec. Chem. Mag.* 25: 30–32.

63 Pitsch, S. (1997). An efficient synthesis of enantiomeric ribonucleic acids from d-glucose. *Helv. Chim. Acta* 80: 2286–2314.

64 Furusawa, K., Ueno, K., and Katsura, T. (1990). Synthesis and restricted conformation of 3′,5'-*O*-(di-*tert*-butylsilanediyl)ribonucleosides. *Chem. Lett.* 97–100.

65 Serebryany, V. and Beigelman, L. (2002). An efficient preparation of protected ribonucleosides for phosphoramidite RNA synthesis. *Tetrahedron Lett.* 43: 1983–1985.

66 Seela, F. and Winter, H. (1995). Synthesis of 6-substituted purine N^7-(2-deoxy-β-D-ribonucleosides) via anion glycosylation and anomerization during the N^7/N^9-glycosyl transfer. *Nucleosides Nucleotides* 14: 129–142.

67 Holý, A. (1971). A synthesis of 2'deoxy-L-uridine. *Tetrahedron Lett.* 12: 189–192.

68 Turner, J.J., Hoos, J.S., Vonhoff, S., and Klussmann, S. (2011). Methods for L-ribonucleotide sequence determination using LCMS. *Nucleic Acids Res.* 39: e147.

69 Maasch, C., Buchner, K., Eulberg, D. et al. (2008). Physicochemical stability of NOX-E36, a 40mer L-RNA (Spiegelmer) for therapeutic applications. *Nucleic Acids Symp. Ser.* 52: 61–62.

70 Bonilla, J.V. and Srivatsa, G.S. ed. (2011). *Handbook of Analysis of Oligonucleotides and Related Products.* Boca Raton, FL: CRC Press.

Index

Oligonucleotide-Based Drugs and Therapeutics: Preclinical and Clinical Considerations for Development, First Edition. Edited by Nicolay Ferrari and Rosanne Seguin.

b

e

i

m

q

r

u

v

Printed and bound by CPI Group (UK) Ltd, Croydon, CR0 4YY
18/09/2026
14964995-0001